THE NUCLEUS

NEW PHYSICS FOR THE NEW MILLENNIUM

THE NUCLEUS
NEW PHYSICS FOR THE NEW MILLENNIUM

Edited by

F. D. Smit
National Accelerator Centre
Faure, South Africa

R. Lindsay
University of the Western Cape
Bellville, South Africa

and

S. V. Förtsch
National Accelerator Centre
Faure, South Africa

Kluwer Academic / Plenum Publishers
New York, Boston, Dordrecht, London, Moscow

Library of Congress Cataloging-in-Publication Data

The nucleus: new physics for the new millennium/edited by F.D. Smit, R. Lindsay, and S.V. Förtsch.

p. cm.

Includes bibliographical references and index.

ISBN 0-306-46302-4

1. Nuclear physics—Congresses. I. Smit, F. D. II. Lindsay, Robert Bruce, 1900– III. Förtsch, S. V. IV. Proceedings of the International Conference on The Nucleus: New Physics for the New Millennium (1999: Faure, South Africa)

QC770.N825 1999
539.7—dc21

99-055325

Proceedings of the International Conference on The Nucleus: New Physics for the New Millennium, held January 18–22, 1999, at the National Accelerator Centre, Faure, South Africa

ISBN 0-306-46302-4

233 Spring Street, New York, N.Y. 10013

http://www.wkap.nl/

10 9 8 7 6 5 4 3 2 1

A C.I.P. record for this book is available from the Library of Congress

Printed in the United States of America

These proceedings are dedicated to John Sharpey-Schafer on the occasion of his sixtieth birthday.

John Francis Sharpey-Schafer

Sharpey - A Student's View

Of John Sharpey-Schafer, what can you say?
His knowledge is boundless, increasing each day.
He always inspires us,
His wit never tires us,
And he greets you with a cheerful "Hey Hey"

As a student in winter, if you wanted advice
He was often in Bormio, skiing on ice.
But when he came in to land
His help was on hand
And his sun tan was really quite nice.

Then John (having itchy feet)
In summer would move on to Crete,
And - funding permitting
We sometimes went with him
And came back with badly burned feet

On experiments he was always with it,
On night shifts he'd make coffee and sip it.
He was there for our sake
(Though not always awake)
And was certainly with us in spirit !

We did Erbium, Holmium, Thullium, Thallium,
Putting on gates and finding new bands in 'em.
And month after month,
Year after year,
Dysprosium 152 ad nauseam.

Happy birthday John! May you have many more.
For your 70th we'll be back for sure.
But - 2010 ?
We can't wait 'till then
So we're planning some experiments at Faure.

So, to echo the words of a wiley
young man by the name of Mark Riley...
To us, you're the best.
And I would have guessed
That on that point we all agree.

Michael Bentley

PREFACE

These proceedings contain the contributions to the Conference "The Nucleus: New Physics for the New Millennium" held at the National Accelerator Centre (NAC) in South Africa from 18-22 January 1999. With this as a theme, the invited speakers spent part of the time putting forward their views of what developments are likely to take place in their topics of interest in the next few years.

The conference was organised to honour John Sharpey-Schafer on the occasion of his 60th birthday. Probably the most important lesson to learn from John's career, is his ability to stay positive and use whatever opportunities are available. For example, during the early eighties it required a major effort to ensure that research activities at Liverpool grew while the Daresbury Nuclear Structure Facility was being built, and use had to be made of overseas facilities such as the Niels Bohr Institute in Denmark.

When Daresbury started to deliver beams, albeit rather slowly, John got stuck in and contributed mightily to probably the golden era of high-spin physics, despite the problems with the laddertron, the beam energies and Thatcherism. When Daresbury was unexpectedly shut down, he jumped at the possibilities of the European collaborations. And last but by no mean least, he became director of NAC in 1996 and started up a high-spin state programme from scratch. That is typical of John. He seized the new opportunity of high energy beams from the cyclotron to do front line physics.

Several aspects made this conference different if not unique: Summer weather in January; accommodation on the beach front; lunch with the zebras and ALL the delegates joining John at his house. All of which by all accounts went down very well.

The conference highlighted the extent to which science has become an international collaboration, where e-mail and airlines allow for a conference to be held at the Southern part of Africa. Nuclear physics, especially the high-spin experimental field has now joined particle physics in becoming so expensive that international collaboration is a must and not a choice. Globalisation is most likely to increase in the next millennium with implications for us as scientists as well. Research equipment is becoming increasingly more expensive.

The foreign delegates enjoyed themselves so much one would most probably quite easily find support amongst them for the idea of developing research facilities in exotic locations like South Africa. What at first may seem to be far-fetched has some really positive spinoffs. It is in everyone's interest to develop science everywhere in order to draw as many of the brightest minds to the field. However, this requires a need for good science to be available in as wide a geographical distribution as possible. If most scientists have John's penchant for exotic locations, this should prove no hardship! At the extra cost of an airline ticket an environment is created in which even the most educated people born in the host location can have a scientific career at home. Without such opportunities, once educated, the best emigrate to bright futures in developed countries resulting in a loss to their own countries.

Hopefully the conference at NAC will give a major boost to the expansion of links between NAC and the rest of the world. The NAC has already a good infrastructure and this fertile ground could well be where this idealistic view of science for the new millennium could grow. This would enable NAC to continue its role in the educational development of all South Africans; a task which John has worked for tirelessly since his appointment and which is clearly as important as the publication of papers or new discoveries in nuclear physics.

F.D. Smit,
R. Lindsay,
S.V. Förtsch

The Conference Delegates

The Setting

CONTENTS

I Opening Speaker

NEW IDEAS IN NUCLEAR STRUCTURE PHYSICS

Neil Rowley

Institut de Recherches Subatomiques
23 rue du Loess
F-67037 Strasbourg Cedex 2
France

Abstract: Just over a century after the discovery of radioactivity by Becquerel and of new elements by the Curies, we have begun to master the techniques of performing experiments with radioactive beams. And, of course, the search for other new elements goes on. At the same time the exploitation of γ-ray arrays of increasing power pushes observation limits lower and lower in the search for new phenomena at high spins. Ancillary devices, capable of giving cleaner channel selection, also permit the study of higher-mass systems and of more exotic charge-to-mass ratios even with stable beams.

A personal selection of some of the topics of interest will be presented, as well as some comments on how relatively simple detection systems can still make major contributions to our understanding of nuclear structure and reactions.

1 INTRODUCTION

It is an honour to be asked to present the opening physics talk at this conference, timed to coincide roughly with the 60th birthday of our old friend John Sharpey-Schafer. I have known John for many years now and have always had enormous respect for him both as a physicist and as a person. I could mention various important discoveries in high-spin physics in which John was intimately

The Nucleus: New Physics for the New Millennium
Edited by Smit et al., Kluwer Academic / Plenum Publishers, New York, 2000.

involved, such as the discoveries of band terminations and superdeformations, but I would also like to remind you that it was John who convinced the nuclear physics community of the need to do proper, clean spectroscopy using arrays of Compton-suppressed γ-ray detectors, an idea which ultimately led to the creation of such devices as EUROBALL and GAMMASPHERE.

Perhaps even more importantly he has played a tremendous rôle in attracting a multitude of bright young people into our field of research, both through his lectures in Liverpool and through his general enthusiasm for his chosen subject, manifested at many international conferences, schools and seminars. I must say that I am delighted to see how many ex-students of John are still in the field of nuclear structure research and that so many of them are here in the audience today. Of course those students who did not remain in the field have gone out into important positions in industry and teaching etc. equipped with the excellent education that a doctorate in experimental nuclear physics provides.

Having attracted this talent into research in the physics department in Liverpool, John did not forget them, not only helping them greatly with their projects but also looking after their personal and social well-being through, for example, extra-curricular sessions in some of Liverpool's more interesting nightspots as well as through bracing walks in the mountains of Snowdonia.

With the latter in mind, I would like to take you on a trip, a fairly random walk, through the hills and valleys – not of course those of North Wales but those of the N-Z plane. The landmarks we shall note on the way are rather subjective, being a collection of topics which I have heard talked about at recent conferences or recent visits to other laboratories. I offer my apologies to anyone who feels that their work should have been included if it is not.

There were a couple of advantages to being a theorist in Liverpool, as I was from 1975-78. One, of course, was that one did not need to do experiments. But the second was that from the 7th floor of the Chadwick tower we had wonderful views over the city; between the cathedrals, across the River Mersey and out to Snowdonia itself. Liverpudlians will tell you that if you cannot see Snowdonia, then it is raining...and of course if you can see Snowdonia, then it is going to rain! Usually if one could see anything at all, it was just the summit of Snowdon itself sticking out above the mist and clouds.

And so it was with the nucleus at the end of the last century, hidden beneath its clouds of electrons. Fortunately Nature provided us with isotopes of uranium which generate a radioactivity that penetrated these clouds for Bequerel to realise that something interesting lay beneath them. As she often is, Nature was kind in giving these isotopes – long enough lived to have survived since the supernova explosion which created them and spewed out much of the matter from which the Solar System was built, yet with lifetimes still short enough to give significant and easily measurable radiations.

One should perhaps interject at this stage that it is the study of nuclear structure and reactions which has given us the deep understanding we have of such astrophysical processes; where the matter of which we are made came from, the evolution of stars – the whys and wherefores of the Hertzsprung-

Russel diagram – and how the Sun produces the energy on which our existence depends. One of the major goals of nuclear physics research over the coming years will be to further extend this base of knowledge through the exploration of nuclei far from the line of β-stability which play a determining rôle in nature of our Universe. For example, understanding the production mechanisms of proton-rich stable isotopes will give considerable insights into the exotic cosmological conditions in which they are forged. Similarly the properties of certain key nuclei on the neutron-rich side of stability, waiting points on the precipitous r-process path up the side of the valley, are essential to a complete understanding of the creation of the elements more massive than iron, including the uraniums themselves (see Fig. 1).

2 SOME HISTORY AND SOME GIANTS

There are many modern versions of the sentiment but I believe that it was Samuel Taylor Coleridge (1772 – 1834) who first said:

A dwarf sees further than the giant when he has the giant's shoulders to mount on.

I would like to mention some of the giants on whose shoulders we stand:

A CENTURY OF NUCLEAR PHYSICS

1896: Becquerel discovers radioactivity

1898: The Curies discover the elements Po and Ra

1911: Rutherford finds the nucleus and sees how big it is

1932: Chadwick discovers the neutron

1935: Joliot and Joliot-Curie make the first artificial radioactive isotope by bombarding ^{27}Al with α-particles

1936: Fermi makes lots of them using neutrons:
$[\mathrm{A}, \mathrm{Z}] + n \rightarrow [\mathrm{A}+1, \mathrm{Z}] + \gamma \rightarrow [\mathrm{A}+1, \mathrm{Z}+1] + e^- + \bar{\nu}$

1939: Hahn, Meitner and Frisch discover fission (neutron-induced)

1955: Mayer and Jensen invent the shell model

1956-7: Bardeen, Cooper and Schrieffer invent pairing

1962: Polikanov discovers fission isomers (^{242}Am); the first superdeformed nuclei

I would have liked to stop there because I would not like John to think that I consider anything too much after his PhD as history. However, since it plays such an important part in our everyday lives, I cannot omit:

1975: Bohr, Mottelson and Rainwater are awarded the Nobel Prize for the collective model and its relation to single-particle structure

nor the important rôle in the game played by Nilsson and Strutinsky.

Of course when Becquerel discovered radioactivity he had no real idea of what it was, though he did eliminate many of the more conventional possibilities. The Curies still did not understand what radioactivity was but they did realise that it was associated with the transmutation of the elements.

In order to understand better the nature of the nucleus, one has to cross the English Channel and go to Manchester to participate in the experiments of Geiger and Marsden. These lead to Rutherford's explanation of the backscattered α-particles in terms of an atomic nucleus of incredible smallness. This was an interesting example of international collaboration since Rutherford did indeed cross the Channel with the α-particle source purchased for these experiments from the Curies. The first, but not the last time a British experiment would be done with a beam provided by France.

The next step in our history takes us the 45 miles down the East Lancs Road from Manchester to Liverpool. The journey takes a surprising 21 years. The reason for this of course is the neutrality of the neutron which made it difficult for Chadwick to detect. Following this discovery, things rapidly begin to fall into place. Heisenberg is happy and the existence of isotopes is understood, since the mass number A is recognised as the sum of the numbers of neutrons and protons $A = N + Z$ in the nucleus.

Shortly after, Joliot and Joliot-Curie make the first artificial radioactive isotope but soon Fermi makes lots of them, exploiting the freshly discovered neutron. None of this of course is destined to take us far from β-stability.

Impressed by Fermi's success in producing new isotopes, Hahn, Meitner and Frisch set out to produce transuranic elements through neutron capture and β-decay. Their failure is of course the great success of the discovery of fission[1] and all that that entailed.

Not only did the neutron prove to be an extremely powerful experimental tool, it also provided a key to the theoretical understanding of many problems of nuclear structure. The single-particle shell model of Mayer and Jensen follows with the understanding of the two-body correlations leading to pairing phenomena coming from physicists working in an entirely different field. Many pairing phenomena are, however, unique to nuclei and such cross-fertilisations provide an essential argument for maintaining a strong research base in our subject as a part of the overall fabric of science.

3 BACK TO THE FUTURE

I have mentioned the above history since John asked me to talk about the future. I say this without irony since we all know, as scientists, that to extrap-

[1] Of course both induced and spontaneous fission do produce neutron-rich fragments far from stability and their spectroscopy has been [2] and will continue to be a major theme of research.

olate from a single point (the present) is impossible, whereas to extrapolate from two points is merely extremely unreliable. So let us revisit the problems mentioned above and see where we are today and where we might be going.

3.1 Nuclear radioactivity; so what's new?

Since the earlier days of nuclear physics when the classification of α-, β- and γ-rays emerged, later to be follwed by induced and spontaneous fission, many new modes of nuclear decay have been observed. Relatively recently, for example the extraordinary ground-state decays via emission of heavy clusters such as $^{12,14}C$, ^{24}Ne etc., giving us fascinating insights into many-particle correlations in heavy systems and the profound rôle of shell structures.

I would, however, like to mention in a little more detail a type of decay which takes us right out to the ridge of nuclear stability, the proton drip line [3], which is proving to be an extremely powerful spectroscopic tool [4].

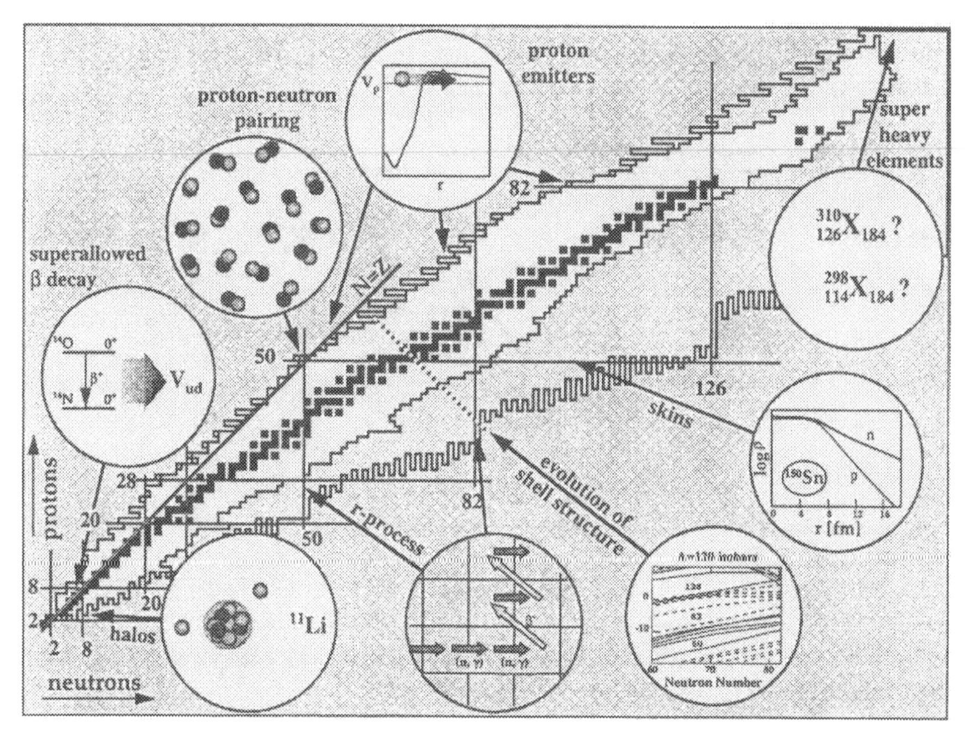

Figure 1 The nuclear $N-Z$ plane, showing many new phenomena either under exploration or which may shortly become accessible with the advent of radioactive beams [1].

Fig. 1 indicates the position of the proton emitters and the inset their decay mechanism, tunneling through the Coulomb, and possibly centrifugal barriers. By definition they lie on or near the proton drip line; a stiff hike up from the valley bottom to overlook the abyss of particle instability. Once past the drip line, the last proton in the nucleus finds no unoccupied state which lies below the top of the potential barrier and leaves the nucleus unimpeded in a time of around 10^{-22} s. The last proton in an emitter, however, although still having positive energy, must tunnel through the barrier and consequently the nucleus has a much longer lifetime. This can be of the order of milliseconds for a proton with high angular momentum l. Indeed a measurement of the proton energy and lifetime allows one to extract l and thus study the single-particle structure of these unbound levels.

An excellent example can be found in the review article of Woods and Davids [4]. Fig. 2 shows the ground-state-to-ground-state decay of ^{167}Ir to ^{166}Os via proton emission. Note, however, that a metastable state of ^{167}Ir can also decay to the ground state of ^{166}Os. The lifetimes for these two decays allow angular momentum assignments of $l = 0, 5$ respectively, permitting spin and parity assignments of the two states in the parent nucleus. Alongside this, however, ^{167}Ir (both ground state and isomer) can give rise to a chain of four α-decays down to the closed-neutron-shell nucleus ^{151}Tm ($N = 82$). Again lifetimes show that all these decays are $l = 0$, permitting spin assignments as well as the localisation of the ground states and isomers in all four daughter nuclei. Note especially the inversion between ^{159}Ta and ^{155}Lu.

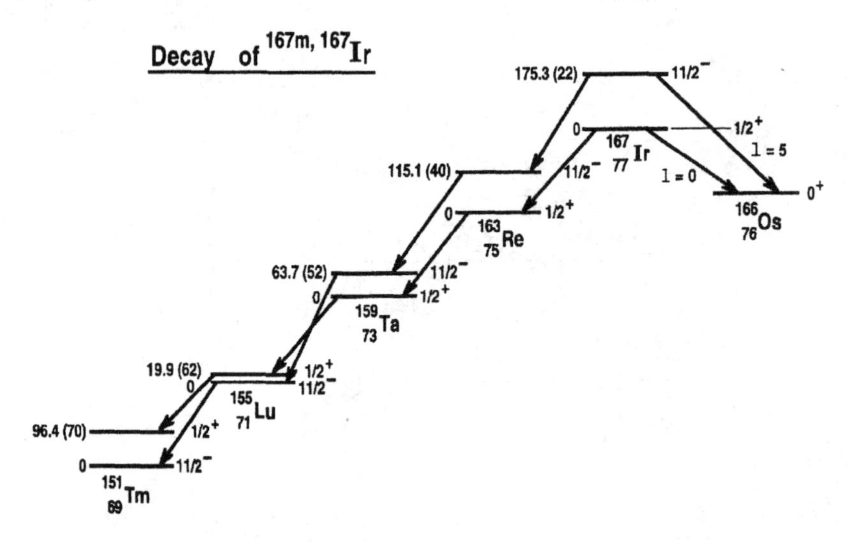

Figure 2 The interplay of proton decay and α-decay in this single example leads to a mine of nuclear structure information [4].

Much work remains to be done in this region of the $N - Z$ plane using such techniques. In addition one is starting to resolve the kind of fine structure in proton emission (long-since observed in α-decay) where the proton may leave the daughter nucleus in excited states. The search for simultaneous 2-proton emission is also underway with the interesting question as to whether the spatial correlations emerging from such experiments will allow us to talk of di-protons.

Moving down from the proton drip line, nuclei decay by β^+ emission or by electron capture. Fermi transitions, especially for $N = Z$ nuclei (see e.g. Ref.

[5]), are of particular interest. In addition, due to the steepness of the valley wall, much energy may be available for the weak decay, to the extent that a considerable part of the strength of the Gamow-Teller resonance may fall into the β-decay window (see e.g. Ref. [6]). Measured distributions of this strength will provide a fierce challenge to theoretical calculations of nuclear structure in this region. The high decay energy also means that further decays (γ-decay and proton emission) may follow providing a wealth of nuclear structure information through the exploitation of new detectors designed for the appropriate coincidence measurements.

3.2 The alchemists

The Curies' search for unknown elements was certainly arduous and demanded outstanding dedication to their task. At least, however, these elements existed to be discovered. If we are interested in seeing more, we must set about making our own! Most of the recent progress has come from the SHIP at GSI, where the new elements Z =110, 111 and 112 have been produced (see e.g. Ref. [7]).

An enormous effort is currently being put into taking one step further than this at the Joint Institute for Nuclear Research in Dubna, where particle-microamp beams of ^{48}Ca are being used to bombard a ^{244}Pu target and I am confident that the next superheavy element will be discovered in this way[2].

Of course future radioactive beams of sufficient intensity, especially those rich in neutrons, may open up new shipping lanes to the superheavy island of stability; lower intensities possibly being compensated by large enhancements of the fusion cross section [9] arising from strong coupling to neutron-transfer channels etc.

3.3 How big and how heavy?

Rutherford taught us that the nuclear radius varies according to $R = r_o A^{1/3}$, with $r_o \approx 1.1$ fm. The nuclear mass is approximately given by A ($= N + Z$) atomic mass units. This is powerful information but of course much more interesting is how these properties deviate from this apparent simplicity. The semi-empirical Bethe-Weiszäcker formula tells us that macroscopic properties play an important rôle in determining the nuclear binding energy and hence its mass. Of course pairing and shell effects are also crucial.

[2] A week or so after this talk was given, the discovery of $Z = 114$ was officially announced in Dubna. The reaction was ^{48}Ca + ^{244}Pu $\rightarrow ^{292}_{114}$X*. A single event was registered apparently corresponding to the evaporation of three neutrons to form a $^{289}_{114}$X residue. After 30.4 s, this nucleus emitted an α-particle of energy 9.71 MeV, followed by a second α-particle of energy 8.67 MeV 15.4 min later and a third α-particle of energy 8.83 MeV after 1.6 min. The final product ^{277}Hs spontaneously fissioned after 16.5 min. The very long lifetime of the $Z = 114$ isotope leads one to believe that the evaporation residue lies very close to the fabled island of stability [8].

The reader may wonder at the author's clairvoyance, or may suspect that he had a little inside information before his talk.

Figure 3 shows dramatically these latter two effects through the neutron separation energy S_n and the 2-neutron separation energy S_{2n} for the calcium isotopes. The curve for S_n shows clearly the effects of pairing and that for S_{2n} shows the effects of crossing the neutron shells at $N = 20$ and $N = 28$. Thus a single piece of information, the nuclear mass, recounts interesting stories of nuclear structure. This is just as well since for many exotic nuclei this may be the only information accessible in the relatively near future! Enormous leaps are, however, being made in providing such information. One might, for example, cite the Schottsky Mass Spectrometer experiments at GSI, where vast swathes of the $N - Z$ plane have been tackled and mass measurements obtained [10].

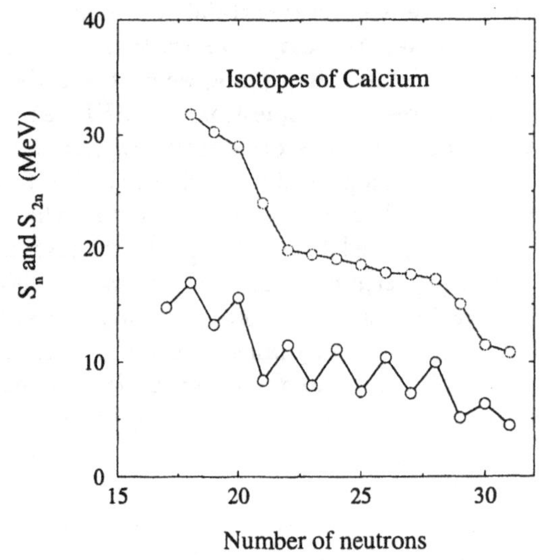

Figure 3 The neutron separation energy S_n and the 2-neutron separation energy S_{2n} for the calcium isotopes show beautifully the effects of pairing and of shell closures.

One can go one stage further than separation energies and define the pairing energy directly through $P_N = |M_{A+1} - 3M_A + 3M_{A-1} - M_{A-2}|/4$. A study of this property shows that the pairing strength itself may change abruptly at shell closures, though a full analysis of the effect also requires [11] a detailed knowledge of the corresponding nuclear deformation. Again, fortunately, this is a quantity which can be obtained from experiments involving relatively few atoms using laser-spectroscopy techniques [12]. These give a measure of the nuclear rms charge radius which in turn depends on its deformation. Recent experiments at Jyväskylä [13] (the first to successfully handle refractory elements on-line) have even yielded information on the differences in radii between nuclei in their ground and meta-stable states.

The increasing sophistication of techniques involving trapped ions of exotic nuclei should allow us in the future to push such measurements further and further from the stability line.

3.4 Breaking up is hard to do; the still-neutral neutron

As commented early, the neutron remained a mystery for many years due to its lack of charge. This property still makes it difficult to detect but large modern arrays of efficient neutron detectors such as the franco-belgian DEMON array make possible experiments which would have left Chadwick breathless.

One could cite, for example, the use of this detector system as a nuclear clock and thermometer in the measurement of the multiplicity and energy spectrum of the neutrons from a fission event. This idea has recently been applied to the phenomenon of bi-modal fission [14] of ^{226}Th (^{18}O + ^{208}Pb at the Strasbourg Vivitron) where the fission-fragment mass distribution displays two distinct components, one symmetric and the other asymmetric. This shows the existence of two separate valleys in the fragment-fragment potential energy and the associated neutrons give valuable information on the timescales relating to the propagation down these.

A real *tour de force*, however, is the exploitation of this multidetector to the measurement of the size of neutron halos (see Fig. 1). Here [15], two neutrons emitted from a halo during a break-up reaction are detected in coincidence so that the technique of interferometry (as used to measure the size of distant stars) can be employed to measure the size of the halo; and this despite the very low intensity beams of ^{11}Li and ^{14}Be etc. available.

4 GAMMA-RAY SPECTROSCOPY

The discovery of a superdeformed fission isomer by Polikanov in 1962 [16] suggested that maybe such states existed in other nuclei and that one should think of other means of populating them. This was eventually achieved using the EUROGAM detector in Daresbury in 1986 [17] by the creation of ^{152}Dy at very high angular momentum, an experiment in which John Sharpey-Schafer played an active part. Since then a plethora of superdeformed rotational bands have been discovered in many other nuclei in the same and different regions of mass. More importantly than this, some of these bands have been shown to have some remarkable properties; a C_4 or $\Delta I = 4$ staggering, identicity of bands [18] in different nuclei etc.

One obstacle which remained for many years was to establish how these bands decay to the known ground-state bands. Without this information, these structures were left "hanging in the air" with no firm assignments of their spins and with no absolute measure of their energies. The breakthrough of finding the linking transitions was made in 1996 [19, 20] in ^{194}Hg and ^{194}Pb. This opened up many new possibilities for the future; a genuine spectroscopy of the second potential well, an experimental determination of the pairing strength in the bands if links are found in adjacent nuclei etc.

Of course high-spin physics is not simply concerned with superdeformations and other topics of interest can be found in the splendid GAMMASPHERE booklet [21] for which we should congratulate Mark Riley and his colleagues. Figure 4, taken from this booklet, highlights a number of the phenomena in question. Of course much of the physics shown on this figure also stems from EUROGAM (both in Daresbury and in Strasbourg) and from earlier arrays such as TESSA as well as work pursued with EUROBALL III in Legnaro and to be further developed with EUROBALL IV, again in Strasbourg. While some of this physics is well established, much is still in its infancy. Problems for the future will almost certainly include the coupling of charged-particle devices to such γ-ray arrays, not simply to provide a gate by which to clean up the spectra of particular residues but to explore the details of reaction mechanisms, molecular resonances and fission etc. The coupling of appropriate recoil devices will also permit the observation of the excited states of very heavy nuclei produced weakly in fusion-evaporation experiments such as those on ^{254}No at the Argonne Laboratory and in Jyväskylä [24, 25].

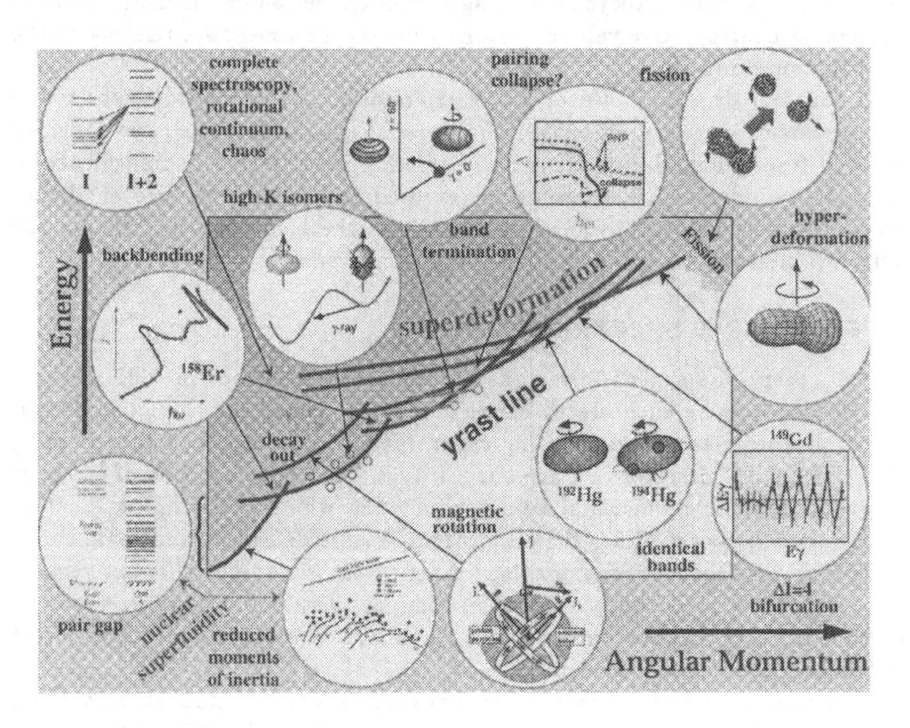

Figure 4 The angular momentum world of the nucleus [21].

The future will also almost certainly have surprises up its sleeve, like the totally unexpected proton decay of the highly deformed band in ^{58}Cu [22] or the curious "shears" bands corresponding to "magnetic rotations" [23]. It may even herald the arrival of the ultimate γ-ray devices exploiting "tracking" techniques.

I will say no more about γ-ray physics, since I suspect that we will hear much more about it from other (more qualified) speakers at this conference. I just leave you with the thought that on a board at the poster session I noticed a copy of a recent paper on fission isomers "Experimental Evidence for *Hyperdeformed*

States in U Isotopes [26]" which transports us neatly back to the beginning of this section...

5 EPILOGUE: DOES SIZE MATTER?

The future of nuclear physics seems bright, though we have hinted at the dominance of fearsome arrays of detectors and radioactive beams. Do laboratories which do not possess such capabilities have a rôle to play in this future?

I think that the answer is clearly yes and would like to give two examples to support this. The first is the relatively modest GAREL+ array which we have exploited in Strasbourg to partially fill the breach left by EUROGAM before the return of EUROBALL. By coupling its 14 large-volume Ge detectors to the Recoil Filter Detector (Berlin/Krakow/Strasbourg) and the Betatronc conversion-electron spectrometer (CSNSM Orsay/ ISN Grenoble), much interesting physics became accessible. Indeed you will hear a couple of contributions on GAREL+ during this meeting.

My second example comes from the southern hemisphere. The Nuclear Physics Laboratory at the Australian National University in Canberra has similarly exploited its small CESAR array to become world leaders in the study of high-K isomeric states through the application of imaginative timing techniques. Also, in the field of my own theoretical work, the use of a relatively simple velocity filter has allowed the group to lead in the field of "fusion barrier distributions" [9].

I am, therefore, confident that the new AFRODITE array at the National Accelerator Centre here in Faure, coupled to its ancillary detectors but more importantly, coupled to the great enthusiasm and experience of John Sharpey-Schafer, can also play an important rôle in the future of nuclear physics research.

Acknowledgments

The author is grateful to the many colleagues who generously offered him figures for use in his talk.

References

[1] W. Nazarewicz, B. Sherrill, I. Tanihata, and P. Van Duppen, Physics with radioactive beams, Nuclear Physics News 6 (1996) 17

[2] I. Ahmad and W.R. Phillips, Gamma-rays from fission fragments, Reports on Progress in Physics 58 (1995) 1415

[3] W. Gelletly and P.J. Woods, Neutron-deficient nuclei studied with stable and radioactive beams, Phil. Trans. R. Soc. London A 356 (1998) 2033-2062

[4] P.J. Woods and C.N. Davids, Nuclei beyond the proton drip-line, Annu. Rev. Nucl. Part. Sci. 47 (1997) 541-90

[5] C. Longour et al., Phys. Rev. Lett. 81 (1998) 3337

[6] Ch. Miehe et al., Proc. of Int. Symp. on New facets of spin giant resonances in nuclei, Tokyo, Japan 1997, Eds. H.Sakai, H. Okamura and T. Wakasa, (World Scientific, 1998) 140

[7] S. Hofmann, Recent results from SHIP, Proc. of the VI Int. School-Seminar on Heavy Ion Physics, Dubna, September 1997, Eds. Yu. Ts. Oganessian and R. Kalpakchieva (World Scientific, Singapore, 1998) 385

[8] Yu. Ts. Oganessian et al., Synthesis of superheavy nuclei in the ^{48}Ca + ^{244}Pu reaction, submitted to Phys. Rev. Lett.; Preprint JINR, E7-99-53

[9] M. Dasgupta, D.J. Hinde, N. Rowley and A.M. Stefanini, Measuring barriers to fusion, *Annu. Rev. Nucl. Part. Sci.* 48 (1998) 401-61

[10] H.Geissel, T.Radon et al., CP455, ENAM 98: Proc. 2nd Int. Conf. on Exotic Nuclei and Atomic Masses, Eds. B.M. Sherrill, D.J. Morrissey, and Cary N. Davids, 1998, The American Institute of Physics 1-56396-804-5/98;
T. Radon et al., Phys. Rev. Lett. 78 (1997) 4701

[11] O. Burglin and N. Rowley, Details of nuclear masses away from stability: Pairing and deformation effects, Int. Symp. on Exotic Nuclear Shapes, Debrecen, May 1997, Heavy Ion Phys. 6 (1997) 189

[12] J. Billowes and P. Campbell, "High-resolution laser spectroscopy for the study of nuclear sizes and shapes", J.Phys. G: *Nucl. Part. Phys.* 21 (1995) 707.

[13] J.M.G. Levins et al., Phys. Rev. Lett. 82 (1999) 2476

[14] A. Kelic et al., submitted to Phys. Rev. and Strasbourg Preprint IReS 98-23;
M.G.Itkis et al., Proc. of INPC 98 Conference, Paris, August 1998, Nucl.Phys. A in press

[15] F.M. Marqués, M. Labiche, N.A. Orr et al., submitted to Phys. Rev. Lett., LPC Report LPCC-99-13

[16] S.M. Polikanov, Soc. Phys. JETP 15 (1962) 1016

[17] P.J. Twin et al., Phys. Rev. Lett. 57 (1986) 811

[18] B. Haas, Identical superdeformed bands, Prog. Part. Nucl. Phys. 38 (1997) 1

[19] T.L. Khoo et al., Phys. Rev. Lett. 76 (1996) 1583

[20] A. Lopez-Martens et al., Phys. Lett. 380B (1996) 18

[21] GAMMASPHERE Highlights Booklet, Ed. M.A. Riley, published by the Gammasphere Users Executive Committee (1998). See http://www-gam.lbl.gov

[22] D. Rudolf et al., Phys. Rev. Lett. 80 (1998) 3018

[23] R.M. Clark et al., Phys. Rev. Lett. 78 (1997) 1868

[24] P. Reiter et al., Phys. Rev. Lett. 82 (1999), 509.

[25] M. Leino et al., XXXIII Zakopane School of Physics, Zakopane, Poland, 1-9 Sept. 1998 and to be published; R.-D. Herzberg et al., Verhandl. DPG 34 (1999) 153.

[26] A. Krasznahorkay et al., Phys. Rev. Lett. 80 (1998) 2073

II Delegate Contributions

STRUCTURE OF LIGHT MASS (EXOTIC) NUCLEI AS EVIDENCED BY SCATTERING FROM HYDROGEN

K. Amos[A], P. J. Dortmans[A], and S. Karataglidis[B]

[A] School of Physics, University of Melbourne, Parkville, Victoria 3052, Australia.
[B] TRIUMF, 4004 Wesbrook Mall, Vancouver, British Columbia, V6T 2A3, Canada.

Abstract: Microscopic optical model potentials generated by full folding of realistic two-nucleon (NN) interactions with nuclear structure specified by large basis shell model calculations have been constructed. With those (nonlocal) potentials, predictions of light mass nuclei–hydrogen scattering result that agree well with observations of cross sections and analyzing powers.

1 INTRODUCTION

A topic of current interest is the specification of the structures of exotic nuclei such as the neutron/proton rich isotopes of light mass nuclei. Many of these nuclei can be formed as radioactive beams and experiments made to determine their scattering from, and reactions with, stable nuclei. The scattering of such exotic nuclei from hydrogen targets is of particular interest as that scattering data should be sensitive to properties of the ground state of these nuclei. Analysis of that data is feasible also as inverse kinematics equates the process to the scattering of energetic protons from them as targets.

It is now possible to predict observables from elastic and inelastic proton-nucleus (pA) scattering at intermediate energies [1] in a manner consistent with that employed for electron scattering. To do so, three basic aspects of the system under investigation are required. Where possible, these properties must be determined independently of the pA scattering system being studied. First, the description of the nucleus (i.e. one body density matrix elements, OBDME) should be determined from large scale structure calculations which describe well the ground state properties (and low excitation spectra if pertinent) of the nucleus in question. The second aspect is the choice of nucleon bound state wave functions. They, with the given OBDME, can be assessed by their use in fitting elastic electron scattering form factors. The final ingredient is the complex, energy and density dependent, effective NN interaction that exists between the incident and struck nucleon. This effective interaction, which we suppose has central, tensor and two–body spin–orbit components each having a radial variation that is a sum of Yukawa functions, is defined so that it reproduces accurately (momentum space) half–off–shell NN t– and g–matrices associated with realistic NN potentials.

With all three ingredients specified, energy dependent, complex, and non-local optical potentials have been formed for the scattering of 65 to 800 MeV protons from any nucleus. With those nonlocal optical potential, solutions of the Schrödinger equations yield differential cross sections, analyzing powers and spin rotations for elastic scattering in very good agreement with data [1]. We stress that they are predictions. No *a posteriori* adjustment to any of the details of the calculations has been made.

The effective NN interaction in nuclei

We consider a realistic microscopic model of pA reactions to be one that is based upon NN t–matrices whose on–shell values are consistent with measured NN scattering data to and above the incident energies of interest, and whose properties off of the energy shell are consistent with data such as NN bremsstrahlung. For energies below the pion threshold, the Paris, Bonn, and Hamburg (OSBEP) interactions [2] satisfy those requirements quite well. Above that threshold, no meson exchange model accounts well enough for the resonances and flux loss effects to match the Arndt phase shifts. Recently, however, by supplementing the OBEP model interactions with NN optical potentials [3] those phase shifts could be matched to 2.5 GeV with (NN) optical potentials that are smooth functions of energy, and are consistent with both the known resonance characteristics and the known profile function of very high energy NN scattering. Such we have found to be appropriate starting interactions to determine effective NN interactions within the nuclear medium when energies are above the pion threshold when minimal relativity is considered.

Considering energies below pion threshold, the NN t–matrices are solutions of Lippmann–Schwinger equations. However, if the struck nucleon is embedded in a nuclear medium, calculations of the NA optical potential should be based instead upon medium modified NN g–matrices. We take those to be solutions of Brueckner–Bethe–Goldstone (BBG) equations in which allowance is made for Pauli blocking and average fields upon the scattering. Details of the calculations have been given previously [4]; the results being tables of complex numbers for each NN channel, for each incident energy, at each Fermi momentum value, and for a selected set of relative momenta.

At energies to over 200 MeV, the medium effects that differentiate the g- from the t-matrices are quite severe [4]. That is especially so for the on-shell values. Furthermore these effects are quite complex and cannot be represented at all reasonably in any simple function of the density itself.

The shell models of structure

For light mass nuclei ($A < 16$), all significant structure calculations have been made using the shell model program OXBASH [5]. With very light mass nuclei (such as the He and Li isotopes), in the main we have used the matrix elements of Zheng *et al.* [6] in those calculations. Complete $(0 + 2 + 4)\hbar\omega$ and also $(0+2+4+6+8)\hbar\omega$ model space calculations have been considered for $A \leq 6$. For $6 < A > 16$, complete $(0 + 2)\hbar\omega$ space calculations have been made using a standard (MK3W) set of potentials. The structure of exotic nuclei (^{6,8}He, ^{9,11}Li) have been determined in this way as well. More details of these shell model structures of light mass nuclei have been published recently [7]. The results of the calculations are OBDME, and, for the the most recent studies [6, 7], the single nucleon bound state wave functions.

The pA optical potentials

When effective interactions as described above are folded with the target OBDME and proper account taken of the antisymmetry of the the pA wave function, complex nonlocal spin dependent optical potentials result, with the form

$$
\begin{aligned}
U(\vec{r}_1, \vec{r}_2; E) &= \delta(\vec{r}_1 - \vec{r}_2) \sum_n \zeta_n \int \varphi_n^*(\vec{s}) v^D(\vec{r}_{1s}, E; \rho[k_f(r_{1s})]) \varphi_n(\vec{s})\, d\vec{s} \\
&\qquad + \sum_n \zeta_n \varphi_n^*(\vec{r}_1) v^{Ex}(\vec{r}_{12}, E; \rho[k_f(r_{12})]) \varphi_n(\vec{r}_2) \\
&\Rightarrow U_D(\vec{r}_1, E) + U_{Ex}(\vec{r}_1, \vec{r}_2; E)
\end{aligned}
$$

where v^D and v^{Ex} are appropriate combinations of the NN ST channel elements of the effective interaction, $\varphi_j(\vec{r})$ are the single nucleon bound state wave functions and ζ_n are the shell occupancies of the target nucleus. In the past, the leading term has been used alone (the $g\rho$ or $t\rho$ approximation), or the nonlocal (exchange) elements have been approximated by 'equivalent' local interactions. Neither is a satisfactory approach for the analyses of data from the scattering of intermediate energy protons. Indeed when we used the $g\rho$ form with our effective interaction, the cross sections and analyzing powers that result are markedly changed from those given by our complete calculations and which reproduce observed results very well [1].

2 RESULTS OF CALCULATIONS

Using the complete, nonlocal optical potential generated microscopically as described above, we have made predictions of differential cross sections, analyzing powers, and spin rotations for the elastic scattering of 65 and 200 MeV protons from 50 nuclei ranging from ^{3}He to ^{238}U. All of those predictions are in very good agreement with observation for data forward of 60° scattering angle typically (for which cross sections are of order 1 mb/sr and greater).

Scattering from stable nuclei

Results of our calculations of proton scattering from many stable nuclei have been presented in detail elsewhere [1]. Herein to illustrate, sample results for 65 MeV from a set of nuclei (^{7}Li to ^{64}Zn) are given in Fig. 1 (cross sections and analyzing powers on the left and right respectively). We stress that the calculated results presented are predictions; the data were included in these figures AFTER the curves shown were defined. A complete discussion of the crucial role of using a medium modified NN effective interaction is explained in recent literature [1].

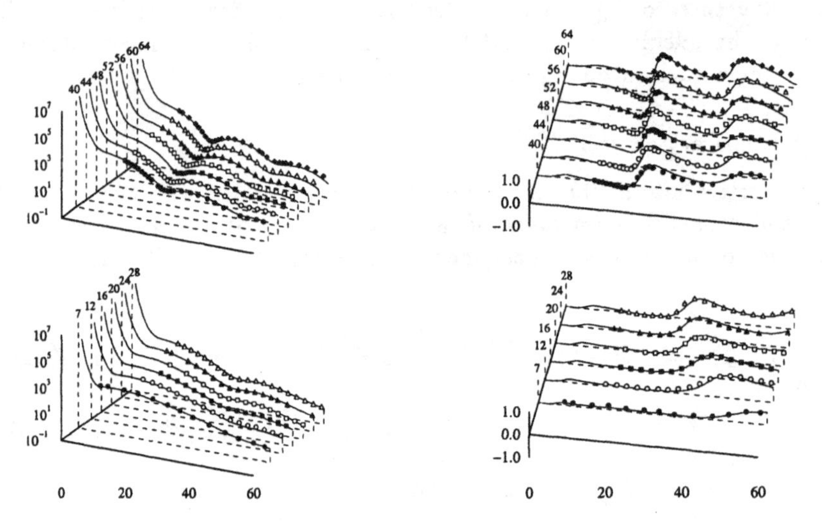

Figure 1 $\frac{d\sigma}{d\Omega}$ and A_y of 65 MeV protons scattering from diverse nuclei

Scattering of ^{6,8}He, and ^{9,11}Li from Hydrogen

The results from our analyses of proton scattering from ^{3,4}He and ^{6,7}Li [7] indicate that we have appropriate shell model descriptions of these light nuclei. Such shell models should be appropriate also for the exotic light mass nuclei. Indeed already the $(0+2)\hbar\omega$ model space structure has been used with success in studies of proton scattering from ^{9,11}Li [8]. Also, and very recently, the low excitation spectrum of ^{11}Li seems to have been identified [9] and, although no spin-parity assignments have yet been made, the negative parity states of our spectrum match likely entries from that experiment. For the other 'exotic' nuclei to be considered, ^{6,8}He, the $(0+2+4)\hbar\omega$ model space structure has been used. In fact the ground state of ^{6}He we take as the isobaric analogue of the $0^{+};1$ (3.56 MeV) state in ^{6}Li. The structure calculations were made using the G matrix interaction of Zheng *et al.* as input to the shell model code OXBASH. Calculations that use single-particle wave functions from the shell model calculations do not coincide with a "halo" structure. We estimate the effects of an existent halo by varying those single particle wave functions.

In all cases then we first specify the single particle bound states by Woods-Saxon (WS) wave functions. Those which gave good reproduction of the elastic electron scattering form factors of ^{6}Li were used for the ^{6,8}He calculations while those which reproduced the elastic electron scattering form factors of ^{9}Be were used in the calculations for ^{9}Li and ^{11}Li. With such wave functions, we consider the nuclei to be of "non–halo" type.

In these analyses, the ^{8}He and ^{9}Li results act as controls. Since the single neutron separation energies are 2.137 MeV and 4.063 MeV for ^{8}He and ^{9}Li, respectively, we consider ^{8}He to be an example of a neutron skin while ^{9}Li we believe is a simple core nucleus. We artificially ascribe a halo to these nuclei to ascertain if the procedure and data are sensitive enough to detect the flaw. To specify "halo" nuclear properties, we adjust the WS potentials such that the relevant neutron orbits are weakly bound. This guarantees an extensive neutron distribution. For ^{6,8}He, the 0p-shell binding energy was set to 2 MeV, which is close to the separation energy (1.87 MeV) of a single neutron from ^{6}He. For ^{9}Li and ^{11}Li, the halo was specified by setting the binding energy for the WS functions of the $0p_{\frac{1}{2}}$ and higher orbits to be 0.5 MeV. The neutron density profiles for ^{6}He, ^{8}He, ^{9}Li, and ^{11}Li obtained from the present shell model calculations are shown in Fig. 2. Therein the dashed and solid lines portray, respectively, the neutron profiles found with and without the halo conditions

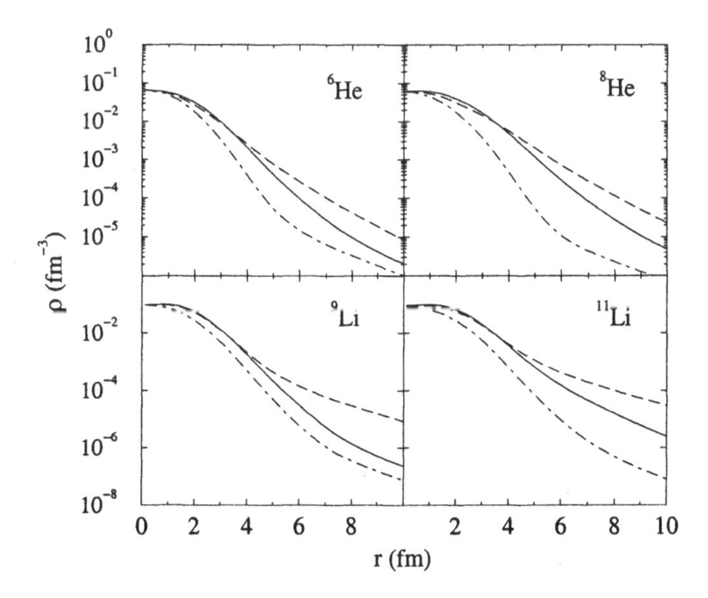

Figure 2 The proton and neutron (non-halo and halo) densities of some exotic nuclei

being implemented. The dot-dashed line in each case represents the proton density. As the folding process defines the optical potentials, we expect that the internal ($r < r_{\rm rms}$) region influences the predictions of differential cross sections, notably at large scattering angles.

With the structures defined above, calculations in inverse kinematics were made of the scattering of 72 MeV per nucleon ^{6,8}He and of 60 - 62 MeV per nucleon ^{9,11}Li ions from hydrogen targets. The core single particle wave functions were those used in our calculations of scattering from the other He and Li isotopes. Now, however, there are more loosely bound neutrons and we have little data other than the scattering from hydrogen to help ascertain details of these. When the OBDME and wave functions discussed above were used in calculations of proton scattering, the differential cross sections that are compared with the data in Fig. 3 were obtained. The predictions obtained by using

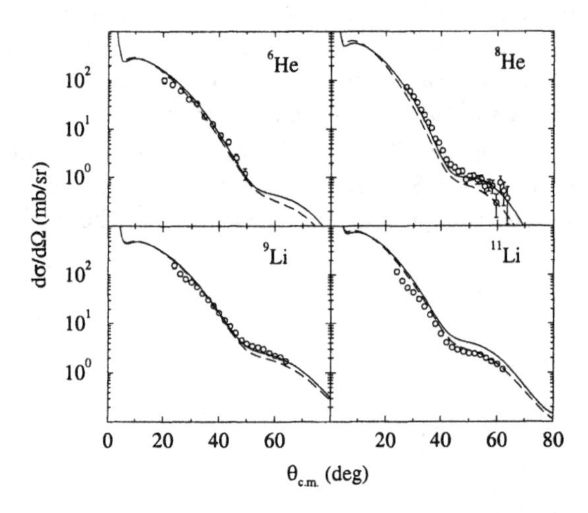

Figure 3 $\frac{d\sigma}{d\Omega}$ for the scattering of $72A$ MeV ^{6,8}He and of $62A$ MeV ^{9,11}Li from hydrogen.

the no-halo (basic shell model) structure information are displayed by the solid curves, while those found by using the enforced neutron 'halo' forms are depicted by the dashed curves. Our results are very suggestive of a neutron 'halo' for ^{11}Li given that the match to data came with the $0p_{1/2}$, $1s$–$0d$, and the $0f$–$1p$ shell neutrons in the ground state being described by WS wave functions with 0.5 MeV binding energies. Likewise the comparisons indicate that ^{9}Li and ^{8}He do not have more extensive neutron distributions than the neutron skins that standard shell model calculations give.

It is of note that the halo nature of the nucleus is manifest in these differential cross sections at relatively large momentum transfer values. As such those cross sections are not particularly sensitive to the details of the wave function at the large radii (the traditional halo region). Thus the current scattering data reflect the 'depletion' (or no) of neutron strength in the interior of these nuclei from that we expect of non-halo constructs. At these energies, there are variations in cross section predictions caused dominantly by the neutron probability amplitudes at large radii. But they occur in the vicinity of the hadron–Coulomb interference regime. For 60 to 70A MeV light ions this is the region between 5 and 10° C. of M. scattering angle.

Scattering of 800 MeV protons from ^{12}C

Radioactive beams are planned at higher energies and our results [1] indicate that the microscopic model approach can be used with confidence at 200A MeV. We demonstrate next that such should also be the case at 800A MeV.

We have used boson exchange model NN interactions [2], OSBEP and BCC3 specifically, modulated by NN optical potentials [3], to specify NN t- (and g-) matrices at 800 MeV. With those t-matrices, the SM97 phase shifts to 2.5 GeV are fit extremely well. Coordinate space effective interaction forms that map very well the associated 800 MeV t– and g–matrices have been determined and then used in a full folding model to specify the complex and nonlocal optical potentials for 800 MeV protons incident on ^{12}C. The structure of the target used in that folding was determined from a large space (complete $(0+2)\hbar\omega$) shell model calculation which, in the past, gave electron scattering form factors in very good agreement with measured values. Thereby all quantities required in the folding process have been preset to make solution of the associated non-local pA Schrödinger equations predictive of the pA scattering phase shifts and so of the differential cross sections and analyzing powers. The differential cross section data from 800 MeV protons scattering off of ^{12}C are shown in Fig. 4. Clearly with just the basic OBEP used to specify our optical potentials (the

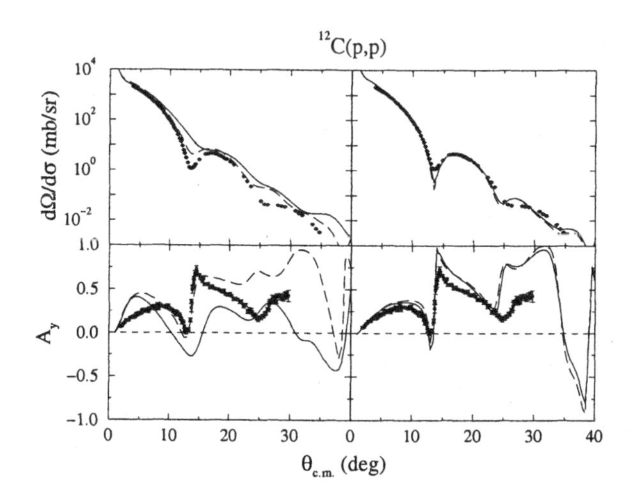

Figure 4 $\frac{d\sigma}{d\Omega}$ and A_y for 800 MeV protons scattered from ^{12}C.

left hand panel) neither result matches observation, although the BCC3 interaction gives better predictions than the OSBEP case. The BCC3 interaction accounts for effects of the Δ resonance and so innately is a 'more realistic' base interaction for the energy regime above 300 MeV. However, when either of the interactions are modulated by an (NN) optical potential and the proton-^{12}C optical potential defined by a full folding of the effective interactions specified by the attendant t- and g-matrices, our p$-^{12}$C cross section and analyzing power predictions are in very good agreement with the data to scattering angles of 25° (by which the magnitudes have fallen to less than 0.1 mb/sr). The effects of the modulation of both the OSBEP and the BCC3 models are very noticeable with the cross sections and even more so with the analyzing powers. Only with the modulations that tune OSBEP and BCC3 against the SM97 data set has satisfactory reproductions of that analyzing power structure been found.

3 CONCLUSIONS

Our predictive theory of $p-A$ scattering permits analyses of radioactive beam–hydrogen scattering to assess conjectured matter profiles of those exotic nuclei. We are confident that such is the case for all energies in the range $60A$ to $800A$ MeV. The available scattering data from hydrogen confirm that ^{11}Li is a halo nucleus, while the analysis of the scattering data correctly determines that both ^{8}He and ^{9}Li are not. The low-angle scattering results also suggest that ^{8}He is a neutron skin nucleus, as found from breakup reactions. The available scattering data for ^{6}He from hydrogen are not extensive enough to discriminate between the halo and non-halo scenarios as yet.

References

[1] P. J. Dortmans, K. Amos, and S. Karataglidis, *J. Phys. G* 23, 183 (1997); P. J. Dortmans, K. Amos, S. Karataglidis, and J. Raynal, *Phys. Rev. C* 58, 2249 (1998).

[2] M. Lacombe *et al.*, *Phys. Rev. C* 21, 861 (1980); R. Machleidt. *Adv. Nucl. Phys.* 19, 189 (1989); L. Jäde and H.V. von Geramb, *Phys. Rev. C* 55, 57 (1997); *ibid.* C 56, 1218 (1998).

[3] H.V. von Geramb, K.A. Amos, H. Labes, and M. Sander, *Phys. Rev. C* 58, 1948 (1998).

[4] P. J. Dortmans and K. Amos, *J. Phys.* G17, 901 (1991); *Phys. Rev C* 49, 1309 (1994).

[5] B. A. Brown, A. Etchegoyen, and W. D. M. Rae, *computer code OXBASH-MSU*, MSUCL Report No. 524, 1986 (unpublished).

[6] D. C. Zheng *et al.*, *Phys. Rev. C* 52, 2488 (1995).

[7] S. Karataglidis, B. A. Brown, P. J. Dortmans, and K. Amos, *Phys. Rev. C* 55, 2723 (1997); P. J. Dortmans, K. Amos, and S. Karataglidis, *Phys. Rev. C* 57, 2433 (1998).

[8] S. Karataglidis *et al.*, *Phys. Rev. Lett.* 79, 1447 (1997).

[9] M. G. Gornov, *et al.*, *Phys. Rev. Lett.* 81, 4325 (1998).

LIFETIME MEASUREMENTS AND DIPOLE TRANSITION RATES FOR SUPERDEFORMED STATES IN ^{190}Hg

H. Amro,[1,2] R. V. F. Janssens,[2] E. F. Moore,[1] G. Hackman,[2] S. M. Fischer,[2] I. Ahmad,[2] M. P. Carpenter,[2] B. Crowell,[3] T. L. Khoo,[2] T. Lauritsen,[2] D. Nisius,[2] J. Timar,[4] A. Wilson[4]

[1]North Carolina State University, Raleigh NC 27695, and Triangle Universities Nuclear Laboratory, Durham NC 27708-0308, USA, [2]Argonne National Laboratory, Argonne, IL 60439, USA, [3]Fullerton Community College, Fullerton CA 92833, USA, [4]Oliver Lodge Laboratory, University of Liverpool, Liverpool, United Kingdom

Abstract: The Doppler-shift attenuation method was used to measure lifetimes of superdeformed (SD) states for both the yrast and the first excited superdeformed band of ^{190}Hg. Intrinsic quadrupole moments Q_0 were extracted. For the first time, the dipole transition rates have been extracted for the inter-band transitions which connect the excited SD band to the yrast states in the second minimum. The results support the interpretation of the excited SD band as a rotational band built on an octupole vibration.

1 INTRODUCTION

The first excited SD band in the ^{190}Hg nucleus has a rather unusual behavior, when compared with excited SD bands in other nuclei of the region [1, 2, 3, 4]. In particular, this SD band decays entirely into the yrast SD band, instead of decaying directly towards the normal deformed states. The inter-band transitions, probably of E1 character, linking this band to the yrast SD band were observed [2, 3] allowing for the excitation energy, spins and possible parity to be determined relative to the yrast SD band. At the point of decay into the yrast SD band, the excitation energy of this excited band is 911 keV, i.e. lower than would be expected for a two quasiparticle excitation. The dynamic moment of inertia $\Im^{(2)}$ for the excited band is essentially flat as a function of rotational frequency and about 20% larger than that of any SD band in the A=190 region of superdeformation. When this excited band and its decay pattern were

The Nucleus: New Physics for the New Millennium
Edited by Smit et al., Kluwer Academic / Plenum Publishers, New York, 2000.

observed for the first time, it was proposed that this could correspond to a rotational band built on an octupole-vibrational phonon in the SD well [3].

The results from the present work provide answers to some of the remaining questions regarding this excited band. First, the dipole transition rates for the linking transitions were determined. Second, the quadrupole moments, Q_0, were extracted for both the excited and the yrast SD band.

2 LIFETIMES AND TRANSITION RATES

Superdeformed states in ^{190}Hg were populated with the ^{160}Gd(^{34}S,4n) reaction at a beam energy of 159 MeV. The beam was provided by the 88-Inch Cyclotron at LBNL. A ^{160}Gd target (1.17 mg/cm^2) evaporated onto a thick (13 mg/cm^2) Au backing was used. The γ rays were detected by the Gammasphere array, which consisted of 87 Compton-suppressed Ge detectors at the time of the experiment. A total of 1.1×10^9 events with fold ≥ 4 was recorded. The data were sorted into a number of spectra gated by triple coincidence windows placed on combinations of transitions in the SD bands. Spectra corresponding to each of the angular rings of Gammasphere were constructed individually. In order to derive lifetimes and transition rates from the data, a Doppler shift attenuation method (DSAM) analysis was performed. Linear fits of $\overline{E_\gamma}$ versus $cos(\theta)$ were performed using the first-order Doppler shift expression $\overline{E_\gamma} = E_{\gamma_0}[1+\beta_0 F(\tau)cos(\theta)]$ to extract the $F(\tau)$ values and unshifted γ-ray energies E_{γ_0}. The $F(\tau)$ values are presented in Fig. 1 as a function of the γ-ray energy for transitions in both bands and for the four linking transitions. The intrinsic quadrupole moments Q_0 of the two SD bands were extracted from the experimental F(τ) values using the computer code FITFTAU [6] which requires a velocity profile for the recoiling residues and a model for γ-ray decay of the SD cascade. The velocity profiles of the recoiling Hg ions in the target and the Au backing were calculated using the code TRIM, version 1995, by Ziegler [7]. The following assumptions are made for the decay cascade: (1) the Q_0 values are the same for all SD levels within a band, (2) the sidefeeding into each SD state is approximated by a single rotational cascade (with the number of transitions in the sidefeeding cascade proportional to the number of transitions in the main band above the state of interest), having the same $\Im^{(2)}$ moment as the main band, and controlled by a sidefeeding quadrupole moment Q_{sf} (assumed to remain the same throughout an entire SD band), and (3) a one-step delay at the top of all feeder cascades was parameterized by a single lifetime T_{sf}. A χ^2 minimization using the fit parameters Q_0, Q_{sf}, and T_{sf} was then performed to the measured $F(\tau)$ values. Results from the fit are illustrated in Fig. 1. It is clear that the Q_0 moments for both SD bands are the same within their respective error bars. Using the lifetimes derived from the present analysis and the branching ratios from ref. [4], absolute transition rates for the dipole inter-band transitions band transitions were calculated under the assumption of either $E1$ or $M1$ character using the expressions $B(E1) = 6.288 \times 10^{-16} B.R./\tau E_\gamma^3$ and $B(M1) = 5.687 \times 10^{-14} B.R./\tau E_\gamma^3$, respectively. Here $B.R.$ is the out-of-band

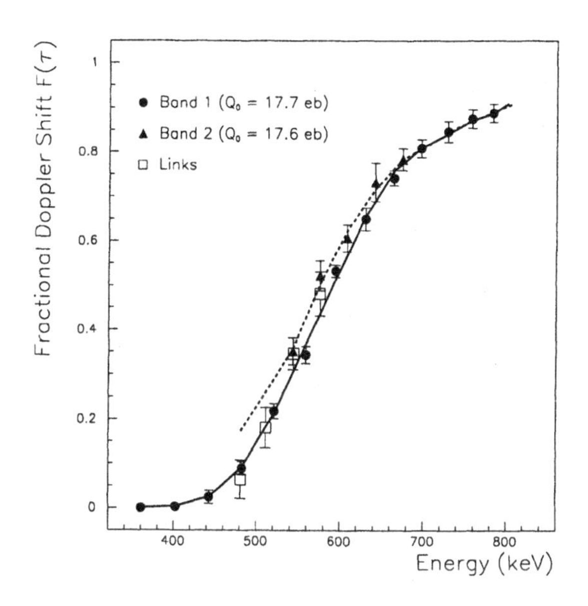

Figure 1 Measured fractional shifts $F(\tau)$ for γ-rays in the two SD bands (1 and 2) and for the inter-band transitions in ^{190}Hg The. $F(\tau)$ values for the linking transitions are plotted at the γ-ray energies of the corresponding in-band transitions of band 2. The solid and dashed curves represent the best fits corresponding to the values of the Q_0 moment given in the figure.

branching ratio, E_γ is the γ-ray energy in MeV, and τ is the mean lifetime of the SD level in seconds. The results are presented in Table 1.

Table 1 Information obtained for the dipole transitions linking band 2 and band 1. The reduced matrix elements assuming pure $E1$ or $M1$ radiation are given in terms of Weisskopf units.

Transition energy (keV)	*Inter-band branching ratio*	*Lifetime of level (fs)*	$B(E1)$ *(W.u.* $\times 10^{-3}$*)*	$B(M1)$ *(W.u.)*
757	0.23 (8)	110 (20)	1.5 (6)	0.16 (6)
812	0.29 (10)	130 (30)	1.2 (5)	0.13 (5)
864	0.50 (16)	100 (20)	2.3 (9)	0.25 (9)
911	0.67 (27)	110 (20)	2.5 (11)	0.26 (12)

3 DISCUSSION

$B(M1)$ values of the magnitude seen in Table 1 have been observed between strongly coupled signature-partner bands in odd-A SD nuclei [3, 4]. However, in the even-A ^{190}Hg nucleus, the yrast SD band is most likely based on the fully paired vacuum state. Any excited SD band based on a quasiparticle excitation would be very unlikely to decay to the yrast band through $M1$ transitions of such large strength. Conversely, while the observed $B(E1)$ rates are orders of magnitude larger than those seen in deformed nuclei, they are of the same order as those reported for $E1$ inter-band transitions in nuclei with a substantial dipole moment arising from octupole collectivity, as is the case, for example, in some nuclei of the actinide region [8]. Therefore, the measured $B(E1)$ rates can be viewed as strong evidence in support of the interpretation of the excited SD band as a rotational band associated with the lowest octupole vibrational mode. In the calculations of ref. [5], the latter is proposed to be the $K = 2, \alpha = 1$ octupole excitation, although there is considerable mixing with other low-lying octupole excitations in the region of frequency where the band is observed experimentally [3, 4]. It should be pointed out that further support for the assignment of a $E1$ character to the inter-band transitions comes from the analogy with the situation in ^{194}Hg where inter-band transitions have recently been observed between bands 3 and 1 and have been shown to have $E1$ character [9]. In addition, the direct comparison of the quadrupole moments of the two bands as shown in Figure 1, indicates a difference in the quadrupole moments of $\Delta Q_0 = 0.1 \pm 1.9$ eb. If the difference in $\Im^{(2)}$ moments between the two bands ($\Delta\ \Im^{(2)} \simeq 15\hbar^2$/MeV) was simply due to a difference in deformation, then at a rotational frequency of $\hbar\omega \simeq 0.28$ MeV, the expected difference in the quadrupole moment would be $\Delta Q_0 > 5$ eb. Such a difference is well outside the limits imposed by the data. This is consistent with the assumption of equal quadrupole moments used in the RPA calculations of Nakatsukasa *et al* [5]. In summary, the fact that (i) no differences in deformation are found between the two SD bands, and (ii) that the absolute transition rates for the linking dipole transitions are large favors the interpretation of the excited SD band as a rotational band built on an octupole vibration.

This work is supported by the US Department of Energy, Nuclear Physics Division, under contracts no. W-31-109-ENG-38, and DE-FG05-88ER40411.

References

[1] M. W. Drigert *et al.*, Nucl. Phys. A530, (1991) 452 .

[2] B. Crowell *et al.*, Phys. Lett. B333, (1994) 320.

[3] B. Crowell *et al.*, Phys. Rev. C51, (1995) R1599.

[4] A. N. Wilson *et al.*, Phys. Rev. C54, (1996) 559.

[5] T. Nakatsukasa, K. Matsuyanagi, S. Mizutori and Y. R. Shimizu, Phys. Rev. C53, (1996) 2213. (1994) 782.

[6] E. F. Moore *et al.*, in *Proceedings of the Workshop on Gammasphere Physics*, Berkeley, CA, 1995, edited by M. A. Deleplanque, I.-Y. Lee, and A. O. Macchiavelli (World Scientific, Singapore, 1996), p. 137.

[7] J. F. Ziegler, J. P. Biersack, and U. Littmark, *The Stopping and Range of Ions in Solids* (Pergamon, New York, 1985); J. F. Ziegler (private communication).

[8] P. Butler and W. Nazarewicz, Rev. Mod. Phys. 68, (1996) 349.

[9] G. Hackman *et al.*, Phys. Rev. Lett. 79, (1997) 4100.

INTERACTIONS BY INVERSION

B. Apagyi and B. Báthory

Department of Theoretical Physics
Technical University of Budapest
H–1521 Budapest Hungary

apagyi@phy.bme.hu

Abstract: A method of model independent determination of the interaction potentials from elastic scattering data is presented. The procedure consists of a phase shift analysis (using, alternatively, a gradient method and a random search procedure to minimize the χ^2 value) and an inversion of the phase shifts to potential (by using the modified Newton-Sabatier theory). The method is illustrated on synthetic data imitating the elastic scattering of ^{12}C nuclei by the ^{32}S target at fixed energy of $E_{cm} = 26.02$ MeV.

1 INTRODUCTION

Ab initio determination of interaction potentials between nuclei at positive energies is a difficult task. In principle it requires the solution of the many-body problem of the underlying clusters and the knowledge of the effective two-body interaction between the constituent nucleons. It is well known that both problems can be accomplished only approximately, by using proper model assumptions. The resonating group method or the generator coordinate method are such examples in conjunction with oscillator models.

On the other hand, quantum inversion theories have been extensively used in the last decade, and by now they become a general tool for the determination of the interactions between nuclei or nucleons. The inversion methods provide an accurate prescription of how a set of phase shifts can uniquely be converted to potentials which describe the scattering of the system in question at fixed energies or angular momenta. If the set of phase shifts is derived in a model independent way then the inversion potential can be considered as obtained directly from the measured scattering data.

The Nucleus: New Physics for the New Millennium
Edited by Smit et al., Kluwer Academic / Plenum Publishers, New York, 2000.

2 THEORY

2.1 *Phase shift analysis of heavy-ion elastic scattering data*

There are various methods for extracting phase shifts from the elastic scattering angular distribution. An example is the controlled random search method which provides a global minimum of the corresponding χ^2 value in the parameter space of $S_l = \exp(2i\delta_l)$. We shall adopt here a method of Chiste *et al* [1] that is based on the observation that the error square value, $\chi^2 = (\sigma_{th} - \sigma_{exp})^2/\Delta\sigma^2_{exp}$, is of fourth order in the parameters S_l so that, by writing

$$S_l^n = S_l^{n-1} + \alpha \partial \chi^2_{n-1} / \partial S_l^{n-1}, \tag{1}$$

one easily gets the fourth order equation

$$\chi^2_n = A\alpha^4 + B\alpha^3 + C\alpha^2 + D\alpha + \chi^2_{n-1} \tag{2}$$

from which the optimal step difference α can be obtained simply by differentiation at each iteration number n. The coefficients A, B, C and D in the above equation are known analytical expressions depending on the set $\{S_l^{n-1}\}$. The iteration is continued until a required precision ($\chi^2_n/N_{exp} < 1$) is attained. One should start from different points of the parameter space in order to check that the result called 'experimental phase shifts' corresponds to the global minimum in the parameter space. Alternatively, one may combine Chiste's method with a random search procedure.

2.2 *Modified Newton-Sabatier inversion method*

The modified Newton-Sabatier scheme has been introduced by Münchov and Scheid [2] using the physical assumption that most interactions of quantum physics are known beyond a certain radial distance r_0. For example, this interaction is the repulsive Coulomb potential in the heavy ion collisions. Also they showed that the proofs of existence and uniqueness of the solutions of the method hold if an infinite number of phase shifts are involved in the procedure. In practice, however, the modified theory proved also useful when a finite set of data was used as input. This practical property makes the modified Newton-Sabatier method applicable to invert experimental phase shifts, especially at low energies where only few partial waves are important.

In the course of the inversion method one solves the Regge-Newton equations

$$\varphi_l(\rho) = \varphi_l^0(\rho) - \sum_{l'=0}^{l_{max}} c_{l'} L_{ll'}(\rho)\varphi_{l'}(\rho) \tag{3}$$

for the coefficients c_l, the knowledge of which provides the inversion potential

$$U(\rho) = U^0(\rho) - \frac{2}{\rho}\frac{d}{d\rho}\sum_{l=0}^{l_{max}} c_l \varphi_l^0(\rho)\varphi_l(\rho)/\rho. \tag{4}$$

In the above equations L is a given matrix $L_{ll'}(\rho) = \int_0^\rho \varphi_l^0(\rho')\varphi_{l'}^0(\rho')d\rho'/\rho'^2$ with $\varphi_l^0(\rho) = \rho j_l(\rho)$ being the regular solutions belonging to the reference potential U_0 and the experimental phase shifts enter as input via the asymptotic form of the solution functions φ_l as $\varphi_l(\rho > kr_0) \propto \rho[j_l(\rho) - \tan\delta_l^{exp} n_l(\rho)]$.

3 RESULTS

We have tested the method on synthetic data generated by a Woods-Saxon potential model simulating $^{32}S+^{12}C$ elastic scattering at $E_{cm} = 26.02$ MeV. The WS parameters are as follows: $V_r = -15$ MeV, $r_{0r} = 1.3$ fm, $a_r = 0.4$ fm, $V_i = -1$ MeV, $r_{0i} = 1.8$ fm, $a_i = 0.8$ fm, $r_{0C} = 1$ fm. The cross section data $\sigma^e(\theta_i)$ have been generated in the range of $\theta_{cm} = 1 - 179$ degree with step of 1 degree and also errors of $1 - 10$ per cent have been attributed to the data.

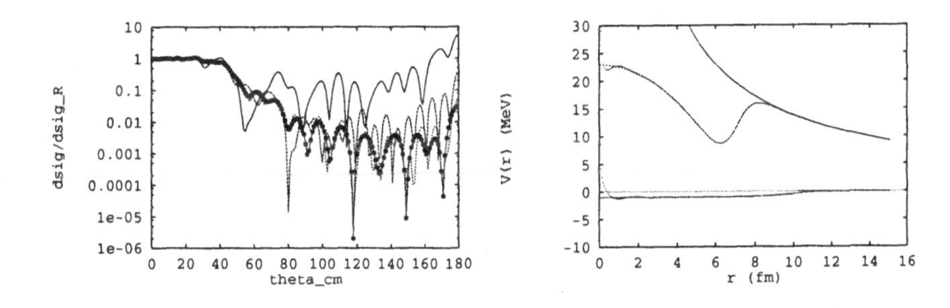

Figure 1

Figure 1 (on the left) shows as dots the 'experimental' cross section data together with the error bars, as a function of θ_{cm}. As dashed lines are shown the results of the phase shift analysis obtained at three different stages of the search characterized by $\chi^2 = 10^6, 10^2, 10^{-2}$ when $l_{max} = 45$ is used.

Figure 1 (on the right) shows as dashed lines the real and imaginary potential obtained by the inversion calculation using the code BIC [3] in conjunction with the set of S_l-matrix provided by the previous search at the smallest χ^2 value. As orientation, also the point-like Coulomb potential (dashed line) and the exact potential (full line) are depicted on the figure.

In conclusion one can say that with a careful analysis of experimental scattering data it may be possible to derive model independent interactions by using a subsequent inversion of the extracted phase shifts.

Acknowledgments

The authors wish to thank Prof. R. Lichtenthäler for useful discussions about phase shift analysis. Work supported by OTKA grants U31206 and T17179.

References

[1] V. Chisté, R. Lichtenthäler, A. C. C. Villari, and L. C. Gomes, Phys. Rev. C 54, 784 (1996).

[2] R. G. Newton, J. Math. Phys. 3, 75 (1962); P. C. Sabatier, J. Math. Phys. 7, 1515 (1966); M. Münchow and W. Scheid, Phys. Rev. Letters 44, 1299 (1980).

[3] B. Apagyi and G. Endrédi, Budapest Inversion Code (BIC), unpublished (1995).

K=0+ EXCITATIONS IN DEFORMED NUCLEI

A. Aprahamian

Department of Physics
University of Notre Dame
Notre Dame, IN 46556, USA

ani.aprahamian.1@nd.edu

Abstract: New measurements of level lifetimes for members of $K^\pi = 0^+$ bands will be presented in three nuclei. Each case highlights a different characteristic of the $K^\pi = 0^+$ excitations.

1 INTRODUCTION

The main focus of this conference is new physics of the nucleus for the next millenium. A perfect time to assess the extent of our understanding for one of the most abundant excitations in the spectra of deformed nuclei, namely $K=0^+$ excitations, and to carve out a piece of the future in order to determine their character.

$K^\pi = 0^+$ excitations have not been understood very well either experimentally or theoretically for essentially the same reasons. The data on $K^\pi = 0^+$ excitations and their absolute decay probabilities have been sparse and the few existing data points have been difficult to characterize in any systematic way. A compilation of all the known absolute transition probabilities from the first excited $K^\pi = 0^+$ bands [1] reveals enormous changes in collectivity as a function of A. For example, the B(E2) values from the 2^+ member of the first excited $K^\pi = 0^+$ band varies from 90 to 0.31 W.u. across the Gd isotopes (A=154-158) indicating a dramatic change in structure of the band.

Collective $K^\pi = 0^+$ excitations can be constructed within the framework of various theoretical models. In the unified geometric model [2] of the nucleus, quadrupole shape oscillations around an equilibrium shape result in $K^\pi = 0^+(\beta)$ and $K^\pi = 2^+(\gamma)$ excitations where the β type of oscillation maintains axial symmetry while the γ type of vibration breaks axial symmetry and

The Nucleus: New Physics for the New Millennium
Edited by Smit et al., Kluwer Academic / Plenum Publishers, New York, 2000.

has a projection of $K^\pi = 2^+$ on the symmetry axis. Two quanta of vibrational excitations would then include $\gamma\gamma$ ($K^\pi = 0^+$ and $K^\pi = 4^+$), $\beta\gamma$ ($K^\pi = 2^+$) and $\beta\beta$ ($K^\pi = 0^+$) types of vibrations. At the present time, the viability of the β type of vibrational mode for nuclei is in question. Specifically, the association of the first excited $K^\pi = 0^+$ excitations with this vibrational mode due to the large observed variations in collectivity. There is to date no clear evidence of a $\beta\beta$ type of vibration, and one example [3, 4] of a $\gamma\gamma$ ($K^\pi = 0^+$) type of excitation.

In this short paper, we present three new measurements of lifetimes for levels within $K^\pi = 0^+$ bands in ^{178}Hf [1], ^{166}Er [5, 3] and ^{158}Gd nuclei [6, 7] that reveal a different aspect of the character of these excitations. In ^{178}Hf, there is an indication of an excited $K^\pi = 0^+$ band at 1772 keV connected with collective transitions to the first excited $K^\pi = 0^+$ band at 1199 keV. In studies of ^{166}Er, there is some evidence for the third excited $K^\pi = 0^+$ band exhibiting the degree of collectivity expected from a β type of vibrational excitation. Finally in ^{158}Gd where strong collective transitions between an excited $K^\pi = 0^+$ band at 1452 keV and the γ seem to result from bandmixing effects [6] and not from a collective excitation built on the γ band. In this same nucleus, thirteen excited $K^\pi = 0^+$ bands have been observed [7] by a (p,t) transfer reaction.

2 EXPERIMENTS AND RESULTS

New measurements in three deformed nuclei are discussed below.

2.1 A new nuclear collective mode?

The^{178}Hf nucleus has been studied rather extensively by a variety of spectroscopic techniques. The spectrum shows five $K^\pi = 0^+$ bands below an excitation energy of 2 MeV. Lifetime measurements of levels in the ^{178}Hf nucleus were carried out using the GRID technique [8, 9, 10]. The first excited $K^\pi = 0^+$ band is at 1199.4 keV. The lifetime of the 0^+ bandhead level is not known. The 2^+ member of this band at 1276.7 keV had a previously measured lifetime of 8.8$\pm$3.5 ps determined by Coulomb excitation [11, 12] resulting in B(E2: $2^+_{K^\pi=0^+_2} \rightarrow 0^+$) and B(E2: $2^+_{K^\pi=0^+_2} \rightarrow 4^+$) values of 0.06 and 0.38 W.u., respectively. This measurement had previously led to the conclusion that the first excited $K^\pi = 0^+_2$ band is not collective. In this work we report on the lifetime of the 4^+ member of the same band at 1450.4 keV. The resulting B(E2: $4^+_{K^\pi=0^+_2} \rightarrow 6^+$) value range is 0.9$\rightarrow$13 W.u. typical for a transitions between a single-phonon vibrational excitationand the ground state. The extracted range of B(E2) values depopulating this level indicates that the 4^+ state at 1450.4 keV is indeed a member of this $K^\pi = 0^+$ band connected to the 2^+ member of the band by a collective 173.7 keV transition[11]. These values fall within the range of expected collectivities of β type of vibrational excitations. The 0^+ and 4^+ members of this $K^\pi = 0^+$ band at 1199 keV were populated in the (p,t) reaction [13] but not the 2^+ member. The present character of this band remains a puzzle. The most important result concerns the highest energy $K^\pi = 0^+$ band known in ^{178}Hf at 1772.2 keV. The extracted B(E2: $2^+_{K^\pi=0^+_5} \rightarrow 4^+_{g.s.}$) is $\leq$1 W.u., while the transitions to the first excited $K^\pi = 0^+_2$ band at 1199.4 keV are highly collective. This is the first evidence for observation of collective transitions between two excited $K^\pi = 0^+$ bands in any nucleus where a high-lying $K^\pi = 0^+_5$ band at 1772.2 keV is seen to primarily decay to the first excited $K^\pi = 0^+_2$ band at 1199.4 keV via transitions of collective strength. values

2.2 A beta vibrational excitation?

The lifetimes of levels in the ^{166}Er nucleus have been studied with the (n,n'γ) reaction [5, 3] and Coulomb excitation to yield evidence for the only observation of a $K^\pi = 0^+$ $\gamma\gamma$ vibrational excitation to date. More recently, it has been shown [3] that it is the third excited $K^\pi = 0^+$ band at 1934 keV that shows the degree of collectivity expected from a β type of vibrational excitation and not the first.

2.3 Simple bandmixing?

The ^{158}Gd nucleus has been recently studied [7] by the (p,t) reaction at the University of Munich Q3D spectrograph with a 27MeV proton beam. The results show 13 excited $K^\pi = 0^+$ bands below an excitation energy of approximately 3 MeV. A recent study of the same nucleus with the ultra-high resolution capabilities of the GAMS4 spectrometer and the GRID technique [6] have shown a $K^\pi = 0^+$ excitation that is connected via collective transition probabilities to the γ band. The results point to a $K^\pi = 0^+$ band whose properties can be understood in terms of an existing bandmixing calculation [6, 14] and not as a collective excitation built on the γ band.

3 DISCUSSION

New measurements in three nuclear systems have been presented to elucidate in each case a different aspect of the character of observed $K^\pi = 0^+$ excitations. In ^{178}Hf, an excited $K^\pi = 0^+$ band is shown to be connected to the first excited $K^\pi = 0^+$ band. Until the puzzling characteristicts of this latter excited $K^\pi = 0^+$ band can be understood, the new $K^\pi = 0^+$ band can only be seen as a new type of collective excitation built on the first excited $K^\pi = 0^+$ band. Studies of ^{166}Er nucleus point to the existence of a β type of vibrational excitation but it is not the first excited $K^\pi = 0^+$ band. In the same nucleus, there is evidence for a $K^\pi = 0^+$ $\gamma\gamma$ type of vibrational excitation. Studies of $K^\pi = 0^+$ bands in the ^{158}Gd nucleus reveal yet another aspect of the character of these excitations that can be understood in terms of existing bandmixing calculations. Also, several new $K^\pi = 0^+$ bands have been observed by a new (p,t) measurement whose characters are yet to be determined.

4 CONCLUSIONS

There is now a growing data set of newly measured level lifetimes for members of several $K^{\pi} = 0^{+}$ bands in a number of deformed nuclei. Perhaps it is time for a theoretical revisitation to the study of these $K^{\pi} = 0^{+}$ excitations in the coming millenium. The application of modern theoretical techniques may provide the insight necessary to determine the nature of the numerous $K^{\pi} = 0^{+}$ bands observed in the spectra of deformed nuclei.

Acknowledgments

The author acknowledges research support from the NSF and NATO under contracts PHY-9402761 and 91057 and collaborations with her colleagues R. de Haan, S.R. Lesher, H. Börner, E.R. Marshalek, V. Zamfir, and R.F. Casten.

References

[1] Aprahamian, A. *et al.*, *Journal of Phys. G*, in press, (1999)

[2] Bohr A and Mottelson B R *Nuclear Structure Vol. II*, (Reading: W A Benjamin) (1975)

[3] Garrett, P.E. *et al.*,*Phys. Rev. Lett.* 78, 4545 (1997).

[4] Fahlander, C. *et al Phys. Lett.* B 388 475 (1996)

[5] Garrett, P.E. *et al.*, *Phys. Lett.* B 400, 250 (1997)

[6] Börner, H. *et al.*, *Phys. Lett.* B, in press, (1999)

[7] Lesher, S.R. *et al.*, to be published, (1999)

[8] Börner, H. and Jolie, J., *J. Phys. G: Nucl. Part. Phys.* 19, 217 (1993) and all the references therein.

[9] Dewey, M.S. *et al.*, *Nucl. Instrum. Meth.* A 284, 151 (1989)

[10] Kessler, E.G. *et al.*, *J. Phys. G: Nucl. Phys.* 14, 167 (1988)

[11] Browne, E., NDS 72, 221 (1994).

[12] Ronningen, R.M. *et al.*, *Phys. Rev. C* 15, 1671 (1977)

[13] Sheline R.K., *et al.*, *Pramana 41*, 151 (1993)

[14] Greenwood, R.C., *et al.*, *Nucl. Phys. A*, 304, 327 (1978).

COMPETITION BETWEEN BETA AND DOUBLE BETA DECAY IN ^{48}Ca AND ^{96}Zr

M. Aunola, J. Suhonen and T. Siiskonen

Department of Physics, University of Jyväskylä,
P. O. Box 35 (Y5), FIN-40351 Jyväskylä, Finland

matias.aunola@ux.phys.jyu.fi

Abstract: Highly forbidden beta decays of ^{48}Ca and ^{96}Zr are studied and their relative importance as compared to the double beta decay of these nuclei is evaluated. ^{48}Ca and ^{96}Zr are the only naturally occurring nuclei in which these processes can occur simultaneously. Although usually ordinary beta decay overwhelms double beta decay unless the former is energetically forbidden, in these cases the high degree of forbiddenness and small release of kinetic energy makes the half-lives of these modes comparable to each other.

In the case of ^{48}Ca the partial half-lives for the highly-forbidden transitions to three lowest lying states ($J^\pi = 6^+, 5^+, 4^+$) of ^{48}Sc are calculated [1]. Results for this decay are shown in Table 1. We find the decay to be dominated, as expected, by the unique fourth-forbidden transition to the excited $J^\pi = 5^+$ state of ^{48}Sc. A very significant enhancement of the non-unique transitions to the $J^\pi = 6^+$ ground state and $J^\pi = 4^+$ excited state can be seen. The partial half-life is found to be $1.1^{+0.3}_{-0.3} \cdot 10^{21}$ years which is in reasonable agreement with the previous results [2]. The obtained half-life is approximately 25 times longer than the measured double-beta-decay half-life of $(4.3^{+2.4}_{-1.1}[\text{stat}] \pm 1.4[\text{syst}]) \cdot 10^{19}$ years [3]. Thus beta decay competes with the double beta decay only weakly.

The partial half-lives for the highly-forbidden transitions from the ground state of ^{96}Zr to the three lowest-lying ($J^\pi = 6^+, 5^+, 4^+$) of ^{96}Nb are calculated by using both the harmonic oscillator and the Woods–Saxon mean-field wave functions [4]. Results have been compiled into Table 2. We find the decay to be dominated by the unique fourth-forbidden transition to the excited 5^+ state of ^{96}Nb. The theoretical beta-decay half-life of ^{96}Zr is found to be $(1.1^{+0.5}_{-0.4}) \cdot 10^{20}$

The Nucleus: New Physics for the New Millennium
Edited by Smit et al., Kluwer Academic / Plenum Publishers, New York, 2000.

years which is comparable to the half-life of $[2.1^{+0.8(stat)}_{-0.4(stat)} \pm 0.2(syst)] \cdot 10^{19}$ years of the two-neutrino double beta decay measured in counter experiments by the NEMO collaboration [5]. Our result indicates that the β^- decay of ^{96}Zr is a sizable contaminant of the high-precision geochemical double-beta-decay experiments that are being planned in order to study the possible time dependence of the weak-interaction constant [6].

Table 1 Theoretical partial half-lives $t^{\beta^-}_{1/2}$ for the transitions $^{48}\mathrm{Ca}(0^+) \rightarrow {}^{48}\mathrm{Sc}(J^\pi)$ in the harmonic-oscillator (HO) and Woods–Saxon (WS) mean-field basis for FPBP and FPKB3 interactions.

J^π_f	$t^{HO}_{1/2,FPBP}$	$t^{HO}_{1/2,FPKB3}$	$t^{WS}_{1/2,FPBP}$	$t^{WS}_{1/2,FPKB3}$
6^+	$1.3 \cdot 10^{31}$ y	$9.8 \cdot 10^{30}$ y	$1.7 \cdot 10^{29}$ y	$1.5 \cdot 10^{29}$ y
5^+	$1.1 \cdot 10^{21}$ y	$8.3 \cdot 10^{21}$ y	$1.2 \cdot 10^{21}$ y	$9.2 \cdot 10^{20}$ y
4^+	$5.2 \cdot 10^{26}$ y	$5.0 \cdot 10^{26}$ y	$8.8 \cdot 10^{23}$ y	$9.1 \cdot 10^{23}$ y

Table 2 Theoretical partial half-lives $t^{\beta^-}_{1/2}$ for the transitions $^{96}\mathrm{Zr}(0^+) \rightarrow {}^{96}\mathrm{Nb}(J^\pi)$ in the harmonic-oscillator mean-field basis for different sets of single particle energies, namely Woods–Saxon (WS) and the adjusted Woods–Saxon (AWS) sets.

J^π_f	$t^{\beta^-,(WS)}_{1/2}$	$t^{\beta^-,(AWS)}_{1/2}$
6^+	$3.2 \cdot 10^{32}$ y	$8.4 \cdot 10^{30}$ y
5^+	$1.2 \cdot 10^{20}$ y.	$1.0 \cdot 10^{20}$ y
4^+	$1.0 \cdot 10^{25}$ y	$1.2 \cdot 10^{24}$ y

References

[1] M. Aunola, J. Suhonen and T. Siiskonen, submitted to Europhys. Lett. (1998)

[2] E.K. Warburton, Phys. Rev. C 31 (1985) 1896

[3] A. Balysh et al., Phys. Rev. Lett. 77 (1996) 5186

[4] M. Aunola and J. Suhonen submitted to Phys. Rev. Lett. (1998)

[5] R. Arnold, C. Augier, J. Baker et al., submitted to Nucl. Phys. A (1998)

[6] A.S. Barabash, JETP Lett. 68 (1998) 1

IN-BEAM GAMMA SPECTROSCOPY OF VERY NEUTRON-RICH NUCLEI AT GANIL

F. Azaiez[1], M. Belleguic[1], O. Sorlin[1], S. Leenhardt[1], C. Bourgeois[1], C. Donzaud[1], J. Duprat[1],D. Guillemaud-Mueller[1], A.C.Mueller[1], F. Pougheon[1], M.G.Saint-Laurent[2], M. J. Lopez[2], N. L. Achouri[2], J. M. Daugas[2], M.Lewitowicz[2], F. De Oliveira[2], H. Savajols[2], Yu-E. Penionzhkevich[3], Yu. Sobolev[3], C. Borcea[4], M. Stanoiu[4], J. C. Angelique[5], S. Grevy[5], I. Deloncle[6], J. Kiener[6], M. G. Porquet[6], A. Gillibert[7], J. E. Sauvestre[8], Z. Dlouhy[9], W. Shuying[10].

[1]IPN-Orsay (France), [2]GANIL Caen (France), [3]FLNR-KINR Dubna (Russia), [4]IAP Bucarest (Roumania), [5]LPC Caen (France), [6]CSNSM Orsay (France), [7]DAPNIA-CEA (France), [8]DAM-CEA (France), [9]RPI-REZ (Czech Republic), [10]GSI Darmstadt (Germany)

azaiez@ipno.in2p3.fr

Abstract: The structure of nuclei far from stability can be investigated by means of in-beam gamma spectroscopy using both stable and radioactive beams. Two such experiments have been recently performed at GANIL. i) Secondary beams of neutron-rich Ge, Zn and Ni isotopes around N=40 were produced from the fragmentation of a primary ^{86}Kr beam, and analyzed by means of the LISE3 spectrometer. For the transitions between the 0^+ ground state and the first 2^+ state, B(E2) values have been extracted from Coulomb excitation cross-section measurement. Some of the obtained results related to the effectiveness of the N=40 spherical sub-shell closure are discussed. ii) In-beam γ-spectroscopy of neutron-rich nuclei around ^{32}Mg produced by fragmentation of ^{36}S have been performed recently. Gamma decay of relatively higher excited states have been measured in a large number of exotic light nuclei. Preliminary results obtained in a number of neutron-rich nuclei around N=20 are presented.

The Nucleus: New Physics for the New Millennium
Edited by Smit et al., Kluwer Academic / Plenum Publishers, New York, 2000.

1 INTRODUCTION

One of the most challenging goals of experiments with radioactive beams is to determine how the structure of atomic nuclei changes near the drip lines as the binding energies of single particles approach zero. Because of the relatively low intensity of the existing radioactive beams, the structure of very exotic nuclei has been investigated mainly through beta-decay and isomer-decay studies. Only recently, elastic and inelastic secondary reactions, induced by radioactive beams on stable targets, have been used in order to extract nuclear structure information. Among them, inelastic scattering in inverse kinematics and Coulomb excitation are known to provide valuable nuclear structure information such as the energy and the collectivity of the first excited states of nuclei. Inelastic proton scattering to the first excited 2^+ state of the doubly magic ^{56}Ni nucleus is a nice example of such experiments with radioactive secondary beams [1]. On the other hand, the large cross section of the Coulomb excitation process at intermediate incident energies is a key factor to overcome the difficulties imposed by the weak intensity of presently available radioactive beams. This has been recently demonstrated in Coulomb excitation experiments of unstable beams of ^{11}Be [2, 3], ^{32}Mg [4] and neutron rich Ar and S isotopes [5, 6]. Coulomb excitation of secondary radioactive beams as well as a novel method based on in-beam γ-spectroscopy of exotic nuclei produced by projectile fragmentation have been recently used at GANIL. The description of the two experiments together with the obtained results on the spectroscopy of neutron rich nuclei around N=40 and N=20 will be presented.

2 NEUTRON-RICH NUCLEI AROUND N=40

Calculations in the framework of Hartree Fock Bogoliubov (HFB) and Relativistic Mean Field (RMF) theories suggest that for neutron-rich nuclei, a better description would be obtained with a rounded potential that can be schematically simulated by the harmonic-oscillator potential [7]. The increase of the N=40 gap arises naturally from this approach. The ^{68}Ni$_{40}$ nucleus is known to exhibit a high 2^+ energy of 2.03 MeV [8] in contrast to its neighboring isotopes ^{66}Ni [8] and ^{70}Ni [16] which have lower 2^+ energies of 1425 keV and 1259 keV, respectively. The sudden increase of the 2^+ energy at N=40 suggests that ^{68}Ni is spherical and can be considered as a good core to modelize more neutron-rich Ni isotopes up to ^{78}Ni [10]. The unknown reduced transition probability B(E2: 0^+ to 2^+) of ^{68}Ni should then be measured in order to confirm this shell effect. In this respect, it would be interesting to compare this B(E2) value to that of the doubly magic ^{56}Ni, which was found to be 600 (120) e^2fm^4 [1]. For heavier N=40 isotones, proton excitations in the fp-shell increase the collectivity to B(E2)=1600(140) e^2fm^4 for ^{70}Zn [11] and 2130(60) e^2fm^4 for ^{72}Ge [11].

Another feature of the N=40 nuclei is that they exhibit a $0^+{}_2$ state at relatively low excitation energy. For ^{68}Ni and ^{72}Ge, this state is below the 2^+ state, and is therefore an E0 isomer since it cannot decay to the $0^+{}_1$ ground state by γ-emission. The structure of the $0^+{}_2$ isomer in ^{68}Ni is not yet understood and needs further experimental investigations. It could be viewed, on the one hand, as a two quasi-particle excitation of the core [8]. On the other hand, the self-consistent HFB calculations using Gogny [12] and Skyrme [17] effective interactions interpret it as a shape isomer with a large quadrupole deformation of β_2 larger than 0.4 [12]. The examination of a Nilsson diagram, where a shell gap is clearly visible at such a deformation parameter, also supports this interpretation. Whether the predicted shape-isomer corresponds to the known $0^+{}_2$ level or to a more excited $0^+{}_3$ state is still debated. In addition to the 2^+ level, information about excited states of ^{68}Ni have been obtained by Broda et al. [8] from multi-nucleon transfer reactions. In this study, a 5^- isomer at 2.847 MeV was discovered and interpreted as due to a neutron particle-hole ($g_{9/2}$,$p_{1/2}$) excitation. Since this isomer has a long half life of $T_{1/2}$=0.86ms, it can be used as a secondary beam in order to induce coulomb excitation to study its configuration. This holds true for the $0^+{}_2$ isomer with $T_{1/2}$=220ns[18], even if a fraction of it can be lost during the flight time in the spectrometer. Recently, an experimental program has started at GANIL in order to address the opened questions concerning the spherical gap at N=40 and the nature of isomers in ^{68}Ni. In the following, we will present preliminary results from a Coulomb excitation experiment, performed at GANIL, with neutron-rich nuclei in the vicinity of ^{68}Ni.

The Coulomb excitation of secondary beams has been induced by the Coulomb field of a thick target placed in the center of a large gamma-array detector. The nuclei ^{76}Ge, ^{70,72}Zn and ^{68}Ni were produced at an energy of about 50MeV/u by the fragmentation of a ^{86}Kr beam at 65 MeV/A. They were identified event-by-event in two large-area (25cm^2) silicon detectors mounted at a distance of 50 cm from the secondary lead target. Two clover Ge-detectors were placed around the implantation detector in order to measure the γ-rays originating from the decay of isomers produced in the fragmentation of the primary beam and transmitted by the LISE3 spectrometer. The scattered fragments were detected up to an angle of 3 degrees in the laboratory frame. At these small deflection angles, the Coulomb inelastic contribution strongly dominates the total cross section. The mean production rate of the fragments was about 100 particles/second for ^{76}Ge, ^{72}Zn and 20 particles/second for ^{68}Ni and ^{70}Zn. The lead target of 220 mg/cm^2 thickness was surrounded by the γ-array of 70 BaF_2 detectors of Chateau de Cristal, mounted in the 4π geometry. The diameter of each crystal is about 9 cm with a length of 14 cm. The γ-rays of interest, subsequent to the coulomb excitation, are emitted in flight with a velocity of v/c=0.3. Consequently, the Doppler effect induces a broadening of the γ-lines. By placing the BaF_2 detectors at a distance of 35 cm from the lead target, the opening angle viewed from the target is reduced, keeping the broadening to a reasonable value of 15% at 1 MeV.

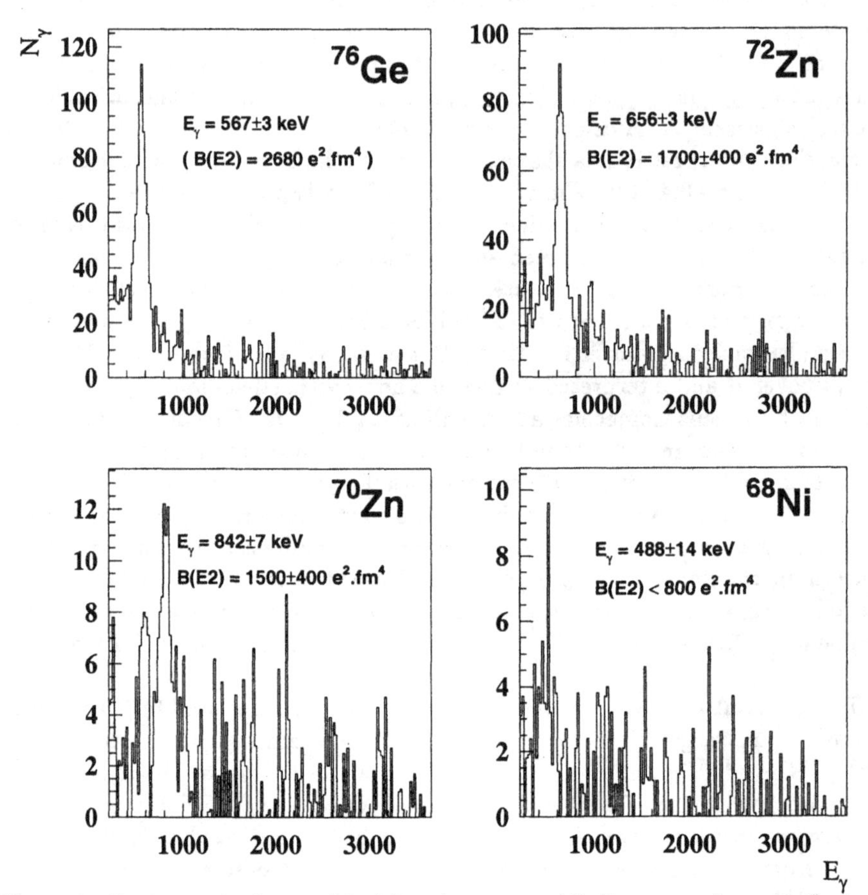

Figure 1 Background subtracted and Doppler corrected BaF_2-spectra from the Coulomb excitation of ^{76}Ge, ^{70,72}Zn and ^{68}Ni.

In the spectra of figure 1, background subtraction and Doppler shift corrections as a function of the emission angle of the γ-rays have been applied. The ^{76}Ge nucleus is a well-known case of a deformed nucleus, exhibiting a low 2^+ state (563 keV) and a strong excitation probability, B(E2) = 2680 e^{2}fm^4. It has been used as a calibration measurement in order to determine B(E2) values of ^{70,72}Zn and ^{68}Ni. Even if the spectrometer was not optimized for ^{70}Zn, an excitation probability B(E2) = 1500(400) e^{2}fm^4 has been determined for the 2^+ state at 885 keV. The agreement between the present value and the one measured at low energy [11], where the coulomb excitation is the dominant fraction of the 2^+ excitation, is a clear indication of the validity of other extracted B(E2) values from our data. The experimental B(E2) values [15] are summarized in figure 3 which includes also values for neighboring isotopes. The B(E2) values of the Zn isotopic chain indicate that the collectivity is increasing at N=40, the maximum of collectivity occurring at N=42. Using the prescription of Raman et al. [16] for deformed nuclei, a deformation parameter of β_2=0.23 is found. But it seems that the collectivity for N=40 isotones is more of a vibrational type. This is reflected by the low-energy level schemes of ^{70}Zn, ^{72}Ge, and ^{74}Se which feature E(4$^+$)/E(2$^+$) ratios very close to 2.0. This shows that protons are quickly washing out the N=40 effect due to the large number of 2qp-excitations in the fp shell above Z=28. For the coulomb excitation of ^{68}Ni, very few counts are found around the expected energy E(2$^+$) = 2.03 MeV. Therefore, only an upper limit of B(E2) = 800 e^{2}fm^4 could be derived.

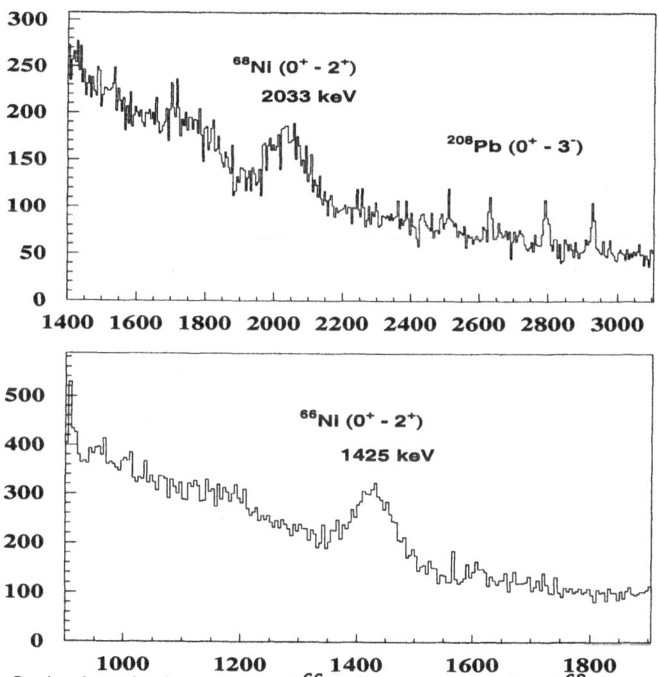

Figure 2 Coulomb excitation spectra of ^{66}Ni (lower spectrum) and ^{68}Ni (upper spectrum), obtained after Doppler correction using the segmentation of the Clover detectors. The four sharp lines in the ^{68}Ni represent the coulomb excitation to the 3$^-$ state of the ^{208}Pb target (the four lines correspond to the four different angles of the segments).

More recently, this experiment has been repeated using more favorable conditions: (i) A neutron-rich primary beam of ^{70}Zn, closer to the nuclei of interest, has been used and the production rate of ^{68}Ni has been increased by almost three orders of magnitude. (ii) Instead of the BaF$_2$-array, a higher resolution γ-array consisting of four segmented clover Ge detectors has been used as four faces of a cube around the lead target. The immediate consequense was a much higher resolving power for the γ-ray detection. This is nicely demonstrated in the spectra of figure 2, where gamma-lines corresponding to the 2^+ to 0^+ transitions in ^{66}Ni and ^{68}Ni are clearly visible. Despite a relatively smaller efficiency (5.5% at 1.3 MeV) of the four segmented clover detecors, the increase of the secondary beam intensities allowed an unambiguous determination of the B(E2: 0^+ to 2^+) value in ^{66}Ni and ^{68}Ni. The obtained values have been added to the systematics in figure 3. This indicates for the first time the effectiveness of the N=40 spherical shell effect in the Ni isotopes.

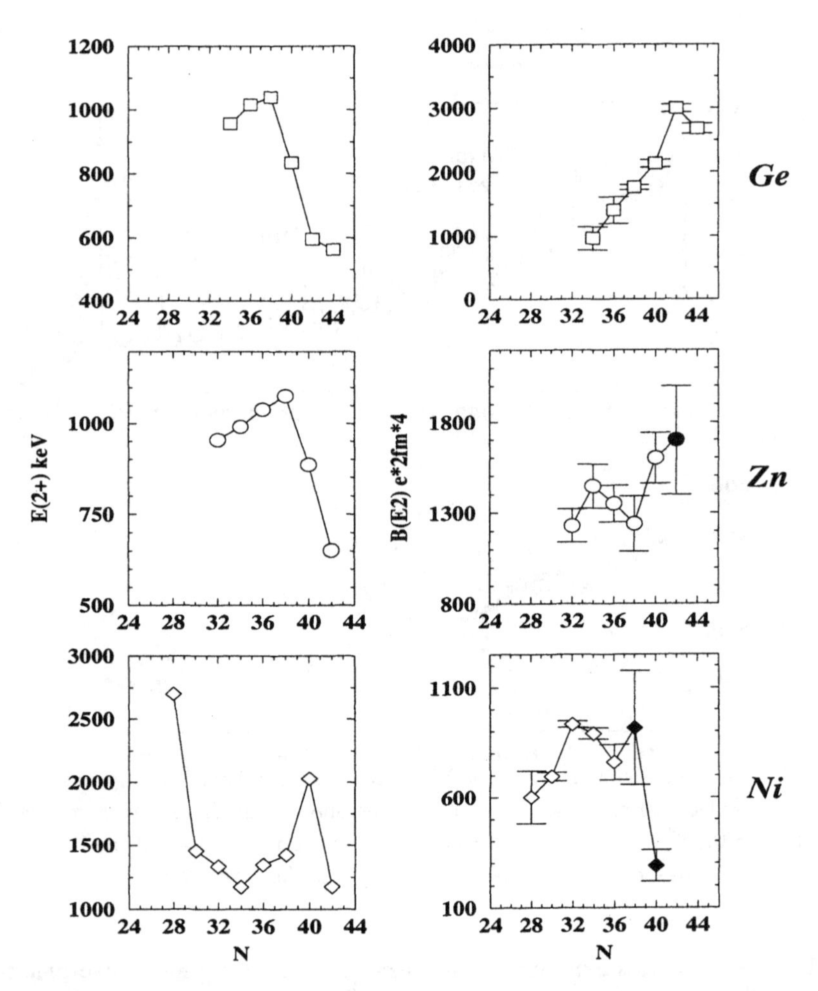

Figure 3 Systematics of the 2^+ energies and B(E2) values in the Ge,Zn and Ni isotopes. New results from our experiments are indicated by filled symbols.

3 NEUTRON RICH NUCLEI AROUND N=20

From the study of the structure of light neutron-rich nuclei, it has been recently suggested that some major shell-gaps are weakened when large isospin values are encountered. The typical cases of ^{32}Mg and ^{44}S, where a large quadrupole collectivity has been found [4, 5, 6] have brought some evidence for such a shell-gap weakening at large neutron excess. Though, information on the excitation energies of the first 2^+ states and on the B(E2) values of the 0^+ to 2^+ transitions is not sufficient to fully understand the structure of these nuclei. For instance the measurement of the $E(4^+)/E(2^+)$ ratio should shed some light on the origin of the large quadrupole collectivety observed. In order to bring more spectroscopic information on ^{32}Mg and neighboring nuclei, a novel experimental method has been used. This method is based on the production of very neutron-rich nuclei in relatively higher excited states, through the projectile fragmentation process, and on the detection of their in-beam γ-decay. Such an experiment aiming for the measurement of the $E(4^+)/E(2^+)$ ratio in $^{30-32}$Mg, $^{26-28}$Ne and $^{20-22}$O has been recently performed at GANIL. A ^{36}S beam, at 77MeV/u was used on a 2.77 mg/cm^2 Be target. The target was located at the entrance of the SPEG spectrometer which was used to analyze the different fragments produced in the reaction. At a proper $B\rho$ = 3.4 T.m setting of the spectrometer, many neutron rich exotic nuclei have been produced and identified. The produced nuclei are identified in a time of flight versus energy-loss plot. The energy-loss is measured in an ionization chamber at the focal plane of the SPEG spectrometer, whereas the time of flight of the fragments is given by the time difference between the RF pulse from the accelerator and the signal from a plastic scintillator located at the SPEG focal plane. It is worth pointing out that most of the produced nuclei are TERRA INCOGNITA for nuclear spectroscopy and thus γ-spectroscopy of these nuclei (such as $^{22-23}$O, $^{27-28}$Ne , $^{32-33}$Mg) is completely unknown. Gamma-spectroscopy for all the produced exotic nuclei is obtained by performing coincidences between the analyzed fragments and γ-rays emitted in flight during their decay to the ground state. For that purpose a highly efficient (25 % at 1.33MeV) γ-array consisting of 74 BaF_2 crystals (the same used for the Coulomb excitation experiment) was used around the target covering symmetrically the upper and lower hemispheres (roughly 80 % of the solid angle around the target is covered). This array is supposed to provide γ-fragment as well as γ-γ-fragment coincidences. The latter is needed to build-up a level scheme for each fragment. In addition to the BaF_2 array, four 70 % high resolution Ge detectors were used at the most backward angles (in between the two hemispheres) in order to help identifying some more complex BaF_2 spectra (see spectra below). Even though the analysis of the data is in progress, enough results are obtained to date to prove the feasibility and the power of the method. In the following, part of the results obtained in even-even nuclei are presented. After gating on the proper fragment and on the true γ-fragment coincidences (subtracting the random coincidences contribution), some Doppler corrected γ-spectra are presented and commented. No result has been obtained yet on the gamma angular distribu-

tion and correlation for angular moment assignment. Therefore our discussion will be based on the assumption that the strongest γ-line in a spectrum of an even-even nucleus represents the 2^+ to 0^+ transition. This is a fairly valid assumption for light nuclei where no low-lying octupole states are expected. Nevertheless this has to be proven yet.
i) The obtained γ-spectra of ^{22}O from both the BaF_2 and Ge detectors is presented in figure 4. If we assume, from an intensity argument, that the γ-line observed for the first time at 3.1 MeV represents the 2^+ to 0^+ transition, this extends the systematic of the 2^+ transition energies of Oxygen isotopes up to $N = 14$. One can see in figure 5, that Oxygen isotopes exhibit the lowest 2^+ energy at half-occupancy of the $d_{5/2}$ state (N=12) just like the Ne and Mg isotopes do (the energy scale of figure 5 is to be multiplied by a factor 4 for the Oxygen isotopes). Whether it will continue to follow the same trend up to N=16 or not is a key point to understand why the last bound Oxygen isotope seems to be ^{24}O [17, 18, 19].

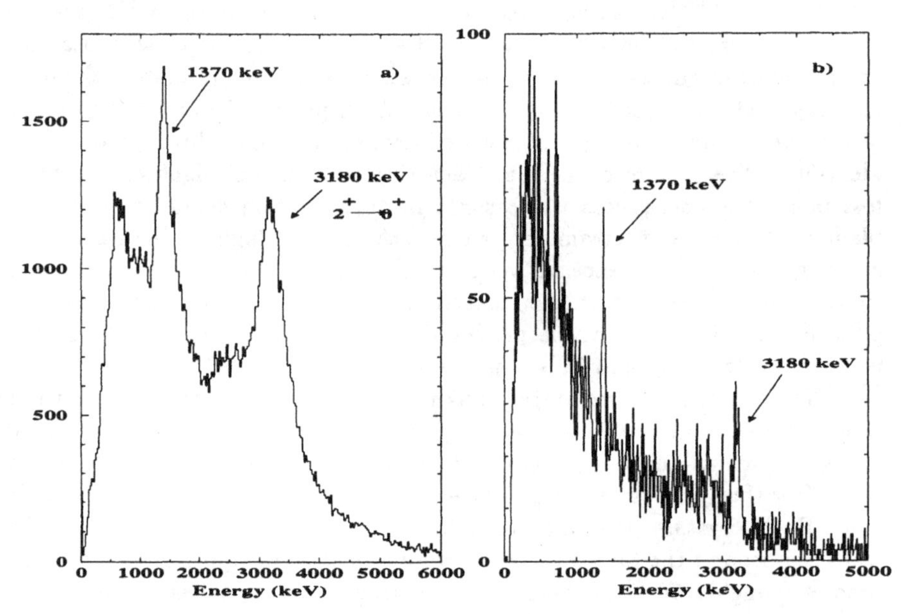

Figure 4 BaF_2 (a) and Ge (b) spectra of the in-beam γ-decay of ^{22}O.

ii) The BaF_2 and Ge spectra of ^{28}Ne are shown in figure 6. This figure illustrates the importance of the high efficiency for this experiment. While the Ge spectrum does not show any significant line, the BaF_2 spectrum exhibits a quite convincing structure at 1.3 MeV. This γ-line is very likely to represent the 2^+ to 0^+ transition in ^{28}Ne which shows for the first time that, approaching N=20, the 2^+ energies in the Ne isotopes decrease dramatically. One can see in figure 5 that the 2^+ energy drops from around 2 MeV in ^{24}Ne and ^{26}Ne to 1.3 Mev in ^{28}Ne (it is worth pointing out that the 2^+ excitation energy of ^{26}Ne has been already measured in a β-decay experiment at GANIL [20]). This behavior is presumably a sign of shell structure change for neutron rich Ne isotopes similar to the one observed a long time ago in the Mg isotopes [21].

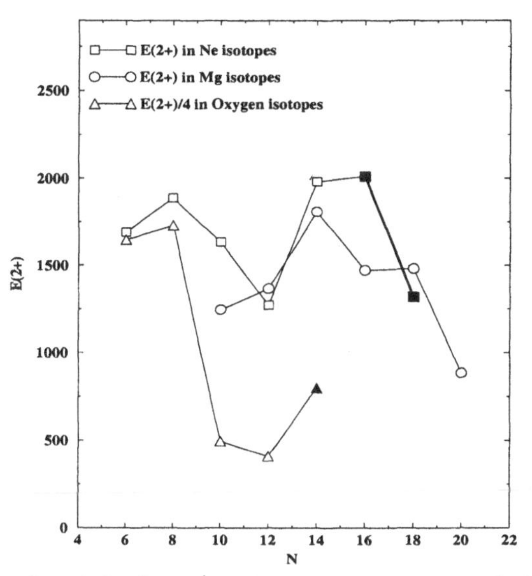

Figure 5 Systematics of the first 2^+ excitation energies in the O, Ne and Mg isotopes. The labels are Triangles for O, squares for Ne and circles for Mg (filled symbols indicate the newly measured values in our experiment).

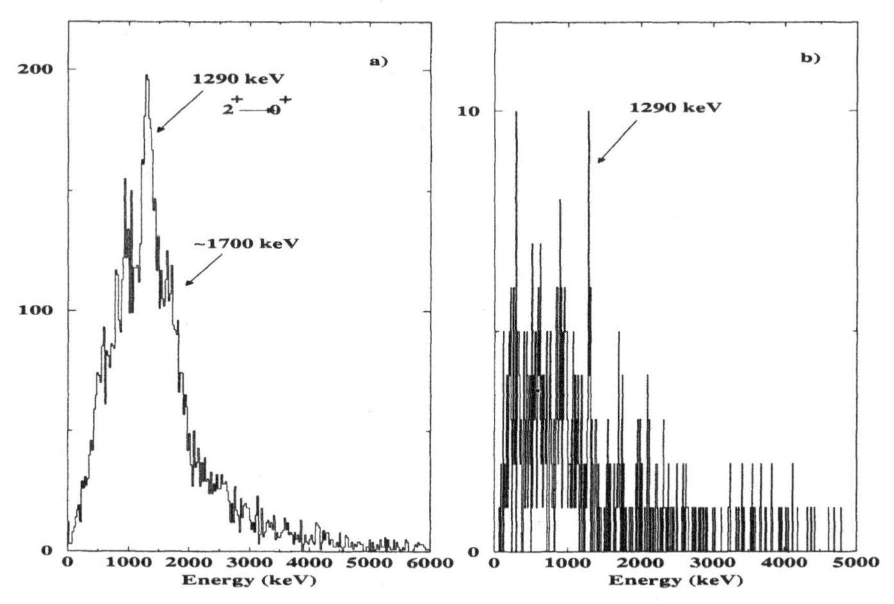

Figure 6 BaF_2 (a) and Ge (b) spectra of the in-beam γ-decay of ^{28}Ne

iii) In figure 7, BaF_2 spectra and Ge spectra of ^{32}Mg are shown. Like for many other produced fragments, the gamma spectra (Ge and BaF_2 spectra) of ^{32}Mg exhibit more than one line. For all these fragments, γ-angular distribution and γ-γ coincidence between BaF_2 detectors has to be analyzed in order to deduce a level scheme. This type of analysis is in progress for ^{32}Mg and reveals that the two lines: the 885Kev (the well known 2^+ to 0^+ transition in ^{32}Mg [21]) and the 1.4 Mev newly observed line, are in coincidence. The nature (multipolarity) of the 1.4Mev γ-ray is not yet extracted from the data. Though, it is likely to be either the 4^+ to 2^+ transition or a transition from a second $2_2{}^+$ to a first $2_1{}^+$ state. It is worth pointing out that Monte-Carlo Shell Model calculations [22], made prior to our experiment, reproduce very well the excitation energy of $2_1{}^+$ states observed in $^{26,28}Ne$ and ^{32}Mg. Furthermore these calculations, in agreement with the observed new excited state in ^{32}Mg, report a 4^+ state at 1.4 MeV above the 2^+ state in this nucleus. In any case these new experimental results will shed more light on the physics underlying the so-called shell-effect quenching at the neutron-rich side of the valley of stability.

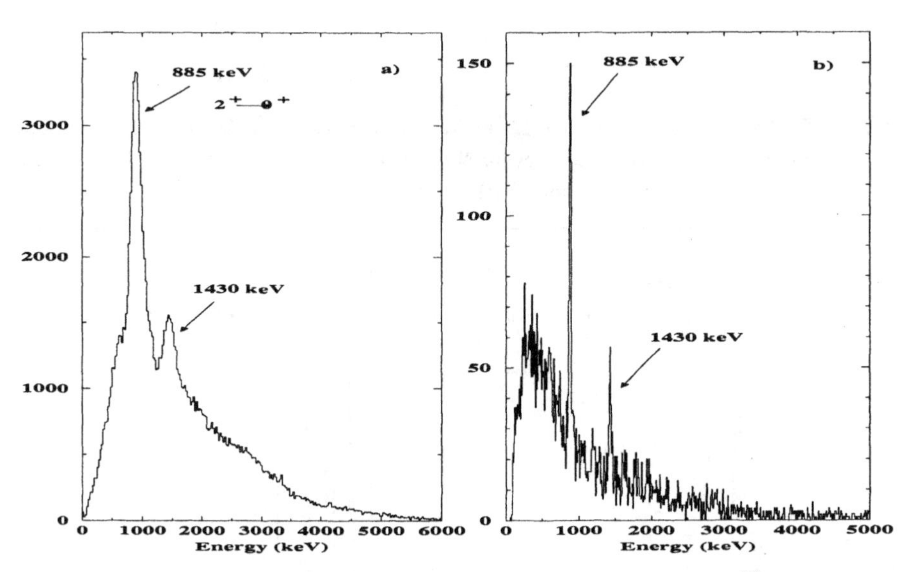

Figure 7 BaF_2 (a) and Ge (b) spectra of the in-beam γ-decay of ^{32}Mg

Beside the importance of the obtained preliminary results, this experiment shows that the in-beam-γ-spectroscopy from fragmentation reactions is very promising for exploring nuclear structure far from stability. It also highlights the needs for a dedicated gamma detection system from the point of view of efficiency, resolution and Doppler Broadening reduction. These three features are the basic requirements for which EXOGAM [23] was built and thus make it the ideal gamma detection system for such experiment.

References

[1] G. Kraus et al., *Phys. Rev. Lett.* 73 , 1773 (1994)
[2] R. Anne et al., Z. Phys. A 352 (1995) 397.
[3] T. Nakamura et al., Phys. lett. B 394 (1997) 11.
[4] T. Motobayashi et al., Phys. Lett. B 346, 9 (1995)
[5] H. Scheit et al., Phys. Rev. Lett. 77, 3967 (1996)
[6] T. Glasmacher et al., Phys. Lett. B 395, 163 (1997)
[7] J. Dobaczewski et al., Phys. Rev. Lett. 72, 981 (1994)
[8] R. Broda et al., Phys. Rev. Lett. 74, 868 (1995)
[9] R. Grzywacz, Phys. Rev. Lett. 81, 766 (1998)
[10] H. Grawe et al., Prog. in Part. and Nucl. Phys. 38, 15 (1997)
[11] P. H. Stelson and F. K. McGowan, Nucl. Phys.A 32, 652 (1962)
[12] M. Girod et al., Phys. Rev. C 37, 2600 (1988)
[13] P. Bonche et al., Nucl. Phys. A 500, 308 (1989)
[14] M. Bernas et al., Phys. Lett. B 113, 279 (1982)
[15] S. Leenhardt, thesis work, I.P.N Orsay
[16] S. Raman et al., Phys. Rev. C 43, 556 (1991)
[17] D. Guillemaud-Mueller et al., Phys.Rev C41, 937 (1990)
[18] M. Fauerbach et al., Phys. Rev. C53, 647 (1996)
[19] O. Tarasov et al, Phys. Lett. B409, 64 (1997)
[20] A. T. Reed et al. private communication
[21] D. Guillemaud-Mueller et al., Nucl. Phys. A426, 37 (1984)
[22] T. Otsuka et al., Nuclear Structure '98, Gatlingurg,tennessee, USA (1998)
[23] F. Azaiez and W. Korten., Nucl. Phys. News, Vol. 7, No. 4, (1997)

EXPERIMENTAL STUDIES OF THE NUCLEON-NUCLEON INTERACTION AT KVI WITH 190 MeV PROTONS

Jose C.S. Bacelar

Kernfysisch Versneller Instituut
Zernikelaan 25, 9747AA Groningen
The Netherlands

bacelar@kvi.nl

Abstract: The real and virtual photon emission during interactions between few-nucleon systems have been investigated at KVI with a 190 MeV proton beam. Here I discuss the results of the proton-proton system and proton-deuteron capture. Predictions of a fully-relativistic microscopic-model of the NN interaction are discussed. For the virtual photon processes, the nucleonic electromagnetic response functions were obtained for the first time and are compared to model predictions.

INTRODUCTION

The study of photon emission during nucleon-nucleon collisions provides a sensitive and unobtrusive way to study the nucleon-nucleon interaction. The photon couples directly to the electromagnetic currents associated with the dynamics of the collision. Recent accurate measurements [1, 2] of this process provided exciting developments [3, 4, 5, 6, 7] in the theoretical attempts to describe the interacting nucleons. Most of the precise data is associated with proton-proton collisions, for obvious experimental simplicities.

At KVI we have performed [2] the most accurate measurements to date on the reaction $pp \to pp\gamma$. These data are discussed in detail in reference [2]. The proton beam is polarized, with an energy of 190 MeV. A liquid H_2 target [8] is used to reduce background. The forward angle hodoscope SALAD (Small Angle Large Acceptance Detector) [2] was used to measure the energies and the tracks of the two protons, within scattering polar angles of 6^0 and 28^0. It consists of two wire chambers (with a total of five wire planes) and two stacked planes

The Nucleus: New Physics for the New Millennium
Edited by Smit et al., Kluwer Academic / Plenum Publishers, New York, 2000.

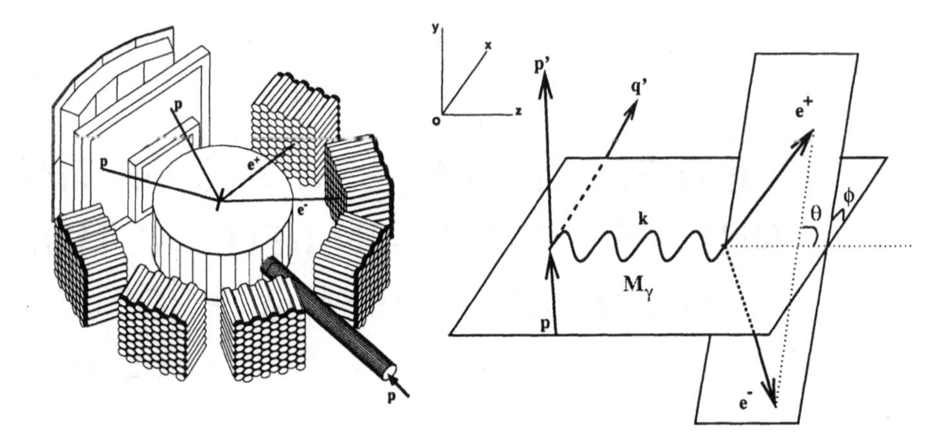

Figure 1 Left: Schematic view of the experimental setup consisting of SALAD (two wire-chambers and 50 plastic scintillators) and TAPS (384 hexagonal BaF_2 crystals). Not drawn are the thin plastic scintillators placed in front of each BaF_2 crystal. An example of a ppe^+e^- event originating from the liquid-hydrogen target is indicated. On the right panel the chosen coordinate system to describe the dynamics of these events is given.

of scintillator detectors (see fig. 1). This detector is equiped with a specially designed trigger system [9] which allows it to deal with the unwanted 14 MHz rate of the elastic channel at these small angles. The photon was detected with TAPS [10], an array of 384 BaF_2 crystals which for the data discussed here were placed in the median plane around the target in blocks of 8x8 crystals (see fig. 1). The experimental setup shown in this figure is common to all the experiments discussed in this paper. In the figure a characteristic event for the reaction $pp \rightarrow ppe^+e^-$ is depicted.

PROTON-PROTON BREMSSTRAHLUNG

The real photon bremsstrahlung data [2] is well reproduced by a fully-relativistic microscopic-model [5, 6, 7] of the nucleon-nucleon interaction in some regions of the phase space. The model, based on the Fleischer-Tjon potential [11], includes the off-shell dynamics of the interacting protons. It includes virtual Δ-isobar excitations, meson exchange currents, negative energy states ($\overline{p}$) and rescattering diagrams explicitly. This model also reproduces the measured anisotropies. For regions of the phase space where at least one of the protons scatter at small angles (less than 8^0), the model overpredicts the data by as much as 20%.

During this experiment we measured [12, 13] for the first time the virtual photon emission process, which in the laboratory means the measurement of the positron-electron pair. Such an event is shown in fig. 1. The added information provided by the virtual photon is the decomposition of the electromagnetic response of the nucleon in components related to the polarizations of the photon. The six nucleonic response functions are related to specific angular distributions of the two leptons. Following reference [14] the reaction cross section for the virtual bremsstrahlung process, $pp \rightarrow ppe^+e^-$, is proportional to the following amplitude:

$$|A|^2 = \frac{1}{M_\gamma^2} \left\{ W_T \left(1 - \frac{\ell^2}{2M_\gamma^2} \sin^2\theta\right) + W_L \left(1 - \frac{\ell^2}{k_0^2} \cos^2\theta\right) + \frac{\ell^2 \sin^2\theta}{2M_\gamma^2} \left(W_{TT} \cos 2\phi + W'_{TT} \sin 2\phi\right) + \frac{\ell^2 \sin 2\theta}{2k_0 M_\gamma} \left(W_{LT} \cos\phi + W'_{LT} \sin\phi\right) \right\}, \quad (1)$$

where M_γ and k_0 are the invariant mass and the energy of the virtual photon, respectively. The polar (θ) and azimuthal (ϕ) angles of the momentum-difference vector, ℓ, of the two leptons are shown in fig. 1. The six independent response functions (W_i) determining the reaction cross section are defined as

$$\begin{aligned} W_T &= M_x M_x^* + M_y M_y^*, & W_L &= \frac{M_\gamma^2}{k_0^2} |M_z|^2, \\ W_{TT} &= M_y M_y^* - M_x M_x^*, & W_{LT} &= -2\frac{M_\gamma}{k_0} \Re(M_z M_x^*), \\ W'_{TT} &= -2\Re(M_x M_y^*), & W'_{LT} &= -2\frac{M_\gamma}{k_0} \Re(M_z M_y^*). \end{aligned}$$

Here, M_i refers to the spatial coordinates of the covariant $pp \rightarrow ppe^+e^-$ amplitude [14], for which the photon momentum is chosen parallel to the z-axis and the incoming proton momentum lies in the xz-plane (see fig. 1).

The cross-section for the virtual bremsstrahlung process as a function of the invariant mass of the photon, as well as its angular distribution are plotted in fig. 2. The data are compared with predictions of the fully-relativistic microscopic-calculation [7] (solid line) as well as a low energy (LET) calculation [14, 15] (dashed line). The LET calculation uses an expansion procedure for the on-shell T-matrix in order to account for the off-shell dynamics of the interacting protons. Furthermore, it only takes into account the rescattering contributions, meson-exchange currents and the virtual Δ-isobar excitation by enforcing charge and current conservation. The microscopic calculation overpredicts the virtual photon data by approximatelly 30% over the full acceptance of the experimental setup. The LET calculation on the other hand reproduces the data rather well. We note that the cross-section depicted in fig. 2, is integrated over the leptonic angles and is therefore a measure of the transverse (W_T) and longitudinal (W_L) nucleonic response functions. The interference terms cancel out in the integration over the leptonic phase-space (see equation 0). The different amplitudes of the orthogonal cosine and sine functions of the leptonic dihedral angle ϕ (see fig. 1 and equation 0) were extracted from the data. They are directly related to each of the four interference nucleonic response functions: W_{TT}, W_{TL}, W'_{TT} and W'_{TL}. The experimentally extracted response functions are shown in fig. 3. They are integrated over the full ac-

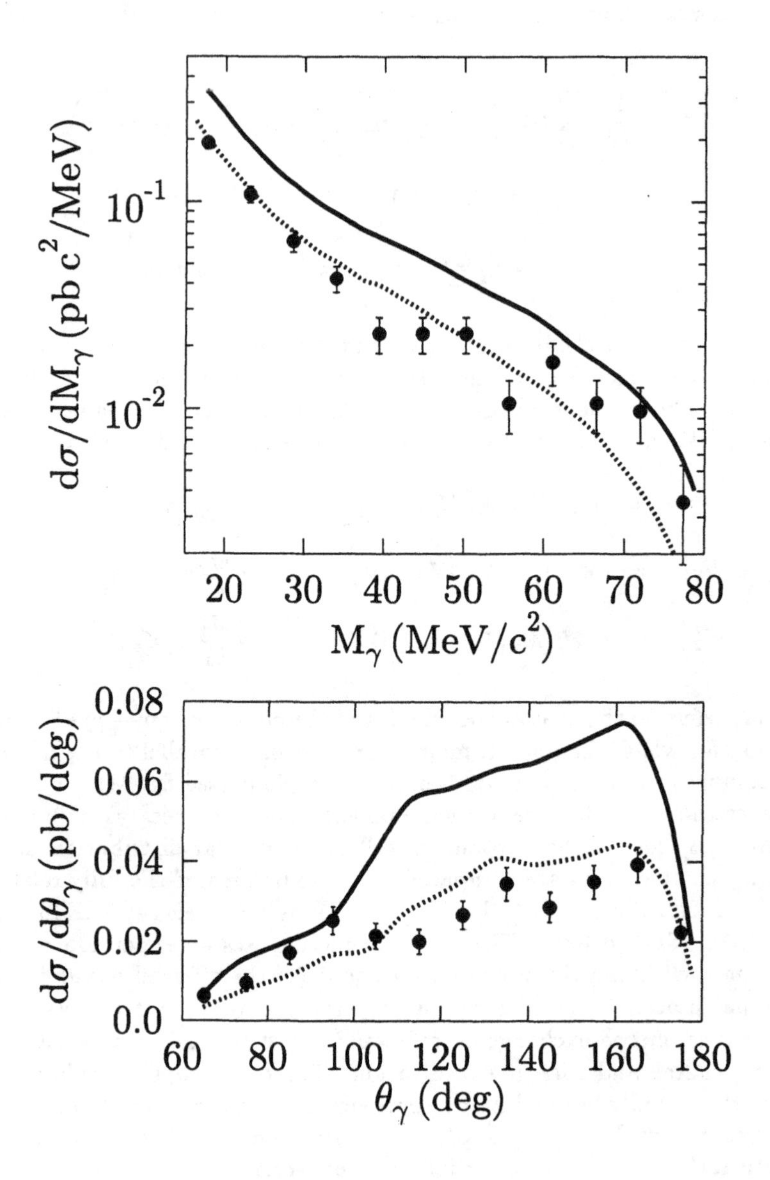

Figure 2 Top: differential cross section as a function of the invariant mass (M_γ) of the virtual photon integrated over the entire detector acceptance. Bottom: virtual-photon angular distribution in the laboratory frame for invariant masses integrated from 15 to 80 MeV/c^2. The solid lines present the results of a microscopic model and the dotted lines the results of a LET calculation. The calculations are discussed in the text.

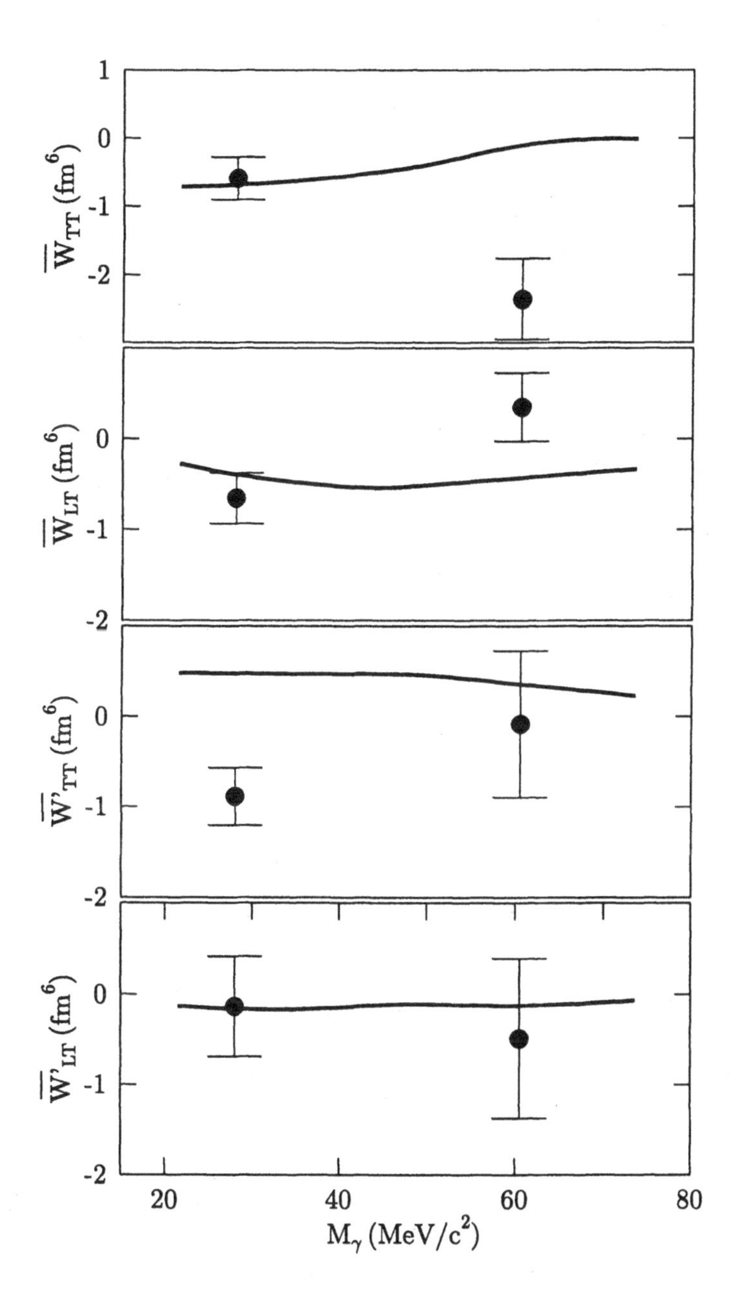

Figure 3 The experimental values of the proton electromagnetic interference response functions $\overline{W_{TT}}$, $\overline{W_{LT}}$, $\overline{W'_{TT}}$ and $\overline{W'_{LT}}$. These are integrated over the whole phase-space covered by the experimental setup. The data points represent the average values for two windows in the invariant mass of the virtual photon: 15-40 MeV/c^2 and 40-80 MeV/c^2. The solid lines represent the predictions of a fully-relativistic microscopic model of the NN interaction.

ceptance of our experimental setup, and shown for two regions of the invariant mass of the photon. The predictions of the fully relativistic calculations are also shown as solid lines. Within the accuracy of the data the model predictions are in general good. These response functions are sensitive to the different diagrams included in the calculations, and with an accurate dataset one will be able to disentangle the effects of individual diagrams.

PROTON-DEUTERON CAPTURE PROCESS

Another reaction investigated was proton-deuteron scattering. All exit channels were measured. Here I will only discuss the process $pd \rightarrow {}^3$He both for real as well as virtual photon capture. The experimental setup is the same as in the proton-proton studies. The differential cross-sections as function of the photon angle are shown for the real and virtual photon capture in fig. 4. For the real photon capture our data are compared with previously published data from IUCF [16] measured at a proton beam energy of 150 MeV, and an experiment performed at TSL [17] at 176 MeV. The data is compared with the predictions of an LET calculation discussed in ref. [18]. The dashed line is a calculation with the wave function of the excited ^{3}He state given by the Argonne V18 potential, and the nominal magnetic moment. The solid line is obtained by modifying the self-energy of the ^{3}He, which is equivalent to increasing the magnetic moment by a factor of 1.4.

FUTURE PLANS AT KVI

At KVI this whole program is going to enter its second phase, whereby the Plastic Ball detector [19], originally developed at GSI, is used as a photon/dilepton spectrometer. This detector has an almost complete coverage of 4π for the electromagnetic radiation, and increases the efficiency of the experiments reported here by a factor of 25. It is highly granular, with a total of 654 phoswich elements. At KVI this detector is being modified to include an inner-shell of Cherenkov-detectors which will provide a highly selective trigger for dilepton events. The electromagnetic response functions of the proton will be studied with a precision such that the effects of different diagrams entering in the nucleon-nucleon interaction will be disentangled.

Acknowledgments

I would like to acknowledge all my KVI collaborators which made this experimental program possible: M.J. van Goethem, M.N. Harakeh, M. Hoefman, H. Huisman, N. Kalantar-Nayestanaki, H. Löhner, J.G. Messchendorp, R. Ostendorf, S. Schadmand, R. Turissi, M. Volkerts, H.W. Wilschut and A. van der Woude. In particular J.G. Messchendorp who analysed all the data presented here. Also the support by the TAPS collaboration in the operation of the Two-Arm Photon Spectrometer at KVI is acknowledged. I would like to thank A.Yu. Korchin, G. Martinus, R. Timmermans and O. Scholten for providing the computer codes for the calculations and for useful discussions. The work provided by the cyclotron and polarized-ion source group

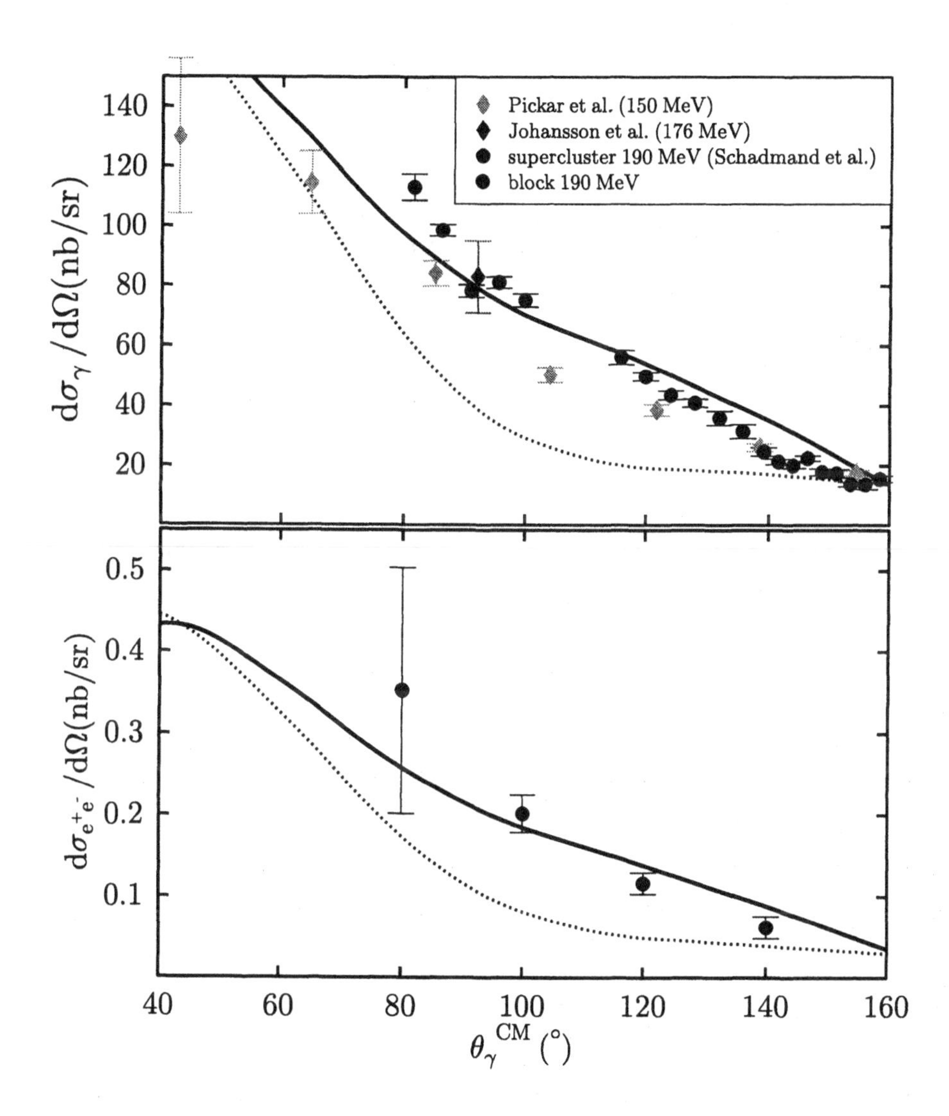

Figure 4 Measured differential cross-sections for the real photon (top) and virtual photon (bottom) $p + d \rightarrow {}^3$He capture processes. The filled dots are from the present work. Calculations with an LET model are shown as solid and dashed lines and are discussed in the text.

in delivering a high-quality beam was essential for the success of the experiment. This work was supported in part by the "Stichting voor Fundamenteel Onderzoek der Materie" (FOM) with financial support from the "Nederlandse Organisatie voor Wetenschappelijk Onderzoek" (NWO), by the "Granting Agency of the Czech Republic", by GSI, the German BMBF, and by the European Union HCM network under contract HRXCT94066.

References

[1] K. Michaelian et al., Phys. Rev. D41, 2689 (1990).
[2] N. Kalantar-Nayestanaki, Nucl. Phys. A631, 242c (1998). 48 (1993) 229.
[3] F. de Jong and K. Nakayama, Phys. Lett. B385, 33 (1996).
[4] J.A. Eden and M.F. Gari, Phys. Lett. B347, 187 (1995); Phys. Rev. C53, 1102 (1996).
[5] G.H. Martinus, O. Scholten and J.A. Tjon, Phys. Rev. C56, 2945 (1997).
[6] G.H. Martinus, O. Scholten and J.A. Tjon, Phys. Lett. B402, 7 (1997).
[7] G.H. Martinus, Ph.D. Thesis, University of Groningen (1998), and to be published.
[8] N. Kalantar-Nayestanaki et al., Nucl. Instr. Meth. Phys. Res.A417 215 (1998).
[9] S. Schadmand et al., Nucl. Instr. Meth. Phys. Res. A423 174 (1999).
[10] A.R. Gabler et al., Nucl. Instr. Meth. Phys. Res. A346, 168 (1994).
[11] J. Fleischer and J.A. Tjon, Nucl. Phys. B84, 375 (1975); Phys. Rev. D21, 87 (1980).
[12] J.G. Messchendorp et al., Nucl. Phys. A631, 618 (1998).
[13] J.G. Messchendorp et al., Phys. Rev. Lett. (1999) in press.
[14] A.Yu. Korchin and O. Scholten, Nucl. Phys. A581, 493 (1995).
[15] A.Yu. Korchin, O. Scholten and D. Van Neck, Nucl. Phys. A602, 423 (1996).
[16] M.J. Pickar et al., Phys. Rev. C35, 37 (1987).
[17] R. Johansson et al., Nucl. Phys. A641, 389 (1998).
[18] A.Yu. Korchin and O. Scholten, Nucl. Phys. (1999) in press.
[19] A. Baden et al., Nucl. Instr. Meth. Phys. Res. 203, 189 (1982).

STUDY OF MASS A=130 FISSION FRAGMENTS IN THE $^{14}N + ^{238}U$ REACTION AT 10 MeV/A*

D. L. Balabanski[1], A. Minkova[1], G. I. Rainovski[1], M. Danchev[1], N. Erduran[2], B. Akkus[2], M. Bostan[2], G. Gurdal[2], M. Yalçinkaya[2], R. Beetge[3], R. W. Fearick[3] G. Mabala[3], D. Roux[3], W. Whittaker[3], B. R. S. Babu[4],J. J. Lawrie[4], S. Naguleswaran[4], R. T. Newman[4], J. V. Pilcher[4], F. D. Smit[4], and J. F. Sharpey-Schafer[4]

[1] Faculty of Physics, St. Kliment Ohridsky University of Sofia, BG-1164 Sofia, Bulgaria
[2] Department of Physics, University of Istanbul, Vezneciler-Istanbul, Turkey
[3] Department of Physics, University of Cape Town, 7701 Cape Town, South Africa
[4] National Accelerator Centre, 7131 Faure, South Africa

"Dimiter Balabanski" <mitak@phys.uni-sofia.bg>

Abstract: Fission fragments in the mass A≈130 region, lying close to the β-stability line, were produced in the $^{14}N + ^{238}U$ fusion-fission reaction at 10 MeV/A. The ^{14}N beam was delivered by the Separated Sector Cyclotron at the National Accelerator Centre, Cape Town, South Africa. The γ-rays of the fission fragments were detected by the AFRODITE spectrometer. The X-ray LEPS detectors were used for element selection and γ-X-ray coincidences for fragment identification.

1 INTRODUCTION

An intense effort has been made to understand the roles of the shell gaps and active orbitals which support the stabilization of the nuclear shapes at different deformations in the mass A≈130 region. Yet, experimental data is lacking

*This work was partially supported by the Research Fund of the University of Istanbul through grants UP-8/270598 and UP12/040199.

for the heavier isotopes, which lie close to the β stability line and beyond it. This provides the necessity to push back the borders of the unknown and investigate these nuclei. These studies will provide new information and will allow one to understand many structural effects which are already established in this mass region, *i.e.* the shape competition between prolate and oblate bands which was observed in the heaviest investigated iodine isotopes [1, 2, 3]. At higher angular momenta in these nuclei a competition between the collective and single-particle states was observed, which was interpreted in terms of abrupt band-terminations [4, 2, 5]. Significant asymmetry in γ-direction was also reported in this region, which is based on the observed signature inversion in the $\pi h_{11/2} \otimes \nu i_{13/2}$ and $\pi h_{11/2} \otimes \nu h_{11/2}$ bands in odd-odd nuclei [7]. All these effects need to be traced to higher masses.

The nuclei at the β-stability line and beyond it, cannot be produced in heavy-ion fusion-evaporation reactions, which are the common tool to populate high-spin states in nuclei. Therefore, other reaction mechanisms need to be utilized in these cases. In recent experiments at Gammasphere and at Euroball, spontaneous fission of ^{248,252}Cf was used to study the excited states of fission fragments in the vicinity of the N=82 neutron shell [8, 9].

Here we report an experiment which aims to populate excited states in A$\geq$125 iodine nuclei in the ^{14}N + ^{238}U reaction at 140 MeV (10 MeV/A).

2 REACTION MECHANISM AND EXPERIMENTAL TECHNIQUE

There are two fission mechanisms which contribute significantly to the total cross-section, when bombarding a heavy target (e.g. ^{238}U) with a high-energy beam well above the Coulomb barrier: low-energetic fission and deep-inelastic processes. As a result of the former mechanism an asymmetric mass distribution of the fragments is obtained. In this case the nucleus is excited by predominantly peripheral colisions. The process is dominated by shell effects and the heavy fragment is a neutron-excessive nucleus, having approximately 50 protons and 82 neutrons. Yet, the energy of the system is high enough to allow the evaporation of a few nucleons, and the mass of this fragment is shifted towards the line of stability. These interactions account for about 20-30% (or even more) of the total fission cross-section. On the other hand, lighter nuclei are formed through high excitation fission processes, involving larger momentum and energy transfer, with smaller impact parameters and large overlap of the incident ions and the target nucleus. The width of the mass distribution for each of these processes is known to be about 5 mass units [10]. Thus, the interplay between the target-projectile combination and variation of the projectile energy provides a possibiliy to move the centroid of the fragment mass distribution throughout the (N,Z)-plane and, eventually, access nuclei, which cannot be produced in spontaneous fission, or through symmetric fission, which has a maximal yield for Z/2 and N/2; Z and N are the sums of the protons and neutron of the target and the projectile, respectively.

In a previous study, Yu *et al.* investigated the reaction ^{12}C + ^{238}U at 20 MeV/A [10] and report considerable yields for the A$\geq$125 iodine isotopes. We have chosen a ^{14}N projectile in order to somewhat enhance the production of odd-Z isotopes.

The ^{14}N beam was delivered by the Separated Sector Cyclotron at the National Accelerator Centre, Faure (Cape Town), South Africa. A thick ^{238}U target ($\approx$65 mg/cm^2) was used for this experiment. The γ-decay and the X-rays of the fission fragments were detected with the AFRODITE spectrometer [11], which for this experiment consisted of 8 Compton suppressed Clover detectors (4 of them at forward and the other 4 at backword angles with respect to the beam) and of 7 LEPS detectors. All the LEPS detectors were in the plane perpendicular to the beam (90° plane). The target was tilted at 45° with respect to the 90° plane, in a way that 3 LEPS detectors monitor the front side of the target, three of them the back side and one the bottom of the target. The scattering chamber had mylor windows. The treshold of the LEPS detectors was put as low as possible in order to detect X-rays in the mass A$\approx$130 region, which were used for element identification in this experiment. This allowed us to have a detection limit for the X-rays as low as 8 keV. We used 50 μm brass absorbers for the LEPS detectors, in order to reduce the $_{92}$U L X-rays, having energies of E_{L_α}=13.5 keV, E_{L_β}=17.2 KeV and E_{L_γ}=20.2 keV. The intensity of the L_β- and L_γ-rays is 16% and 2.5% of the total X-ray intensity of $_{92}$U, but, as intense flushes of X-rays occur during the slowing down of the fission fragments in the target, they were overloading the LEPS detectors. It should be noted that the intensity of the L X-rays in this case was much larger compared to that of the K X-rays. This indicates that L-electrons are ionized predominantly during the slowing down of the fragments. Nevertheless the X-rays of interest were detected reasonably well.

A typical event in this experiment had a multiplicity >3 in the LEPS detectors and mucltiplicity $\geq$2 in the Clover detectors. Fivefold events were required during the experiment with multiplicity $\geq$2 in the Clover detectors.

3 RESULTS AND DISCUSSION

A total projection spectrum, detected by the LEPS detectors of the AFRODITE array, revealing the X-rays which were observed in this experiment is presented in fig. 1. The intense X-rays of $_{74}$W and to a less extend of $_{82}$Pb are due to the W collimators and Pb material which are used in the set-up. Some extra intensity of the $_{79}$Au X-rays may be due to Au-plated contacts of the LEPS detectors. The spectrum demonstrates the general trend of decreasing intensity of the X-rays with the increase of the atomic number throughout the rare-earth region. The somewhat smaller X-ray intensity for $_{52}$Te and $_{53}$I is due to the attenuation of the X-rays in the low-energy region.

The X-ray efficiency of the LEPS detectors of the array provides a condition for element selection. After gating on the corresponding X-rays, γ-ray spectra, detected by the Clover detectors, were projected. γ-rays, which correspond to some $_{52}$Te, $_{53}$I and $_{54}$Xe isotopes, were identified in these spectra. The results are presented in fig. 2. At this stage of the analysis it is not possible to determine absolute yields, but the results suggest that the distribution of

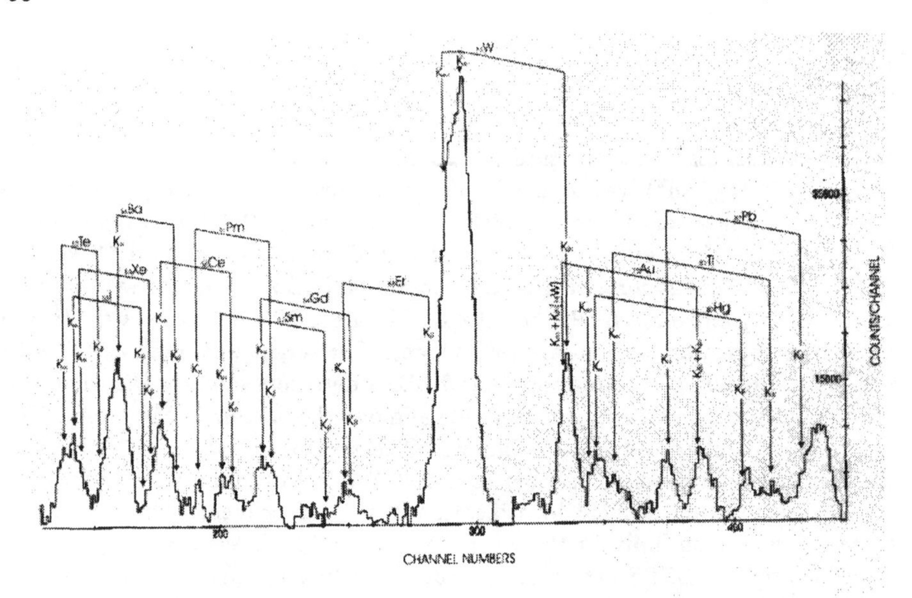

Figure 1 A total projection X-ray spectrum detected by the LEPS detectors of the AFRODITE array in the ^{14}N + ^{238}U reaction at 140 MeV. Arrows indicate the K_α and K_β lines for most of the observed elements.

the heavy fragment is somewhat shifted towards lighter masses, which might be due to the evaporation of a few neutrons.

In summary, we have carried out an experiment to populate excited states of neutron-excessive nuclei in the mass A≈130 region, utilizing the ^{14}N + ^{238}U reaction at 140 MeV. The γ- and X-rays of the fission fragments which were produced in the reaction were detected by the AFRODITE spectrometer and used for isotope identification. The study demonstrates the applicability of the method. A shortcoming of this experiment is the hindrance of the X-ray element selection, due to the strong contamination of the $_{92}$U X-rays. An improvement

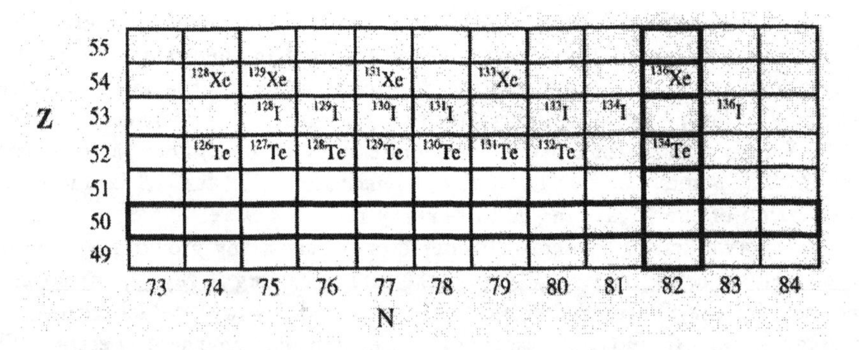

Figure 2 Part of the nuclear chart, displaying the identified $_{52}$Te, $_{53}$I and $_{54}$Xe isotopes in the ^{14}N + ^{238}U reaction at 140 MeV.

of the technique should be achieved by direct fragment identification (*i.e.* by adding solar cells to the array), which will improve the selectivity and the quality of the γ-spectra.

Acknowledgments

Travel grants for D.L.B., A.M., G.G. and M.Y. from the Centre for Turkish-Balkan Physics Research and Applications are greatfully acknowledged. The work is also supported in part by the Bulgarian National Science Fund.

References

[1] F. Dönau and U. Hagemann. *Zeitschrift für Physik A*, 293:31, 1979.

[2] Y. Liang, D.B. Fossan, J.R. Hughes, D.R. LaFosse, T. Lauritsen, R. Ma, E.S. Paul, P. Vaska, M.P. Waring, and N. Xu. *Physical Review C*, 45:1041-1057, 1992. And references therein.

[3] G. Rainovski, D.L. Balabanski, G. Lo Bianco, G. Falconi, N. Blasi, D. Bazzacco, G. de Angelis, D.R. Napoli, M.A. Cardona, A.J. Kreiner, and H. Somacal. Submitted to *Physical Review C*, 1999.

[4] D.L. Balabanski, G. Rainovski, N. Blasi, G. Lo Bianco, S. Signorelli, D. Bazzacco, G de Angelis, D.R. Napoli, M.A. Cardona, A.J. Kreiner, and H. Somacal. *Physicsl Review C*, 56:1628-1632, 1997.

[5] E.S. Paul, J. Simpson, S. Araddas, C.W. Beausang, M.A. Bentley, M.J. Joycet, and J.F. Sharpey-Schafer. *Journal of Physics G*, 19:913, 1993.

[6] Y. Liu, J. Lu, Y. Ma, S. Zhou, and H. Zheng. *Physical Review C*, 54:719-730, 1996.

[7] T. Komatsubara, K. Furano, T. Hosoda, J. Mukai, T. Hayakawa, T. Morikawa, Y. Iwata, N. Kato, J. Espino, J. Gascon, N. Gjørup, G.B. Hagemann, H.J. Jensen, D. Jerrestam, J. Nyberg, G. Sletten, B. Cederwall, and P.O. Tjøm. *Nuclear Physics A*, 557:419c-438c, 1993.

[8] I. Ahmad and W.R. Phillips. *Reports on Progress of Physics*, 58:1415, 1995.

[9] J.H. Hamilton, A.V. Ramayya, S.J. Zhu, G.M. Ter-Akopian, Yu.Ts. Oganessian, J.D. Cole, J.O. Rasmussen, and M.A. Stoyer. *Progress on Particle and Nuclear Physics*, 35:635, 1995.

[10] Y.W. Yu, C.H. Lee, K.J. Moody, H. Kudo, D. Lee, and G.T. Seaborg. *Physical Review C*, 36:2396-2402, 1987. Prentice-Hall, 1992.

[11] J. F. Sharpey-Schafer in the *Structure of Vaccuum & Elementary Matter, Widerness, South Africa, March 10–16, 1996*, edited by H. Stöcker, A. Gallmann, and J. H. Hamilton (World Scientific, Singapore, 1997) p. 656.

THE ISAC RADIOACTIVE BEAM FACILITY AT TRIUMF: PRESENT STATUS AND FUTURE PLANS

G.C. Ball[1], R. Baartman[1], J. Behr[1], P. Bricault[1], L. Buchmann[1], J.M. D'Auria[2], P. Delheij[1], M. Dombsky[1], G. Dutto[1], D. Hutcheon[1], K.P. Jackson[1], R. Kiefl[3], R. Laxdal[1], P. Levy[1], J.-M. Poutissou[1], P. Schmor[1], G. Stanford[1], and T. Ruth[1]

[1]TRIUMF, 4004 Wesbrook Mall, Vancouver, BC, Canada, V6T 2A3
[2]Simon Fraser University, 8888 University Way, Burnaby, BC, Canada, V5A 1S6
[3]University of British Columbia, Vancouver, BC, Canada, V6T 2A3

ball@triumf.ca

Abstract: A new radioactive beam facility of the ISOL type, ISAC, is being built at TRIUMF to provide intense beams of both low-energy (<60 keV) and accelerated radioactive ions up to 1.5 MeV/u. The facility is scheduled for completion by the fall of 2000. With further energy upgrades anticipated in the next five-year plan, Canada is positioning itself as a leader in the next generation of radioactive beam facilities. This report presents a general description of the facility, an overview of the initial experimental program, the present status of the project and the future plans for ISAC-II.

1 INTRODUCTION

In June 1995, the Canadian federal government funded the construction of a new radioactive beam facility at TRIUMF called ISAC (Isotope Separator and ACcelerator), which will provide intense beams of accelerated radioactive isotopes by the fall of the year 2000 with energies up to 1.5 MeV/u and with a possible extension towards higher energies (6.5 MeV/u) in the future. The artist's conceptual view of the new ISAC facility at TRIUMF is presented in Fig. 1.

The Nucleus: New Physics for the New Millennium
Edited by Smit et al., Kluwer Academic / Plenum Publishers, New York, 2000.

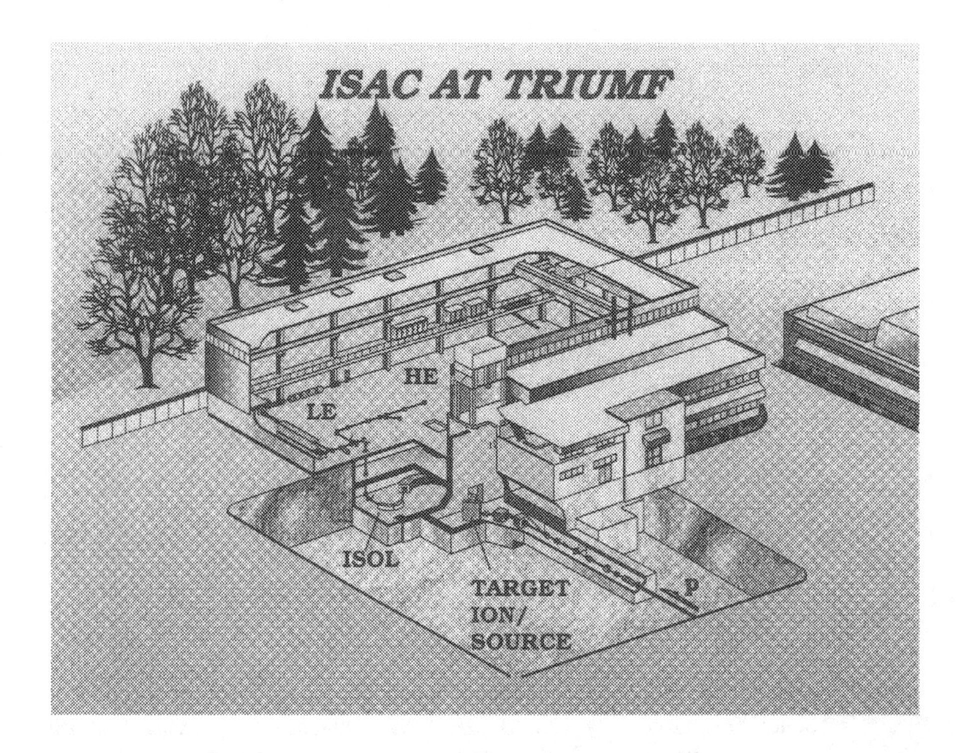

Figure 1 Sketch of the ISAC-1 building layout.

ISAC will produce radioactive beams by the ISOL method using very intense beams of protons at 500 MeV to bombard thick production targets. The facility is being constructed with a capability to handle proton beam intensities on target of up to 100 μA. Protons from the TRIUMF main cyclotron are extracted into a new beam line which feeds the new ISAC building. In the target hall, two target stations will receive the proton beam alternatively to increase the efficiency of ion beam delivery. The targets will be initially of the type proven at TISOL, the 1 μA test facility which has been in operation since 1987[1], or similar targets used at CERN-ISOLDE.

The target is coupled to the ion source via a small transfer tube. Several types of ion sources will be available; the first operation will use a surface ionization source to produce beams of alkali elements. In addition, this ion source should also produce beams of elements such as Sr, Ga, Yb, and others. Development has started on a compact microwave ECR ion source to produce beams of volatile elements. At present there are only conceptual plans to develop other sources, such as FEBIAD or Laser ion sources.

The target area will be the most activated location of the new ISAC facility. To handle the components of the proton beam line near the production target and the first elements of the ion beam production, it was decided to use the concept of a double vacuum enclosure with a shielding plug (2 m in length) above each component. Services, connections and vacuum couplings are outside the high radiation field region where a hands-on operation is possible. The changeover of components can be done remotely via a high-precision crane.

After exiting the ion source, the ion beam will be mass analyzed, by a low resolution pre-separator followed by a high acceptance mass analyzer with a resolving power of $\frac{\Delta M}{M} \simeq \frac{1}{10000}$. The whole assembly sits on a HV platform to allow the operation of the mass separator at constant momentum, independent of the actual beam energy. The mass separated beam is then directed vertically from the underground mass separator floor, up approximately 8 m to the experimental hall located at the grade level. Alternatively, the beam can be deflected to the TRINAT (TRIUMF Neutral Atom Trap) facility, located on a well-shielded intermediate floor. Once at ground level in the new ISAC experimental hall, the beam will be directed either to the RFQ/DTL accelerators for experiments with accelerated beam (HE), or to the experimental areas planning to use the unaccelerated radioactive beams. The cw RFQ accelerator will take a singly charged ion beam with $q/A > \frac{1}{30}$ from 2 keV/u to 150 keV/u. After the RFQ, the beam will be stripped, and the selected charge directed to a cw DTL which will accelerate an ion beam with $q/A > \frac{1}{6}$ to the desired energy in the range from 0.15 to 1.5 MeV/u. The RFQ will be operated in continuous duty cycle mode at 35 MHz. As this is the most challenging part of the accelerator chain, it was decided to do an early test of the accelerating structure with a stable ^{14}N beam and 7 of 19 rf rings. This was successfully accomplished in June, 1998. The entire acceleration system including matching sections between the RFQ and DTL are expected to be ready for commissioning by the beginning of 2000. It is anticipated that radioactive beams at full energy will be available in the HE experimental area by the fall of 2000.

2 EXPERIMENTAL PROGRAM

A wide range of radioactive beams at low energy ($\leq$ 60 keV) will be provided for a number of experimental programs including: 1) the test of weak interaction symmetries by trapping radioactive atoms in a magneto-optic trap; 2) tests of CVC and the Standard Model from precise measurements of the intensities for superallowed Fermi $0^+ \rightarrow 0^+ \beta$-decay; 3) a systematic study of ground state moments for nuclei far from stability using low-temperature nuclear orientation methods; 4) condensed matter studies of small structures and interfaces in semiconductors and superconductors by β-NMR with a polarized ^{8}Li beam; and 5) the production of pure radionuclides of importance in nuclear medicine. The accelerated radioactive ion beams with masses A<30 and energies from 0.15–1.5 MeV/u will be used primarily for research in nuclear astrophysics. The HE experimental area will house a state of the art recoil separator (DRAGON) and a general purpose scattering facility. Figure 2 gives a general layout of the planned experimental facilities in the ISAC hall. A brief overview of some of the experimental programs is presented below.

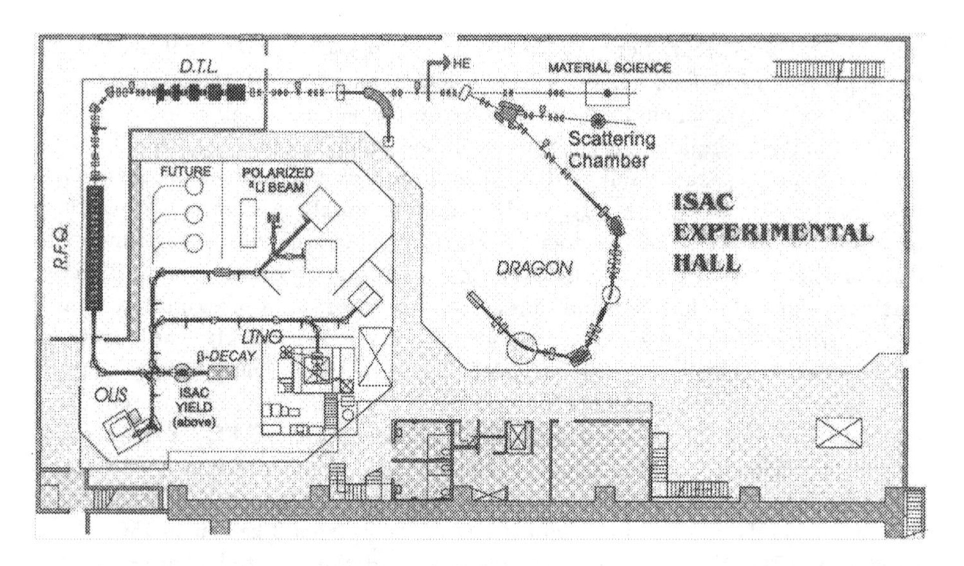

Figure 2 Planned layout of scientific stations in the ISAC experimental hall.

2.1 Fundamental Symmetry Tests

The TRINAT facility traps radioactive atoms produced at the on-line separator TISOL in a magneto-optic trap (MOT)[2]. The atoms are suspended in space, so unperturbed nuclear recoils from β-decay can be detected, and therefore the ν momentum can be deduced, making new correlation measurements possible. The atoms (and hence nuclei) can be polarized, allowing improvements in measurements of the degree to which parity is maximally violated in the weak interaction. In autumn 1998 the trap was moved to ISAC.

Being a purely weak process, nuclear β-decay is directly sensitive to physics at the energy scales characterized by the mass of the lefthanded vector bosons, M_w=80.4 GeV/c^2. The $\beta - \nu$ correlation experiments in the $0^+ \rightarrow 0^+$ decay of ^{38m}K ($t_{\frac{1}{2}}$=0.925 sec) at accuracy 0.01 are sensitive to masses of new scalar exchange bosons of ($\sim(0.01)^{-\frac{1}{4}} M_w$ $\sim$260 GeV/c^2) (for standard electroweak coupling), complementary, for example to direct searches at HERA. Limits on a possible scalar interaction are poor, both from beta decay and from particle physics, and a 1% measurement of the $\beta - \nu$ correlation coefficient a would be competitive. Measurements of spin correlation variables in ^{37}K to $\leq 10^{-3}$ accuracy will be necessary to set limits on right-handed W's complementary to direct searches at Fermilab and HERA.

The feasibility of these experiments has now been demonstrated. Using a double trap system, 10,000 atoms/sec of ^{38m}K have been collected and moved by magneto-optic forces to a detection trap, free of background decays of untrapped atoms. The momentum of the β^+ is measured in a ΔE-E telescope of a double-sided Si strip detector backed by a plastic scintillator, and the position and time-of-flight of the Ar recoils is measured in coincidence in a microchannel plate. Neutral Ar recoils are measured at back-to-back angles with the β^+, and a uniform electric field collects essentially the entire angular distribution of charged Ar recoils. About 500,000 coincident events have been recorded so far and are under analysis[3].

For the second phase of this program, high polarization of the trapped ^{37}K nuclei is required and a scheme based upon optical pumping of a hyperfine transition with circularly polarized light is being developed (initially with stable ^{41}K atoms).

2.2 *Tests of CVC and the Standard Model*

Precise measurements of the rate for superallowed Fermi $0^+ \rightarrow 0^+\beta$-decays provide demanding and fundamental tests of the properties of the weak interaction. In particular, since the axial vector decay strength is zero for such decays the intensities are directly related to the weak vector coupling constant. Presently nine transitions have been determined with sufficient precision to confirm the conserved vector current (CVC) hypothesis at the level of $3\text{x}10^{-4}$. These data together with the muon lifetime also provide the most accurate value for the up-down quark mixing matrix element of the Cabibbo-Kobayashi-Maskawa (CKM) matrix, V_{ud}. However, the resulting unitarity test of the first row of the CKM matrix differs from unity by more than two standard deviations[4]. Since there is no known reason to question the uncertainties in the small calculated Coulomb and radiative corrections that have been applied to these data, additional measurements are required to examine this discrepancy.

At ISAC a program has been initiated to measure the half-lives and branching ratios for T_z=0 nuclei with $A \geq 62$. These data together with accurate Q-values will test the theoretical calculations for analogue symmetry breaking which are predicted to be much larger for these nuclei. The nucleus ^{74}Rb has been chosen for the initial experiment since it should be the easiest one to produce. The experiments will be carried out using techniques that were perfected by members of the collaboration in previous high-precision beta-decay half-life measurements[5]. Although the measurements are simple in principle, great care must be taken to achieve the required precision (~0.05%). In addition, since the Q-value is large, several significant branches to excited states will need to be accurately determined. The measurements will be very demanding because of the short half-life of ^{74}Rb (~69 ms) and limited beam intensity.

2.3 *Nuclear Astrophysics*

The description of the energy production in astrophysical sites like stars or in explosive events like novae, supernovae, and X and γ rays bursts and the understanding of the generation of heavy elements in the universe relies on a comprehensive understanding of experimental and theoretical nuclear physics involving many unstable nuclei which will be reachable at ISAC. Typical information required in a network of linked nuclear reactions in explosive stellar events are the capture cross sections, mainly for (p,γ), (n,γ), (α,γ) at the energies involved (of the order of $\leq$ 1.5 MeV/u), and the β-decay characteristics of unstable nuclei for a large range of nuclei extending to the limit of stability. Although a very large number of data are required, some reactions are key either because they are slow or are related to astronomically observable isotopes. Theory can be used to make order of magnitude predictions for other reactions. For the reactions involved in the low mass region, for example, in the CNO cycle and NeNa cycle, a small number of nuclei constitute waiting points which critically control the evolution of stars. They involve nuclei participating in the rapid proton capture (rp) process occurring in novae and x-ray bursts and controlling the production of proton-rich elements up to mass 100 and beyond.

ISAC has been optimized for this program by providing high intensities of low mass ($A<30$) nuclei at energies continuously variable between 0.15 and 1.5 MeV/u. Since most of the capture reactions of interest involve proton or alpha captures, the studies will be done in reverse kinematics by bombarding a high pressure windowless gas target with radioactive beams and extracting the reaction products with high selectivity in a recoil mass fragment separator (DRAGON).

The initial physics program will focus on the ^{21}Na(p,γ)^{22}Mg reaction which by passes the production of ^{22}Na in novae. The observations of ^{22}Ne abundance in meteorites and the absence of ^{22}Na gamma rays in the recent data from γ-ray observatories in orbit are not consistent with the current reaction network parameters[6]. The measurement of ^{21}Na(p,γ) will be followed by studies of a few key reactions, such as ^{19}Ne(p,γ), ^{18}F(p,γ), ^{18}Ne(α,γ), involved in the breakout path from the CNO cycle toward the Na-Ne cycle which sets up the onset of the rp path towards higher mass elements. Ultimately, DRAGON has been designed to access the ^{15}O(α,γ) reaction which is the key breakout path from the CNO cycle for supernovae. This benchmark study will require very high ^{15}O yields and a separation capability of the recoil separator at the 10^{15} level between ^{15}O beam ions and ^{19}Ne recoil ions.

As part of the nuclear astrophysics program, a number of nuclear structure studies are envisaged using a new sophisticated elastic scattering facility with modern pixel Si detectors of high granularity. Of particular interest are (α,p) reactions on waiting point nuclei in the rp process.

3 FIRST LOW-ENERGY RADIOACTIVE BEAM FROM ISAC

The construction of the ISAC facility is proceeding on schedule. In the fall of 1998 all of the components required to produce and deliver a radioactive ion beam to the TRINAT facility and to the precision β-decay lifetime experiment were completed and tested with stable ^{39}K, ^{40}Ca beams from the ISAC target ion source. A CaO (compressed pellet) target was used to produce the first radioactive beams, ^{37}K, ^{38m}K, from ISAC on November 30, 1998. With the proton beam current limited to $\leq 1\ \mu$A, the initial radioactive beam intensities were comparable to those obtained from TISOL. The ^{37}K, ^{38m}K beams were delivered to TRINAT for several days during this initial development period. Following the completion of the hot cells and other remote handling systems, the experimental program with low-energy radioactive beams from ISAC is scheduled to begin in the spring of 1999.

4 ISAC-II

When the intense beams of exotic radioactive nuclei are available from ISAC-I, outstanding opportunities will exist to attack problems in the areas of nuclear astrophysics, fundamental symmetries, condensed matter physics, nuclear medicine, and applied engineering. If in addition it were possible to accelerate these ions to energies above the Coulomb barrier, the facility becomes particularly valuable for research into nuclear structure physics and nuclear reactions.

In response to TRIUMF's request for input into the next Five-Year Plan (2000–2005), the Canadian nuclear physics community made a strong submission to have ISAC-I extended. This upgrade called ISAC-II would be designed to accelerate radioactive ions with masses $A \geq 150$ to energies of $E \geq 6.5$ MeV/u. A charge state booster such as an electron-cyclotron-resonance (ECRIS) or electron beam (EBIS) ion source would be used to provide a q/A of at least 1/30 required to accelerate all masses $A \leq 150$ with the existing RFQ. A new DTL would be built to accelerate the ion beams from 0.15 to 0.4 MeV/u, the optimum stripping energy for $30 \leq A \leq 150$. Finally, a superconducting linac with many short modules similar to those operating reliably at other heavy ion labs would boost the energy to $6.5 \leq E \leq 15$ MeV/u for $\frac{1}{7} \leq \frac{q}{A} \leq \frac{1}{3}$. The construction of ISAC-II would be carried out with a minimal interruption to the experimental program with ISAC-I. In addition, by installing superconducting modules downstream of the ISAC-I DTL, beams of $E \sim 5$ MeV/u for $A < 60$ could be available in the existing ISAC building by the end of 2002.

5 NUCLEAR SCIENCE WITH ISAC-II

Until now, our understanding of nuclei has been limited by our ability to only make nuclei which lie near the valley of stability. Large regions of the N-Z plane with extreme ratios of Z/N have been inaccessible. The properties of both proton-rich and neutron-rich nuclei are expected to be quite different from those near the stability line. Accelerated radioactive ion beams will allow us to study these nuclei in detail and learn how isospin and neutron density affect the nucleon-nucleon and spin orbit force, what the form of the pairing interaction is for weakly bound nuclei, how the nuclear shell structure evolves with N/Z and what if any are the magic numbers at the limits of stability. These are all questions of key importance to our understanding of the nucleus as a many-body quantum system.

Letters of intent have been submitted to TRIUMF to relocate two Canadian experimental facilities at ISAC-II, namely: the 8π spectrometer and the CRL/Laval 4π multidetector array. The ISAC-II physics interests of Canadian and international collaborations include: (1) nuclear structure physics from heavy-ion fusion-evaporation reactions with proton and neutron-rich projectiles; (2) spectroscopy of exotic nuclei with resonance reactions; (3) nuclear shell structure of nuclei far from stability from one and two-nucleon transfer reactions with radioactive beams in inverse kinematics; (4) exploiting the isospin degree of freedom in the study of heavy-ion reaction dynamics with light ($A<40$) ions at $E\geq6.5$ MeV/u; (5) Coulomb excitation of neutron-rich nuclei of importance to the rapid neutron capture (-r) process and where the weakening of the shell structure of nuclei is predicted; (6) the mechanism of heavy-ion reactions with neutron-rich projectives leading to the production of heavy elements; and (7) an extension of the ISAC-I nuclear astrophysics program to higher masses using the DRAGON recoil separator and General Purpose Scattering facility currently under development.

Radioactive ion beams with $A\leq150$ and energies ≤15 MeV/u are required to study all of the exciting nuclear structure and nuclear reaction physics addressed in the (November 1997) US White Paper "Scientific Opportunities with an Advanced ISOL Facility". It is evident that ISAC-II would be a world-class facility capable of accessing most of this physics. Many topics could be explored using the full range of nuclear techniques developed for studies with stable beams. The choice will depend not only on the interests of the Canadian nuclear physics community and the international users of ISAC-II but also on the availability and/or development of the instrumentation required to exploit these exotic beams.

6 CONCLUSION

With the construction of ISAC at TRIUMF, Canada is positioning itself in the new millenium as a leader in the next generation of radioactive beam facilities. When combined with a strong diversified program in nuclear physics, nuclear astrophysics, condensed matter research and biosciences and with the proposed ISAC-II upgrades, the facility holds the promise to provide a unique scientific opportunity in North America for many years to come.

References

[1] K. Oxorn *et al.*, Nucl Inst Meth **B26** (1987) 143; M. Dombsky *et al.*, Nucl Inst Meth A295 (1990) 291.

[2] J.A. Behr *et al.*, Phys Rev Lett 79 (1997) 375.

[3] J.A. Behr *et al.*, Trapped Charged Particles and Fundamental Physics, ed., D.H.E. Dubin, AIP 1998 (to be published).

[4] J.S. Towner and J.C. Hardy, Proc of the V International Symposium on Weak and Electromagnetic Interactions in Nuclei, Santa Fe, NM, June 1998, to be published and references cited therein.

[5] V.T. Koslowsky *et al.*, Nucl Inst Meth A401 (1997) 289.

[6] M. Politano *et al.*, Ap J. 448 (1995) 807.

THE PION-NUCLEUS INTERACTION AND PIONIC ATOMS

R. C. Barrett[†] and Y. Nedjadi[††]

[†]Physics Department, University of Surrey
Guildford, GU2 5XH, UK
[††]Nuclear and Astrophysics Laboratory
University of Oxford, Keble Rd
Oxford, OX1 3RH, UK *

r.barrett@surrey.ac.uk y.nedjadi1@physics.ox.ac.uk

Abstract: We have derived a relativistic impulse approximation to calculate the relativistic DKP pion-nucleus optical potentials, using the Duffin-Kemmer-Petiau (DKP) wave equation. We give a systematic comparative analysis of our DKP relativistic pion-nucleus optical potential with typical non-relativistic models available in the literature. We compare these models by looking at their respective prescriptions for the real and absorptive, S-wave and P-wave, isoscalar and isovector, pieces of the pion-nucleus optical potentials.

1 INTRODUCTION

In the conventional treatment of the pion-nucleus system the Klein-Gordon (KG) equation is solved using some assumed pion-nucleus interaction [1]. In this work an alternative relativistic approach based on the Duffin-Kemmer-Petiau (DKP) equation has been used [2].

Although the KG-based approach has been widely adopted, arguments based on the non-conserved vector current [3], the pion sub-structure [4] and the smoothness of the pion propagator [5] would favor the DKP equation over the KG one. Furthermore the first-order relativistic DKP equation is free from the ambiguities associated with the inclusion of interactions in the KG equation and it can easily be used for coupled channel systems.

*Partial funding provided by EPSRC grant GR/J95867.

Important features of the empirical pion-nucleus potential, such as a strong S-wave repulsion and a modified Kisslinger P-wave component, can be shown to arise naturally from the dynamics of the DKP equation.

Friedman, Kälbermann and Batty [6] demonstrated that analysing the pionic level shifts and widths using the DKP equation results in a marginal improvement over the fits obtained with the Klein-Gordon equation. However their analysis involved purely phenomenological pion-nucleus potentials and the authors chose to insert potentials in the DKP equation in a rather artificial way so as to make the resulting Klein-Gordon equivalent equation have a Lorentz-Lorenz-Ericson-Ericson (LLEE) parameter with a value of 2. Here the real part of the pion-nucleus interactions is not phenomenological and such artificial prescriptions are not necessary.

2 THE DKP EQUATION IN PION-NUCLEUS PHYSICS

The DKP equation for a pion interacting with a nucleus can take the form

$$(i\beta^\mu \partial_\mu - m_\pi)\psi = (U_s(r) + \beta^0 U_v^0)\psi \tag{1}$$

where U_s and U_v^0 are some Lorentz scalar and time-like vector DKP pion-nucleus potentials.

The first component of the DKP spinor (ϕ) obeys the Klein-Gordon equation

$$\left(\nabla^2 + (E^2 - m^2)\right)\phi = 2m\,(U_{S-wave} + U_{P-wave})\,\phi \tag{2}$$

where

$$U_{S-wave}(r) = U_s + U_v^0 + \frac{1}{2m}(U_s^2 - U_v^{0^2}) + \frac{E-m}{m}U_v^0 \tag{3}$$

$$U_{P-wave}(r) = \frac{1}{2m}(1+\alpha(r))\nabla \frac{\alpha(r)}{1+\frac{1}{3}\xi\alpha(r)}\nabla \quad \text{with} \quad \alpha(r) = \frac{U_s}{m} \quad \text{and} \quad \xi = 3. \tag{4}$$

The DKP approach therefore generates a non-local term similar to the Kisslinger potential modified to account for the LLEE effect in pion-nucleus scattering [11].

The DKP model prescribes a value of 3 for the EELL parameter (ξ); ξ is not well determined by pionic atoms and low-energy pion-nucleus scattering data. Its theoretical estimates vary in the range $1.2 \leq \xi \leq 2.6$ [1]. A value of 3 for the EELL parameter corresponds to smoothing the short-range behaviour built-in the Kisslinger potential. The same kind of effect was also found in Dirac nucleon-nucleus scattering : relativistic propagators simulate nuclear medium effects.

3 CONSTRUCTING THE DKP OPTICAL POTENTIAL : 3 STEPS

- The experimental $\pi - N$ amplitudes are represented as a set of Lorentz-invariant t-matrices :

$$\mathcal{T} = SII + S'IP + V\gamma_\mu\beta^\mu + V'\gamma_\mu P\beta^\mu + T\sigma_{\mu\nu}\Sigma^{\mu\nu}, \tag{5}$$

where I, P and β^{μ} are the DKP unit, projector and β matrices respectively while I and γ_{μ} are the Dirac unit and γ matrices; $\sigma_{\mu\nu}$ and $\Sigma^{\mu\nu}$ are the Dirac and DKP anti-symmetric tensors respectively. The matrix elements of these Lorentz-invariant $\pi - N$ amplitudes between the free DKP (π) and Dirac particle (ψ) spinors are equated to the experimental center of mass two-body scattering amplitude between Pauli spinors (χ) for the nucleon :

$$\chi^{\dagger}\mathcal{F}\chi = \frac{M}{4\pi\sqrt{s}}\bar{\psi}\bar{\pi}T\pi\psi, \tag{6}$$

where M is the nucleon mass, $\sqrt{s}$ is the pion-nucleon centre of mass energy, and

$$\mathcal{F} = b + c\,\mathbf{k}\cdot\mathbf{k}' + d\,\boldsymbol{\sigma}\cdot(\mathbf{k}'\wedge\mathbf{k}). \tag{7}$$

- The t-matrices are then folded with the relevant Lorentz nuclear densities

$$\rho_s(r) = \sum_{\alpha}^{occ}\psi_{\alpha}^{\dagger}(r)\gamma^{0}\psi_{\alpha}(r) \quad \text{and} \quad \rho_v(r) = \sum_{\alpha}^{occ}\psi_{\alpha}^{\dagger}(r)\psi_{\alpha}(r). \tag{8}$$

- The absorptive potentials assume a quadratic dependence on the nuclear densities with complex parameters to be determined from a global fit.

4 ANALYSIS OF THE DKP RELATIVISTIC MODELS

- The empirical features of low-energy pion-nucleus interactions are found to favor the SVT and SV′T models.

- The SVT model produces an S-wave component smaller than those in non-relativistic impulse approximations. The SV′T model yields a very strong S-wave repulsion in the interior. This strong repulsion may indicate an alternative mechanism for explaining the problem of the missing S-wave repulsion.

- The SVT and the SV′T models produce a P-wave component similar to those obtained by non-relativistic models. The P-wave component is not so sensitive to the model dependence of the scalar density.

- The SVT and SV′T models produce the same effective isovector S-wave potential, only a few MeV smaller than those in non-relativistic impulse approximations.

- The SVT and SV′T models yield an identical P-wave term which roughly coincides with the ones obtained in non-relativistic impulse approximations.

5 CONCLUSION

We have derived a relativistic impulse approximation to calculate the relativistic DKP pion-nucleus optical potentials. The empirical pion-nucleus interaction discriminates between the alternative relativistic models in favor of the SVT and SV′T models.

The real and absorptive, S-wave and P-wave, isoscalar and isovector, pieces of the DKP pion-nucleus optical potentials either reproduce or improve on their counterparts in traditional non-relativistic models.

We are currently re-analysing experimental results on strong level shifts and widths in pionic atoms over the whole of the periodic table. Once we have a satisfactory description of pionic atoms, we plan to look at pion-nucleus elastic and inelastic scattering.

Acknowledgments

We thank Dr E. D. Cooper for useful discussions. The financial support of EPSRC through GR/J95867 is gratefully acknowledged.

References

[1] J. A. Carr, H. McManus and K. Stricker-Bauer, *Phys. Rev.* C25 (1982) 952;
E. Friedman, *Phys. Lett.* **B207** (1988) 381.

[2] R. J. Duffin, *Phys. Rev.* 54 (1938) 1114;
N. Kemmer, *Proc. Roy. Soc.* 166 (1938) 127;
Proc. Roy. Soc. A 173(1939) 91;
G. Petiau, *Acad. R. Belg. Cl. Sci. Mem. Collect.* 16 No. 2

[3] E. Fishbach, M. M. Nieto and C. K. Scott, *Phys. Rev.* D7 (1973) 207.

[4] R. A. Krajcik and M. M. Nieto, *Am. J. Phys.* 45 (1977) 818 and references therein.

[5] E. D. Cooper and N. B. De Takacsy, *Phys. Lett.* **B220** (1989) 17;
E. D. Cooper, private communication.

[6] E. Friedman, G. Kälbermann and C. J. Batty, *Phys. Rev.* C34 (1986) 2244.

[7] Y. Nedjadi and R. C. Barrett, *J. Math. Phys.* 35 (1994) 4517.

[8] Y. Nedjadi and R. C. Barrett, *J. Phys.* A27 (1994) 4301.

[9] Y. Nedjadi, S Ait-Tahar and R. C. Barrett *J. Phys. A* 31 (1998) 3867.

[10] G. Kälbermann, *Phys. Rev.* C33 (1986) 1814;
G. Kälbermann, *Phys. Rev.* C34 (1986) 2240.

[11] M. Ericson and T. E. O. Ericson, *Ann. Phys.* 36 (1966) 323.

NUCLEAR STRUCTURE PHYSICS AT YALE AND BEYOND

C.W. Beausang

Wright Nuclear Structure Laboratory,
Yale University,
New Haven, CT06520,USA

Abstract: Some of the new experimental equipment, available at the Wright Nuclear Structure Laboratory including the YRAST Ball array and some of its auxiliary detectors are described. Aspects of the nuclear structure physics program at the WNSL is reviewed. Several new results are presented including the investigation of low lying 0^+ states in ^{154}Sm and the observation of a possible new magnetic rotational band in ^{205}Rn is reported.

1 YRAST BALL AND ITS AUXILIARY DETECTORS

It is with great pleasure that I am able to attend this conference in honor of John Sharpey-Schafer. A little over a year and a half ago John attended our workshop at Yale and presented, as he always does, a wonderful talk on the past present and future program at NAC. Romantic names such as AFRODITE made us jealous both of this powerful detector system and that he lived in a country that made such puns possible! Since arriving in South Africa I can also see that the country itself is as beautiful as the mythical goddess herself! In this paper I will present some results from Yale, where we strive to have a Y in the name of each piece of apparatus. As you will see this is not always such an easy task!

The Nucleus: New Physics for the New Millennium
Edited by Smit et al., Kluwer Academic / Plenum Publishers, New York, 2000.

The Nuclear Structure program at the Wright Nuclear Structure Laboratory has recently been revitalized by the addition of several new researchers and by the construction and implementation of several new major experimental tools. The new equipment includes the YRAST Ball array and its suite of auxiliary detectors.

The YRAST Ball (Yale Rochester Array for SpecTroscopy) [1] is a powerful new γ-ray spectrometer. The mechanical structure supports up to 28 Compton suppressed Ge detectors, including up to 9 Clover Ge detectors around the target position. Currently the array consists of four segmented Clover detectors, each with approximately 150% relative efficiency, one 70% coaxial Ge and nineteen 20-25% coaxial Ge detectors. These detectors give the array a total photopeak efficiency of $\sim$ 2%, making the YRAST Ball the largest university based spectrometer in the world.

A variety of auxiliary detectors are available for use with the YRAST Ball. These include four LEPS detectors, a thirty two element BGO sum-energy multiplicity filter and an array of $\sim$ 12 neutron detectors. In addition, an array of solar cells for heavy ion detection has been developed and a state of the art recoil distance plunger for lifetime measurements has been constructed.

Solar-cells, which operate without bias, can provide a cost effective solution for the detection of energetic charged particles. For example, heavy-ions following fission, scattered beam, α-particles and perhaps β-rays may be detected using such devices.

Recently, we have exploited this fact at Yale by constructing a prototype array consisting of eight 1 cm^2 solar cells [2]. The cells are arranged as elements of a 3 x 3 matrix and placed at backward angles at a distance of about 3 cm from the target, the beam passing through the empty center element of the matrix. This array has been successfully coupled with the YRAST Ball, to detect fission fragments from both source (^{252}Cf) and beam induced fission and also to detect backscattered beam following Coulomb excitation. For example, Figure 1 shows the total γ-ray spectrum recorded in YRAST Ball following the ^{14}N + ^{197}Au reaction. The large background, from fission induced γ-rays, has been removed in the bottom spectrum by subtraction of the γ-ray spectrum measured in coincidence with fission events recorded in the solar-cell array. The geometric efficiency of this prototype array is on the order of 10%.

Encouraged by these results we are constructing a larger array of solar cell detectors [2]. The new array will consist of 56 1 cm^2 solar cells arranged in two half boxes (plus some additional end-pieces). The boxes close over the target position giving a total geometric coverage of almost 95% of 4π. Each half box consists of 28 cells which will be electrically connected to produce 21 signals. Clearly, when used as a fission veto or selector only one half box (the back half) is required because of the kinematics of the reaction mechanism. The new array will be operational by early 1999.

Construction of a new, state of the art recoil distance plunger device has just been completed. The New Yale Plunger Device (NYPD) is now available

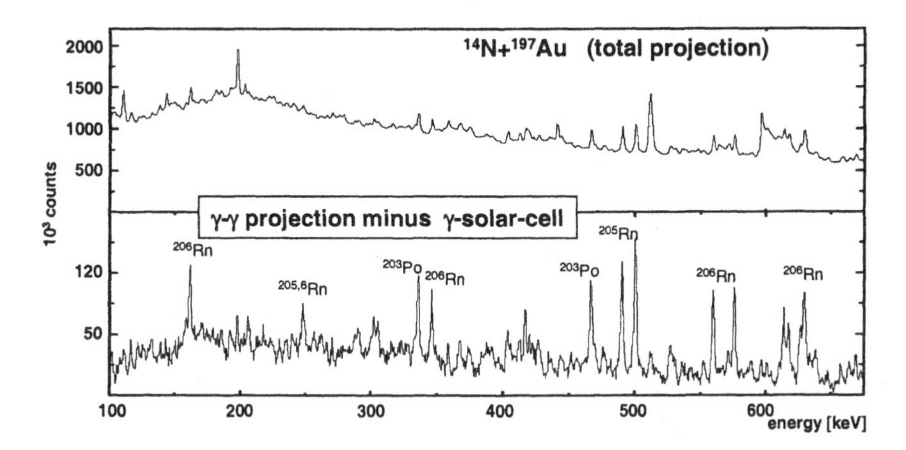

Figure 1 The large background in the top spectrum, due to fission events, has been removed in the lower panel by subtraction of the γ-ray spectrum measured in coincidence with the prototype solar-cell array.

for use in lifetime measurements [3]. The first experiments are scheduled for February 1999.

The design of the NYPD is based on the very successful Köln plunger designed by A. Dewald. The device, designed for use with large Ge detector arrays such as Gammasphere and YRAST Ball, has a minimum of material near the target chamber. A stable mechanical guidance system moves the target and stopper foils and ensures that the foils remain parallel at all times. In addition, a capacitance measurement and feedback mechanism keeps the distance between the foils constant and compensates for thermal expansion caused by beam heating effects. The feedback mechanism stabilizes the distances to better than 0.1 μm. To open up new frontiers in lifetime measurements (e.g. to study nuclei populated by deep-inelastic, Coulex or fission reactions), it is possible to couple the NYPD to an array of charged particle detectors, for example the Rochester University CHICO array or our solar-cell detectors.

In addition to YRAST Ball, several other programs have been developed at WNSL or are under active consideration. A moving tape collector has been implemented for beta-decay studies of short lived isotopes. Here, radioactive nuclei produced via heavy-ion fusion evaporation reactions are implanted on a mylar tape. Direct beam irradiation of the tape is prevented by a small plug placed directly behind the target. Advantage is taken of small angle straggling in the target to extract most (efficiency around 70%) of the recoils around the plug. The tape periodically transports the isotopes to a counting station where decay γ-rays are detected in four YRAST Ball Clover detectors arranged in a close packed geometry. In this case the total photopeak efficiency for γ-ray detection is on the order of 10%, rivaling the efficiency of Gammasphere.

A program to utilize the WNSL Enge Split Pole Magnetic Spectrometer as a gas filled separator is under active development. The split-pole, placed at zero degrees is used to focus recoils of interest onto heavy-ion detectors, e.g. solar cells, while rejecting beam and fission fragments. Early tests of the system have been successful and have provided proof of principle. Transmission efficiencies for fusion evaporation residues of $\sim$ 30% have been achieved. A new target chamber and mechanical support structure for several Clover Ge detectors are being designed.

In the above, I have described some of the new experimental apparatus available at the Wright Nuclear Structure Laboratory. To date we have utilized this apparatus to pursue a varied physics program in both high- and low-spin nuclear structure physics. Physics topics which we have concentrated on include the investigation of low-spin collective modes, the study of phase transitions or phase coexistence in nuclei, the evolution of structure in neutron rich nuclei and the investigation of high-spin properties of light actinide nuclei. In the remainder of this paper I will describe some of our early results from Yale and Beyond.

2 THE NATURE OF EXCITED $^{+}$0 STATES IN ^{154}Sm

Despite years of investigation, the nature of low lying 0^+ states in deformed even-even nuclei remains poorly understood. The traditional interpretation of the first excited 0^+ states as a β-vibration [4] has recently been the subject of intense discussions [5]. Indeed recent experimental evidence and theoretical studies are consistent with this state being a member of an anharmonic double-γ vibrational multiplet.

Recently, the nature of the first excited 0^+ states in ^{152}Sm has been investigated. It has been shown that ^{152}Sm provides a very special case, since it happens to lie in the middle of a very sharp shape transition between the spherical ^{150}Sm and the deformed ^{154}Sm [6]. The strongly reduced transition probability, B(E2), between the 0_2^+ and the 2_1^+ in ^{152}Sm can be explained by a shape mixing between deformed and spherical 0^+ states, the deformed state forming the ground state in ^{154}Sm, a spherical state similar to the ground state in ^{150}Sm.

To better understand this shape transition and to check the consistency of the interpretation the nature of low lying 0^+ states in ^{154}Sm was recently investigated at Yale [7] via Coulomb excitation utilizing an ^{16}O beam at 65 MeV. The beam was incident on a 3 mg/cm^2 ^{154}Sm foil backed by a 15 mg/cm^2 Au layer to stop the recoils.

The emitted γ-rays were detected using YRAST Ball while back scattered beam particles were detected using the solar-cell array. By selecting γ-rays in coincidence with backscattered beam particles, forward going ^{154}Sm nuclei were identified, allowing a Doppler Shift Attenuation Method lifetime analysis to be performed. Lifetimes of 1.3 (3) ps and 0.61 (4) ps were determined for the 0_2^+ and 2_γ^+ levels in ^{154}Sm, respectively. The B(E2: $0_2^+ \rightarrow 2_1^+$) = 12 (2) Wu, extracted from these data, is consistent with expectations for a β-vibration.

3 HIGH-SPIN STATES OF NUCLEI IN THE LIGHT ACTINIDE REGION

The light actinide region, lying to the northwest of the doubly magic ^{208}Pb nucleus has long been a fertile ground for the study of shell model states. As expected for nuclei close to the doubly magic shell closure at $Z = 82$ and $N = 126$ their near ground state excitation spectra are made up of single particle excitations. However, at higher excitation energies and spins a variety of collective rotational behaviors are manifest. For example, superdeformed states ($\beta \sim 0.5$) are well established in Hg, Tl, Pb and Bi [8, 9] nuclei and an example has recently been observed in Po [10].

Another exotic phenomenon, shears or magnetic rotation is also manifest in the mass 190-200 region [11, 12, 13, 14]. In this model, in a nearly spherical nucleus, the single-particle (hole) angular momentum vectors of high-j protons and neutrons are initially oriented almost perpendicular to each other so that the total angular momentum vector is not aligned along a principal axis of the nucleus, but instead is tilted. Higher angular momentum states are by the slow closing of these angular momentum vectors or shears. Such a mechanism gives rise to characteristic features of the resulting 'rotational' bands, namely enhanced magnetic dipole transitions and large B(M1)/B(E2) ratios which decrease with rotational frequency.

Such exotic shape coexistence phenomena are predicted to persist as one moves to nuclei further away from ^{208}Pb. However, surprisingly little information is available on high spin properties of At, Rn, Fr and Ra nuclei, particularly odd or doubly-odd nuclei, with $Z > 84$ and $N < 126$. With this at mind we have decided to pursue a program at Yale (and Beyond) to investigate this difficult but sometimes fascinating region. Early experiments carried out at Yale and at the University of Jyväskylä have concentrated on the high spin structure of nuclei in the vicinity of ^{205}Rn.

Two experiments were performed to populate high-spin states in ^{205}Rn. The first, carried out at the University of Jyväskylä Accelerator Laboratory, utilized the ^{40}Ar +^{170}Er reaction at a beam energy of 183 MeV. The gamma-rays were detected using the Jurosphere array which consisted of 15 large Eurogam Phase I [9] ($\sim$ 70% relative efficiency) and 10 TESSA-type [16] ($\sim$ 25% relative efficiency) Compton-suppressed Ge detectors, total photopeak efficiency $\sim$ 1.5% at 1.3 MeV. To discriminate against the very large fission background encountered in this reaction ($\sigma_{fusion} \sim 5$ mb, $\sigma_{fission} \sim 300$ mb) fusion evaporation residues were detected, in coincidence with prompt γ-rays detected at the target, using the RITU gas filled separator [17] (transmission efficiency $\sim$ 30% for this reaction).

A second experiment, carried out at Yale utilized the ^{14}N + ^{197}Au reaction at beam energies ranging from 90 to 110 MeV. For this experiment the recoils were stopped in the thick ($\sim$ 20 mg/cm^2) Au target. The gamma-rays were detected using YRAST Ball.To generate cleaner gamma-ray spectra, fission products were detected using our prototype solar cell array, see Figure 1.

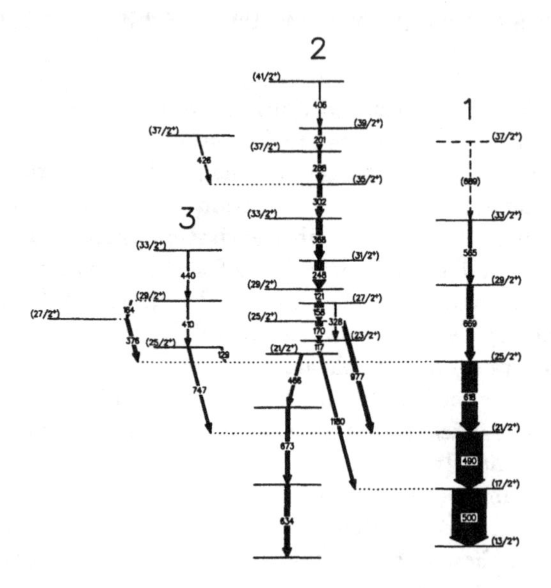

Figure 2 The level scheme of ^{205}Rn.

Prior to these experiments no gamma-ray transitions were known in ^{205}Rn. Our level scheme is presented in Figure 2. The transitions were assigned to Rn based on X-ray coincidences and to ^{205}Rn based on the results of our excitation function measurements using YRAST Ball while the level scheme has been deduced mostly using the Jurosphere data.

Two cascades, labeled bands 1 and 2, respectively, dominate the high-spin portion of the decay. To illustrate the quality of the data, spectra gated by the 689 keV and 288 keV transitions are presented in Figure 3. Spin and parity assignments are tentative and are based on our assignment of $13/2^+$ for the bandhead of band 1, which is assigned the $\nu i_{13/2}$ configuration [18].

Band 2 consisting of 10 stretched magnetic dipole transitions with weak, unobserved crossover transitions is of particular interest. Such a band appears similar to the magnetic rotational bands in Pb and in Bi nuclei [12, 13, 14]. Band 2 is proposed as another example of a magnetic rotational band, the first in a Z = 86 nucleus.

There are two likely candidates for the configuration of band 2, the negative parity $\nu i_{13/2} \otimes \pi(h_{9/2} i_{13/2})$ or the positive parity $\nu i_{13/2} \otimes \pi(i_{13/2})^2$ configuration. Both are calculated to have a small oblate deformation of $\epsilon_2 < 0.1$. Using this deformation and the proton and neutron pairing energies, Δ_p and Δ_n obtained from the TRS calculations, TAC calculations were performed. The results in-

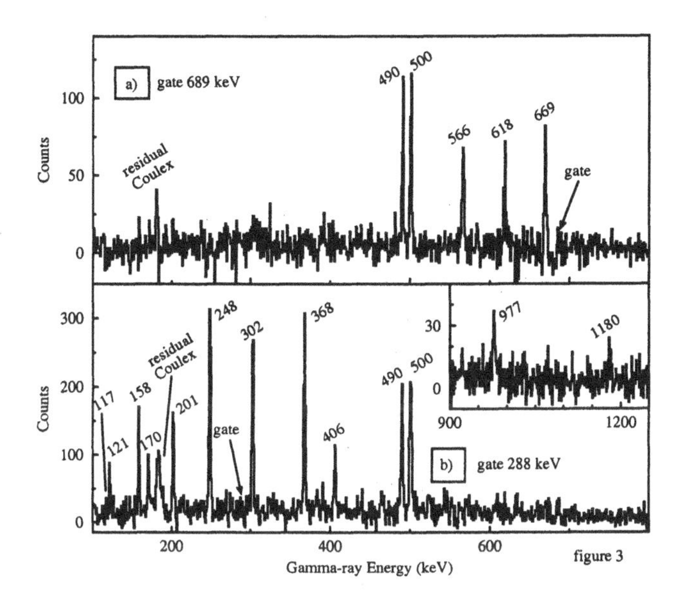

Figure 3 Spectra gated on the 689 and 288 keV transitions in ^{205}Rn.

dicate that both of these configurations indeed correspond to tilted or shears bands with large B(M1)/B(E2) ratios. Further, the calculations indicate that both positive and negative parity configurations lie at a comparable excitation energy at about 1.8 MeV relative to the yrast $\nu i_{13/2}$, in excellent agreement with the experimental data. Since the angular distribution and correlation data favor a quadrupole character, most likely E2, for the 977 and 1180 keV transitions linking bands 1 and 2, the positive parity $\nu i_{13/2} \otimes \pi i^2_{13/2}$ configuration is tentatively assigned. Lifetime measurements, to extract the absolute B(M1) values for the transitions in band 2 are necessary to confirm the shears character of this band. In addition conversion electron and gamma-ray polarization measurements are planned to confirm our configuration assignment for band 2.

Encouraged by our results for ^{205}Rn, several other $A \sim 200$ actinide nuclei have recently been studied at Yale and Beyond. We have extended our search for magnetic rotation to the $N = 119$ isotones, ^{203}Po and ^{204} using the Garel array at the IReS Strasbourg and the YRAST Ball array at Yale, respectively. Prior to these experiments no information was available on excited states in ^{204}At). These data are currently under analysis.

A beginning has been made to extend our investigations to heavier Z nuclei. The reaction ^{16}O + ^{197}Au at beam energies of 90 - 100 MeV was used to populate high-spin states in ^{207,208,209}Fr for the first time. The gamma-rays were detected using YRAST Ball while the solar cells were used to discriminate against fission events. A preliminary level scheme for ^{208}Fr is shown in Figure 4.

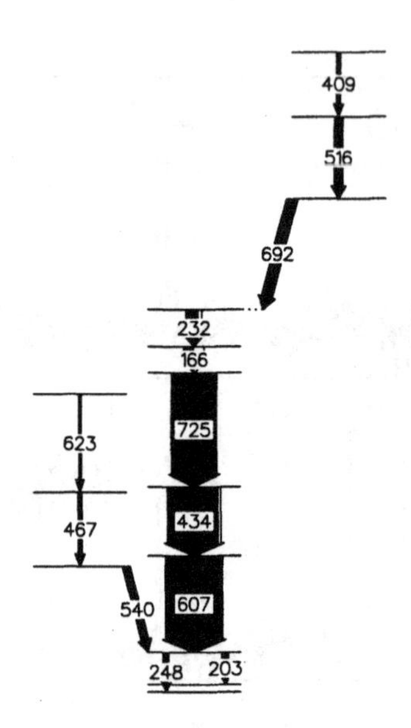

Figure 4 Preliminary level scheme for ^{208}Fr. Assignment to Fr is based on X-ray coincidences, however the assignmnet to $A = 208$ is tentative.

This paper represents the work of many individuals. In particular, the work of John Novak who has been responsible the data analysis on ^{205}Rn is acknowledged. Reiner Krüken is responsible for the ^{154}Sm analysis. Discussions with Stefan Frauendorf and Jing-ye Zhang are gratefully acknowledged. This work is supported by the U.S. D.O.E. under grant number, DE-FG02-91ER-40609.

References

[1] C.W. Beausang, et al., to be published

[2] R. Krüken, et al. to be published.

[3] R. Krüken et al., to be published

[4] A. Bohr and B. Mottelson, Nuclear Structure, (Benjamin, New York, 1969, 1975, and World Scientific, New York, 1998), Vol II.

[5] See for example, R.F. Casten and P. von Brentano, Phys. Rev. C50, R1280, (1994), and Phys. Rev. C51, 3525, (1995).

[6] F. Iachello, N.V. Zamfir, and R.F. Casten, Phys. Rev. Lett. 81, 1191, (1998).

[7] R. Krüken, et al., to be published.

[8] R.V.F. Janssens and T.L. Khoo, Ann. Rev. Nucl. Part. Sci. 41, 321, (1991).

[9] R. Firestone and B. Singh, Nucl. Data. Sheets, 78, 1, (1996).

[10] D.P. McNabb, G.Baldsiefen, L.A.Bernstein, J.A.Cizewski, H.-Q.Jin, W.Younes, J.A.Becker, L.P.Farris, E.A.Henry, J.R.Hughes, C.S.Lee, S.J.Asztalos, B.Cederwall, R.M.Clark, M.A.Deleplanque, R.M.Diamond, P.Fallon, I.Y.Lee, A.O.Macchiavelli, and F.S.Stephens, Phys. Rev. C53, R541, (1996).

[11] S. Frauendorf, Nucl. Phys. A557, 259, (1993).

[12] G.Baldsiefen, M.A.Stoyer, J.A.Cizewski, D.P.McNabb, W.Younes, J.A.Becker, L.A.Bernstein, M.J.Brinkman, L.P.Farris, E.A.Henry, J.R.Hughes, A.Kuhnert, T.F.Wang, B.Cederwall, R.M.Clark, M.A.Deleplanque, R.M.Diamond, P.Fallon, I.Y.Lee, A.O.Macchiavelli, J.Oliveira, F.S.Stephens, J.Burde, D.T.Vo, and S.Frauendorf, Phys. Rev. C54, 1106, (1996),

[13] R.M.Clark, S.J.Asztalos, G.Baldsiefen, J.A.Becker, L.Bernstein, M.A.Deleplanque, R.M.Diamond, P.Fallon, I.M.Hibbert, H.Hubel, R.Krucken, I.Y.Lee, A.O.Macchiavelli, R.W.MacLeod, G.Schmid, F.S.Stephens, K.Vetter, R.Wadsworth, and S.Frauendorf, Phys. Rev. Lett. 78, 1868, (1997).

[14] P.J.Dagnall, C.W.Beausang, R.M.Clark, R.Wadsworth, S.Bhattacharjee, P.Fallon, P.D.Forsyth, D.B.Fossan, G.de France, S.J.Gale, F.Hannachi, K.Hauschild, I.M.Hibbert, H.Hubel, P.M.Jones, M.J.Joyce, A.Korichi, W.Korten, D.R.LaFosse, E.S.Paul, H.Schnare, K.Starosta, J.F.Sharpey-Schafer, P.J.Twin, P.Vaska, M.P.Waring, and J.N.Wilson. J. Phys. G20, 1591, (1994).

[15] P.J. Nolan, Nucl. Phys. A520, 657c, (1990).

[16] P.J. Nolan, D. Gifford, and P.J. Twin, Nucl. Inst. and Meth. A236, 95, (1985).

[17] M. Leino, J.Aysto, T.Enqvist, P.Heikkinen, A.Jokinen, M.Nurmia, A.Ostrowski, W.H.Trzaska, J.Uusitalo, K.Eskola, P.Armbruster, and V.Ninov, Nucl. Inst. and Meth. B99, 653, (1995).

[18] J.R. Novak, et al. to be published.

[19] See S. Frauendorf in Proceedings of the Conference on Physics from Large Gamma-ray Detectors, Berkeley, 1994, (LBL-35607), Vol. II, Proceedings of the Workshop on Gammasphere Physics, Berkeley, (World Scientific, Singapore, 1995, p 52) and Z. Phys. A356, 263 (1996).

PHOTOACTIVATION OF ^{180M}Ta AT ASTROPHYSICALLY RELEVANT ENERGIES

D. Belic[1], C. Arlandini[2], J. Besserer[3], J. de Boer[3], J. Enders[4], T. Hartmann[4], F. Käppeler[2], H. Kaiser[4], U. Kneissl[1], M. Loewe[3], H. J. Maier[3], H. Maser[1], P. Mohr[4], P. von Neumann-Cosel[4], A. Nord[1], H. H. Pitz[1], A. Richter[4], M. Schumann[2], S. Volz[4], A. Zilges[4]

[1] Institut für Strahlenphysik, Universität Stuttgart, D–70569 Stuttgart, Germany
[2] Forschungszentrum Karlsruhe, Institut für Kernphysik III, D–76021 Karlsruhe, Germany
[3] Sektion Physik, Ludwig-Maximilians-Univ. München, D–85748 Garching, Germany
[4] Institut für Kernphysik, Technische Univ. Darmstadt, D–64289 Darmstadt, Germany

Abstract: A possible connection of the isomeric state in ^{180}Ta to its ground state band via low-lying intermediate states (IS) was investigated by means of photoactivation using bremsstrahlung. A depopulation of the isomer has been found down to a bremsstrahlung endpoint-energy of 1 MeV but not at the lowest bombarding energy of 0.8 MeV. Therefore an IS slightly lower than 1 MeV has to be assumed. Further IS were observed at about 1.2, 1.5, 2.1, 2.5, and 2.8 MeV, respectively.

1 INTRODUCTION

^{180}Ta is the rarest isotope in nature occuring with a relative abundance of only 0.012 %. ^{180}Ta is the only nucleus found being stable in an isomeric state with a halflife of $t_{1/2} \geq 1.2 \cdot 10^{15}$y, while its ground state decays with a halflife of 8.1h (see fig.1). The nucleosynthesis of ^{180}Ta remains a puzzle [1]. An s-process origin of ^{180m}Ta is only possible if the synthesized nuclei are not destroyed by a depopulation of the isomer, e.g. by photoactivation in a stellar photon bath. The photoactivation process of ^{180m}Ta has been subject

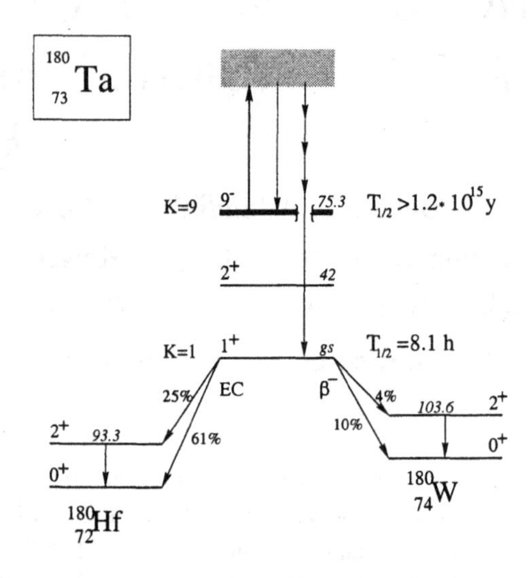

Figure 1 Level and decay scheme of ^{180}Ta

of several experimental investigations using real photons (bremsstrahlung) [2] or virtual photons in Coulomb-excitation experiments with heavy ion- [3, 4] or p- and α-beams [5, 6]. In the bremsstrahlung experiment performed at the S–DALINAC at Darmstadt [2] the photoactivation could be detected down to an endpoint-energy of 2.8 MeV. Two photoexcited intermediate states, at roughly 2.8 and 3.6 MeV excitation energy, were deduced from the measured yield curve. However, the energies of both states are too high for an effective population at s-process temperatures. The aim of the present collaboration was to search for lower-lying photoexcited states in bremsstrahlung experiments of improved sensitivity.

2 EXPERIMENTAL DETAILS

The experiments used a total beam time of 6 weeks at a new irradiation facility at the Stuttgart DYNAMITRON accelerator. Typical electron currents of 200–460 μA bombarded a watercooled bremsstrahlung production target. The distance between radiator target and sample was about 9 cm. In total 150 mg of Ta-oxide material enriched to 5.6 % in ^{180m}Ta were available, corresponding to 6.7 mg of ^{180m}Ta. Alternating with this enriched target sheets of natural Ta-metal with a thickness of $\approx$ 0.2 mm were irradiated for about 2 halflives (12–16h). Furthermore, another target consisting of 150 mg of natural Ta-oxide, prepared in exactly the same way as the enriched target, was activated at some energies to verify the enrichment factor. After the irradiation the radiation from the activated samples was measured offline by two well-shielded, high-resolution, low-energy germanium photon (LEP) detectors. Fig.2 shows examples of such spectra. The low-energy photon energy spectra show charac-

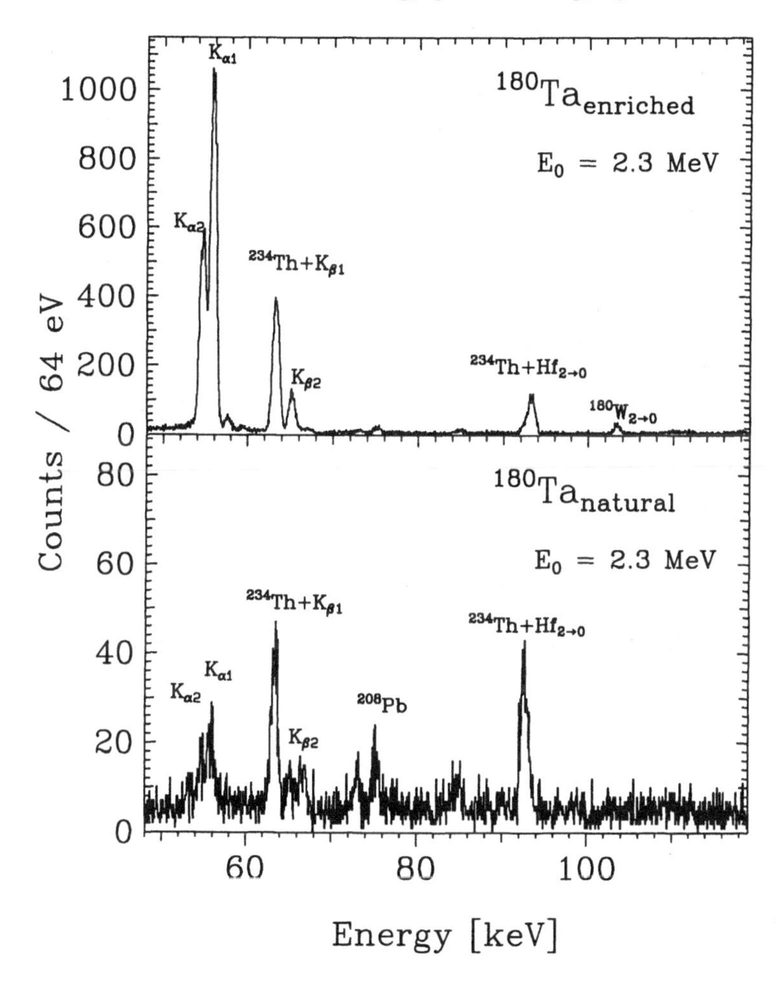

Figure 2 Low-energy γ- and X-ray spectra of photoactivated enriched (upper part) and natural metallic (lower part) Ta-target. The bremsstrahlung endpoint-energy for both targets was $E_0 = 2.3$ MeV. Note that the scales differ by a factor of 12.

teristic X-rays and nuclear γ-transitions which are characteristic for the decay of the ^{180}Ta ground state. They are dominated by the well-resolved K_{α_1} and K_{α_2} lines of Hf. The corresponding K_{β_1} and K_{β_2} peaks are also visible. Unfortunately, there are accidental overlaps of background lines such as nuclear γ-transitions following the β-decay of ^{234}Th with the K_{β_1} line and the $2^+ \to 0^+$ γ-transition in the residual nucleus ^{180}Hf. In the spectra of the enriched sample even the $2^+ \rightarrow 0^+$ transition in ^{180}W is observed, in spite of its weak feeding in the β^--decay of the ^{180}Ta (see fig. 1). The ratio between the properly normalized intensities of the Hf-X-rays of both spectra represents the ratio between the quantities of ^{180m}Ta in the enriched and natural metallic target, which amounts to 37. Another proof for observing the ground state decay is its halflife ($t_{1/2}$=8.1h). It could be precisely confirmed in a time differential measurement covering about 8 halflives. To determine the strength of the IS transition, it is necessary to measure the photon flux accurately. We, therefore, used different methods to monitor the photon beam. Besides measuring the electron current on the radiation target and using an ionisation chamber to obtain the integrated flux, a nuclear resonance fluorescence (NRF) setup with a mixed target (Al/Cu/LiF) was used to determine the photon energy distribution. Additionally, an ^{115}In-target was used as a flux-monitor [7], because Indium has a longlived isomer (E=336 keV, $t_{1/2}$=4.5 h), which can be populated by real photons of energies higher than $\approx$ 940 keV. Therefore In-targets were irradiated simultaneously with the Ta-targets and their activation was also measured offline. Finally these results for the photon flux can be compared to Monte Carlo simulations with the codes GEANT and EGS, which are currently being performed.

3 RESULTS

In fig.3 the yields measured using the enriched sample are presented on linear scale as a function of the bremsstrahlung endpoint-energy to identify the energies of the low-lying IS. Each break in the yield curve indicates the existence of an IS at this energy.

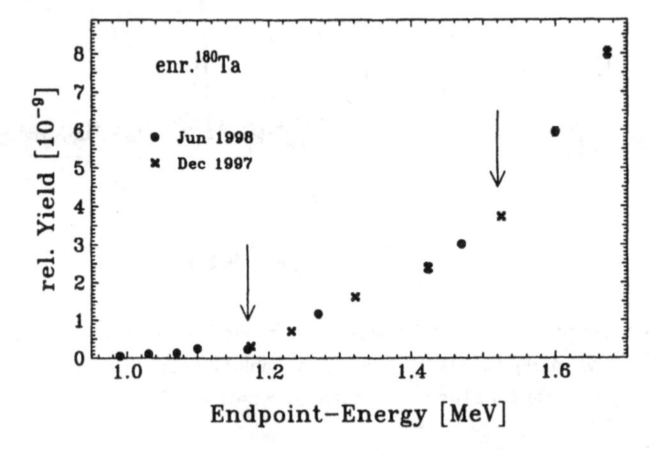

Figure 3 The yields of the Hf X-rays normalized to the charge on the radiator target and the measuring time, measured for the enriched target. The observed breaks in the yield curve are marked by arrows. They indicate IS energies.

We can deduce the energetical positions of the IS. Low-lying IS can be detected at $\leq$1.0 MeV and at about 1.2 MeV and 1.5 MeV. Furthermore, IS were found at about 2.1 MeV, 2.5 MeV, and 2.8 MeV in the measurements with the natural metallic target.

The integrated cross section σ_γ can be calculated from the yield Y, if the photon energy distribution N_{BS} is known:

$$Y(E_0) = \int_0^{E_0} \sigma_\gamma(E_\gamma) \cdot N_{BS}(E_0, E_\gamma) \cdot dE_\gamma \quad (1)$$

As the analysis is not yet finished, these cross sections have not yet calculated.

Systematic measurements were performed with bremsstrahlung endpoint-energies in the range of E_0= 0.8–3.1 MeV (fig.4). At bombarding energies above 1.5 MeV metallic Ta-sheets could be used. The enrichment factors between enriched, metallic and natural samples were also measured as expected. A depopulation of the isomer was observed down to 1 MeV. The availability of the enriched Ta sample, together with the new setup, has improved the sensitivity of the present study roughly by a factor of 5000 compared to previous experiments [2].

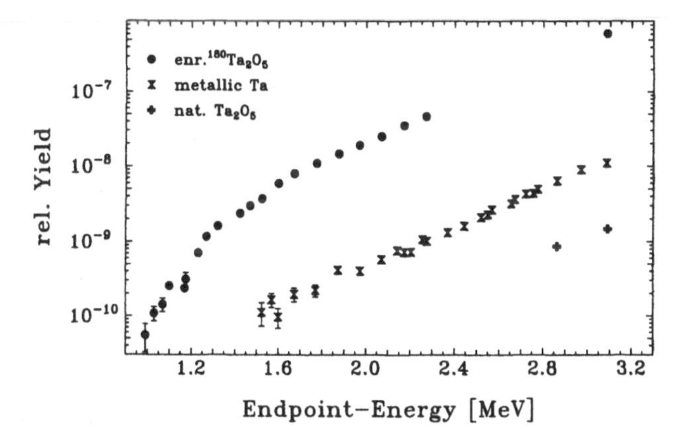

Figure 4 Yields of the Hf X-rays of one LEP-detector, normalized to the charge accumulated on the radiator target and to the measuring time, in a logarithmic scale.

An attempt to measure directly the most interesting IS at low energies in a nuclear resonance fluorescence experiment (E_0=1.5 MeV) failed. The cross section to populate an IS below 1 MeV and its consequences for astrophysics will be discussed in a future paper.

Acknowledgments

The authors wish to thank Prof. Dr. G. Sletten and Prof. Dr. C. Günther for lending us the LEP-detectors used in the experiments. The support by the Deutsche Forschungsgemeinschaft under Contract Nos. Ri 242/12-1, Kn 154/30 and by the Munich Tandem Accelerator Laboratory are gratefully acknowledged.

References

[1] Zs. Németh *et al.*, *Astrophys. J.* 392, 277 (1992).

[2] C. B . Collins *et al.*, *Phys. Rev.* C 42, R1813 (1990).

[3] C. Schlegel *et al.*, *Phys. Rev.* C 50, 2198 (1994).

[4] P. von Neumann-Cosel *et al.*, *Nucl. Phys.* A621, 278c (1997).

[5] M. Schumann *et al.*, *Nucl. Phys.* A621, 274c (1997).

[6] M. Schumann *et al.*, *Phys. Rev.* C 58, 1790 (1998).

[7] P. von Neumann-Cosel *et al.*, *Phys. Lett.* B 266, 9 (1991).

MIRROR NUCLEI AND ODD-ODD $N = Z$ NUCLEI IN THE $f_{7/2}$ SHELL

M. A. Bentley[a], C. D. O'Leary[a,b], D. E. Appelbe[b,c],
R. A. Bark[d], D. M. Cullen[b], S. Ertürk[b,e], A. Maj[f] and D. D. Warner[g]

[a] School of Sciences, Staffordshire University, College Road, Stoke-on-Trent U.K.
[b] Oliver Lodge Laboratory, University of Liverpool, Liverpool L69 7ZE, U.K.
[c] Department of Physics, McMaster University, Hamilton, Canada L8S 4M1
[d] Department of Nuclear Physics, Australian National University, Canberra, Australia
[e] Nigde Universitesi, Fen-Edebiyat Fakültesi, Fizik Bölümü, Nigde, Turkey
[f] Niewodniczanski Institute of Nuclear Physics, 31-342 Krakow, Poland
[g] CLRC Daresbury Laboratory, Warrington WA4 4AD, U.K.
M.A.Bentley@staffs.ac.uk

Abstract: The phenomena of mirror-symmetry and cross-conjugate symmetry have been investigated through a study of nuclei near $N = Z$ at the centre of the $f_{\frac{7}{2}}$ shell. Two pairs of mirror-nuclei, ${}^{49}_{25}\mathrm{Mn}/{}^{49}_{24}\mathrm{Cr}$ and ${}^{47}_{24}\mathrm{Cr}/{}^{47}_{23}\mathrm{V}$, have been studied up to their band termination states. Coulomb energies have been deduced and are shown to be an extremely sensitive probe of the spatial correlations of valence particles. A comprehensive new level scheme has been determined for the odd-odd $N = Z$ nucleus ${}^{46}_{23}\mathrm{V}$, and the results yield new information on the $T = 1$ (and possibly $T = 0$) np-pairing modes.

1 INTRODUCTION

In this presentation, we will examine two fundamental symmetries of nuclear structure. The concept of mirror symmetry is well known, in which two nuclei with conjugate numbers of neutrons and protons show virtually identical energy level schemes. Assuming a charge-symmetric nuclear interaction, the small differences between energy levels have been traditionally interpreted entirely in terms of Coulomb effects. Studies of this kind therefore have the potential to exploit the well-known nature of the Coulomb interaction to investigate the spatial behaviour of the active "valence" nucleons [1, 2, 3, 4]. The second (and less well known) symmetry is cross-conjugate symmetry. This symmetry can be identified *only* in a pure isolated j–shell (i.e. in a spherical single-j shell model). In such a scenario, every nucleus within this isolated shell will have a cross conjugate partner (with identical energy level structures) following a simultaneous exchange of protons for neutrons and particles for holes. Clearly there are few mass regions where this symmetry is expected to hold to any significant extent. However, nuclei towards the centre of the $f_{\frac{7}{2}}$ shell have wavefunctions originating almost entirely from $j = \frac{7}{2}$[5], and cross conjugate symmetry may play an important role[6]. Here we examine four nuclei around the mid-shell nucleus ^{48}Cr - two mirror-pairs and two cross-conjugate pairs - and make detailed comparisons of their energy level structures up to the $J^{\pi} = \frac{31}{2}^{-}$ band-termination states.

The phenomenon of neutron-proton pairing is of considerable current interest. In odd-odd $N = Z$ nuclei, np-pairing is expected to play a major role due to the relative proximity of the neutron and proton Fermi levels. A number of experimental and theoretical investigations have been carried out for such nuclei in an attempt to firmly establish evidence for the different modes of np pairing e.g.[7, 8]. The first mode ($T = 1$) is equivalent to pairing between like particles and allows for pairs coupled with opposite orbital and spin angular momentum vectors. The second mode ($T = 0$) is only available to np pairs, and at low spin allows for aligned spin vectors but anti–aligned orbital angular momenta. This latter type of correlation is of particular interest and despite recent experimental investigations [7, 8], the signature for this effect is still the subject of much debate. Here we present some very tentative evidence for such a signature in the odd–odd $N = Z$ nucleus ^{46}V, as well as an investigation of $T = 1$ pairing correlations. The $T = 1$ states in ^{46}V are compared with their analogue states in ^{46}Ti, and the comparisons described in terms of Coulomb effects.

2 EXPERIMENTAL DETAILS

The experimental work was carried out at the Niels Bohr Institute using the PEX apparatus. In this work the apparatus consisted of four Euroball cluster detectors[9] coupled to a 31–element silicon ball [10] and a 15 element neutron array [11]. This allowed the coincident measurement of the gamma rays, charged particles and neutrons emitted in the reaction. In the experiment a 500 μg/cm^2 enriched ^{24}Mg target was bombarded with a ^{28}Si beam at 87 MeV. The nuclei of interest were populated via the ^{24}Mg(^{28}Si, $2pn$)^{49}Cr, ($p2n$)^{49}Mn, (αp)^{47}V, (αn)^{47}Cr and (αpn)^{46}V reactions. Further details of the experiment and data analysis can be found in references [3, 4, 12].

3 MIRROR-SYMMETRY AND CROSS-CONJUGATE SYMMETRY

The four nuclei ^{49}Cr, ^{49}Mn, ^{47}V and ^{47}Cr were investigated and their main yrast structures have extended up to the band–termination state. For all four nuclei this is $J^{\pi} = \frac{31}{2}^{-}$ – the maximum spin allowable in the isolated $f_{\frac{7}{2}}$ shell. The detailed level schemes can be found in references [3, 4]. These four nuclei correspond to one proton or neutron added or subtracted from the mid–shell nucleus ^{48}Cr. The pairs of nuclei ^{49}Cr/^{49}Mn and ^{47}V/^{47}Cr are mirror pairs and assuming a charge–symmetric nuclear force their level schemes should be identical, with small differences attributable to the Coulomb effect. Interestingly, the pairs ^{49}Cr/^{47}V and ^{49}Mn/^{47}Cr are cross–conjugate pairs and assuming a charge independent nuclear force and an isolated j–shell, their energy levels should also be identical. Thus in this very simplistic picture, all four nuclei should exhibit virtually identical behaviour. In reality, mirror symmetry is expected to be far better than cross–conjugate symmetry (due to contributions to the wavefunctions from other shells which destroys cross–conjugate symmetry).

For a mirror pair, the difference in energy between equivalent states of the same spin is expected to arise almost entirely from Coulomb effects. This is known as the Coulomb Energy Difference (CED) and is plotted in figure 1 for the $A = 49$ and $A = 47$ mirror pairs.

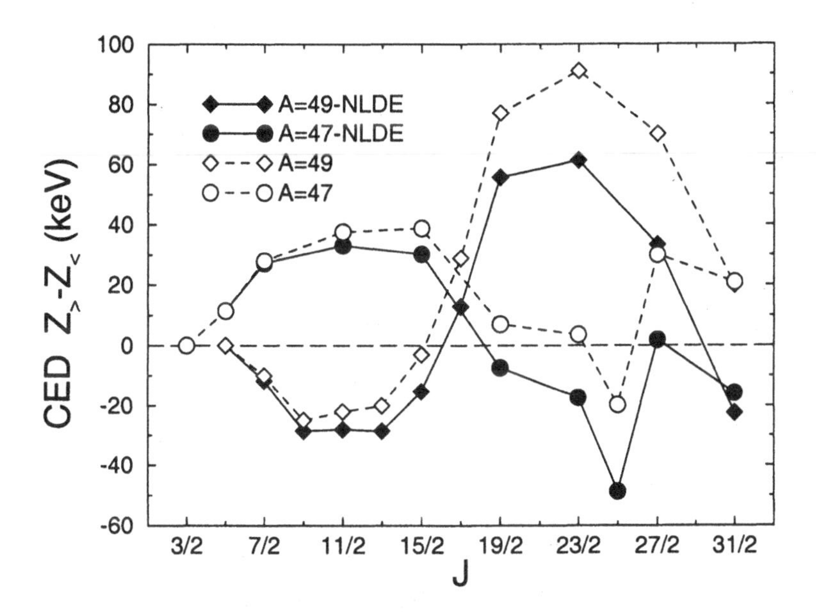

Figure 1 (Dashed lines, open symbols) The Coulomb Energy Difference (CED) for the $A = 49$ and $A = 47$ mirror pairs. The CED is defined as the excitation energy of a state in the low–Z member of the pair subtracted from the energy of the equivalent state in the high–Z member. (Solid Lines, filled symbols) As above, but having subtracted the liquid–drop correction as described in the text.

The CED for the $A = 49$ mirror pair shows a sharp rise at around $J^{\pi} = \frac{17}{2}^{-}$. This has been attributed [1, 3] to a rotational alignment of a pair of protons in ^{49}Cr, thus reducing their spatial overlap hence reducing the Coulomb self energy of the pair. In ^{49}Mn the same effect occurs, though blocking arguments require that a pair of neutrons must align - thus there is no Coulomb effect. At higher spin, the CED reduces towards zero as the band–termination is approached. One explanation for this is that the band-terminating state corresponds to a full non–collective alignment of *all* valence particles (protons and neutrons). Thus all the valence protons are fully aligned in *both* nuclei, and there will no longer be any significant difference in the Coulomb energy between the two nuclei. The observed CED therefore seems to be an extremely sensitive probe of the spatial correlations of the valence particles. The $A = 47$ mirror pair are the cross–conjugate partners of the $A = 49$ mirror nuclei, and their underlying structural behaviour should be very similar. However, the Coulomb effects must have the opposite sign as we have exchanged protons/neutrons and particles/holes. The data is shown in figure 1, and to some extent the general trends of the CED for the $A = 47$ mirror pair bear out these cross–conjugate symmetry arguments. However, the two sets of CEDs are not symmetrical about the CED$= 0$ axis as the above arguments would suggest.

One other effect which can contribute to the CED comes from the macroscopic (liquid drop) charge distribution. Recent calculations [13] show that for all of these nuclei the shape is expected to change from a reasonably deformed prolate shape ($\epsilon_2 = 0.2$) at low spin towards an almost spherical shape at high spin. Due to their different proton numbers, this results in a Coulomb effect of a different magnitude in each member of the mirror pair, thus contributing to the CED. We have calculated the magnitude of this effect by incorporating the calculated deformation parameters into the deformed liquid drop model of Larssen [14]. This liquid–drop effect on the CED has been subtracted from the CED data, and this corrected data should yield information on only the microscopic effects. This is shown in figure 1, and it can be seen that the CED now follows closely the cross–conjugate symmetry arguments described above. It is very gratifying (and perhaps a little surprising) that the mirror symmetry is good enough to allow such a complete understanding of the Coulomb effects and particularly that cross–conjugate symmetry seems to hold to a significant extent.

4 NP-PAIRING MODES IN ODD-ODD N=Z ^{46}V

The odd–odd $N = Z$ nucleus ^{46}V has been studied in detail, and a new extensive level scheme has been produced – see figure 2. This scheme has been discussed in detail in a recent publication [12], and in this presentation we concentrate on two specific aspects associated with *np* pairing.

The ground state $J^\pi = 0^+$ is a $T = 1$ state, whilst the strongly-populated excited states have almost exclusively been identified as $T = 0$. However, a sequence of states (labelled C) has been identified with $J^\pi = 2^+$, 4^+ and 6^+. These states have virtually identical energies to the yrast band ($T = 1$) states in the neighbouring even–even nucleus ^{46}Ti. We therefore identify these new states as having $T = 1$ and as the isobaric analogue states of the ^{46}Ti yrast sequence. Assuming a charge–independent nuclear force, the energies of these two sets of states will be identical and, as with mirror symmetry, small differences may be attributable to Coulomb effects. Thus we can evaluate a

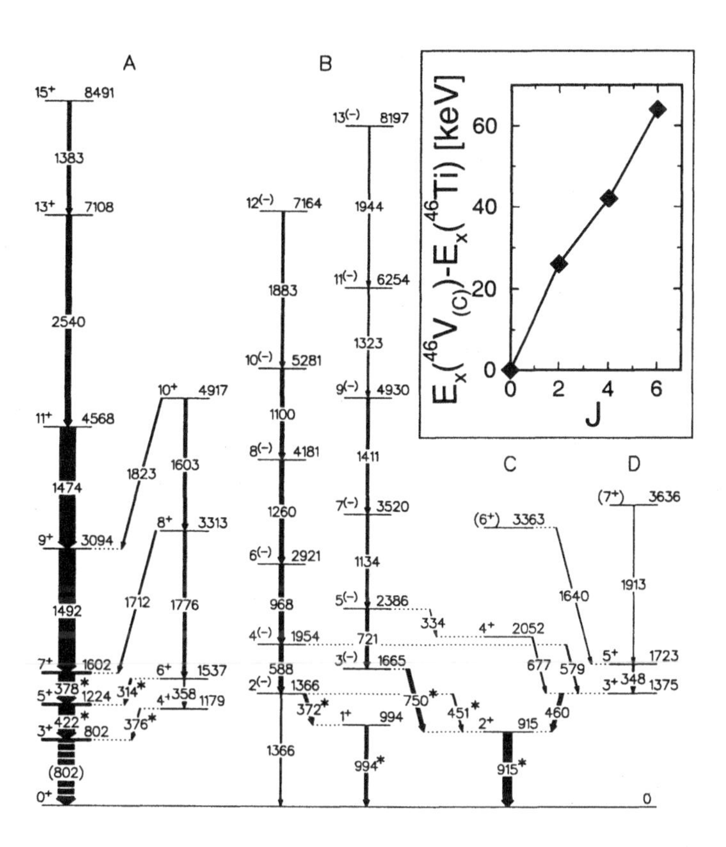

Figure 2 The deduced energy level scheme of ^{46}V. Details can be found in reference [12] and tentative spin/parity assignments are indicated by parentheses. The insert shows the difference in excitation energy between states in band C and their isobaric analogue equivalent states in ^{46}Ti.

CED defined as the difference in energy between states in band C and their equivalent isobaric analogue states in ^{46}Ti. This is shown in the insert of figure 2, and indicates a gradual rise in the CED with increasing spin. Following the work on mirror nuclei, we now have a more quantitative understanding of how particle alignments influence the measured Coulomb effects. Following a comparison with effects seen in mirror nuclei, the CED data for these $T = 1$ states implies that an alignment of protons is taking place in ^{46}Ti and *not* in ^{46}V - again yielding a Coulomb effect. The first rotational alignment in ^{46}Ti is calculated [15] to be an alignment of a $T = 1$ proton-proton pair. In ^{46}V, however, the $T = 1$ states are expected to comprise of both $T = 1$ *np* pairs and $T = 1$ *pp* or *nn* pairs. Thus it seems that in ^{46}V the alignment effect is not due exclusively to a *pp* pair and is most likely to be associated, at least partly, with the alignment of an *np* pair. From this we can are beginning to see that studies of CEDs between such analogue states my be a useful technique for probing the nature of pairing correlations at low spin.

Another structure of interest in ^{46}V is that labelled B in figure 2. This $T = 0$ structure appears to be a rotational band consisting of two signatures with little signature splitting. Experimentally it has not been possible in this work to firmly assign the parity of this structure (though the $J = 1$ state at 994 keV has definite positive parity). We have tentatively assigned negative parity to the remaining states based on the systematics of these light $f_{\frac{7}{2}}$–shell nuclei in which excitations from the $d_{\frac{3}{2}}$ shell are known (e.g. [2, 16]) to yield opposite parity deformed structures at low energy. However, the assignment of negative parity is not certain, and if the structure was positive parity, then there is an interesting possibility of a signature of $T = 0$ pairing.

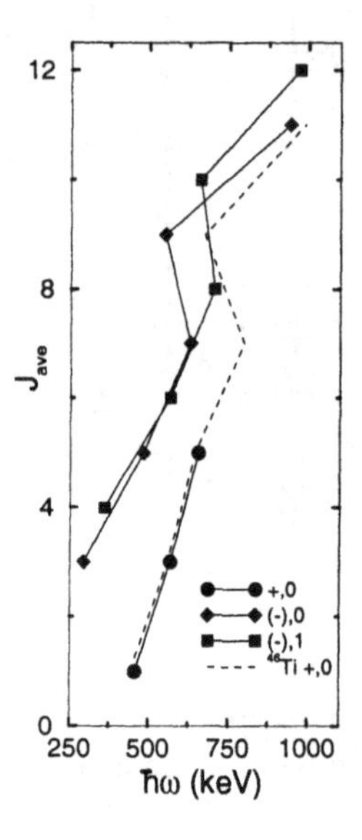

Figure 3 The average spin (J_{ave}) versus rotational frequency for the structure labelled B in ^{46}V (squares and diamonds). The dashed line indicates a similar plot for the yrast structure in ^{46}Ti and the circles correspond to their analogue states in band C of ^{46}V.

A spin *vs.* rotational frequency plot for these bands is shown in figure 3, where it can be seen that there is a rotational alignment occurring in both signatures at a similar frequency to the yrast band of ^{46}Ti. This implies that an $f_{\frac{7}{2}}$ pair alignment is occurring. If the structure was positive parity, in the absence of *np* pairing this alignment would be blocked in this odd–odd $f_{\frac{7}{2}}$ system. However, if $T = 0$ *np* pairing was present, then the two odd particles would be paired and the blocking effect would no longer occur, thus allowing the alignment to take place. It should be stressed that although this would certainly be a strong signature of this elusive type of pairing, the most likely scenario is that the structure is negative parity, in which case the observed behaviour can be explained by the excitation from the $d_{\frac{3}{2}}$ shell. However, when amongst friends at a conference it is sometimes more entertaining to be speculative! A direct measurement of the parity of this structure is planned for future work.

Acknowledgments

We wish to thank A. Poves, G. Martinez-Pinedo, S. M. Lenzi and J. A. Cameron for very helpful discussions. This work was supported by the United Kingdom Engineering and Physics Sciences Research Council. Finally, I (MAB) would like to add my own special thanks to John Sharpey-Schafer, to whom this conference has been dedicated on the occasion of his 60th birthday. Like so many others, John has been an inspiration to me, and he is the only reason I am doing nuclear physics research today.

References

[1] J. Cameron *et al.* Phys. Lett. B 235 (1990) 239

[2] J. Cameron *et al.* Phys. Lett. B 319 (1993) 58

[3] C. D. O'Leary *et al.* Phys. Rev. Lett. 79 (1997) 4349

[4] M. A. Bentley *et al.* Phys. Lett. B 437 (1998) 243

[5] G. Martinez-Pinedo, A. P. Zuker, A. Poves and E. Caurier. Phys. Rev. C 55 (1997) 187

[6] J. A. Cameron *et al.* Phys. Rev. C 58 (1998) 808

[7] D. Rudolph *et al.* Phys. Rev. Lett. 76 (1996) 376

[8] S. M. Vincent *et al.* Phys. Lett. B 437 (1998) 264

[9] J. Eberth *et al.* Prog. Part. Nucl. Phys. 28, (1992) 495

[10] T. Kuroyanagi *et al.* Nucl. Instrum. Methods A 316 (1992) 289

[11] S. E. Arnell *et al.* Nucl. Instrum. Methods A 300, (1991) 303

[12] C. D. O'Leary *et al.* Submitted to Phys. Lett. B (1999)

[13] A. Afanasjev and I. Ragnarsson. (private communication)

[14] S. E. Larsson Physica Scripta 8 (1973) 17

[15] J. A. Sheikh, D. D. Warner and P. Van Isacker. Phys. Lett. B 443 (1998) 16

[16] P. Bednarczyk *et al.* Phys. Lett. B 393 (1997) 285

STUDY OF THE DOUBLY-ODD NUCLEUS ^{66}As

Alison Bruce

University of Brighton
Brighton
BN2 4GJ
UK

alison.bruce@brighton.ac.uk

1 INTRODUCTION

The study of odd-odd N=Z nuclei is particularly interesting as it is in these nuclei that the dominance of neutron-proton pairing would imply the reappearance of a pairing gap similar to that in an even-even nucleus. An example of this has recently been observed in the N=Z=37 nucleus ^{74}Rb [1] where the lowest T=1 states mirror the structure of their isobaric analogue ^{74}Kr. In that study, however, insufficient states were seen to determine whether a pairing gap could be observed in the T=0 states also. In contrast to this, no evidence of T=1 states (except for the ground state) was observed in a recent study of ^{62}Ga [2] where it was the T=0 states which were populated. The 1st excited T=0 state is at 571 keV in ^{62}Ga and at ~1 MeV in ^{74}Rb, which suggests a gradual increase in the excitation energy of the T=0 states in this region. In the intermediate odd-odd N=Z nucleus ^{66}As it might therefore be expected that both T=0 and T=1 states could be observed below 1 MeV. This might therefore be the ideal nucleus in which to study the relative excitation energies and structures of the T=0 and T=1 states in this region. A recent study [3] of isomeric states in ^{66}As using a projectile fragmentation reaction has recently been carried out but as yet no prompt decaying radiation has been observed in this nucleus. We have therefore undertaken a study of the yrast and near-yrast states in ^{66}As.

The Nucleus: New Physics for the New Millennium
Edited by Smit et al., Kluwer Academic / Plenum Publishers, New York, 2000.

2 EXPERIMENTAL DETAILS AND PROCEDURES

The nucleus ^{66}As was studied at the Oak Ridge National Laboratory using the ^{40}Ca(^{32}S,αpn)^{66}As reaction at a beam energy of 105 MeV. PACE calculations suggest that the nucleus of interest, ^{66}As, is populated with a cross-section of $\sim$ 5 mb in this reaction which is $\sim$ 1% of the total cross-section. The main reaction channels produce ^{69}As, ^{68}Ge and ^{66}Ge with predicted cross-sections of 72 mb (25%), 49 mb (17%) and 61 mb (21%) respectively. Therefore, the recoiling nuclei were analysed using the RMS and a split anode ionisation chamber. The settings of the RMS were adjusted so that two charge states (q = 16 and 17) of mass 66 nuclei could be measured on the focal plane of the separator simultaneously. Gamma-rays were collected using an array of 5 clover detectors and 6 suppressed germanium detectors situated at the target position. Figure 1 shows a gamma-ray spectrum measured in coincidence with mass (A) = 66 recoils detected at the focal plane of the RMS. The gamma-rays which are labelled, and indeed all the strong transitions in the spectrum, can be associated with $^{66}_{32}$Ge from previous studies of that nucleus [4].

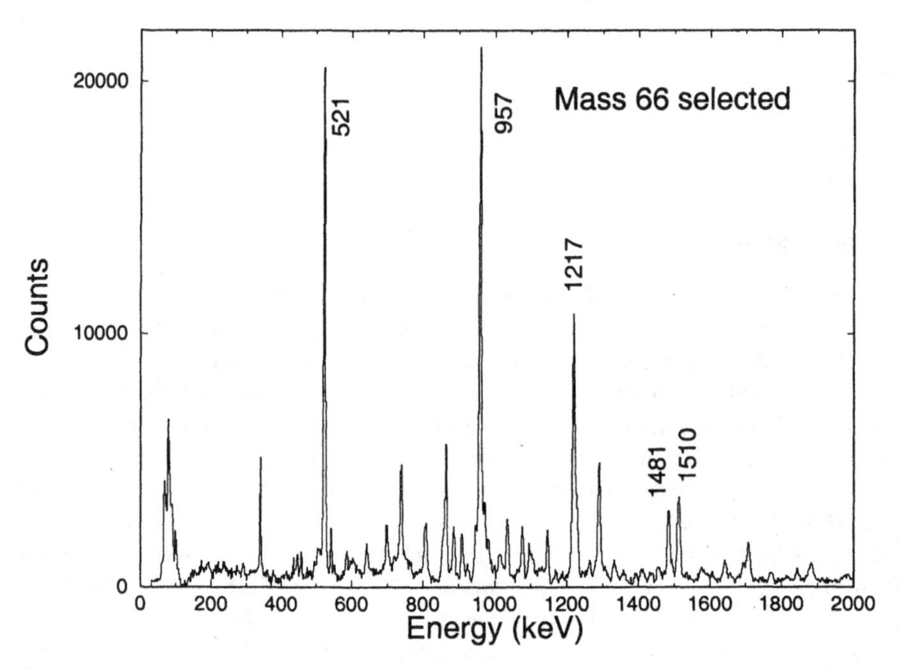

Figure 1 The spectrum of gamma-rays seen in coincidence with mass 66 recoils

In order to get some Z-selection it is necessary to try and use the information from the ion chamber. Indeed, this was the first experiment using this ion chamber. Unfortunately the velocity of the recoils in this experiment was quite low (v/c = 3.5 %) so there was not much Z separation in the ion chamber. However it has been possible to produce two mass gated spectra which correspond to two different Z slices and these are shown in figure 2. The bottom

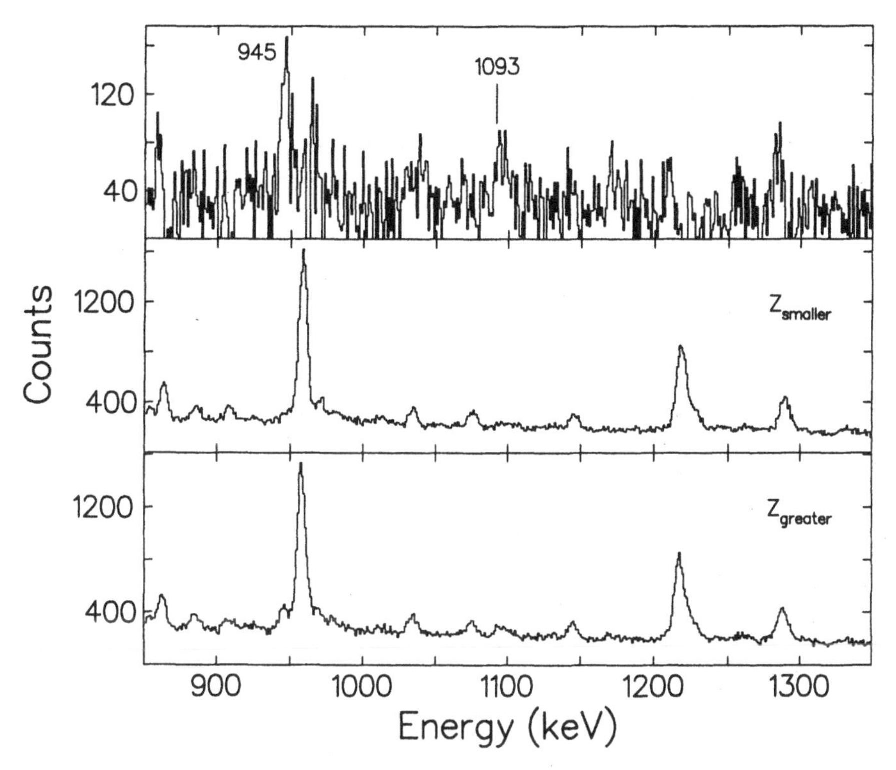

Figure 2 The spectrum of gamma-rays seen in coincidence with mass 66 recoils and bottom) $Z_{greater}$ middle) $Z_{smaller}$. The top spectrum is the difference between the two.

portion of this figure shows the spectrum obtained by gating on $Z_{greater}$ in the ion chamber and the middle portion shows the spectrum obtained by gating on $Z_{smaller}$. The top portion of the figure shows the difference spectrum between these two after suitable normalisation has been carried out.

It is evident that there are two peaks which stand out in this spectrum and these have been labelled at 945 and 1093 keV. Careful examination shows that these peaks are in the $Z_{greater}$ spectrum but not the $Z_{smaller}$ spectrum, and that they are in coincidence with each other. Therefore these transitions could possibly be the yrast transitions from $^{66}_{33}$As. It was however mentioned in the previous paragraph that two charge states (q=16 and 17) of mass 66 nuclei were measured on the focal plane of the separator simultaneously and further analysis shows that the gamma-rays measured at 945 and 1093 keV are observed only in coincidence with nuclei in one of the charge states.

Further analysis reveals that the transitions are in fact the two lowest transitions in the ground-state band of $^{70}_{34}$Se and are seen because of an A/q ambiguity between mass 66, charge state 16 (A/q = 4.125) and mass 70, charge state 17 (A/q = 4.117). Although disappointing in terms of the overall aim of the experiment, it is confirmation that there is some Z-selection in the data.

3 DISCUSSION

It is interesting then to speculate on why there is no evidence of prompt gamma-rays from ^{66}As in the A = 66 gated matrix. ^{66}As is predicted to be populated with a cross-section ~ 1/20 of that of ^{66}Ge and, inspection of the spectrum shown in figure 1 suggests that it should be possible to see transitions with ~ 1/20 of the counts of transitions in ^{66}Ge. The problem might be one of identification but, it has been shown in figure 2 that there is some Z selection so that should not be the case.

The second option then is that all the intensity feeds through the known isomers [3] and therefore no strong gamma-ray transitions would be expected to be observed in this measurement. Grzywacz et al., [3] identified two isomers in this nucleus, one at an excitation energy of 1357 keV which has a half-life of 1.9 μs and the other at an excitation energy of 3024 keV with a half-life of 17 μs. The spins and parities of these states are predicted to be 5^+ and 9^+ respectively and as such these states will probably be part of the yrast sequence for this nucleus. The vast majority of the population in the fusion evaporation reaction will follow the yrast sequence and therefore will populate the higher-lying of these 2 isomers. In this case the strength feeding the isomeric state would be severely fragmented and the non-observation of any line corresponding to ^{66}As could be understood.

Acknowledgments

This work is a collaboration with the following people: D.D.Warner, C.Baktash, C.J.Barton, M.A.Bentley, R.A.Cunningham, E.Dragulescu, L.Frankland, T.N.Ginter, C.J.Gross, R.C.Lemmon, B.MacDonald, C.D.O'Leary, J.Simpson, S.M.Vincent, C.H.Yu and V.Zamfir.

References

[1] D.Rudolph et al., Phys. Rev. Letts. 76 (1996) 376.

[2] S.M.Vincent et al., in *Proceedings of the international conference on nuclear structure at the extremes*, Lewes, UK June 1998, to be published in J.Phys. G. and Phys. Letts. B437 (1998) 264.

[3] R.Grzywacz et al., Phys. Letts. B429 (1998) 247.

[4] U.Hermkens et al., Zeitscrift fur Physik A343 (1992) 371.

TEST OF PARITY DOUBLING IN ^{223}Ra AND ^{223}Th

N. Amzal[1*], P.A. Butler[1], G. D. Jones[1], B. Gall[3], F. Hannachi[2], D. Hawcroft[1], R-D. Herzberg[1], T.H. Hoare[1], F. Hoellinger[3], P.M. Jones[1], C.F. Liang[2], P. Paris[2], D.P. Rea[1] and N. Schulz[3].

[1] Oliver Lodge Laboratory, University of Liverpool, L69 7ZE - U.K.
[2] C.S.N.S.M., Bat.104, F-91405, Orsay Campus - France
[3] IReS Strasbourg, F-67037, Strasbourg cedex - France

Abstract: The level structures of the odd-A nuclei ^{223}Ra and ^{223}Th have been investigated by means α-γ and α-e$_{K,L,M}$ angular correlation measurements, in-beam electron conversion and γ-ray spectroscopy. Values of g-factors for members of the $K = 3/2^{\pm}$ bands in ^{223}Ra and $K = 5/2^{\pm}$ bands in ^{223}Th have been extracted from the measurement of $B(M1)/B(E2)$ branching ratios and compared to theoretical predictions in order to ascertain wether they support other indications of octupole deformation in these nuclei or not.

1 INTRODUCTION

The region of the periodic table which has shown the best evidence for octupole instability is around N or Z $\approx$ 85 - 92 and N $\approx$ 131 - 139. Strong octupole correlations arise when single-particle states which differ by Δl=Δj=3 lie close

*Corresponding author : na@ns.ph.liv.ac.uk

The Nucleus: New Physics for the New Millennium
Edited by Smit et al., Kluwer Academic / Plenum Publishers, New York, 2000.

to each other and to the Fermi surface [1], [2]. In even-even nuclei, the signature of octupole deformation is the presence of a rotational band of interleaved positive and negative parity levels connected by strong E1 transitions. In odd mass nuclei, a reflection-asymmetric deformation leads to bands characterized by levels almost degenerate in energy with the same spin but opposite parities : *"Parity doublets"*. However, the observation of parity doublets is not sufficient to confirm that octupole correlations exist; octupole deformation predicts that the g-factors, for the positive and negative parity bands, should be identical as they arise from identical intrinsic structure of the rotational bands. In the mass 150 region the odd-mass nuclei showing low-lying near-degenerate opposite parity bands have very different g-factors for the different parities (e.g.[3], [4]), which is indicative that they are reflection-symmetric. In the mass region around Z=88, N=134 there are very few systematic data on the behaviour of g-factors in odd-mass nuclei. We report here measurements of gyromagnetic ratios in ^{223}Ra (for the first time in an odd-neutron nucleus in this mass region) and in ^{223}Th (a model example of reflection-asymmetry).

2 THE ^{223}Ra STUDY

2.1 Motivation

In ^{223}Ra two parity doublets $1/2^{\pm}$ and $3/2^{\pm}$ have been identified which is a key indicator of octupole correlations. The decoupling parameters of the $K = 1/2^{\pm}$ bands and the magnetic dipole moments for the $K = 3/2^{\pm}$ bandheads also support a reflection-asymmetric shape for this nucleus [2]. Octupole deformation in the strong coupling limit predicts that the quantity $(g_K - g_R)/Q_0$ (where g_K is the intrinsic g-factor, g_R the rotational g-factor and Q_0 the intrinsic quadrupole moment) should be equal for both positive and negative parity bands. We have carried out measurements of this parameter for the $K = 3/2^{\pm}$ bands in ^{223}Ra, in order to ascertain whether they support other evidence that stable octupole deformation exists in ^{223}Ra at low excitation energies.

2.2 Experimental method

The low spin states of ^{223}Ra (figure 1) have been populated by α-decay from ^{227}Th ($T_{1/2} \approx$ 19 days) which was itself produced via β^- decay of an 10 μCi ^{227}Ac source ($T_{1/2} \approx$ 27years) . A compact system has been designed in order to measure α-γ and α-$e_{K,L,M}$ angular correlations : a Be-windowed germanium detector and a Si(Li) detector were used to detect γ-rays and internal conversion electrons respectively, while α-particles were detected with 5 Silicon PIN diodes positioned at 180°,155°,135°,115° and 90° relative to the Ge detector. Energy calibrations of the Ge and Si(Li) were done using ^{152}Eu and ^{133}Ba γ-ray sources and a ^{207}Bi electron source. The ^{227}Ac itself was used for the energy calibration of the PIN diodes. A complete description of the experimental setup together with the analysis and results is given in [5], [6].

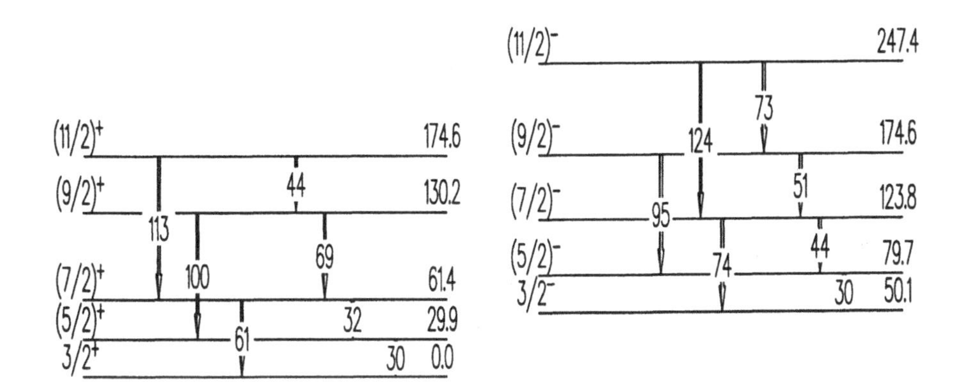

Figure 1 Partial decay scheme for ^{223}Ra. Level and transition energies are in keV.

2.3 Experimental results and discussion

Values of the parameter $(g_K - g_R)/Q_0$ have been extracted for the $K = 3/2^{\pm}$ parity doublet bands from the measurement of the branching ratio for each pair of E2 and (M1+E2) intra band transitions from the same level spin I. The deduced values of $(g_K - g_R)/Q_0$ are shown in figure 2. These values are similar for the positive and the negative parity bands as expected for stable octupole deformation. Theoretical calculations using Woods-Saxon cranking codes [7] have been carried out. These calculations show that reflection asymmetry with $\beta_3 \approx 0.05$ is required in order to reproduce the values deduced experimentally. The results of this study support other evidence that, among the odd N nuclei, ^{223}Ra provides one of the best example of octupole deformation at low excitation energies.

3 THE ^{223}Th STUDY

3.1 Motivation

The nucleus ^{223}Th has been often cited as the best case of parity doubling in atomic nuclei, on account of the near degeneracy of the lowest four rotational bands (both signatures of $K = \frac{5}{2}^+$ and $\frac{5}{2}^-$ bands), and the strong $E1$ transitions connecting the bands of opposite parity. The decay scheme of this nucleus, taken from Dahlinger [8] is shown in figure 3. While the branching of $E1$ and $E2$ transitions have been measured up to high spin, there is very little information on the $M1$ transition intensities which connect the signature partners in each parity band except for the lowest transitions. Comparison of the $B(M1)/B(E2)$ branching ratios will provide a crucial test of the hypothesis that the parity doubling arises from octupole deformation of this nucleus.

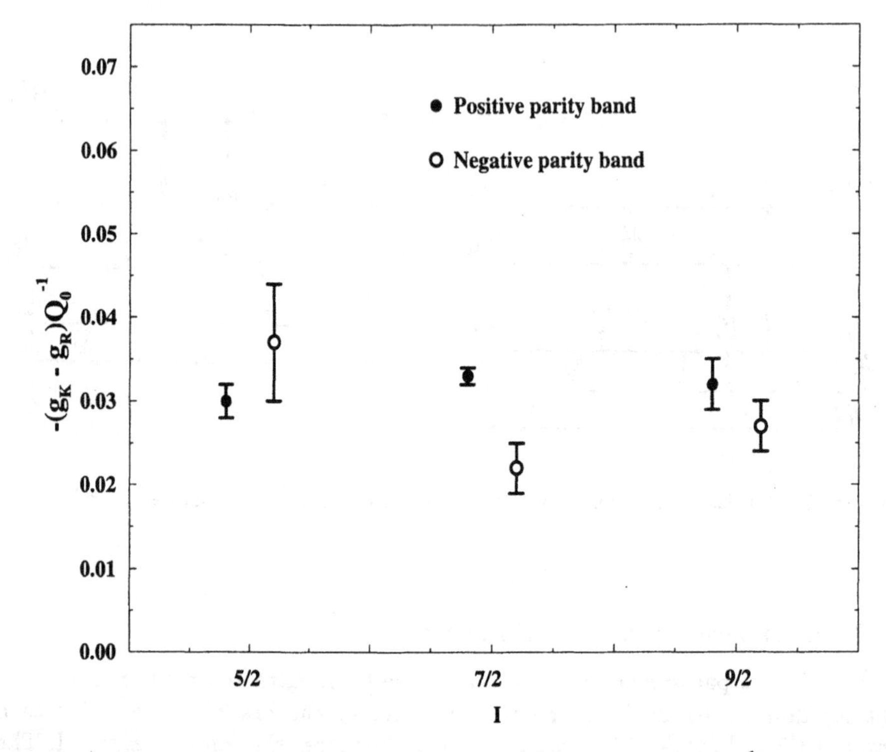

Figure 2 The value of the parameter $-(g_K - g_R)/Q_0$ (in units barns^{-1}) deduced for members of the $K = 3/2^{\pm}$ bands in ^{223}Ra. The transitions are labelled with the spin I of each decaying level and parity π.

3.2 Experimental details

We have carried out measurements of the $B(M1)/B(E2)$ branching ratios for transitions in both positive and negative parity bands in ^{223}Th, using the GAREL+ facility at IReS Strasbourg. The nucleus was produced in the reaction ^{208}Pb(^{18}O,3n) at 86 MeV, using a 250μg/cm^2 target. The strongly converted $M1$ transitions were observed using the Betatronc electron spectrometer and a Si(Li) detector ($\sim$ 1 % efficiency) [9]; the $E1$ and $E2$ γ-ray transitions were detected in 12 large-volume ($\sim$ 70 % relative efficiency) Compton-suppressed Ge detectors and a single LEPS detector. Electron-γ as well as γ-γ coincidences were recorded during 5 days. Energy and efficiency calibrations of the Ge and Si(Li) were done using ^{152}Eu and ^{133}Ba γ-ray sources and a ^{223}Ra electron source.

3.3 Preliminary results and discussion

Values of the quantity $|g_K - g_R|$ have been extracted, for the $K = \frac{5}{2}^+$ and $K = \frac{5}{2}^-$ bands, from the measured intra band $B(M1)/B(E2)$ branching ra-

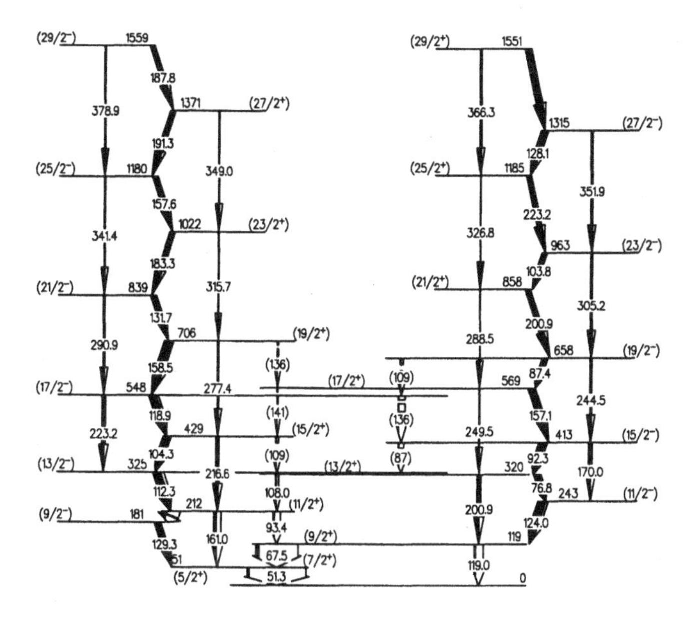

Figure 3 Partial decay scheme for ^{223}Th taken from Dahlinger [8]. Level and transition energies are in keV.

tios. Plots of the values of the parameter $|g_K - g_R|$ for the $\frac{11}{2}^+ \to \frac{9}{2}^+$, $\frac{9}{2}^+ \to \frac{7}{2}^+$, and $\frac{15}{2}^- \to \frac{13}{2}^-$ transitions in ^{223}Th are shown in figure 4. The values extracted by Dahlinger [8] for the $\frac{9}{2}^+ \to \frac{7}{2}^+$ and $\frac{7}{2}^+ \to \frac{5}{2}^+$ are also shown. Within the errors these values are similar within each band and between the two bands of opposite parity which is consistent with the interpretation that ^{223}Th is reflection-asymmetric. However the values determined experimentally do not agree with the theoretical ones obtained using calculations of single particle wave functions in a reflection-asymmetric Woods-Saxon potential assuming that the states have pure $\Omega = 5/2$ configurations. The discrepancy could arise from coriolis mixing with the close lying $K = 3/2^{\pm}$ bands.

4 SUMMARY

The level structure of the odd-A nuclei ^{223}Ra and ^{223}Th have been investigated by means of various experimental techniques including α-γ and α-$e_{K,L,M}$ angular correlation measurements, in-beam electron conversion and γ-ray spectroscopy. Values of g-factors for transitions in both positive and negative parity bands have been extracted from the measurement of $B(M1)/B(E2)$ branching ratios and compared to theoretical calculations performed using Woods-Saxon cranking codes. The deduced values support other indications of octupole deformation in these nuclei.

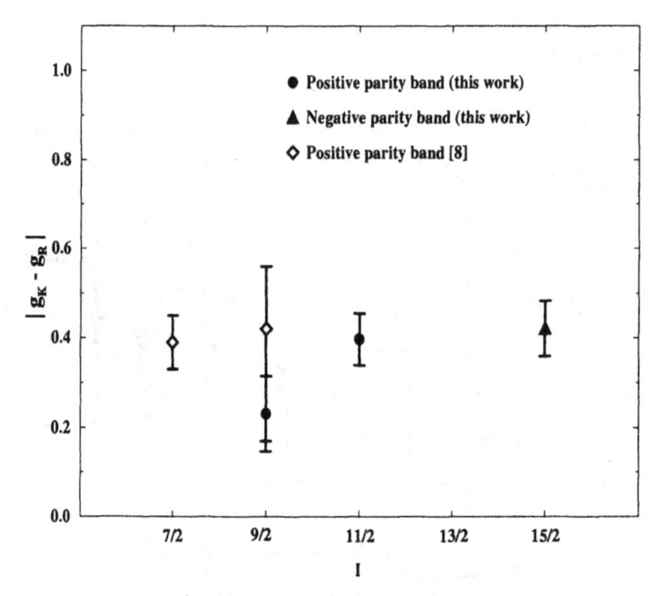

Figure 4 The value of the parameter $|g_K - g_R|$ obtained from intra band $B(M1)/B(E2)$ branching ratios in ^{223}Th plotted against the spin I of the decaying level. Previous measurements from [8] are shown in the cases for which they are available.

Acknowledgments

This work was supported by grants from the U.K. Engineering and Physical Sciences Research Council.

References

[1] P.A. Butler and W. Nazarewicz Rev. Mod. Phys. 68 (1996) 349

[2] I. Ahmad and P.A. Butler Ann. Rev. Nucl. Part. Sci. 43 (1993) 71

[3] W.J. Vermeer *et al.* Phys. Rev. C42 (1990) R1183

[4] C.J. Pearson *et al.* Phys. Rev. C49 (1994) R1239

[5] G. D. Jones *et al.* J. Phys. G: Nucl. Part. Phys. 17 (1991) 713

[6] G. D. Jones *et al.* Eur. Phys. J A 2 (1998) 129

[7] S. Ćwiok *et al.* Comp. Phys. Comm. 46 (1987) 379

[8] M. Dahlinger *et al.* Nucl. Phys. A484, (1988) 337

[9] P. Paris *et al.* NIM A357, (1995) 398

ANGULAR DISTRIBUTION OF PROTONS EMITTED FROM ORIENTED NUCLEI: TOWARD IMAGING SINGLE-PARTICLE WAVE FUNCTIONS

N. Carjan[1], P. Talou[2], D. Strottman[2,3]

[1]Centre d'Etudes Nucléaires de Bordeaux-Gradignan
33175 Gradignan Cedex, France
[2]Theoretical Division and [3]LANSCE, Los Alamos National Laboratory,
Los Alamos, NM 87545, USA

carjan@in2p3.fr, talou@lanl.gov, dds@lanl.gov

1 INTRODUCTION

A major drawback of most quantum mechanical measurements is that they produce only quantities averaged over coordinates or time; nuclear physics is no exception. There is, however, an observable (the angular distribution with respect to the deformation axis of single nucleons emitted from metastable states in oriented nuclei) that is directly related to the spatial distribution of the single-particle wave function, as will be shown in this paper. It will also be argued that measuring probability densities of nucleons in nuclei (not only matrix elements) is more promising and that it should become a new trend in evaluating our understanding of nuclear structure in the future.

2 DENSITY MAPS FROM ANGULAR DISTRIBUTIONS

To calculate the angular distribution with respect to the nuclear symmetry axis of a proton emitted from a metastable state in a deformed nucleus, we have solved numerically the time-dependent Schrödinger equation in two dimensions [1, 2],

$$i\hbar\frac{\partial}{\partial t}\psi(z,\rho,t) = \mathcal{H}(z,\rho)\psi(z,\rho,t) \tag{1}$$

The Nucleus: New Physics for the New Millennium
Edited by Smit et al., Kluwer Academic / Plenum Publishers, New York, 2000.

where the hamiltonian $\mathcal{H}$ is

$$\mathcal{H}(z,\rho) = -\frac{\hbar^2}{2\mu}\left[\frac{1}{\rho}\frac{\partial}{\partial\rho} + \frac{\partial^2}{\partial\rho^2} + \frac{\partial^2}{\partial z^2} - \frac{\Lambda^2}{\rho^2}\right] + V_{pA}(z,\rho). \quad (2)$$

V_{pA} is the potential representing the interaction between the individual proton and the remaining nucleons of the core daughter nucleus. Initial single-proton excited states in ^{208}Pb, as those represented in Fig. 1, were chosen in the frame of a deformed Woods-Saxon potential ($\epsilon = 0$ corresponds to a spherical potential). Knowing $\psi(z,\rho,t)$, one can estimate the tunneling probability as a function of azimuthal angle θ:

$$P_{tun}(t,\theta) = \int_{r_B(\theta)}^{\infty} |\psi(r,\theta,t)|^2 r^2 dr, \quad (3)$$

where $r_B(\theta)$ is the radial position of the potential ridge in the direction θ. The results are shown in Fig. 2 for the $'2h'$ quasi-stationary states at $\epsilon = 0.1$ from Fig. 1.

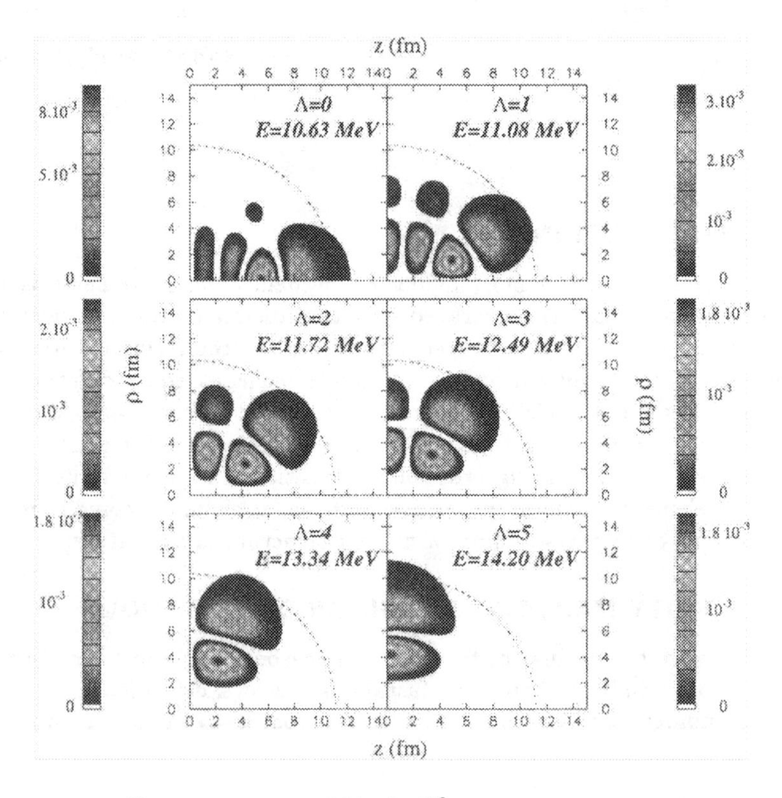

Figure 1 Figure 1 $|\psi(r,\theta,t)|^2$ at $\epsilon = 0.1$.

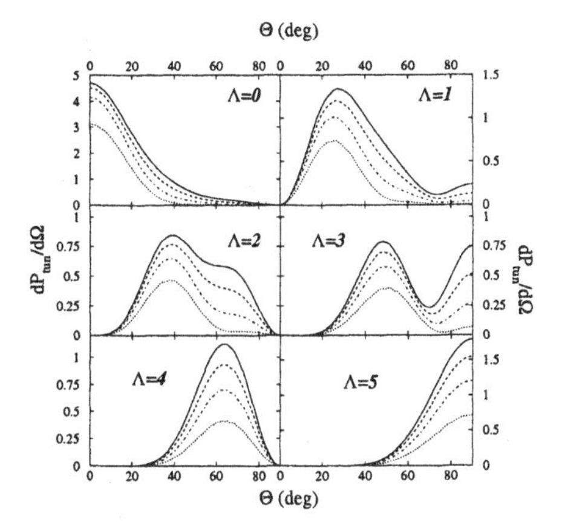

Figure 2 Angular distributions of the $'2h'$ states at $\epsilon = 0.1$.

Except for $\Lambda = 0$, none of these wave functions escapes along the path of minimum barrier ($\theta = 0°$). The main directions of emission are determined by the angles between the branches of the initial wave function and the z axis. The anisotropic barrier acts like a filter modifying the relative intensities of different branches but does not change their directions. Consequently, the emission does not always occur along the nuclear symmetry axis as was intuitively predicted in Ref.[3].

It is therefore possible in most cases to deduce the number of branches in the proton density and their orientations just by inspecting the angular distribution. For more complete information it is, however, necessary to simulate the filter, *i.e.*, to calculate the tunneling through a bidimensional barrier. A detailed study of the connection between angular distributions and density maps is in progress.

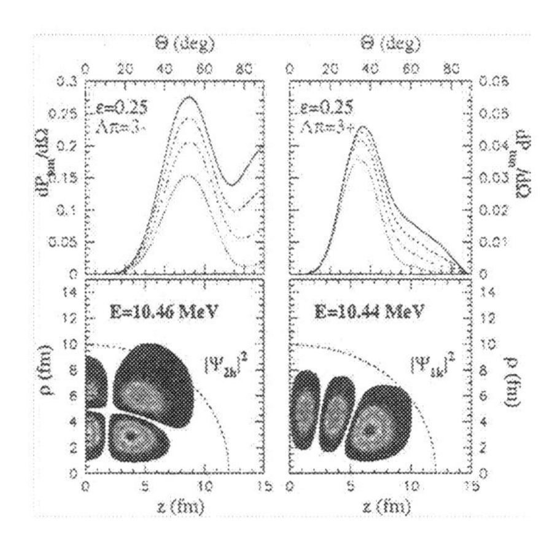

Figure 3 Figure 3 Comparison between density maps and angular distributions for two crossing states at $\epsilon = 0.25$.

The same qualitative results are found at larger deformations (see Fig. 3 for the case of $\epsilon = 0.25$) in spite of a larger difference between the barriers in ρ and z directions. Another aspect that demonstrates the high sensitivity of our method is shown in the last figure: a small deviation from sphericity produces a large anisotropy (Fig. 4).

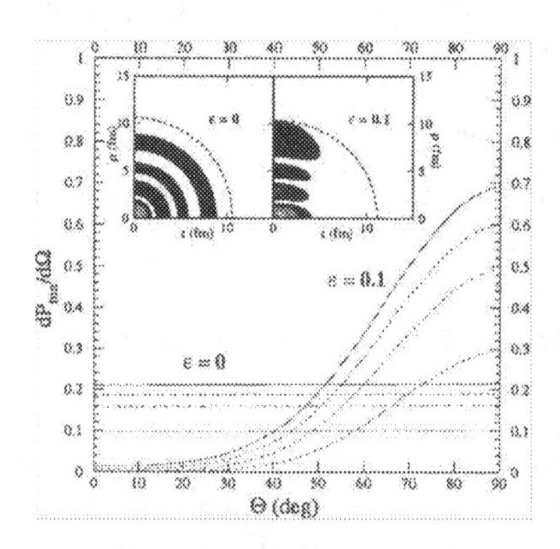

Figure 4 Temporal dependence, $'t' = 5$ to 20 by 5 in units of 10^{-22} sec of the angular distributions for the $'4s'$ state at spherical ($\epsilon = 0$) and slightly deformed ($\epsilon = 0.1$) configurations. The corresponding density maps are inserted for comparison.

3 CONCLUDING REMARKS

Quite accurate maps of single-particle density distributions in deformed nuclei can be obtained by combining a) angular distribution measurements of single nucleons emitted from metastable states in oriented nuclei, and b) numerical simulations of quantum tunneling through an anisotropic coulomb + centrifugal barrier. This would represent the first direct measurement of probability densities of nucleons in nuclei and would allow one to test nuclear models on a deeper and more detailed level.

After this work was completed we found that pronounced anisotropies – such as those calculated above – have been recently observed during a similar process involving α emission from oriented nuclei[4]. Although α decay is easier to measure, proton or neutron decay represents a better tool for exploring in detail the wave functions involved since: a) the modification of the initial wave anisotropy by the tunneling filter is less important, and b) one samples a single particle wave function at a time and not an overlap of the four nucleons entering into the formation amplitude of an α particle, and c) it carries higher angular momenta.

References

[1] P. Talou, N Carjan and D Strottman. To appear in Nucl. Phys. A.
[2] P. Talou, N Carjan and D Strottman. Phys. Rev. C58 (1998) 3280.
[3] D.L. Hill and J.A. Wheeler. Phys. Rev. 89 (1953) 1102.
[4] P. Schuurmans *et al.* Phys. Rev. Lett. 77 (1996) 4720.

PROBING HIGHLY-FRAGMENTED GIANT RESONANCES: COINCIDENCE EXPERIMENTS IN THE NEW MILLENNIUM

J. Carter

Physics Department,
University of the Witwatersrand, PO Wits, Johannesburg, 2050 South Africa

Abstract: Electric isoscalar and isovector giant resonances are excited strongly by inelastic electron and hadron scattering. In particular, the isoscalar giant monopole resonance (ISGMR) and the isoscalar giant quadrupole resonance (IS-GQR) have been found both experimentally and theoretically to be highly fragmented in medium mass nuclei. The use of a coincidence technique to measure the angular correlations of decay products after inelastic excitation of the target nucleus is a powerfull tool for determining the multipolarity of the resonance excited. Only recently with the advent of continuous wave electron accelerators has this been possible for the (e,e′x) reaction. Results will be presented on experiments done at the National Accelerator Centre for ^{40}Ca(p,p′x=p,α) and ^{48}Ca(p,p′n). This work compliments similar experiments performed using electrons at S-DALINAC (Darmstadt) and MAMI A (Mainz) in Germany. It will be shown that decay by a non-zero spin particle to a non-zero spin state in the residual nucleus leads to simple angular correlations from which the relative strengths of the various multipoles excited can be extracted unambiguously. This is unexpected since up to now α-particle decay has been studied, where channel spin plays no role.

The Nucleus: New Physics for the New Millennium
Edited by Smit et al., Kluwer Academic / Plenum Publishers, New York, 2000.

1 FORWARD

I shall be presenting some of the results of an investigation into the fragmentation of giant resonance strength in the doubly magic nuclei ^{40,48}Ca. This work is as a result of the Wits / Darmstadt / NAC / Cape Town / Stellenbosch collaboration. The following people have been involved: C. Bähr, H Diesener, P von Neumann-Cosel, A Richter, K Schweda, S Strauch, A. Staschek (Technical University, Darmstadt), S V Förtsch, J J Lawrie, S J Mills, R T Newman, J V Pilcher, F D Smit, G F Steyn, D M Whittal (National Accelerator Centre, Faure), R W Fearick (University of Cape Town), A A Cowley (University of Stellenbosch) and M N Harakeh (Kernfysisch Versneller Instituut, Groningen).

2 INTRODUCTION

Electric isoscalar and isovector giant resonances are excited strongly by inelastic electron and hadron scattering. In particular, the isoscalar giant-monopole resonance (ISGMR) and the isoscalar giant-quadrupole resonance (ISGQR) have been found both experimentally (see, for example, ref. [1]) and theoretically (see, for example, ref. [2]) to be highly fragmented in medium mass nuclei. The use of a coincidence technique to measure the angular correlations of decay products after inelastic excitation of the target nucleus is a powerfull tool for determining the multipolarity of the resonance excited. Only recently with the advent of continuous wave electron accelerators has this been possible for the (e,e′x) reaction.

Results will be presented of experiments done at the National Accelerator Centre for ^{40}Ca(p,p′x=p,α) and ^{48}Ca(p,p′n). This work compliments similar experiments performed using electrons at the S-DALINAC (Darmstadt) and MAMI A (Mainz) in Germany. It will be shown that decay by a non-zero spin particle to a non-zero spin state in the residual nucleus leads to simple angular correlations from which the relative strengths of the various multipoles excited can be extracted unambiguously. This is unexpected since up to now α-particle decay has been studied, where channel spin plays no role.

3 ANALYSIS OF ANGULAR CORRELATIONS

The inelastic excitation and decay of giant resonances in ^{40,48}Ca were studied in experiments using the K600 magnetic spectrometer of the cyclotron facility at the National Accelerator Centre, Faure. Angular correlations were measured for the reactions ^{40}Ca(p,p′x=p,α) and ^{48}Ca(p,p′n) at E_p = 100 MeV. The inelastically excited ^{40}Ca decays predominantly by the emission of a proton or an α-particle (see Fig. 1). Results for the isoscalar quadrupole strength in ^{40}Ca from the (p,p′α_0) reaction were presented in ref. [3]. By way of example, measured angular correlations and the fitted angular correlation functions are shown in Fig. 2. In this case, because of the 0^+ ground-state spin of the ^{36}Ar residual nucleus after α_0 decay and the 0^+ spin of the α particle, it is possible to extract unambiguously the relative strengths of the E0/E2/E3 components used in the fit [3].

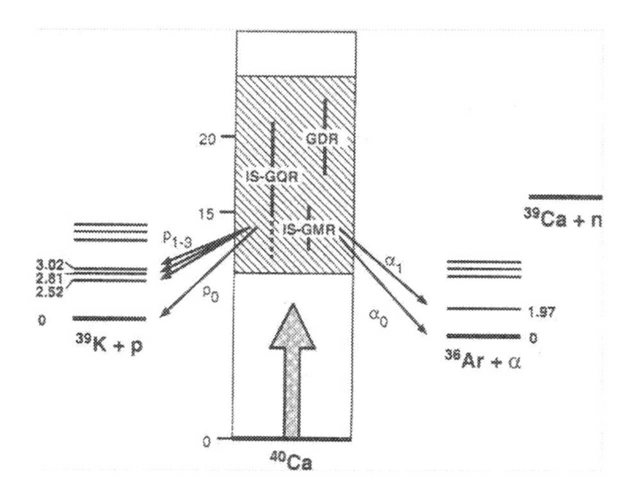

Figure 1 Excitation and decay of ^{40}Ca.

The analysis, however, of the corresponding p_0-decay angular correlations poses a considerably more complicated though not impossible task. Here, p_0 decay to the $3/2^-$ ground state of ^{39}K and the spin 1/2 of the decay proton

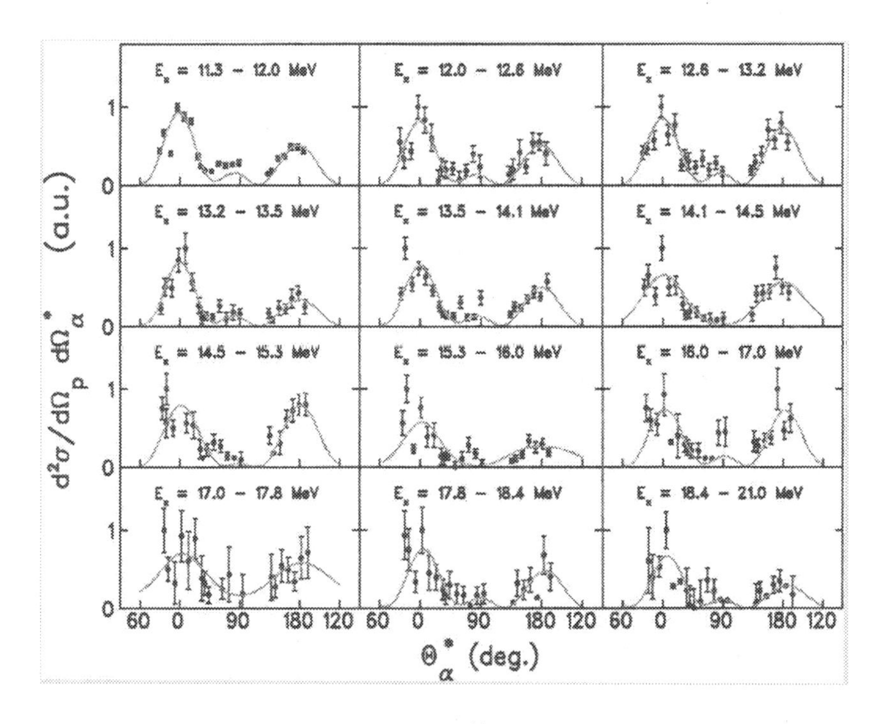

Figure 2 Angular correlations for the reaction ^{40}Ca(p,p$'\alpha_0$) at E_p = 100 MeV and θ_p = 17°.

together with the value of the relative angular momentum of the decay (predominantly $l = 0, 1, 2, 3, 4\ \hbar$) leads to a number of spin-coupling possibilities from a state with $L = 0$, 2 or 3 $\hbar$. The problem is complicated further by the inclusion of the relative phases between the decay angular momenta and their relative strengths (transmission coefficients T_l) as a function of excitation energy in ^{40}Ca. Fortunately, however, it is possible to fix the values of the different parameters. The transmission coefficients [4] can be obtained by using a suitable optical potential for the proton decay. The relative phase between the even/odd l-values is set to $\phi = 180°$, as found in an analysis of ^{40}Ca(e,e$'\alpha_1$) data [5]. Also, the relative strengths E0/E2/E3 determined in the analysis of the corresponding α_0 data can be used. The resulting parameter-free fits to the p_0 angular correlations are shown in Fig. 3. As can be seen, simple, asymmetric structures (reminiscent of the fits to the α_0 channel) are predicted which indeed show good correspondence with the specific features of the data.

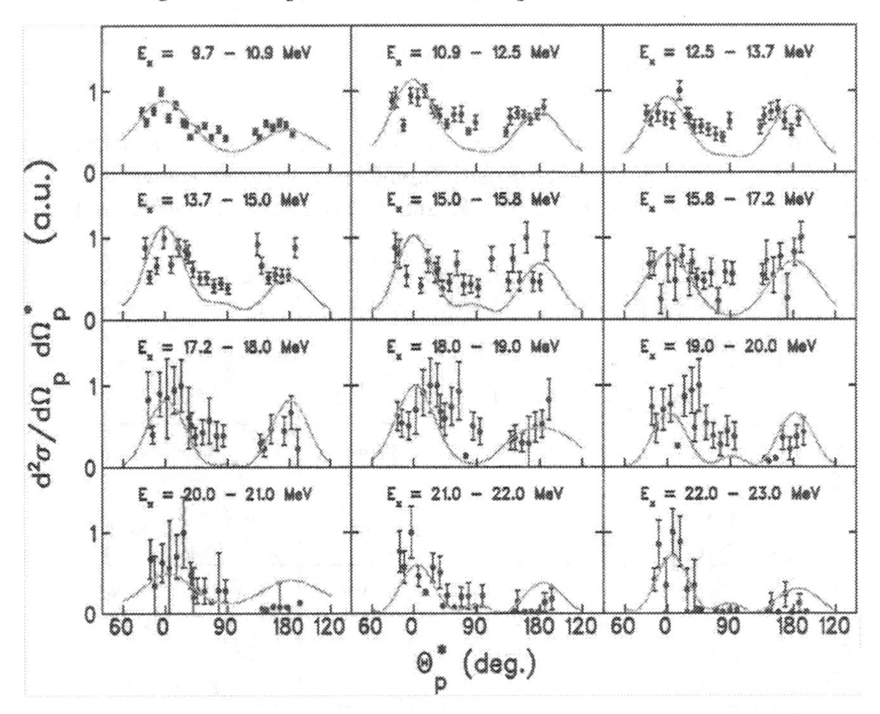

Figure 3 Angular correlations for the reaction ^{40}Ca(p,p$'$p$_0$) at $E_p = 100$ MeV and $\theta_p = 17°$.

Turning now to the doubly magic nucleus ^{48}Ca. The inelastically excited ^{48}Ca decays predominantly by the emission of a neutron (see Fig. 4). An exceptionally good energy resolution was obtained for neutron measurements in the ^{48}Ca(p,p$'$n) experiment at $E_p = 100$ MeV performed recently at the National Accelerator Centre. Further data extraction has led to the first presentation of the measured angular correlations for inelastic proton scattering at

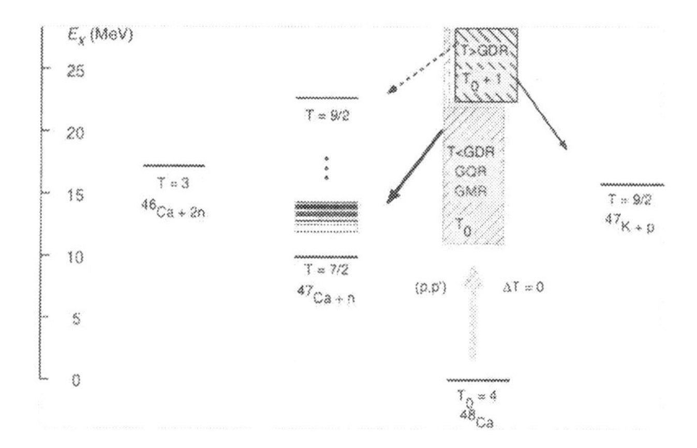

Figure 4 Excitation and decay of ^{48}Ca.

$\theta_p = 21°$ (see Fig. 5). Here, the choice of excitation energy windows is a compromise to maintain good enough statistics while preserving specific features in the data. As can be seen, some angular correlations display strong asymmetry while others are isotropic.

Following the procedures that were so effective in the analysis of the p_0 channel for ^{40}Ca, the n_0 decay was analyzed in the same way. Here, the problem

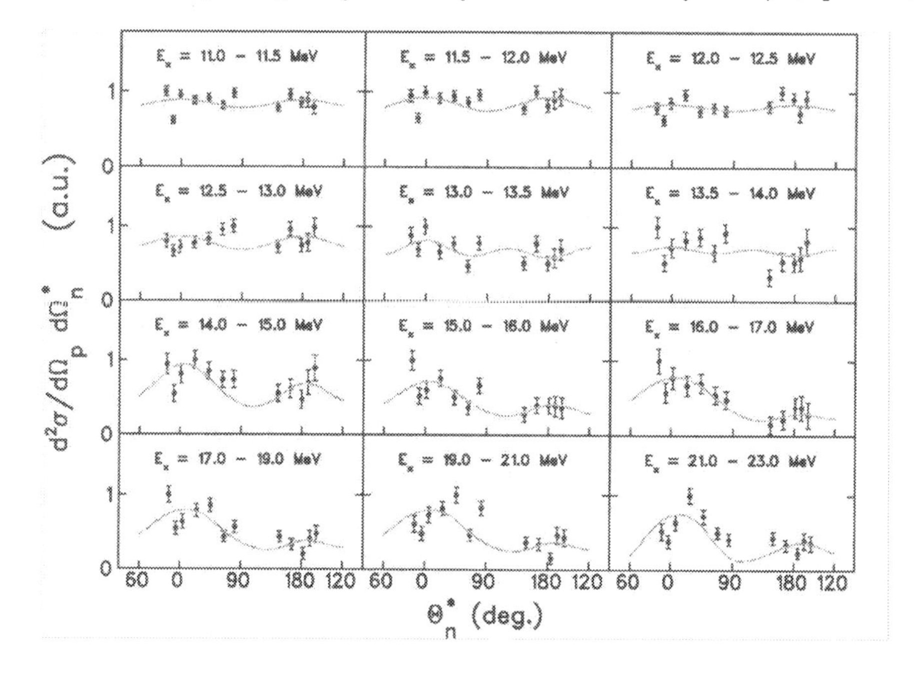

Figure 5 Angular correlations for the reaction ^{48}Ca(p,p'n_0) at E_p = 100 MeV and θ_p = 21°.

is even more complicated than before with the n_0 decay leading to a higher-spin $7/2^-$ ground state in ^{47}Ca. The relative strengths for the E0/E2/E3 excitations in ^{48}Ca were varied until a consistent fit (also to data taken at $\theta_p = 26°$, $31°$ and $36°$) was obtained. In the region E_x = 12 - 14 MeV predominantly E0 strength was found, in agreement with recent state-of-the-art RPA calculations which include coupling to the continuum, 1p1h⊗phonon configurations and ground state correlations induced by these configurations [2]. For the other excitation energy windows E2 was dominant. In all cases, the E3 contribution was only a few percent and was responsible for the strong asymmetries of the angular correlation functions. The use of the odd multipole E1 was investigated but good fits to the data were not obtained. In any case, 1^- states are not populated strongly (see Fujita et al. [6] for inelastic scattering at E_p = 65 MeV) and there was no evidence for such states being excited strongly at E_p = 100 MeV in the present experiment for $\theta_p = 21°$. It should be noted that the data displayed in Figs. 2, 3 and 5 are normalised to unity for each excitation energy window and that most of the cross section lies below E_x = 15 MeV.

4 SUMMARY

The following points have been made:
1) A model-independent multipole analysis of measured angular correlations can be performed for none spin-zero decays (p, n decay, spin 1/2) to resolved final states.
2) Previously, only α-decay (spin 0^+) to 0^+ final states was investigated. It may be worthwhile to look again at fitting existing angular correlation data for p and n decay.
3) New experiments should be designed to obtain good statistics for all resolved decay channels when measuring angular correlations.
4) Many more nuclei could be identified for investigation other than those favourable for α_0-decay. These might include radioactive beams.

References

[1] F. Zwarts, A.G. Drentje, M.N. Harakeh and A. van der Woude, Nucl. Phys. A 439 (1985) 117.

[2] S. Kamerdzhiev, J. Speth and G. Tertychny, Nucl. Phys. A 624 (1997) 328.

[3] J. Carter, A.A. Cowley, H. Diesener, R.W. Fearick, S.V. Förtsch, M.N. Harakeh, J.J. Lawrie, S.J. Mills, P. von Neumann-Cosel, R.T. Newman, J.V. Pilcher, A. Richter, K. Schweda, F.D. Smit, G.F. Steyn, S. Strauch and D.M. Whittal, Nucl. Phys. A 630 (1998) 631.

[4] F. Puehlhofer, computer programme TL, GSI (1979).

[5] M. Kohl, P. von Neumann–Cosel, A. Richter, G. Schrieder and S. Strauch, Phys. Rev. C57 (1998) 3167.

[6] Y. Fujita, M. Fujiwara, S.Morinobu, T.Yamazaki, T. Itahashi, H. Ikegami and S.I. Hayakawa, Phys. Rev. C 37 (1988) 45.

CLUSTERING IN NUCLEI FROM N/Z=1 TO N/Z=2

Wilton N. Catford

Department of Physics
University of Surrey
Guildford GU2 5XH, UK

Abstract: The clustering of nucleons in nuclei is a widespread but elusive phenomenon for study. Here, we wish to highlight the variety of theoretical approaches, and demonstrate how they are mutually supportive and complementary. On the experimental side, we describe recent advances in the study of the classic cluster nucleus ^{24}Mg. Also, recent studies of clustering in nuclei approaching the neutron drip line are described. In the region near N/Z=2, both theory and experiment now suggest that multi-centre cluster structure is important, in particular for the very neutron rich beryllium isotopes.

1 INTRODUCTION: TYPES OF CLUSTERING

Rarely has a topic in nuclear physics attracted as much misunderstanding and debate as clustering behaviour in nuclei. Nuclear clusters are like Schrödinger's cats, being both in and out of existence until they are observed, and in some ways it is even more difficult to untangle their true behaviour from the process of observation. In some instances, such as the spectroscopic factors derived from α-particle transfer reactions, the quantitative data give a sure signature of clustering [1]. More often, the signature is less definite but the cluster model leads to a consistent and intuitive understanding of the data. Taken as a whole, the evidence is compelling that clustering effects are important in nuclei across

The Nucleus: New Physics for the New Millennium
Edited by Smit et al., Kluwer Academic / Plenum Publishers, New York, 2000.

the periodic table, and particularly amongst the lighter nuclei below calcium. The richness of the subject has ensured a steady flow of recent review articles [2, 3, 4, 5].

The existence of α-particle emission from nuclei suggests that pre-formed α-particles might exist inside them. From very early on, however, it has been known [6, 7] that it is overly naive to suppose that any nucleus is composed of real α-particles bound together. In a nucleus such as ^{16}O, say, the α-particles would overlap and effectively tear each other apart through the strong interaction. Still, as we shall see, the tendency towards clustering persists and it is driven by the tight binding of the maximally symmetric system of 2 protons and 2 neutrons coupled to $L = S = 0$; the $(0s_{1/2})^4$ shell model configuration. The Ikeda diagram [8] predicts widespread clustering structure, and in particular α-cluster structure, for states in nuclei at energies near the various cluster separation thresholds.

If we take ^{12}C as an example, it is appealing to imagine three interacting α–clusters dragging past each other in orbital motion, and exchanging nucleons through their strong interaction. For a given nucleon, its wavefunction is not confined to a single α–cluster but is shared between different clusters. Perhaps it is reasonable to think of this as a kind of Bose-Einstein condensate of α-particles. The overall A-body wavefunction must be antisymmetric, and it is the action of the antisymmetrization operator that effectively shares the wavefunction of each nucleon between the cluster mass centres. The average particle density distribution for the properly antisymmetrized intrinsic wavefunction of the nucleus continues to show the cluster-like appearance.

In some theoretical calculations, it is convenient to assume the existence of cluster centres *a priori*, and subsequently to deal with the antisymmetrization either exactly or according to an approximate prescription. Of course, the clusters need not in general be simply α-like. Further, models have recently been developed that allow the general unconstrained A-body problem to be solved using a variational method, wherein the individual nucleons are allowed to move independently with a two-body force acting between them and are not forced into clusters at all. The results of such calculations show a natural tendency towards clustering behaviour in certain cases and mean field behaviour in others. One exciting prediction is that the light neutron rich nuclei such as beryllium, boron and carbon exhibit multicentre cluster structure as neutrons are added out towards the drip-line. This clustering may be a significant factor in determining the structure of neutron haloes in nuclei such as ^{11}Be and ^{14}Be. In any case, the models of halo nuclei which assume a core plus one or two valence nucleons are examples of cluster models, where the core is treated as a cluster.

Clustering behaviour has been seen in a wide variety of guises [9, 10] across the nuclear chart, cf. fig.1, including:

- **Very light nuclei:** in cases such as ^{8}Be $(\alpha + \alpha)$ [11] and ^{7}Li $(\alpha + t)$ [12], or even ^{12}C (3α) [13], there are small subsystems of the nucleus that are quite tightly bound and there is a tendency for this structure to develop and to be competitive in energy with a mean field configuration. Additional neutrons can be added into molecular orbitals in ^{8}Be using the model of Abe [14] and account for the properties of ^{9}Be and ^{10}Be [15, 16],

- Magic core plus orbiting cluster: in nuclei just beyond strong shell closures, the valence nucleons can behave like a cluster orbiting outside of the closed shell nucleus, giving characteristic rotational spectra, which have been observed near the ^{16}O, ^{40}Ca, ^{90}Zn and ^{208}Pb doubly magic closures [17]; this is discussed in the next section,

- Normal core plus nucleon halo: when for example the final neutron is extremely weakly bound within the mean field of the nucleus, then its wavefunction naturally tends to extend substantially beyond the edges of the nuclear potential into the classically forbidden region, particularly if it has a low angular momentum [18]; an even more interesting case is with two nucleons in the halo, such as in Borromean systems [19],

- Large-scale clustering in low-lying levels: complex nuclei have been successfully modelled with the assumption that they comprise two mutually orbiting nuclei of comparable masses, with the predicted spectra showing remarkable agreement from the ground state all the way up to the separation energy [20, 21, 22]; this, despite the fact that the components must surely overlap,

- Large-scale clustering at high excitation: resonances seen at energies near the Coulomb barrier in heavy ion scattering [23, 24, 25], particularly between α-conjugate nuclei, seem to indicate molecular-like states in which the component nuclei orbit at a distance where they just graze against each other [25, 26]; this can also be studied in breakup reactions,

- Complete alpha-particle condensation: an extreme form of clustering behaviour is predicted in some models for $A = 4n$ nuclei, wherein the entire nucleus behaves not as one liquid drop but as a condensation into n separate α-like droplets [13, 27], and when extra excitation energy is added to these nuclei the configuration of the mass centres can unwrap and become spatially extended; the limiting case is represented by linear chain states, which have been predicted over a wide range of masses [28].

The various different models that have been developed are described in the reviews [2, 3, 4, 5]. One theoretical approach is to treat the clusters within the nucleus as real clusters and to solve the many-body scattering problem, for example in the three-body case by solving the three-body Fadeev equations [19]. This can be successful provided that the clusters avoid overlapping. Alternatively, the cluster structure can be assumed but the exchange of nucleons can be allowed, and the wavefunctions properly antisymmetrized. Such models using the resonating group method (RGM) [5, 29] and the generator coordinate

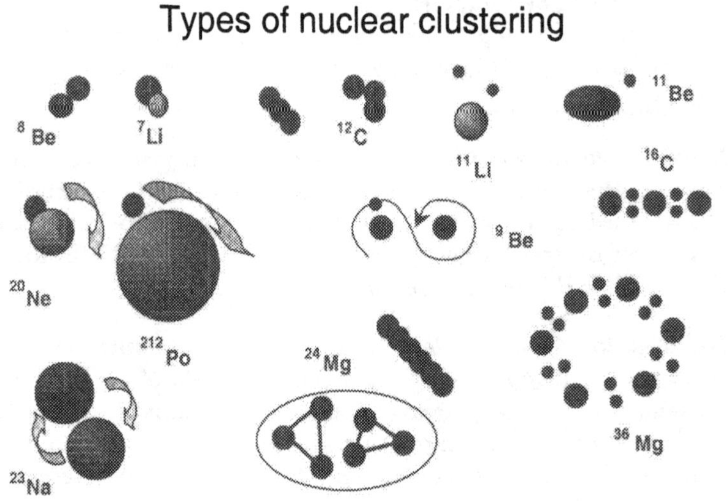

Types of nuclear clustering

Figure 1 A wide variety of different types of clustering behaviour have been identified in nuclei, from small clusters outside of a closed shell, to complete condensation into α-particles, to halo nucleons outside of a normal core.

method (GCM) [30] have been extensively applied to light nuclei and in particular for calculations of astrophysically important processes. Group theoretic approaches have been used to analyse the symmetry properties of cluster states [31]. Approximate two-body models have been applied to nuclei where a cluster orbits a core [32], with the orbital quantum numbers chosen in such a way as to approximately satisfy the Exclusion Principle by avoiding the overlap of the core and cluster wavefunctions. Another model that has been widely applied is the Brink-Bloch α-cluster model [13, 27], in which nucleons are assumed to be bound in harmonic potential wells forming α-like clusters with L=S=0. An effective two-body force acts between the clusters and a fully antisymmetrized A-body Slater determinant wavefunction is derived. The equilibrium positions of the mass centres are determined in a variational calculation. Time dependence can be introduced to this model to give the time-dependent cluster model (TDCM) [33]. Taking this one step further, individual unconstrained nucleons can be treated in a similar fashion with a two-body force acting between independently moving nucleons. This approach has been developed into models known as Fermionic molecular dynamics [34] and antisymmetrized molecular dynamics (AMD) [35]. The quantities that can be calculated in the various models vary from case to case, but energy levels, electromagnetic moments and transition densities have all been very successfully predicted. An exciting recent development in theory, via the AMD model, is the ability to predict systematically the persistence of clustering in systems that differ from the simple $A = 4n, T_z = 0$ α-like nuclei, and this will be discussed below.

2 REVIEW OF SOME KEY RESULTS FROM THE MODELS

2.1 *Positive and negative parity rotational bands*

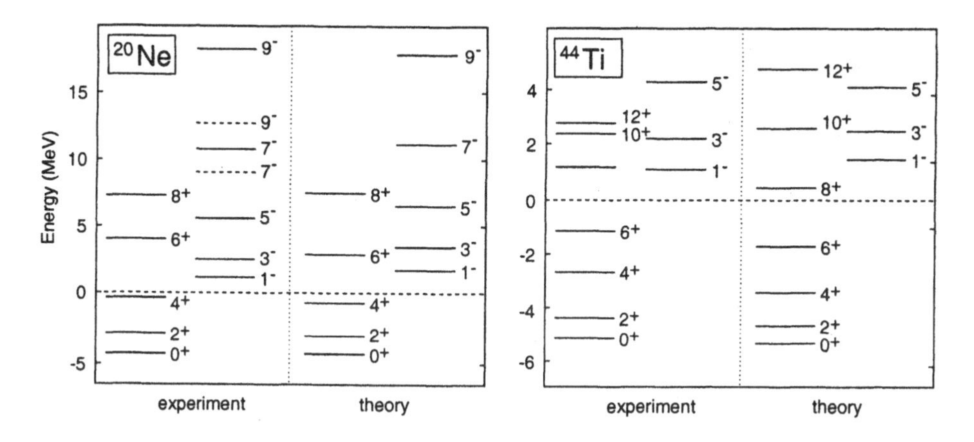

Figure 2 The energy offset between positive and negative parity states in rotational bands, shown here for the ground state bands in ^{20}Ne and ^{44}Ti, supports the cluster model.

In ^{20}Ne for example, there is a negative parity K=0 rotational band with the same moment of inertia as the positive parity ground state band, but displaced upwards in energy. The natural explanation for this in terms of the cluster model was first put forward by Horiuchi and Ikeda [36]. These bands correspond to an α-particle orbiting an ^{16}O core [37]. Assuming that it is possible to start with an α-particle on one side of the core and to pass continuously, via the radial motion of the cluster, to the situation with it on the opposite side of the core, then this implies a symmetric potential energy function that has a barrier at zero separation. The intrinsic solutions centred on the two minima must be properly symmetrized to obtain states of good parity, and the resultant states of different parity have different behaviour in the region of the central barrier. The negative parity state vanishes at zero but the positive parity state does not. The net effect is to shift negative parity states upwards in energy compared to the positive parity states. This behaviour is clearly seen in ^{20}Ne and also in ^{44}Ti, where an α-particle orbits a ^{40}Ca core (see fig. 2, [38]). In fact, if the barrier between the two reflection symmetric states (before parity projection) became infinitely large, then the probability of being at zero would be zero for the positive parity solution also, and there would be no shift between the different parity rotational states in that case. This closely parallels the situation for octupole deformation in nuclei, where intrinsic octupole states can be continuously transformed into their degenerate mirror images. In their review of intrinsic reflection asymmetry, Butler and Nazarewicz [39] point out that the *sd*-shell region provided the first evidence for reflection asymmetry in rotating nuclei, and they include a detailed bibliography for studies of ^{16}O, ^{18}O, ^{20}Ne, ^{24}Mg, ^{28}Si, ^{32}S, ^{40}Ca and ^{44}Ti, concentrating on the octupole aspects.

Another important point to note is that the level scheme of *core + cluster* nuclei is not the only property which is well reproduced by cluster model calculations. Using models such as that of Buck, Dover and Vary [32], electromagnetic transition densities can be calculated correctly, without the need for effective charges such as required in shell model calculations. These calculations have been extended successfully to include α-particle states above the ^{90}Zr and ^{208}Pb double shell closures [17]. Alpha-decay widths and low energy elastic scattering cross sections have been reproduced in a unified treatment for ^{20}Ne and ^{44}Ti [38]. The model can be extended successfully to nuclei in which exotic clusters (those which have been observed in radioactivity, such as ^{24}Ne [40]) orbit outside of a ^{208}Pb core. Buck *et al* successfully reproduced the systematics of exotic decay half-lives [41] and also the excitation energies for rotational bands (including Coriolis antistretching) and electromagnetic transition strengths for collective quadrupole and octupole transitions [42] across 19 even-even actinide nuclei.

2.2 The structure of ^{12}C

The nucleus ^{12}C has long been treated theoretically as a triangular arrangement of α-like structures [13]. Models that presuppose the existence of clusters, such as that of Brink [13], have recently received extra theoretical support from two different directions. In the first instance, the quantum mechanical three-body problem, including the Coulomb force, was solved for real α-particles using a hyperspherical expansion of the Fadeev equations [43]. A phenomenological three-body force was included to account for Pauli and polarization effects. With parameters chosen to reproduce the energy of the three-body (0_2^+, 7.65 MeV) scattering state, the width of the 0_2^+ and the energy of the bound 0_1^+ ground state were reasonably reproduced. The calculations predicted the most likely relative spatial configurations of the three α-particles in the scattering problem. The ground state was an almost pure equilateral triangle with sides of 3.0 fm, whereas the excited (unbound) state was a superposition of several configurations with the main one being a somewhat larger and flattened triangle. The ground state result is quite close to the geometry found in a GCM calculation which used an equilateral triangle with the length of the sides being a fitted parameter [44]. The (0_2^+, 7.65 MeV) state was once thought [45] to be an example of a linear α-particle chain state of the type postulated by Morinaga [46]. Later work showed that the $0_2^+ - 2_2^+$ energy gap was not consistent with a linear 3α linear chain structure [47] and a bent chain was proposed [48]. The three-body Fadeev calculations support this general interpretation although they imply a hinge angle of around 100° in the dominant configuration rather than the 150° favoured by earlier calculations [49].

Another recent theoretical treatment of ^{12}C is that of Kanada-En'yo [50]. She used the antisymmetrized molecular dynamics (AMD) model [2, 35], extended to perform the variational part of the calculation after projecting states of good parity and angular momentum. In these calculations, the relative geometry of the individual nucleons was not constrained at all *a priori*, i.e no clustering was imposed on the nucleus. The predicted spectrum of levels and their transition strengths show several advantages over cluster calculations, in particular for the relative properties of the 0_1^+ ground state and the 0_2^+ state. An analysis of the wavefunctions obtained [50] shows that the α-clusters are significantly dissociated in the ground state, but the results also support a well-developed 3α structure for other rotational bands.

2.3 Dissolution of alpha particles

As discussed above, ^{20}Ne displays a rotational band spectrum that indicates an ^{16}O+α structure. Recently, within the framework of the AMD model, it has become possible to study theoretically the degree of clustering in the different states of the band [35]. These new calculations support many of the earlier deductions [51] concerning reduced clustering in the ground state band towards its termination at spin 8^+. A cranking method was used to study states along the yrast line in ^{20}Ne, starting from the overall minimum energy solution for the 0^+ ground state. Nucleon density distributions were then calculated for the rotating intrinsic states before parity projection. For the lower states in the band, the structure was found to be prolate and approximately axially symmetric, consistent with an ^{16}O+α structure. Furthermore, the ^{16}O part of the mass density comprised a tetrahedral arrangement of four α-like mass centres arranged as predicted by conventional cluster models of ^{16}O. The negative parity states showed particularly well-developed clustering. For spins of 7^- and above, the ^{16}O+α structure tended to dissolve and other structures including oblate shapes and ^{12}C$\alpha\alpha$ cluster configurations began to appear. Although they have certain deficiencies, for example the negative parity energies are not well reproduced, the calculations give a fascinating insight into the evolution of clustering with excitation energy and spin. At the higher excitation energies they help to delineate the limits of applicability for cluster models. At the lower energies the AMD results strongly support the validity of assuming a cluster structure, and hence they support the predictions of other quantities that can be made most effectively using the cluster models.

2.4 Large-scale or di-nuclear clustering

Calculations using an α-cluster model for a large nucleus sometimes give solutions that show a tendency for the nucleus to be grouped into two or more substantial substructures. This generalises the behaviour seen for ^{20}Ne in the AMD. Indeed, the calculations of Marsh and Rae [52] using the Brink model [13] show that the ground state of ^{24}Mg can be viewed as two ^{12}C nuclei in juxtaposition. In a certain sense, therefore, it is not too surprising that the low-lying levels of nuclei such as ^{24}Mg can be modelled as two interacting ^{12}C nuclei. What is truly remarkable, however, is that such a model [21, 22] succeeds with the component nuclei taking on their free-space properties, even though they presumably overlap considerably.

In the approach of Buck *et al* [21], ^{24}Mg is modelled as a bound state of two ^{12}C nuclei which can be internally excited. The internal energy levels and the electromagnetic transition strengths between them are taken to be those for real, free ^{12}C nuclei. Coupled channels calculations are performed to deduce the level scheme up to a maximum energy given by the sum of the Coulomb barrier and the ^{12}C+^{12}C separation energy. Almost the entire experimentally known T=0 spectrum is reproduced, both for positive and negative parity states. The calculated electromagnetic strengths for quadrupole transitions between the low-lying positive parity states are also in good agreement with experimental values, without the need to introduce effective charges. The model has been extended to describe ^{23}Na in terms of ^{12}C and ^{11}B, each with their free-space properties [20]. Excellent agreement is achieved with experimental E2 and M1 transition strengths, and the model has also been extended successfully to predict polarization observables [53]. Whether such models have a physical significance beyond their computational convenience is not clear.

2.5 *Natural clustering tendency in neutron rich systems*

When additional neutrons are added to the ^{8}Be core of two α-clusters, molecular states in ^{9}Be and ^{10}Be are formed [15, 16, 54]. Similarly, protons can be added and account for structures in the nuclei ^{9}B and ^{10}B [16, 55]. These approaches can be extended to the more neutron rich isotopes of beryllium and boron [15, 56, 57] and to analogous 3α systems in the carbon isotopes [56]. Thus, Nature edges towards the speculative suggestion of Wilkinson [58] that complete rings of α-clusters might exist, like necklaces, held together by neutron pairs in covalent bonds between the clusters (cf. fig.1). For now, however, necklace states remain pure speculation and even simple chain states of α-clusters (which are predicted to exist at high excitations over a range of nuclei [28]) are not compellingly in evidence experimentally beyond ^{8}Be.

Von Oertzen [16, 56] has estimated the excitation energies of molecular chain states in neutron rich beryllium, boron, carbon and oxygen, where the α-cluster centres are bound together with covalent bonds in analogy with the well-known [15, 54] α: 2n :α states near 6 MeV in ^{10}Be. The chain states are expected to be at higher excitation energies in the more exotic nuclei [16, 56], although other clustering (e.g. ^{12}Be=^{6}He+^{6}He) could be anticipated at lower excitations.

But what of the wider significance of clustering in the lowest energy levels for light, neutron rich nuclei? An exciting development in the study of these nuclei has been the AMD predictions for ground state nucleon densities. Calculations have been performed for isotopic chains extending to the neutron dripline for ground states of lithium and beryllium [59, 60], boron [61] and carbon [62, 63] and have been summarised by Horiuchi [2]. As might be expected, a tendency towards clustering is predicted near the N=Z line and a tendency towards more mean-field behaviour near the magic number N=8. However, for N>8 the clustering appears to return and in some cases to be enhanced. For example, in boron isotopes the proton intrinsic matter distributions show marked and increasing Li-He clustering in the isotopes ^{15,17,19}B. Across isotopic chains, the

AMD calculations have been tested against measured ground state properties and they are found to reproduce well the radii, binding energies and magnetic moments, except in the case of known halo states [2]. Similarly, the excited state energy spectra and the electromagnetic transition strengths between levels are reproduced quite well, except for halo levels.

The structure of excited states, and the evolution of shape within a given nucleus have been studied in more detail in very recent work with the AMD model for ^{10}Be [64] and ^{12}C [50]. The AMD calculations highlight very clearly the differences in structure, for example between the ground state of ^{10}Be and the quartet of molecular states near 6 MeV [64]. They allow the coexistence of cluster structure and mean field structure to be investigated. Otherwise, in the explicitly molecular models [56, 57] the relationship between the lowest states and the excited molecular states requires careful interpretation.

3 RECENT EXPERIMENTAL RESULTS

3.1 Breakup of ^{24}Mg to ^{12}C+^{12}C

The breakup of excited ^{24}Mg, ^{28}Si and ^{32}S nuclei into ^{12}C and ^{16}O fragments, following nuclear excitation processes, has been studied in detail for more than a decade [3]. The widths of the ^{24}Mg breakup states are of order 100–150 keV [65] which is comparable to the expected statistical width [66]. On the other hand the breakup states and the resonances populated in ^{12}C + ^{12}C scattering near the barrier appear to be closely linked [65], and the scattering resonances have been shown to have enhanced partial widths for ^{12}C decay. The scattering resonances are also correlated between different reaction channels and they γ-decay to the ground state of ^{24}Mg [24].

For the breakup states, there has never been a direct partial width measurement. Recently, Freer *et al.* have attempted to obtain such information indirectly, by comparing the probabilities for the decay of ^{24}Mg states into the ^{12}C+^{12}C and ^{16}O+^{8}Be channels [67]. States in ^{24}Mg were populated in the reaction ^{12}C(^{16}O,α)^{24}Mg at several beam energies and the spins were measured using the angular correlations of the sequential breakup. The energy-spin systematics followed the grazing angular momentum for two ^{12}C nuclei (as expected for molecular states). However, it was also noted that the statistical decay probabilities for fission to two ground state ^{12}C nuclei follow similar systematics, peaking as the barrier penetration probabilities balance against the opening of competing channels (cf. Fig.3). In a quantitative analysis, it appeared that some states decayed to the two fission channels with the statistical ratio of probabilities, but that in addition there were several states that favoured the ^{12}C+^{12}C channel. The situation may have been particularly complicated because the breakup states were populated using the (^{16}O,α) reaction, which has a compound nuclear mechanism according to its angular distribution [68]. The mechanism in ^{12}C(^{24}Mg,^{24}Mg*)^{12}C or in transfer reactions may be more selective of true molecular states. Clearly, there is a lot more work still to be done in the measurement of partial decay widths for breakup states.

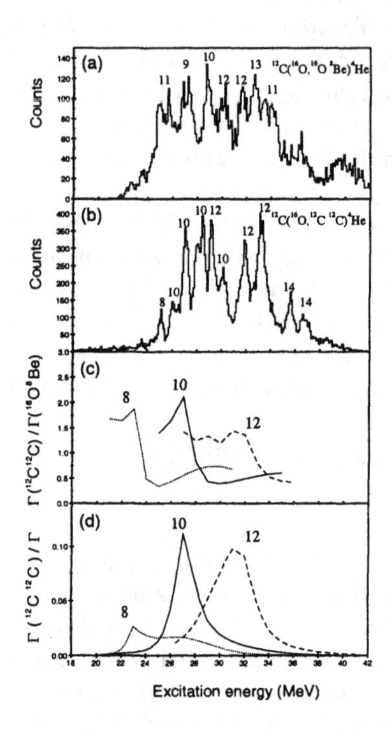

Figure 3 Montage of excitation spectra for breakup states seen in ^{24}Mg depopulated by (a) ^{16}O^{8}Be or (b) ^{12}C^{12}C decay. Also shown, statistical partial width predictions as a function of excitation energy and spin, for (c) ^{12}C^{12}C relative to ^{16}O^{8}Be and (d) for ^{12}C^{12}C as a fraction of the total width.

3.2 Resonances in mutual 3^- *scattering of* $^{12}C+^{12}C$

Chappell *et al.* [71] have recently measured excitation functions and angular correlations for ^{12}C+^{12}C scattering leading to mutual excitation of the ^{12}C*(9.63 MeV; 3_1^-) state. In the Brink-Bloch model, both the ground state and the 3_1^- state have the same intrinsic triangular shape, and the 3_1^- state has three units of angular momentum around its three-fold symmetry axis. The new mutual 3_1^- data complement earlier measurements of the mutual 0_2^+ (7.65 MeV) channel [72] and the $0_2^+3_1^-$ channel [73, 74] and extend over the energy range $E_{\rm cm}$ = 30–45 MeV.

The 3_1^- state decays by α-emission to produce a jet of three α-particles, and these must be reconstructed kinematically in the analysis to flag 3_1^- production. In fact, the experiment needed to measure five-fold coincidences and reconstruct the sixth α-particle prior to the reconstructions of the triple jets. In the excitation function, prominent resonances were observed at $E_{\rm cm}$ = 32.5, 37.5 and 43.0 MeV. The lowest of these corresponds to the energy of the fa-

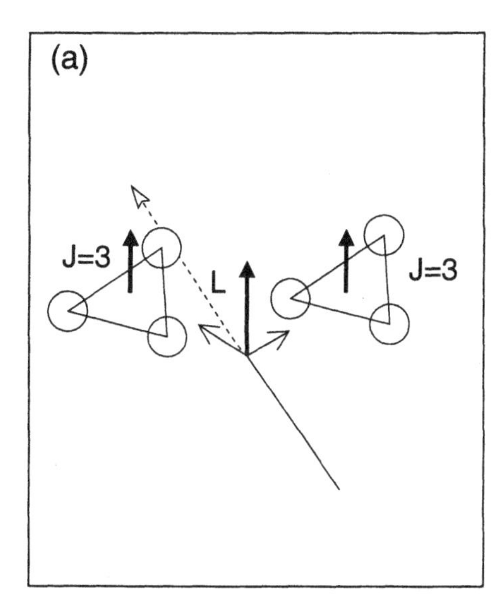

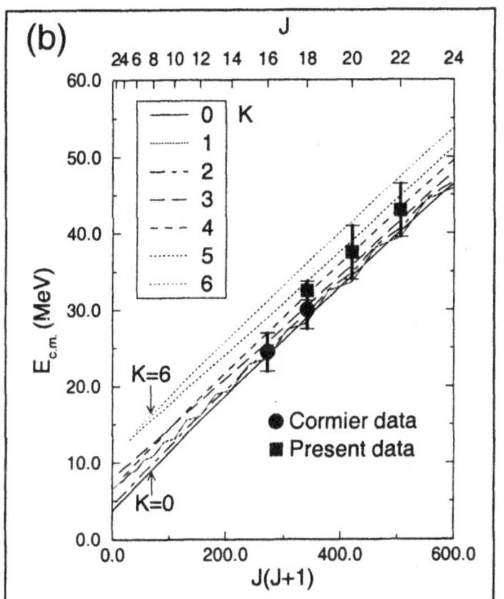

Figure 4 Resonances seen in $^{12}C+^{12}C$ scattering to the mutual 3_1^- channel: (a) the angular correlations indicate fully aligned angular momentum in the final state as shown schematically, (b) the energy-spin systematics are consistent with α-cluster predictions for a very deformed triaxial state in ^{24}Mg.

mous resonance in $0_2^+0_2^+$ scattering [72] that was once believed to correspond to a shape eigenstate resonance in the form of a 6α linear chain [75], but the $3_1^-3_1^-$ channel is found to be much stronger [73]. In view of the triangular structure of the 3_1^- state, the 6α chain hypothesis is untenable. It can be supposed that the strength in the $0_2^+0_2^+$ channel arises through the hinge angle in the 3α 0_2^+ state (see earlier), which gives it a partially triangular structure.

The angular correlation analysis for the $3_1^-3_1^-$ scattering reveals that the final state is fully aligned in angular momentum, as shown schematically in fig. 4(a). The dominant spins for the three resonances appear to be 18, 20 and 22 respectively. These high spins cannot be supported by ^{24}Mg in its normal deformation [71], so the absorptive scattering potential is in principal weaker (since only direct reactions and highly deformed resonances contribute).

Chappell *et al.* discuss their results in the context of the Band Crossing Model [76] using reaction theory, and they use the Brink-Bloch α-cluster model to give a structural interpretation. The Brink model predicts [52] a strongly deformed triaxial state with $^{12}C^{12}C$ structure, which corresponds to that labelled F1 by Flocard *et al.* [77]. When this rotates, it gives rise to a number of bands with different K-values which are close in energy but have very little K-mixing. It was speculated [71] that the $3_1^-3_1^-$ resonances (see fig. 4(b)) may be associated with the K=6 component, and that they extend the sequence of broad resonances first observed by Cormier *et al.* [78] in single and mutual 2_1^+ scattering.

3.3 *Search for gamma-ray decay of breakup states in ^{24}Mg*

The ^{12}C+^{12}C states seen in breakup and resonance reactions appear to show rotational $J(J+1)$ systematics. This could be a genuine structural effect or it could be due to some kinematic selection. To distinguish these possibilities, the observation of decay γ-rays connecting the states would be useful, but this is exceptionally difficult. One experiment has so far succeeded in obtaining results [69]. A well known 10^+ resonance at $E_{\text{cm}} = 16.45$ MeV was populated in ^{12}C+^{12}C scattering. Particles were detected in coincidence using two position sensitive detectors to record their energy and angle. The complete binary kinematics allowed the masses of the particles in the exit channel to be determined, and the ^{12}C+^{12}C channel could be selected. An array of 74 BaF_2 detectors (Château de Cristal) recorded coincident γ-rays. The reaction Q-value was reconstructed and the missing kinetic energy was compared with the γ-ray energy. If the resonance γ-decayed to a lower state in ^{24}Mg, that subsequently decayed to $^{12}C_{gs}+^{12}C_{gs}$, then the compared energies should be equal. The principal complication and source of background comes from events in which the breakup precedes the γ-decay, that is when the resonance decays to one (or even two) excited ^{12}C fragments. This is dominated by the 4.43 MeV, 2_1^+ channel. No simple discrimination of these events, other than by γ-ray energy, can be made. The result from the experiment was quoted as $\Gamma_\gamma/\Gamma = 1.2 \times 10^{-5}$ which, with certain assumptions, corresponds to a transition strength of order 100 W.u. These numbers are comparable to the predictions for a molecular band [69, 70], but because of the low statistics and the background uncertainties it seems most appropriate to interpret them as upper limits. More experiments are definitely required.

3.4 *Two-centre effects in neutron-rich Be isotopes*

The beryllium isotopes, as discussed above, are predicted to exhibit a well developed molecular structure in which valence neutrons bind together two helium-like centres [56, 64]. The molecular states are expected near the threshold for ^{4}He and ^{6}He decay, consistent with the Ikeda diagram [8]. Recently, experimental evidence in support of these predictions has been obtained in a measurement of p(^{12}Be,^{12}Be*)p using a radioactive beam [79]. A beam of 378 MeV ^{12}Be ions was produced at the GANIL laboratory using the LISE3 spectrometer. The fragmentation products from a primary ^{18}O beam were purified to produce a beam of 95% ^{12}Be at 2×10^4 pps. Fragments from the binary breakup of ^{12}Be* were identified by Z and A in E.ΔE telescopes. Measurements of the angles and energies allowed the excitation energy of the parent ^{12}Be to be reconstructed. Evidence was obtained for both ^{6}He^{6}He and α^8He breakup, beginning at an excitation energy just above threshold (close to 10 MeV in each case). The ^{6}He^{6}He spectrum is less complicated because the identical spin zero particles restrict the states to even spin and parity. Unfortunately, no partial width measurements were possible. Remarkably, the relative angular distributions of the ^{6}He^{6}He breakup fragments could be measured sufficiently accurately to suggest spin assignments for several levels. The result was an approximate rotational sequence (fig. 5, [79]) reminiscent of the results for $^{12}C^{12}C$ breakup of ^{24}Mg [80]. The moment of inertia is consistent with two touching ^{6}He spheres and greatly exceeds that for a spherical ^{12}Be nucleus. Now that there is experimental evidence to support the theoretical suggestions of clustering in the neutron rich beryllium isotopes, it is likely that clustering and two-centre effects must be taken into account in all studies of this region. This is relevant in view of the recent measurements of p(^{11}Be,^{10}Be)d spectroscopic factors with a radioactive beam [81], which represent the beginning of transfer reaction spectroscopy far from stability.

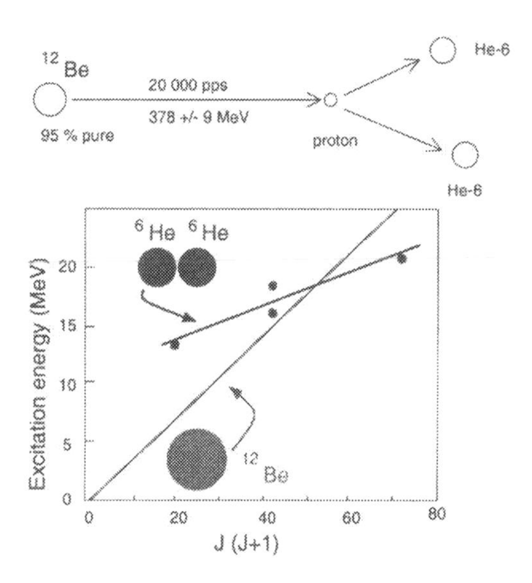

Figure 5 Inelastic scattering of a radioactive beam of ^{12}Be has shown breakup of ^{12}Be into $^{6}He^{6}He$, and the spins of the breakup states are consistent with a rotational band for two touching ^{6}He spheres. The steeper trajectory for a spherical ^{12}Be is shown for comparison.

4 OUTLOOK

Nuclear level schemes reveal an underlying cluster structure in many circumstances. One of the most intriguing continues to be the large-scale structure epitomised by ^{24}Mg as $^{12}C+^{12}C$. Evidence exists that some excited states in ^{24}Mg have a preference (beyond the statistical probability) for decay into two ^{12}C fragments, but the interpretations are open to ambiguities. A key requirement is for partial decay widths such as $\Gamma(^{12}C)/\Gamma$ to be measured for breakup states. The recent experiments using $^{12}C(^{16}O,^{12}C^{12}C)\alpha$ and $^{12}C(^{12}C,^{12}C\gamma)^{12}C$ give some indication of how this might be done. Meanwhile, as experimental techniques advance, $^{12}C+^{12}C$ scattering continues to show remarkable behaviour such as the fully aligned angular momentum in mutual 3_1^- scattering. The recent $3_1^- 3_1^-$ experiment probed ^{24}Mg at high spins, beyond the normal fission limit, in a regime where less ambiguities may exist owing to the lower level density.

Probably the most exciting prospects in cluster studies are amongst the light, very neutron rich nuclei. The AMD calculations have finally taken clustering away from magic shells and the N=Z line, and away from cluster models themselves. The importance of clustering up until now has largely been in the computational power that it provides when the clustering symmetry in the structure is recognised in the formulation of the model. What is somewhat different about the AMD work is that it implies, with no *a priori* assumptions, that it is necessary to take clustering into account in order to get any proper understanding of the structure of light neutron rich nuclei, even in their ground states. The recent results for the ^{6}He^{6}He breakup of ^{12}Be point the way to many exciting experiments in the future, using radioactive beams to explore this region far from stability in a way that was not possible with stable beams.

Acknowledgments

The work described here derives from the author's work over a number of years with the CHARISSA collaboration, and their support and contributions are gratefully acknowledged. Thanks also to Sharpey, for encouragement back in the thesis days.

References

[1] A. Arima and S. Kubono, *in* Treatise on Heavy-Ion Science, Vol. 1, ed. D.A. Bromley (Plenum, New York, 1984) p.617

[2] H. Horiuchi, *in* Correlations and Clustering Phenomena in Subatomic Physics, ed. M. Harakeh *et al.*, Plenum Press (New York) 1997, pp.29-51

[3] M. Freer and A.C. Merchant, J. Phys. G23 (1997) 261

[4] A.H. Wuosmaa *et al.*, Annu. Rev. Nucl. Part. Sci. 45 (1995) 89

[5] K. Langanke, Adv. Nucl. Phys. 21 (1994) 85

[6] J.A. Wheeler, Phys. Rev. 59 (1941) 16 and 27

[7] W. Wefelmeier, Z. Phys. 107 (1937) 332; Naturwiss. 25 (1937) 525

[8] K. Ikeda *et al.*, Prog. Th. Phys. Suppl. Extra Number (1968) 464

[9] R.K. Sheline and K. Wildermuth, Nucl. Phys. 21 (1960) 196

[10] J. Cseh and W. Scheid, J. Phys. G18 (1992) 1419

[11] J.A. Wheeler, Phys. Rev. 52 (1937) 1083,1107

[12] B. Buck *et al.*, J. Phys. G11 (1985) L11; G14 (1988) L211

[13] D.M. Brink, *in* Proc. Int. School of Physics "Enrico Fermi", course XXXVI, Varenna, 1965, ed. C. Bloch (Academic Press, New York, 1966) p.247

[14] Y. Abe, J. Hiura and H. Tanaka, Prog. Theor. Phys. 49 (1973) 800

[15] M. Seya, M. Kohno and S. Nagata, Prog. Theor. Phys. 65 (1981) 205

[16] W. von Oertzen, Z. Phys. **A357** (1997) 355

[17] B. Buck, A.C. Merchant and S.M. Perez, Phys. Rev. **C51** (1995) 559

[18] K. Riisager, Rev. Mod. Phys. 66 (1994) 1105

[19] M.V. Zhukov *et al.*, Phys. Rep. 231 (1993) 151

[20] A. Kabir and B. Buck, Nucl. Phys. **A518** (1990) 449

[21] B. Buck, P.D.B. Hopkins, A.C. Merchant, Nucl. Phys. **A513** (1990) 75

[22] R. Baldock, B. Buck and J.A. Rubio, J. Phys. **G12** (1986) L29

[23] D.A. Bromley *et al.*, Phys. Rev. Lett. 4 (1960) 365 and 515

[24] K.A. Erb and D.A. Bromley, *in* Treatise on Heavy-Ion Science, Vol. 3, ed. D.A. Bromley (Plenum, New York, 1985) p.201

[25] N. Cindro, Riv. N. Cimento 4 (1981) No.6; Ann. Phys. Fr. 13 (1988) 289

[26] H. Feshbach, Proc. European conf. on nuclear physics with heavy ions, Caen, France, 1976. J. de Physique (Colloque) 37 (1976) C5:177

[27] W.D.M. Rae, Int. J. Mod. Phys. **A3** (1988) 1343

[28] A.C. Merchant and W.D.M. Rae, Nucl. Phys. **A549** (1992) 431

[29] K. Wildermuth and Y.C. Yang, A Unified Theory of the Nucleus, (Academic Press, New York, 1977)

[30] P. Descouvemont *et al.*, Nucl. Phys. **A605** (1996) 160

[31] J. Cseh and W. Scheid, J. Phys. **G18** (1992) 1419

[32] B. Buck, C.B. Dover and J.P. Vary, Phys. Rev. **C11** (1975) 1803

[33] W. Bauhoff *et al.*, Phys. Rev. **C32** (1985) 150

[34] H. Feldmeier, Nucl. Phys. **A515** (1990) 147

[35] Y. Kanada-En'yo and H. Horiuchi, Prog. Theor. Phys. 93 (1995) 115

[36] H. Horiuchi and K. Ikeda, Prog. Theor. Phys. 40 (1968) 277

[37] K. Wildermuth, T. Kenellopoulos, Nucl. Phys. 7 (1958) 150; 9 (1958) 449

[38] B. Buck *et al.*, Phys. Rev. **C52** (1995) 1840

[39] P.A. Butler and W. Nazarewicz, Rev. Mod. Phys. 68 (1996) 349

[40] H.J. Rose and G.A. Jones, Nature 307 (1984) 245

[41] B. Buck, A.C. Merchant and S.M. Perez, Phys. Rev. Lett. 76 (1996) 380

[42] B. Buck, A.C. Merchant and S.M. Perez, Phys. Rev. **C58** (1998) 2049

[43] D.V. Fedorov and A.S. Jensen, Phys. Lett. **B389** (1996) 631

[44] M. Dufour and P. Descouvemont, Nucl. Phys. **A605** (1996) 160

[45] H. Morinaga, Phys. Lett. 21 (1966) 78

[46] H. Morinaga, Phys. Rev. 101 (1956) 254

[47] Y. Fujiwara*et al.*, Prog. Th. Phys. Suppl. 68 (1980) 29

[48] H. Horiuchi, K. Ikeda and Y. Suzuki, Prog. Th. Phys. Suppl. 52 (1972) 89

[49] H. Friedrich, L. Satpathy and A. Weiguny, Phys. Lett. **B36** (1971) 189

[50] Y. Kanada-En'yo, Phys. Rev. Lett. 81 (1998) 5291

[51] A. Arima, H. Horiuchi, K. Kubodera and N. Takigawa, Adv. Nucl. Phys. 5 (1972) 345; T. Tomoda and A. Arima, Nucl. Phys. **A303** (1978) 217

[52] S. Marsh and W.D.M. Rae, Phys. Lett. **B153** (1985) 21

[53] A. Kabir and R.C. Johnson, J. Phys. **G18** (1992) 1967

[54] S. Okabe and Y. Abe, Prog. Theor. Phys. 61 (1979) 1049

[55] H. Furutani *et al.*, Prog. Theor. Phys. Suppl. 68 (1980) 193

[56] W. von Oertzen, Z. Phys. **A354** (1996) 37

[57] W. von Oertzen, Il Nuovo Cimento **110A** (1997) 895

[58] D.H. Wilkinson, Nucl. Phys. **A452** (1986) 296

[59] Y. Kanada-En'yo, H. Horiuchi and A. Ono, Phys. Rev. **C52** (1995) 628

[60] A. Doté, H. Horiuchi and Y. Kanada-En'yo, Phys. Rev. **C56** (1997) 1844

[61] Y. Kanada-En'yo and H. Horiuchi, Phys. Rev. **C52** (1995) 647

[62] Y. Kanada-En'yo and H. Horiuchi, Phys. Rev. **C55** (1997) 2860

[63] Y. Kanada-En'yo and H. Horiuchi, Phys. Rev. **C54** (1996) R468

[64] Y. Kanada-En'yo, H. Horiuchi and A. Doté, J. Phys. **G24** (1998) 1499

[65] N. Curtis *et al.*, Phys. Rev. **C51** (1995) 1554

[66] D. Shapira, R.G. Stokstad and D.A. Bromley, Phys. Rev. **C10** (1974) 1063

[67] M. Freer *et al.*, Phys. Rev. **C57** (1998) 1277

[68] M. Freer *et al.*, Phys. Rev. **C51** (1995) 3174

[69] A. Elanique, Thèse, IReS Strasbourg (1997), Report IReS 97-13

[70] H. Chandra and U. Mosel, Nucl. Phys. **A298** (1978) 151

[71] S.P.G. Chappell *et al.*, Phys. Lett. **444B** (1998) 260

[72] A.H. Wuosmaa *et al.*, Phys. Rev. Lett. 68 (1992) 1295

[73] S.P.G. Chappell *et al.*, Phys. Rev. **C51** (1995) 695

[74] A.H. Wuosmaa *et al.*, Phys. Rev. **C54** (1996) 2463

[75] W.D.M. Rae, A.C. Merchant, B. Buck, Phys. Rev. Lett. 69 (1992) 3709

[76] Y. Kondo, Y. Abe and T. Matsuse, Phys. Rev. **C19** (1979) 1356

[77] H. Flocard *et al.*, Prog. Theor. Phys. 72 (1984) 1000

[78] T.M.. Cormier *et al.*, Phys. Rev. Lett. 40 (1978) 924, 38 (1977) 940

[79] M. Freer *et al.*, Phys. Rev. Lett., 82 (1999) 1383

[80] B.R. Fulton *et al.*, Phys. Lett. **B267** (1991) 325

[81] J.S. Winfield *et al.*, J. Phys. **G25** (1999) *in press*

THE DEVELOPMENT AND APPLICATIONS OF SCINTILLATOR-PHOTODIODE DETECTORS

N.M. Clarke

School of Physics and Astronomy, University of Birmingham, B15 2TT, U.K.

Abstract: The advantages of scintillation detectors using photodiodes are compared with those using photomultipliers, and a brief history is presented of their development in the 1960s. These devices have proved valuable in the study of clustering in nuclei via breakup reactions, where the excitation energy in the nucleus can be reconstructed with better resolution than that of the scintillation detector. Combinations of gas filled ion chambers, silicon position sensitive detectors and scintillator-photodiode detectors have been constructed and used recently to observe ^{6}He clustering in states of ^{12}Be.

1 INTRODUCTION

The modern scintillation counter using the photomultiplier(PM) tube was developed by Curran and Baker in 1944 [1]. Despite its poor resolution compared with semiconductor detectors, it continues to be the choice for many industrial and medical applications, and remains in widespread use as anti-Compton shields and for high energy calorimeters. The advantages of scintillators are their availability in large sizes, versatility of shape, short decay times and radiation hardness; many have high density and atomic number [2].

The spectral response of PM tubes is well matched to many scintillators, though the quantum efficiency is poor (20%); the counters suffer from poor stability because of the high gain, and require bulky shields against magnetic

The Nucleus: New Physics for the New Millennium
Edited by Smit et al., Kluwer Academic / Plenum Publishers, New York, 2000.

fields. The photodiode (PD) has a high quantum efficiency (70%) for CsI(Tl) scintillator, and is insensitive to magnetic fields. The PD is more noisy than the PM tube and has no gain, so needs a low-noise high-gain preamplifer, but is very stable, requiring low voltage supplies and needs minimal containment. These aspects are important where packing density, dead space, and transparency to gamma rays and/or neutrons are significant design criteria.

Fig. 1 shows a schematic of a scintillator-photodiode detector (SPD) used by the CHARISSA collaboration and manufactured by Scionix (Netherlands). The detecting volume of the CsI(Tl) crystal is 50x50 mm area by 40 mm deep, capable of stopping ions of $120(Z^2/A)^{0.6}$ MeV per nucleon. The crystal is tapered to a Hamamatsu PD of 18x18mm area. A low noise preamplifier dissipating 180mW is close coupled to the PD; both eV-Products and Hamamatsu devices are used, depending upon the energies of the ions.

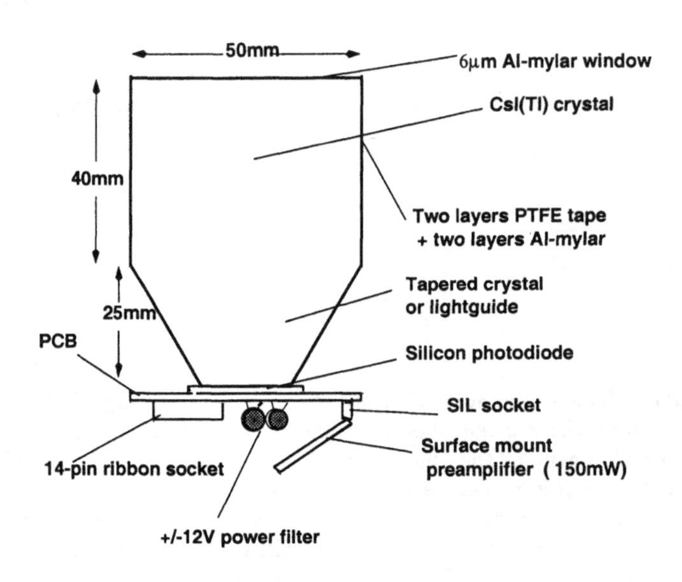

Figure 1 CHARISSA scintillator-photodiode detector

2 PIONEERING RESEARCH ON SPDS 1960-1970

The first SPD detectors used silicon p-n junction diodes made for charged particle detection. At Chicago, Tuzzolino et al. 1962 [3] examined the spectral response of silicon PDs and their gold electrodes, and used 70MeV protons to test an SPD with small crystals of CsI(Tl) and NaI(Tl), finding a resolution Γ of 22%. Further work by Fan, 1964 [4] led to the first application of an SPD in a cosmic ray detector for NASA. Variants of this design flew on IMP and OGO satellites from 1964 to 1974 [5] producing much data on nuclear abundances.

At Harwell, Blamires [6] tested Si-PDs with various electrodes and coupled these to scintillators of ZnS, CsI(Tl) and NaI(Tl), obtaining Γ=8% for the latter with 50MeV protons. Significant progress was made by Bateman [7], who tested the effects of windows and silicone potting on PDs. With a low noise preamplifier, he achieved Γ= 1.96% for a CsI(Tl) -PD detector using 50MeV protons, and showed that Γ was dominated by the noise. Bateman and Ozsan later reduced the noise using a cryogenic PD to obtain Γ=0.85% with 53MeV protons [8]. Kiel describes an SPD which showed noise levels of 200keV and Γ=18.6% for ^{137}Cs γ-rays [9].

3 APPLICATIONS OF SPDS BY THE CHARISSA COLLABORATION

The study of clustering in light nuclei uses the techniques of resonant particle spectroscopy (RPS) [10], with both stable and radioactive nuclear beams (RNB). The fragments are detected with combinations of gas filled ion chamber, silicon position sensitive detector (Si-PSD) and SPD. The Si-PSD yields position signals, and heavy(light) ions are identified via dE signals from the gas(Si-PSD), and E signals from the Si-PSD(SPD) . The energies and angles enable the reconstruction of the relative energy Er between the fragments and hence the excitation energy Ex in the original nucleus. Angular correlations can determine the spins of the states and hence the moment of inertia of the nucleus [11].

Two aspects of SPDs must be considered for RPS; the resolution of the SPD and the light output which is a function of the mass, charge and energy of the ion.

For the Charissa SPDs, the resolution is about 1.5% except for light ions, but the RPS technique produces a compression of energy spread and improved resolution in Er. The value of Er is found from the fragment energies and masses E1,M1 and E2,M2 with relative angle θr :-

$$Er = [E1M2 + E2M1 - 2(E1E2M1M2)^{1/2}cos\theta r]/(M1 + M2) \qquad (1)$$

For symmetric breakup with M=M1=M2, E=E1=E2, the relative energy resolution δEr depends on the angular resolution $\delta\theta$r and the total energy resolution δE as :-

$$\delta Er = E\theta r\delta\theta r \qquad (2)$$

and

$$\delta Er = (\theta r)^2\delta E/4 \qquad (3)$$

Typical values for an RNB experiment are:- M=6, E=200MeV, with θr =0.2 and $\delta\theta$r = 2mm/250mm (mostly beam position) and with δE of about 15MeV (mostly beam energy). These values yield 320keV for eq. [2] and 150keV for eq. [3], demonstrating the reduction in δEr.

A variation of light output with stopping power dE/dx proposed by Birks [12] leads to:-

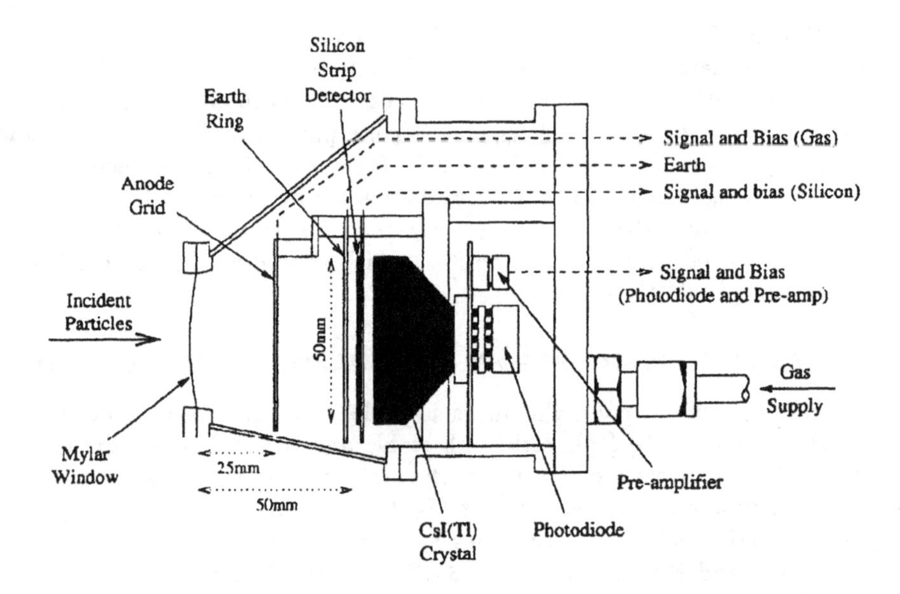

Figure 2 Gas Silicon Scintillator Hybrid Detector

$$L(A, Z, E) = a0 + a1[E - a3AZ^2 ln(1 + E/(a2AZ^2))] \tag{4}$$

where a0-a3 are calibrated using various beams and energies. Horn et al. [13] found a2=a3=0.326 using a CsI-PD whilst Fomichev et al. [14] found a2=0.593 and a3=0.408 using a CsI-PM combination. A comparison of our SPD for ^{4}He, ^{7}Li, ^{12}Be and ^{13}B ions at GANIL with Fomichev's values showed an increase of 10% in a3 improves the fit for ^{13}B ions [15].

At the Australian National University (ANU) a compact SPD 50x50x25mm thick was chosen as the final element for two detector systems. Three hybrid detectors with a gas counter, a 16-strip Si-PSD and an SPD have been constructed (fig. 2) to study the four-body breakup reaction ^{12}C(^{24}Mg, ^{16}O ^{12}C ^{4}He)^{4}He by reconstructing the excited states in ^{16}O [16]; an α and a ^{12}C can be detected simultaneously in separate strips.
The MEGHA array [17] (fig. 3) comprises 5 modules arranged in a Maltese cross; each contains 8 or 9 telescopes consisting of a gas counter, sheet resistive Si-PSD [18], and an SPD; where light ions penetrate to the SPD, the signals are used either as a veto, or to include the ion in the event.

The CHARISSA array at GANIL, France, comprises up to 10 telescopes using either sheet Si-PSDs or strip Si-PSDs, combined with the SPD of fig.1. Freer et al. [19] were able to achieve Γ=700keV for states in ^{12}Be breaking up into ^{6}He +^{6}He ions (fig. 4), despite an 18MeV spread in the beam energy of 378MeV and Γ=1.5% for the SPD. The transparency of the SPD array for neutrons was essential for a study of ^{14}Be using 99 neutron detectors of the DEMON array [20]; a Si-PSD + SPD telescope at 0° identified fragments of ^{10,11,12}Be in coincidence with neutrons. The angular distribution of the neu-

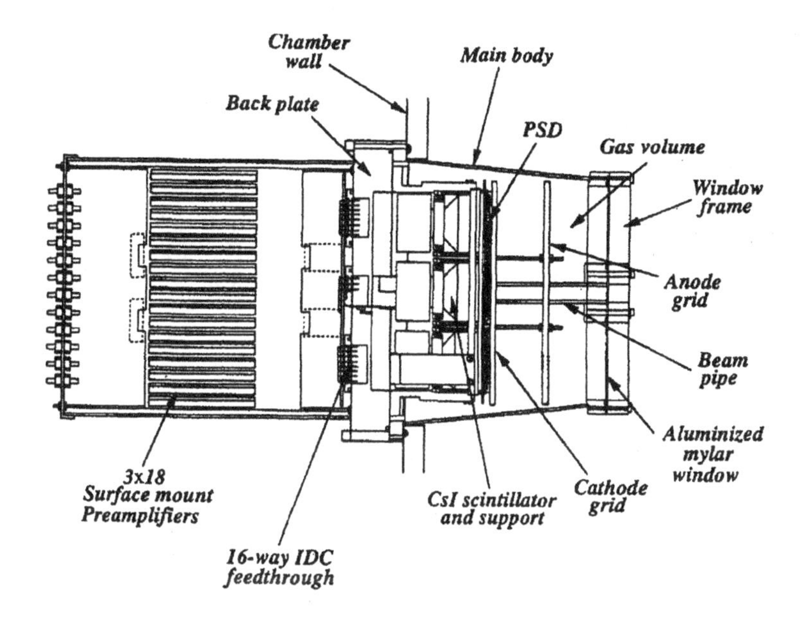

Figure 3 Section through a MEGHA module

trons using C and Pb targets gave momentum distribution widths of 80MeV/c, confirming the halo nature of ^{14}Be.

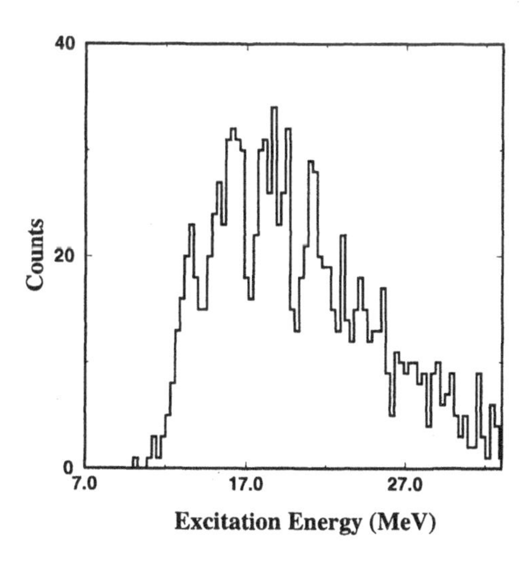

Figure 4 Excitation energy spectrum of states in ^{12}Be

4 SUMMARY

SPDs were first applied to space research, and are now widely used in calorimetry [21], and as particle balls for Ge arrays, and as heavy-ion detectors, where their minimal containment is ideal for neutron detection. SPDs show promise for low level β-spectroscopy, in dark matter research and in combined PET arrays and MRI scanners where their immunity to B fields is essential. New materials for scintillators and photodiodes are available, whilst avalanche and hybrid PDs and silicon drift chambers promise a wealth of new devices beyond the year 2000.

References

[1] S.C. Curran and W.R.Baker, U.S. Atomic Energy Comm. Rep. M.D.D.C.-1296 (1944)

[2] R. L. Heath, R. Hofstadter and E. B. Hughes, Nucl. Instr. Meth. 162(1979) 431; F. D. Brooks, Ibid, 162 (1979) 477

[3] A. J. Tuzzolino, E. L. Hubbard, M. A. Perkins, C. Y. Fan, J. Applied Phys. 33, 1 (1962) 148

[4] C. Y. Fan, Rev. Sci. Instr. 35, 2 (1964) 158

[5] G. M. Comstock, C. Y. Fan, J. A. Simpson, Astrophys. Journal. 146,1 (1966) 51

[6] N. G. Blamires, Nucl. Instr. Meth. 24 (1963)

[7] J. E. Bateman, Nucl. Instr. Meth. 67 (1969) 93, Ibid 71 (1969) 256, 261, 269

[8] J. E. Bateman and F. E. Ozsan, Nucl. Instr. Meth. 108 (1973) 403

[9] G. Keil, Nucl. Instr. Meth. 66(1968) 167

[10] B. R. Fulton and W. D. M. Rae, J.Phys. G: Nucl. Part. Phys. 16 (1990) 333

[11] M. Freer, Nucl. Instr. Meth. Phys. Res. A383 (1996) 463

[12] J. B. Birks, The theory and practice of scintillation counting (Macmillan, New York 1964)

[13] D. Horn et al., Nucl. Instr. Meth. Phys. Res. A320 (1992) 273

[14] A.S. Fomichev et al., Nucl. Instr. Meth. Phys. Res. A344(1994) 378

[15] K. L. Jones, University of Surrey, private communication

[16] M. Shawcross, University of Surrey, private communication

[17] R.L. Cowin et al., Nucl. Instr. Meth. A423 (1999) 75

[18] R.L.Cowin and D L Watson, Nucl. Instr. Meth. A399 (1997) 365

[19] M Freer et al. Phys. Rev. Lett. 82 (1999) 1383

[20] G. Bizard et al., Proc. Int. Conf. New Nucl. Phys. Techn., Ierapetra, Crete 1991, World Scientific p177

[21] E Aker at. al., Nucl. Instr. Meth. Phys. Res. A321(1992)69

PRESENT STATUS OF NEMO 3 : A DOUBLE-BETA DECAY EXPERIMENT WITHOUT ANTINEUTRINO EMISSION

Danielle DASSIE[1] and NEMO Collaboration*

*CEN-Bordeaux-Gradignan[1], France ; CFR-Gif/Yvette, France ; IReS-Strasbourg, France ; Department of Physics-Jyvaskyla, Finland ; FNSPE-Prague, Czech Republic ; INEEL-Idaho Falls, USA ; ITEP-Moscow, Russia ; JINR-Dubna, Russia ; LAL-Or

Abstract: The physics goals of double beta decay experiments are recalled. Various experimental aspects and the expected performances of NEMO 3 are presented.

1 INTRODUCTION

Oscillations of atmospheric neutrinos observed in the Superkamiokande experiment [1] made the problem of neutrino masses and neutrino mixing enter an era of new actuality. An understanding of the physical origin of the masses will however require further experiments. The neutrinoless double beta decay process $(\beta\beta)_{0\nu}$, plays a privileged role in the investigation of the nature of the neutrino and its existence would be a probe for physics beyond the standard model of electro-weak interactions. Neutrinoless double beta decay occurs only when the neutrino is of Majorana type with a finite mass. Besides neutrino mass, the potential of double beta decay experiments also yields information on right-handed W bosons, Majorons, SUSY models etc.

The present contribution will report on neutrino studies by measuring double beta decays with the detector NEMO 3. The NEMO (Neutrino Experiment with MOlybdenum) experiment will be able to study $(\beta\beta)_{0\nu}$ decays of nuclei

The Nucleus: New Physics for the New Millennium
Edited by Smit et al., Kluwer Academic / Plenum Publishers, New York, 2000.

with half-lifes up to 10^{25} years. Using relevant nuclear matrix elements this corresponds to an effective Majorana neutrino mass of the order of 0.1-0.3 eV. The detector is under construction and will be completed by the end of 1999.

2 DOUBLE BETA DECAY

Double beta decay is a transition between nuclei with the same mass number and differing by two units in protons. It is potentially observable when single beta decay to the intermediate nucleus is forbidden by energy conservation or large angular momentum differences.

Double beta decay can occur in several processes:

$$
\begin{array}{llll}
{}^{A}_{Z}X & \longrightarrow & {}^{A}_{Z+2}X & +2e^- + 2\overline{\nu}_e \quad (1) \\
{}^{A}_{Z}X & \longrightarrow & {}^{A}_{Z+2}X & +2e^- \quad (2) \\
{}^{A}_{Z}X & \longrightarrow & {}^{A}_{Z+2}X & +2e^- + \chi \quad (3)
\end{array}
$$

Although rare, decay (1) is allowed as a second effect in the standard model of the electro-weak interaction. It has indeed been firmly established by experiments. Decay channels (2) and (3) are more interesting since lepton number conservation is violated and as a consequence they test physics beyond the standard model. In the $(\beta\beta)_{0\nu}$ decay, the virtual neutrino must be emitted in one vertex and absorbed in the other one. Since in the standard theory the emitted particle is a right-handed antineutrino and the absorbed one a left-handed neutrino, the process requires that the exchanged neutrino is a Majorana particle $\nu = \overline{\nu}$ and that both neutrinos have a common helicity component.
The helicity matching restriction can be satisfied either if the neutrinos have a nonvanishing mass and therefore a "wrong" helicity component proportional to m_ν/E_ν or if there is a right-handed-current weak interaction, but a non vanishing mass is required in any case [2].

The effective Majorana neutrino mass to which $(\beta\beta)_{0\nu}$ experiments are sensitive is generally written in terms of the electron neutrino mixing matrix $U_{e,j}$ as:

$$\langle m_\nu \rangle = \sum_j \epsilon_j m_j U_{e,j}^2 \quad (1)$$

where m_j are the mass eigenstates and ϵ_j phase factors.

The transition rate for $(\beta\beta)_{0\nu}$ events with a Majorana neutrino mass $< m_\nu >$ and neglecting the right-handed-current terms is given as :

$$[T_{1/2}^{0\nu}]^{-1} = G^{0\nu} |M^{0\nu}|^2 \langle m_\nu \rangle^2 \quad (2)$$

where $G^{0\nu}$ is a calculable phase space factor and $|M^{0\nu}|$ is the nuclear matrix element.

It is important to note that the transition rate strongly depends on the nuclear matrix element and various models can lead to different values of $T^{0\nu}_{1/2}$.

Experimentally the three decay modes $(\beta\beta)_{2\nu}$, $(\beta\beta)_{0\nu}$ and $(\beta\beta)_{0\nu\chi}$ are studied by measuring the sum energy of the two electrons. Fig 1 shows schematically the corresponding spectra. In the case of the neutrinoless mode a sharp discrete line at E=$Q_{\beta\beta}$ is expected. For the two-neutrino mode as well as for the Majoron accompanied mode a continuous spectrum will be observed.

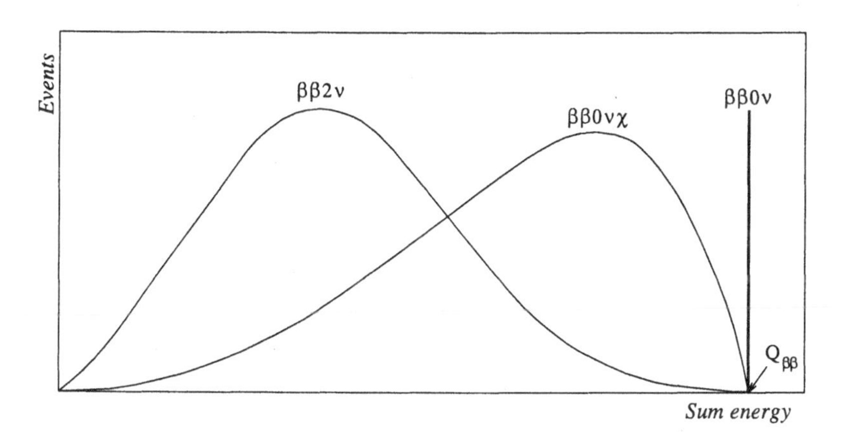

Figure 1 Possible double beta signals

3 THE NEMO EXPERIMENT

The tiny probability associated with a double beta decay event represents a severe experimental challenge. Indeed, a process with a half-life of the order of 10^{25} years must be detected in the presence of inevitable traces of radioisotopes with similar energy release, but having decay rates larger by more than ten orders of magnitude. These impurities, particularly those from uranium and thorium decay chains or those from cosmic origin are a serious source of background.

3.1 NEMO 3 Design

The detector is cylindrical and divided into 20 equal sectors (fig. 2). It consists of a tracking volume filled with helium gas, a thin (50 μm) source foil divides the tracking volume into two concentric cylinders. The surfaces of these cylinders are covered by calorimeters constructed with 1940 large blocks of polystyrene scintillators coupled to low radioactivity photomultiplier tubes. Two types of PMTs are used (3" and 5") depending on the size of the blocks. The tracking system consists of 6180 vertical open octogonal Geiger cells.

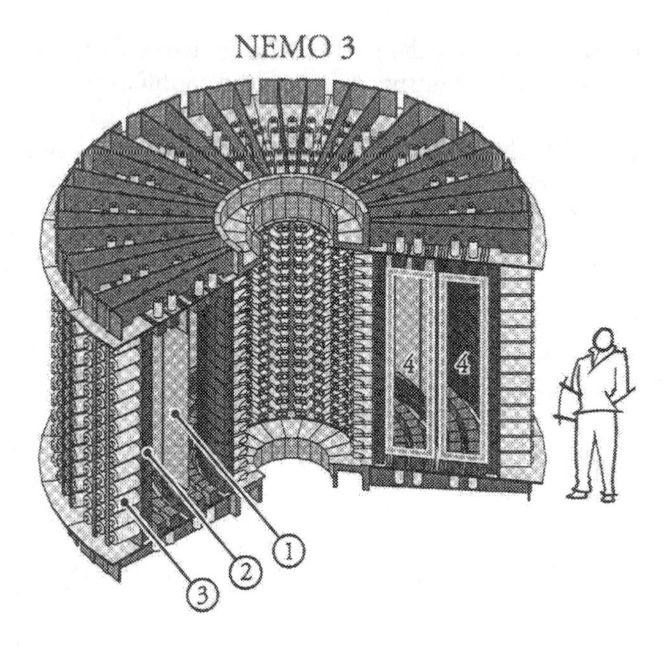

Figure 2 1) cylindrical source foil - 2) scintillator blocks - 3) photomultiplier tubes - 4) tracking volume of drift cells.

NEMO 3 will be able to accomodate up to 10 kg of various double beta decay candidates. To date, attention has been focused on 10 kg of enriched Molybdenum (97% ^{100}Mo). Also available, is one kilogram of each isotope ^{82}Se, ^{116}Cd and ^{130}Te. A solenoid producing a field up to 50 Gauss will surround the detector to reject pair effect events coming from external γ ray background. Finally, external shielding in the form of 20 cm of low activity iron will reduce the γ ray flux and thermal neutrons.

3.2 Background

Of primary importance to all $\beta\beta$ decay experiments is the reduction of radioactivity arising from the uranium and thorium decay chains, in particular the decay of ^{214}Bi and ^{208}Tl which can generate background events which mimic a $\beta\beta$ signal. Careful attention is also paid to the ubiquitous potassium.
Events looking like double beta events and not rejected by time of flight measurements can be created either by gamma rays interacting in the foil or by contamination in the foil.
Gamma rays interacting in the foil can simulate a good event by a two step process (for example double Compton effect or Compton+Möller effect) where two electrons are emitted and photons escape. Beta emitter contamination in the foil can also mimic double beta events when for example a beta particle is emitted simultaneously with another electron (for instance from a converted γ

ray or Möller scattering).
All the materials which have gone into the construction of the detector have been examined with HP Ge detectors at the Fréjus Underground Laboratory or at the C.E.N.B.G. laboratory in Bordeaux. This exhaustive examination of samples caused the rejection of numerous glues, plastics, and metals. The activity in the mechanical pieces which frame the detector are required to be less than 1 Bq/kg. As expected, the radioactive contamination in the experiment is dominated by low radioactivity glass in the PMTs. The total activity of all of the 0.6 tons of PMTs is 800 Bq of ^{40}K, 300 Bq of ^{214}Bi and 18 Bq of ^{208}Tl. These levels are three orders of magnitude below standard PMT levels. There are two sources of plastic scintillators, both with very low levels of radiation. Specifically, from Dubna there are 4 tons with less than 10 Bq of ^{40}K, approximately 5.2 Bq of ^{214}Bi and less than 2 Bq of ^{208}Tl. The one ton of Kharkov scintillators has less than 10 Bq of ^{40}K, 0.7 Bq of ^{214}Bi and 0.3 Bq of ^{208}Tl.
The experimental hall for NEMO 3 is in the Fréjus Underground Laboratory at a depth of 4800 meters of water equivalent. This reduces the cosmic ray muon flux to 4 m^{-2} per day. Vigorous flushing of the air in the hall reduces the radon levels to 10-20 Bq/m^3. The presence of ^{214}Bi decays in the detector from this level of contamination is below that introduced by the PMTs, thereby reducing concerns about additional reductions. Photons of natural radioactivity (<2.6 MeV) are troublesome when interacting with the foil for the $(\beta\beta)_{2\nu}$ and $(\beta\beta)_{0\nu\chi}$ modes. It is not the case for the $(\beta\beta)_{0\nu}$ process and for the chosen $\beta\beta$ emitters, where the energy region of interest $Q_{\beta\beta}$ is above the natural radioactivity limit. In this case the main origin of high energy gamma rays ($\geq$3 MeV) is therefore due to neutron capture inside the detector. Thermal and fast neutrons in the hall are found to be respectively at levels of 1.6×10^{-6} neutrons/s·cm^2 and 4×10^{-6} neutrons/s·cm^2. The effects of these neutrons on the earlier experiment NEMO 2 were studied with an Am-Be source and different formats and levels of shielding. Recent analyses with the software program MICAP is in good agreement with the current understanding of neutron induced backgrounds. In summary thermal neutrons are stopped in the iron shield while fast neutrons go through and are thermalized in the plastic scintillator blocks. The thermalized neutrons are then captured on the copper walls that support the calorimeter producing γ rays up to 8 MeV. Simulations also generate a spectral line at 2.2 MeV suggesting that capture in the scintillators is also significant. However the rates of high energy γ rays interacting with the source foil and producing two electron events or pair creation are expected to be negligible. The magnetic field will be used to study the pair production and confirm the prediction of a negligible contribution. If necessary a paraffin shield surrounding the detector can be added.
As previously stated, the second source of background comes from contamination in the foils. In the $Q_{\beta\beta}$ energy region, there is activity from ^{214}Bi and ^{208}Tl. Of course in the same energy region, the tail of the $(\beta\beta)_{2\nu}$ decay distribution will also play a role. The $(\beta\beta)_{2\nu}$ decays ultimately define the upper half-lives for which the $(\beta\beta)_{0\nu}$ decays can be studied. To insure that the $(\beta\beta)_{2\nu}$ process

indeed defines this limit, maximum levels of ^{214}Bi and ^{208}Tl were calculated. These limits are given in Table 1. For ^{100}Mo it is believed that these limits have been reached, whereas for ^{82}Se with a longer $(\beta\beta)_{2\nu}$ decay half-life, more stringent levels are sought and will require some additional research. Note that the energetic decay of ^{150}Nd ($Q_{\beta\beta}$ = 3.7 MeV) removes concerns of contamination by ^{214}Bi (Q_{β} = 3.3 MeV), but new techniques to enrich Nd will have to be developed for this to be realized.

Isotope	Events/year·10kg			mBq/kg	
	^{214}Bi	^{208}Tl	$\beta\beta2\nu$	^{214}Bi	^{208}Tl
^{100}Mo	0.4	0.4	1.1	0.3	0.02
^{82}Se	0.1	0.1	0.1	0.07	0.005
^{150}Nd	none	0.4	1.1	none	0.02

Table 1: NEMO 3 backgrounds and purity criteria in a 400 keV energy window centered around $Q_{\beta\beta}$.

Currently the plan is to have the ^{100}Mo source foils produced by two different processes. The first is a purification by local melting of solid Mo with an electron beam and drawing a monocrystal from the liquid portion. The monocrystal leaves behind the impurities in the slag of the melt. The crystal is then rolled into a foil for use in the detector. The second purification method is chemical in nature and leaves the Mo in a powder form that is then used to produce foils with a binding paste and mylar strips which have been etched with an ion beam.

4 STATUS OF THE CONSTRUCTION AND EXPECTED PERFORMANCES

Of the 20 sectors required to complete the tracking and calorimeter portions of the detector, 14 have been completed by the end of 1998. The energy resolution for the calorimeter modules associated with the 5" PMTs is 11% at 1 MeV (FWHM), while it is 14.5% for the modules associed with the 3" PMTs. Taking into account the measured energy resolution, Monte Carlo simulations predict a 14% efficiency in the energy window 2.8 to 3.2 MeV. Assuming 10 kg of ^{100}Mo one can expect 2 background events per year in the same energy region.
Compared to all other experiments, NEMO 3 shows very promising performances and has furthermore the possibility of using selected $\beta\beta$ emitters with high $Q_{\beta\beta}$ value and with favorable nuclear matrix elements. It is expected that this experiment will bridge the gap between present and future double beta decay studies using Ge detectors. We plan to start the experiment early in the year 2000.

References

[1] Y. Fukuda et al., Phys. Rev. Lett 81 (1998) 1562.

[2] M. Doi et al., Prog.Theo. Phys. Sup. 83 (1985).

[3] D. Dassié et al., Nucl. Instrum. Meth. A 309 (1991) 465.

[4] R. Arnold et al., Nucl Instrum. Meth. A 354 (1995) 338.

[5] D. Dassié et al., Phys. Rev. D 51 (1995) 289.

[6] R. Arnold et al., Z.Phys. C 72 (1996) 239.

[7] R. Arnold et al., Nucl. Phys. A 636 (1998) 209.

COLLECTIVITY IN MEDIUM MASS N≈Z NUCLEI: ROTATION INDUCED OCTUPOLE CORRELATIONS

G. de Angelis

INFN, Laboratori Nazionali di Legnaro,
via Romea 4, I-35020 Legnaro, Italy
deangelis@lnl.infn.it

Abstract: Medium and heavy nuclei with N≈Z are of special interest due to the fact that protons and neutrons occupy the same orbitals. This leads to a pronounced reinforcement of shell effects which may in some cases result in shapes not observed hitherto. High spin states of the proton rich nuclei in the mass regions A≈70 and 100 have been studied at the GASP and recently at the EUROBALL γ-ray spectrometers. Of the several nuclei populated in the reactions I will discuss here the high spin states of ^{105}In, ^{105}Sn and ^{109}Te as examples for magnetic rotation and rotation induced octupole collectivity. High spin states in the light Ga nuclei have been investigated in order to search for collective modes based on proton neutron pairing correlations. The absence of the blocking effect for rotational structures in odd-odd N=Z nuclei is proposed as an experimental signature.

1 MAGNETIC ROTATION IN THE LIGHT Sn AND In NUCLEI

The recently discovered sequences of $\Delta I = 1$ M1 transitions in the nuclei around ^{200}Pb show very unusual properties (c.f. [1] and [2] and earlier references cited therein): *i)* they are very regular with no signature splitting, similar to high K bands in well deformed nuclei, *ii)* the missing or very small crossover E2 transitions indicate small deformation, *iii)* the B(M1) values are very large (several units of μ_N^2), iv) the ratio of the moment of inertia to the B(E2) value is an order of magnitude larger than for normal or superdeformed nuclei. In short, one observes regular rotational-like sequences in nuclei with very low deformation, which is in contrast to the familiar concept that regular rotational bands appear as a consequence of a substantial deformation of the nucleus. The in-

The Nucleus: New Physics for the New Millennium
Edited by Smit et al., Kluwer Academic / Plenum Publishers, New York, 2000.

terpretation has been given in terms of a novel type of excitation [3], called 'magnetic' rotation. Different from normal rotation, here it is the magnetic dipole that rotates about the angular momentum vector and not the deformed electrical charge distribution. In the Pb region, the angular momentum increase along the band is generated by the simultaneous reorientation of the spins of the proton-particles ($h_{9/2}$ or $i_{13/2}$) and of the neutron-holes ($i_{13/2}$). These angular momentum vectors are essentially perpendicular near the bandheads and align with the total angular momentum of the nucleus for increasing rotational frequency ("shears mechanism").

The Tilted Axis Cranking (TAC) model [3] provides the theoretical framework for a description of magnetic rotation. The appearance of similar bands has been predicted for the A=100 mass region [3]. Here neutron-particles in the $h_{11/2}$ orbital and proton-holes in the $g_{9/2}$ orbital combine with the rotating dipole. High-spin states of the ^{105}Sn and ^{105}In nuclei were populated using the ^{58}Ni + ^{50}Cr reaction at a beam energy of 210 MeV, the beam being delivered by the Tandem XTU of the Legnaro National Laboratory. The GASP array [4], with 40 Compton-suppressed Ge detectors and a multiplicity filter of 80 BGO scintillators, was used for a γ-coincidence measurement. In order to select the reaction channel we have used, in addition to the GASP array, the Si-ball ISIS [5] and the recoil mass spectrometer [6].

Low-energy states in the light tin isotopes originate mainly from excitations of the valence neutrons in the $\nu d_{5/2}$ and $\nu g_{7/2}$ orbitals. All the levels observed in ^{105}Sn up to spin $23/2^+$ are consistent with excitations of 5 valence neutrons into the $d_{5/2}$, $g_{7/2}$, $s_{1/2}$ and $d_{3/2}$ orbitals [7]. At higher excitation energy, starting at spin $I^\pi=29/2^{(-)}$, the level scheme shows a regular sequence of $\Delta I = 1$ transitions, which we tentatively assign M1 character. A comparison of the excitation energy of the dipole bands in the light Sn and In isotopes suggests a negative parity for such a configuration. This changes the previously assumed positive parity assignment [8] and is in agreement with similar structures recently found in heavier Sn isotopes [9]. In the case of ^{105}In, Fig.1, the level scheme above 5.5 MeV of excitation energy shows three regular sequences of $\Delta I = 1$ transitions starting at spin $I^\pi=29/2^-$ which we assume to be M1 in character. For two of them, band 1 and band 2, a negative parity is suggested whereas, based on the decay pattern, a positive parity assignment is preferred for band 3.

We have performed calculations within the TAC model [3] in order to describe the observed bands. Pairing is taken into account for the neutrons using a constant value of Δ_ν=0.80 MeV. For protons the Z=50 shell is closed and pairing is therefore neglected. The equilibrium deformation turns out to be only weakly dependent on the rotational frequency ω. Therefore, we use a constant value of $\epsilon = 0.10$ and $\gamma = 0°$, which are close to the equilibrium values. The most likely configuration of the $\Delta I = 1$ bands consists of $\pi g_{9/2}^{-1}$ coupled to one (negative parity) or two (positive parity) neutrons in the $h_{11/2}$ orbit and one neutron in one of the orbits originating from the mixed $d_{5/2}$ and $g_{7/2}$ multiplets. A further proton hole excitation into the $g_{9/2}$ orbit cancels the magnetic moment leading to a decay characterized by quadrupole transitions (band 4 in Fig.1). In the inset of Fig. 1 the angular momentum, I, of the states of the bandsof ^{105}In as a function of the angular frequency $\hbar\omega$ is reported and compared with the one calculated using the above configurations.

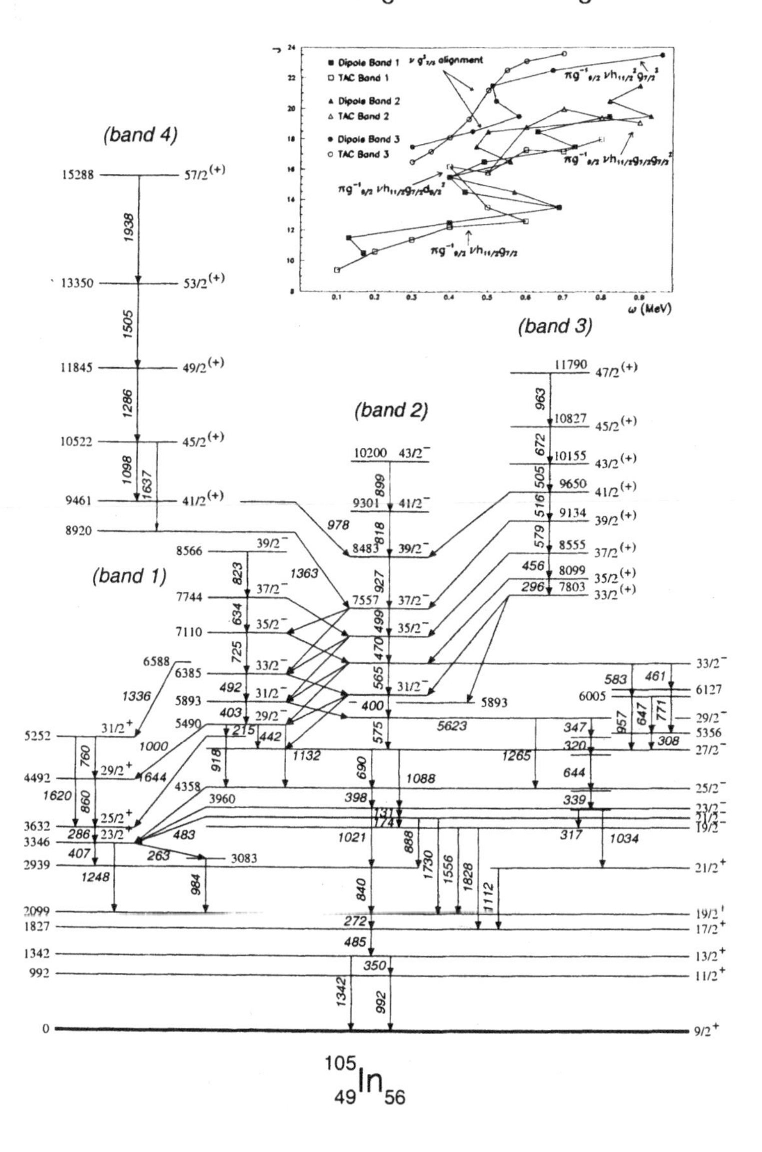

Figure 1 Level scheme of ^{105}In deduced from the GASP data. The γ rays are labeled by energy (in keV). The inset shows the spins of the dipole bands as a function of the rotational frequency. The calculated ones are based on the $\pi g_{9/2}^{-1}\nu[h_{11/2}^{n}(d_{5/2}g_{7/2})^{m}]$ configuration.

2 ROTATION INDUCED OCTUPOLE CORRELATIONS IN THE NEUTRON-DEFICIENT ^{109}Te NUCLEUS

Octupole correlations, generated by the interaction among orbitals near the Fermi surface, which differ by three units of angular momentum, lead to sequences of interleaved states of negative and positive parity connected by strong E1 transitions [10]. The light Te, Xe and Ba nuclei lie in one of the few regions of the nuclear chart where such octupole correlations are expected [11]. In fact, such structures have recently been suggested to be present at high spins in the ^{108,110}Te [12, 13] and ^{114}Xe [14] nuclei.

In this mass region the strongest octupole correlations are expected for the $N = Z$ nucleus ^{112}Ba [11]. There the Fermi surface of both protons and neutrons lies between the $d_{5/2}$ and $h_{11/2}$ orbitals. Moreover at the $N = Z$ line, dynamical proton-neutron octupole coupling may arise. This effect will not be addressed in our work but deserves further investigation. Experimentally, most of such $N = Z$ nuclei cannot be investigated due to the extremely small cross section in fusion evaporation reactions. However, similar strong octupole correlations can arise in Sn- and Te-isotopes also when N is close, although not equal, to Z. Here the ground state quadrupole deformation is smaller than in the Ba-isotopes, but since the $\Omega = 1/2$ state of the high-j $h_{11/2}$ orbital couples strongly to rotation, it will approach the Fermi surface at high angular momentum, resulting in enhanced octupole polarization effects. The quadrupole deformation can play a role similar to that of rotation, since it lowers the $\Omega = 1/2$ state of the $h_{11/2}$ orbit with respect to the Fermi-surface. In particular one can expect octupole correlations to be strong in the 2p-2h (4p-2h) intruder bands in the Sn-(Te-)region. Hence, in the vicinity of neutron number $N = 56$ and proton number $Z = 50 - 52$, rotation and deformation induced octupole correlations can develop.

High-spin states in the neutron deficient nucleus ^{109}Te were populated using the ^{58}Ni+^{54}Fe reaction at 220 MeV, the beam being delivered by the Tandem XTU of the Legnaro National Laboratory. Double and higher fold γ-γ coincidences were acquired with the 4π spectrometer GASP [4]. In order to improve the selectivity of the apparatus for reaction channels involving the evaporation of charged particles, the Si-ball ISIS [5] was mounted inside the inner ball of the GASP spectrometer. The resulting level scheme for the ^{109}Te nucleus is shown in Fig. 2 [15]. The spins and parities of the levels were deduced, where possible, from the analysis of the directional correlation ratios from oriented states (DCO) and of the angular distribution from oriented nuclei (ADO). Special attention was paid to the DCO ratio and ADO analysis using triple coincidence events.

The level scheme below 1.1 MeV of excitation energy is characterized by quasi-neutron excitations involving the $d_{5/2}$ and $g_{7/2}$ orbitals coupled to quadrupole vibrations. The negative-parity states up to $27/2^-$ involve the $h_{11/2}$ neutron orbital most likely coupled to quadrupole vibrational phonons. Above the 3782 keV state two sequences of quadrupole transitions are observed, one of them, the side band, decaying into the ground state band. DCO and ADO

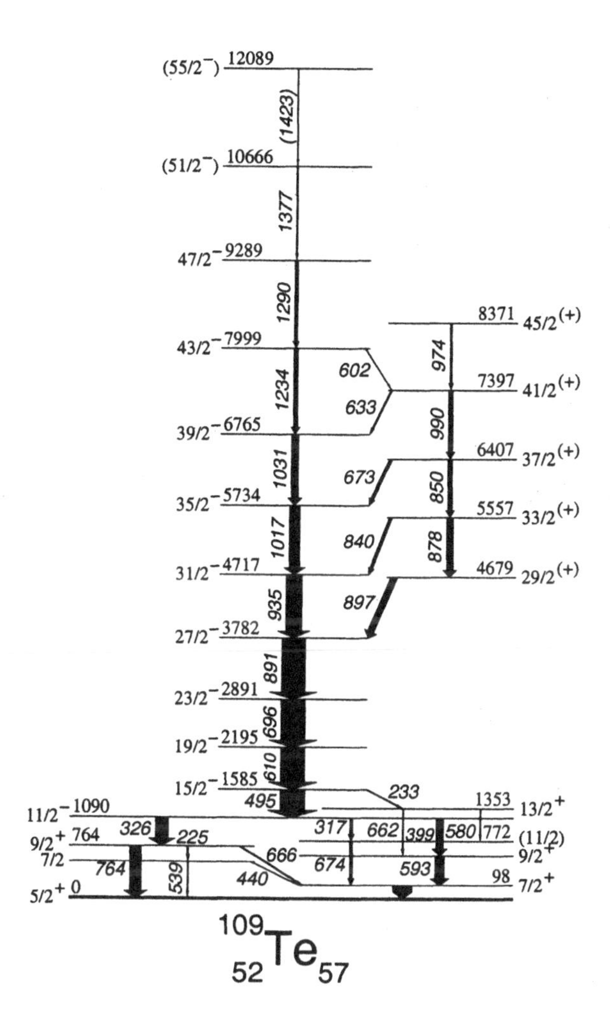

Figure 2 Level scheme of ^{109}Te deduced from the GASP data. The γ rays are labeled by energy (in keV), and the width of the arrows represents the relative intensities.

ratios, extracted for the transitions connecting the side band with the yrast negative parity band, are all consistent with stretched dipoles without mixing. We tentatively assign them an E1 character, resulting in positive parity for the states of the side band. If this assumption is correct, the values of the $B(\mathrm{E1}; I \to I-1)/B(\mathrm{E2}; I \to I-2)$ ratios extracted from the experimental data are of the order of 10^{-6} fm^{-2}. They are similar in magnitude to the $B(\mathrm{E1})/B(\mathrm{E2})$ ratios found in the ^{110}Te nucleus [13] and in the neutron-rich barium nuclei, where octupole correlations are known to play an important role [10, 14]. In the framework of the rotational model, the intrinsic dipole moment D_0 can be extracted from the $B(\mathrm{E1})$ rates [16] as:

$$D_0 \approx \left[\frac{5(I-1)B(\mathrm{E1}; I \to I-1)}{8(2I-1)B(\mathrm{E2}; I \to I-2)}\right]^{1/2} Q_0$$

The intrinsic quadrupole moment can be estimated from Total Routhian surface calculations for ^{109}Te (see below) yielding a value of $Q_0 = 190$ $e$fm^2 ($\beta_2 = 0.15$). The extracted values of the intrinsic dipole moments are D_0=0.057, 0.093, 0.111 $e$fm for the 5557 keV ($33/2^{(+)}$), 6407 keV ($37/2^{(+)}$) and 7397 keV ($41/2^{(+)}$) levels respectively. Such values are similar to those found for ^{110}Te [13] and for the neutron rich Ba region [16]. However they are slightly smaller than those of the Ra-Th region [16].

The onset of octupole deformation at high spins has been investigated in pioneering calculations of total Routhian surfaces (TRS) in the Th-region [17] and later in the heavy Ba-Sm region [18]. However, these calculations were generally restricted to even-even nuclei and pairing correlations were treated in a schematic manner. To investigate more deeply the origin of these band structures, extended self-consistent TRS calculations [20] based on an axially symmetric Woods-Saxon potential and the liquid drop energy from ref. [19] have been performed. In order to address the properties of *odd* nuclei at high spins, the state of the odd particle with simplex quantum number $\pm s$ has been blocked self-consistently. For states in odd nuclei involving the $h_{11/2}$ orbit, we expect states with simplex quantum number $s = +i$ to be the lowest in energy. To avoid the spurious pairing phase transition encountered in the BCS-approximation, we performed approximate particle number projection applying the Lipkin-Nogami projection method [20]. Pairing correlations are treated self-consistently and include both a monopole and doubly-stretched quadrupole pairing force. The shape of the nucleus is optimized with respect to quadrupole, hexadecapole and octupole deformation, assuming axial symmetry.

The TRS calculations for even-even nuclei in the heavy Ba-Sm region show rotation induced shape changes, indicating that at medium spins the magnitude of octupole deformation increases and well separated octupole minima appear [18]. Our calculations - shown in Fig. 3 - yield similar results. At low rotational frequency the $\nu h_{11/2}$ configuration exhibits octupole softness, but the minimum occurs for an intrinsic reflection-symmetric shape. The quadrupole deformation is calculated to be rather small and in fact the observed energy levels in the $h_{11/2}$ negative parity band at low spin are more vibrational in character. With increasing rotational frequency, the surface gets stiffer in the β_3-direction. In the region of $\hbar\omega = 0.5 - 0.6$ MeV, the $g_{7/2}$ neutrons start to align their angular momenta with the rotation and pairing correlations are decreasing. Note that the $g_{7/2}$ orbital is strongly mixed with the $d_{5/2}$. In this rotational frequency region, we observe a sudden increase in the octupole-softness of the TRS and the minimum corresponds to a shape with slight octupole deformation.

The pronounced octupole softness of the routhian surface around $\hbar\omega \approx$ 0.6 MeV can be related to the following mechanism. At low spins, there are states of positive parity based upon the $g_{7/2} - d_{5/2}$ configuration and states of negative parity, involving the $h_{11/2}$ orbital. However, the $(h_{11/2})^2$ neutron alignment in the positive parity structure based upon the $g_{7/2} - d_{5/2}$ orbitals in-

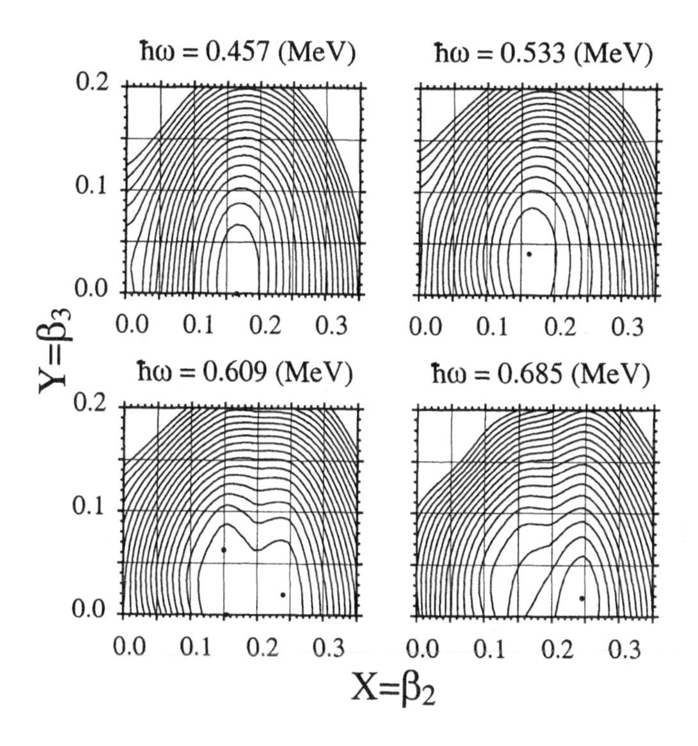

Figure 3 The total Routhian surfaces, for axially symmetric shape, calculated at rotational frequencies ranging from 0.4 to 0.7 MeV in ^{109}Te. At low rotational frequency the surface shows a minimum for a prolate reflection-symmetric shape (β_2 $\approx$0.17). Starting at a rotational frequency of ω $\approx$0.5 MeV, a new minimum at non axially symmetric shape starts to develop. This minimum gets deeper and more octupole deformed at ω $\approx$0.6 MeV. In this rotational frequency region a third minimum, corresponding to the 4p-2h intruder excitation, starts to develop and becomes yrast at the highest frequencies. The distance between the contour lines is 200 keV. For further details see text.

duces enhanced quadrupole deformation. A consequence of the large alignment gain of this configuration is that the negative and positive parity structures approach each other. In the present case, both negative and positive parity band structures are built on three quasi-particle states which start to align with the rotation. Octupole correlations are induced by mixing the aligning $(g_{7/2}-d_{5/2})(h_{11/2})^2$ and $h_{11/2}(g_{7/2}-d_{5/2})^2$ configurations. Due to the Coriolis force, the structures of opposite parity come close in energy with increasing rotational frequency. Note that if no octupole correlations were present, the calculations would predict the negative signature of the positive parity band to be lowest. The experimentally observed positive parity structure has positive signature, but the same simplex quantum number as the $h_{11/2}$ negative parity band. This is another indication of octupole correlations.

3 A NEW SIGNATURE OF PROTON-NEUTRON PAIRING?

An even more interesting aspect of nuclear structure at the N=Z line is the possibility of observing neutron proton correlations in the T=0 channel. Such correlations may lead to a new collective mode based on a neutron-proton pair condensate [21]. The T=1 pairing correlations have been studied extensively and the reduction of the moments of inertia as well as the backbending phenomenon are one of the key signs of superfluidity in atomic nuclei. In contrast, the properties of the T=0 correlations remain to be investigated. The structure of the high spin states of heavy N=Z nuclei, in particular the processes in which high-j particles align their relative orientation, can be used to probe the additional T=0 correlations [22]. In the case of the N=Z=36 ^{72}Kr, the four quasi-particle $g_{9/2}$ alignment has been observed to be significantly delayed in rotational frequency with respect to the heavier Kr isotopes [23]. Since this delay contradicts the extended pairing and deformation self consistent TRS-calculations, it may reflect an additional binding due to the neutron-proton pairing correlations in the T=0 channel. To what extent these correlations affect the band crossing frequencies in nuclei far from stability remains a challenging question. Due to the softness of the potential energy surface, one cannot exclude the possibility that the vibrational motion contributes to the lowering of the ground state.

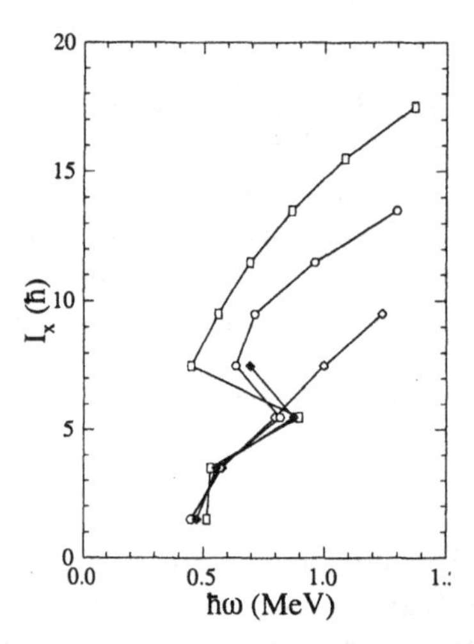

Figure 4 The spin component on rotational axis I_x versus the rotational frequency for the collective bands built on the neutron and proton $g_{9/2}$ structures in the N=62 (filled diamonds), 63 (hexagons), 64 (empty diamonds) and 65 (squares) Ga isotopes.

A stronger experimental signature of the neutron-proton pairing correlations in the T=0 channel can be provided by the absence of the blocking effect in the alignment processes for odd-odd N=Z nuclei. Such alignments are blocked for usual pairs of identical particles due to the Pauli principle but are permitted for pairs of unlike particles.

High-spin states of the N≈Z $^{62,63,64,65}Ga$ nuclei were populated using the ^{32}S + ^{40}Ca reactions at 140 and 150 MeV and using the GASP and EUROBALL γ spectrometers.

In Fig. 4 the spin component on rotational axis I_x versus the rotational frequency is compared for the collective bands built on the $(\pi g_{9/2}\nu g_{9/2})_{9^+}$ structures in the $^{62,63,64,65}Ga$ isotopes. It is noticeable that the first $g_{9/2}$ alignment is observed at $\hbar\omega \approx 0.65$ MeV in the odd-A $^{63,65}Ga$ nuclei and is absent due to the blocking effect in the odd-odd ^{64}Ga. Evidence for an unexpected alignment at $\hbar\omega \approx 0.65$ MeV is observed in the odd-odd N=Z nucleus ^{62}Ga [24]. A preliminary analysis of the Euroball data seems to confirm such an observation. Here one cannot exclude the fact that the lowering of a different structure can cause such an effect.

References

[1] G. Baldsiefen *et al.*, Nucl. Phys.A574 (1994) 521.

[2] M. Neffgen *et al.*, Nucl. Phys. A595 (1995) 499.

[3] S. Fraundorf, Nucl. Phys. A557 (1993) 259c.

[4] D. Bazzacco, Int. Conf on Nuclear Structure at High Angular Momentum, Ottawa, 1992, Vol.2. AECL 10613, p376.

[5] E Farnea *et al.*, LNL-INFN(Rep)-095/95, (1994) 189 Nucl. Instr. and Meth. A400 (1997) 87.

[6] P. Spolaore *et al.*, Nucl. Instr. and Meth. A359 (1995) 500.

[7] R Schubart *et al.*, Z. Physik A352 (1995) 373.

[8] A Gadea *et al.*, Phys. Rev. C55 (1997) R1.

[9] D.G. Jenkins *et al.*, Phys. Lett. B428 (1998) 23.

[10] P.A. Butler and W. Nazarewicz, Rev. Mod. Phys. 68 (1996) 349.

[11] J. Skalski *et al.*, Phys Lett. B238 (1990) 6.

[12] G.J. Lane *et al.*, Phys. Rev. C57 (1998) R1022.

[13] E.S. Paul *et al.*, Phys. Rev. C50 (1994) R534.

[14] S.L. Rugari *et al.*, Phys. Rev. C48 (1993) 2078.

[15] G. de Angelis *et al.*, 437 (1998) 236.

[16] P.A. Butler *et al.*, Nucl. Phys. A533 (1991) 249.

[17] W. Nazarewicz, G.A. Leander and J. Dudek, Nucl. Phys. A467 (1987) 437.

[18] W. Nazarewicz *et al.*, Phys. Rev. C45 (1992) 2226.

[19] W.D. Myers and W.J. Swiatecki, Ann. Phys. (N.Y.), 84 (1969) 395.

[20] W. Satula. R. Wyss and P. Magierski, Nucl. Phys. A578 (1994) 45.

[21] W. Satula and R. Wyss, B393 (1997) 1.

[22] J.A. Sheikh *et al.*, Phys. Rev. Lett. 64 (1990) 376.

[23] G. de Angelis *et al.*, **B415** (1997) 217.
[24] S. Skoda *et al.*, to be published.

COMPETITION OF FISSION WITH THE POPULATION OF THE YRAST SUPERDEFORMED BAND IN ^{194}Pb

I.Deloncle[1],M.-G.Porquet[1],A.Bauchet[1],F.Azaiez[2], M.Belléguic[2]
B.Gall[3],W.Korten[4],A.Astier[5],S.Perriès[5], N.Amzal[6],P.Butler[6]
D.Hawcroft[6],T.Alanko[7],J.Cocks[7], P.Greenlees[7],K.Helariutta[7],
R.Julin[7],S.Juutinen[7], H.Kankaanpaa[7],A.Kanto[7],H.Kettanen[7],
M.Leino[7],P.Mayet[7],M.Muikku[7],P.Rahkila[7],A.Savelius[7],W.Trzaska[7]

[1]CNSM, IN2P3/CNRS, F-91405 Orsay Campus
[2]IPN, IN2P3/CNRS,F-91406 Orsay
[3]Ires, IN2P3/CNRS and Université Louis Pasteur, F-67037 Strasbourg Cedex
[4]DAPNIA/SPhN, CEA Saclay, F-91191 Gif-sur-Yvette
[5]IPN, IN2P3/CNRS and Université C. Bernard Lyon-1, F-69622 Villeurbanne Cedex
[6]Oliver Lodge Laboratory, University of Liverpool, Liverpool L69 3BX, United Kingdom,
[7]JYFL, University of Jyväskylä, FIN-40351 Jyväskylä

deloncle@csnsm.in2p3.fr

Abstract: The ^{194}Pb yrast superdeformed band has been populated in two reactions induced by two different beams delivered by the JYFL cyclotron. These two reactions differ in the asymmetry ratio (i.e. the ratio of the target mass over the projectile one) by more than a factor 3. For the first time, a superdeformed band of the A≈190 mass region has been populated using such a very heavy beam, namely ^{74}Ge.

In the A≈150 mass region, it has been observed [1, 2, 3, 4] that the relative (relatively to the normally deformed states) intensity of the yrast superdeformed (SD) bands increases when the asymmetry ratio (i.e. the ratio of the target mass over the projectile one) of the heavy ion reaction decreases. For this feature an explanation has been proposed [1, 2, 3] involving the competition between the population of SD bands and fission, the latter being inhibited in symmetric reactions. But the A≈150 is not the best mass region for the study of such an interplay, as the fission limit lies far away from the entry points in the (E^*,I) plane. This is not the case in the A≈190 mass region, where it

The Nucleus: New Physics for the New Millennium
Edited by Smit et al., Kluwer Academic / Plenum Publishers, New York, 2000.

has been shown [5, 6] that the highest spin reached in the yrast SD bands is correlated to the fission limit in angular momentum. In particular, the ^{194}Pb nucleus is a good candidate as it presents one of the most intense SD bands of this mass region with a rather low cut-off in angular momentum imposed by fission [6].

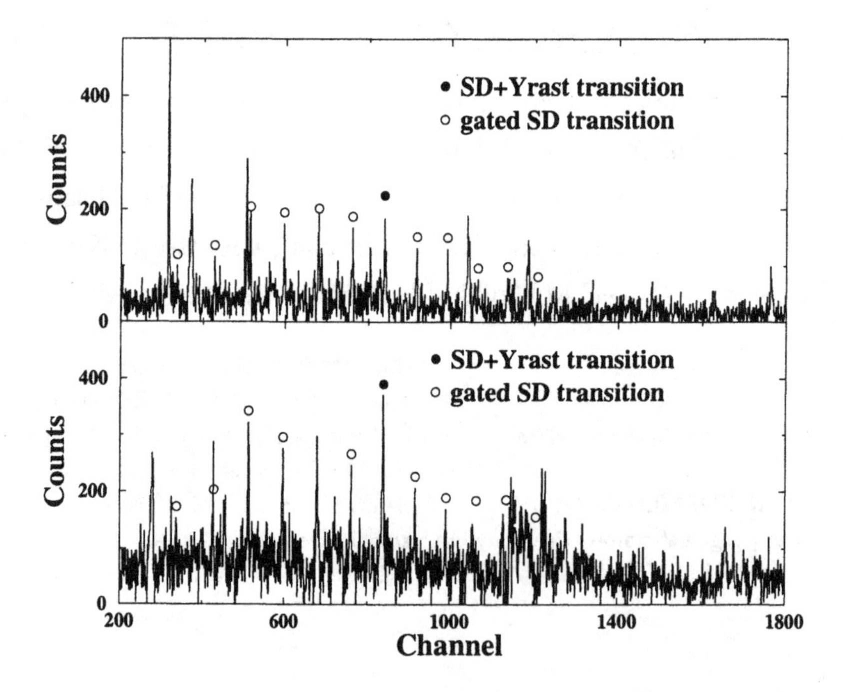

Figure 1 The top spectrum corresponds to the asymmetric reaction and the bottom spectrum to the symmetric one.

The ^{194}Pb nucleus has been produced using two reactions leading to the same compound nucleus (CN) ^{198}Pb. The beams were delivered by the JYFL cyclotron at Jyväskylä. In addition to the 25 Compton suppressed Ge-detectors of Jurosphere, a PPAC detector has been located in the focal plane of the gas-filled separator RITU in order to select the fusion-evaporation channel. The asymmetric reaction (asymmetry ratio of 5.6) was ^{168}Er(^{30}Si,4n) at 145MeV, the symmetric reaction ^{124}Sn(^{74}Ge,4n) reaction at 305MeV. The PACE code predicts $38(\pm4)\hbar$ (resp. $51(\pm7)\hbar$) for the maximum of angular momentum distribution and $60(\pm2.5)$ and $53(\pm4)$MeV for the excitation energy of the CN ^{198}Pb for the asymmetric (resp. symmetric) reaction. The dispersion in the thickness of the self-supported targets (stacks of two $340\mu g/cm^2$ of ^{168}Er foils and ^{124}Sn individual foils of 300,350,375 and $500\mu g/cm^2$) have been taken into account in these numbers given for the mid-target point. There were 16.4×10^7 (resp. 22.2×10^7) Ge-detector events of two- or more fold, 6.6×10^6 (resp. 16.4×10^6) of which had a RITU/PPAC signal in the asymmetric (resp. symmetric) reaction and were chosen to be analyzed. The following spectra of the yrast SD band have been obtained by requiring the presence of at least one of about ten SD transitions (see Fig. 1).

The measurement of the relative intensity of the superdeformed band as compared to normally deformed states ($I_{SD/ND}$) of ^{194}Pb in both reactions has been performed using one-gated background substracted spectra according to the procedure we have already used in [5]. The value of 0.994($\pm$0.1) for the ratio $I_{SD/ND}$ obtained in the case of the asymmetric reaction is in perfect agreement with our previous results in other asymetric reactions [5], whilst this ratio suprisingly falls down to 0.533($\pm$0.06) in the case of the symmetric reaction induced by the Ge beam.

Our surprizing result proves that, as in the A$\approx$150 mass region, the SD bands of A$\approx$190 mass region, have their own excitation curve. Unfortunately, we were not at the maximum in the symmetric reaction. In a forthcoming experiment we will explore this excitation curve in order to reach the highest $I_{SD/ND}$ ratio and try to observe the inhibition of fission.

References

[1] G. Smith et al., Phys. Rev. Lett. 68 (1992) 158

[2] S. Flibotte et al., Phys. Rev. C45 (1992) R889

[3] B. Haas et al., Phys. Lett. B245 (1990) 13

[4] M. Thoennessen and J. R. Beene, MSU CL report, MSUCL-794 (1991)

[5] I. Deloncle et al., J. Phys. G21 (1995) L35

[6] Int. Conf. on Nucl. struct., Crete-Greece 1996 and,
I. Deloncle et al. Z. Phys. A358 (1997) 181

NEW SUPERDEFORMED BANDS IN ^{150}Gd

S. Ertürk[1,2*],
P.J. Twin[1], D.E. Appelbe[1], P. Fallon[3], C.W. Beausang[1†], S. Asztalos[3], R. Krüken[3], T. Lauritsen[4], I.Y. Lee[3], A.O. Macchiavelli[3], and F.S. Stephens[3].

[1] Oliver Lodge Laboratory, University of Liverpool, Liverpool, L69 7ZE United Kingdom.
[2]Nigde Universitesi, Fen-Edebiyat Fakültesi, Fizik Bölümü, Nigde, TURKEY .
[3] Nuclear Science Division, LBNL, Berkeley, California 94720.
[4] Physics Division, Argonne National Laboratory, Illinois 60439.

Abstract: Following an experiment with the GAMMASPHERE γ-ray spectrometer, eight new superdeformed (SD) bands have been observed in the nucleus $^{150}_{64}Gd_{86}$. This brings the total number of SD bands in ^{150}Gd to fourteen, the largest number observed in any nucleus, allowing an examination of the single-particle states close to the nuclear Fermi Surface to be performed. The properties of these bands are discussed in terms of single-particle configurations and are compared with SD bands observed in neighbouring nuclei. The observation of such a large number of excited SD bands has confirmed the SD nature of ^{150}Gd.

*Present Address: Nigde Universitesi, Fen-Edebiyat Fakültesi, Fizik Bölümü, Nigde, TURKEY
†Present Address: W.N.S.L., Physics Department, Yale University, 272 Whitney Avenue, New Haven, CT 06511.

1 INTRODUCTION.

The nucleus ^{150}Gd has been studied extensively since the observation of the first SD band in this nucleus by Fallon *et al.*, [1] in 1989. Until the present study ^{150}Gd has been reported to posses six SD bands [2]. The yrast band, band 1, which was reported by Fallon [1], has been explained in terms of the occupation of two $i_{13/2}$ proton intruders and two $j_{15/2}$ neutron orbitals giving rise to the high-N configuration $\pi 6^2 \nu 7^2$. The quadrupole moment of this band has been measured by Beausang *et al.*, [3] and found to be in good agreement with that predicted by Nazarewicz *et al.*, [5]. Band 2 was reported by Byrski *et al.*, [6], and was found to have almost identical transition energies as those for the yrast band in the Z + 1 nucleus, ^{151}Tb. The explanation behind this observed identicality is based on the properties of the π [301]1/2 natural parity orbital [7]. This orbital shows very little curvature as a function of rotational frequency and thus posses no significant alignment. Furthermore this orbital is flat, hence its occupation does not effect the deformation of the nucleus to any observable degree. Bands 3 and 4 were reported by Beausang [8], following an experiment with the EUROGAM I γ-ray spectrometer [9]. These two bands were observed to lie at the half-points of each other, indicating that they are signature partner bands. Accordingly, these bands were assigned as being based on a single neutron excitation from the second N = 7 intruder (the 7_2) into the [402]3/2 orbital. SD band 5 was reported by Fallon [10] and provided the first evidence for the excitation of a pair of protons in the SD minimum. This band is believed to be based on the excitation of the protons occupying both signatures of the [301]1/2 orbital into the 6_3 and 6_4 orbitals, allowing a comparison to be made between this band and the yrast band in ^{152}Dy [3]. It was also observed [10] that band 5 decays preferentially to band 1 as opposed to the ND states.

This paper reports on the observation of eight new SD bands in the nucleus ^{150}Gd. These bands are understood in terms of single-particle excitations across the N = 86 and Z = 64 SD shell-closures thus confirming the single-neutron and proton level structures in the vicinity of these shell-closures.

2 EXPERIMENTAL DETAILS.

High-spin states in ^{150}Gd were populated following the reaction ^{130}Te(^{26}Mg,6n) ^{150}Gd, at an energy of 149 MeV. The beam, provided by the 88" cyclotron, at the Lawrence Berkeley National Laboratory, was incident upon two stacked self-supporting foils of ^{130}Te each of thickness 500μg cm^{-2} on a gold backing of thickness 500μg cm^{-2}. An additional 100μg cm^{-2} of gold was evaporated on to the front of the target so as to minimise the oxidation of the target material. The resultant γ-rays were detected by the GAMMASPHERE array, which at the time of this experiment comprised 84 large volume Compton suppressed HPGe detectors. Events were collected and stored to magnetic tape when a minimum of six Compton suppressed γ-rays were detected. During the course of this experiment some 1.4$\times$ 10^9 events of fold $\geq$ 6 were collected, which upon unfolding resulted in 4.2$\times$ 10^{10} γ^3 and 4.8$\times$ 10^{10} γ^4 events. The primary exit channel was 6n to ^{150}Gd, carrying ≈66% of the total evaporation cross-section for this reaction. The 5n (^{151}Gd) and 7n (^{149}Gd) channels were found to carry ≈18% and ≈16% of the total evaporation cross-section respectively.

3 RESULTS.

The data were initially sorted into an E_γ-E_γ-E_γ coincidence "cube" for analysis with the Levit8r analysis code [12]. This cube was searched for the existence of new rotational structures with a characteristic energy spacing of $\Delta E_\gamma \approx 50$ keV (that expected for SD bands in this mass region) using the search algorithm incorporated in the Levit8r code [12]. As a result of this search twenty one SD bands were observed in this data, including the six previously known SD bands in ^{150}Gd [2], and the yrast SD band in ^{149}Gd [13]. Eight of the remaining bands have been assigned to ^{150}Gd based on observed coincidences with the low-lying normal deformed (ND) transitions in ^{150}Gd and are reported in this paper. The remaining six bands have been assigned to the nucleus ^{151}Gd, and are reported in a separate publication [14].

Due to the large number of SD bands observed in ^{150}Gd, the labelling scheme for the previously known bands has been rearranged (see Table. I) so as to account for signature partner bands and the remeasured intensities. Although during the course of this work we have been unable to measure the transition quadrupole moments for the "new" bands; the rotational behaviour of these bands along with their SD nature has been inferred from the similarities observed between these bands and rotational structures of confirmed SD nature of adjacent nuclei.

No transitions have been observed that connect these bands to the ND states in ^{150}Gd, hence it is not possible to determine the absolute spins and parities of these SD states. However, it is possible to determine the spins of these states relative to some suitable reference band, while parities may be assigned from the parities of occupied orbitals.

Table 1 The labelling scheme now used to identify the SD bands in ^{150}Gd.

Old Scheme	New Scheme
Band 1	Band 1
Band 2	Band 3
Band 3	Band 4b
Band 4	Band 4a
Band 5	Band 2
Band 6	Band 5

3.1 Bands 1-4

The latest results for these bands are presented for completeness and corresponding spectra are presented in Figure 1. Band 1 has been extended further by the addition of a 1601 keV transition at the top of the band from that reported by Fallon *et al.,* [1]. Band 2, previously labelled band 5, has been extended by the addition of two transitions of energy 1595.9 keV and 1645.5 keV, further, it has been confirmed that band 2 decays into band 1 with the observation of the inter-band linking transitions. The properties of these are discussed in a separate publication [14, 16]. Band 4a, previously known as band 4, Ref. [8], has been extended with the addition of a 1658.4 keV γ-ray at the top of the band, while the two lowest transitions in band 4b (previously known as band 3), the 618 keV and 665 keV γ-rays are now found NOT to be in coincidence with the remaining band members (see Table ??). This reduces the number of transitions in this band by two compared with that measured by Beausang *et al.,* [8] (these two transitions have been identified as normal deformed (ND) transitions in the nucleus ^{149}Gd). Further, Beausang *et al.,* [8] observed an interaction taking place between degenerate levels in Bands 3 and 4a. Our experiment has confirmed this degeneracy and is the subject of a separate publication [14].

3.2 Band 5

The properties of this band have been discussed in Ref. [2], where it was suggested that band 5, was based on the configuration $\pi 6^3 \nu 7^2$, due to the similarity between the $\mathcal{J}^{(2)}$ for band 5 and the yrast SD band in ^{151}Tb [1]. As a result of increased statistics collected during this experiment, the transition energies for band 5 have been remeasured, and its $\mathcal{J}^{(2)}$ moment of inertia recalculated. From Figure 2, it can be observed that the $\mathcal{J}^{(2)}$ for band 5 is almost identical to that for band 1 at high frequencies ($\hbar\omega \geq 0.6$ MeV), while at lower rotational frequencies the $\mathcal{J}^{(2)}$ for band 5 diverges from that for band 1. It is suggested that band 5 is based on a single-neutron excitation from the $\nu 7_2$ into $\nu 7_3$ orbital.

3.3 Bands 6a and 6b

From Figure 3, it can be seen that $\mathcal{J}^{(2)}$ moments of inertia for these two bands are almost identical. It can be observed that band 6a lies at the half-points of band 6b (with a root mean square (r.m.s) difference of 1.24 keV between the measured energies and corresponding calculated half-points over the entire frequency range). This suggests that band 6a and 6b are based on a single-particle excitation into a Nilsson state that exhibits little, or no signature splitting as a function of rotational frequency. As can be observed from Figure 3 the $\mathcal{J}^{(2)}$ for bands 6a and 6b is similar to that for the yrast band of ^{149}Gd, indicating a similar high-N configuration, namely $\pi 6^2 \nu 7^1$ [17]. Furthermore the occupied natural parity orbitals are expected to be flat, with a minimal contribution to the $\mathcal{J}^{(2)}$.

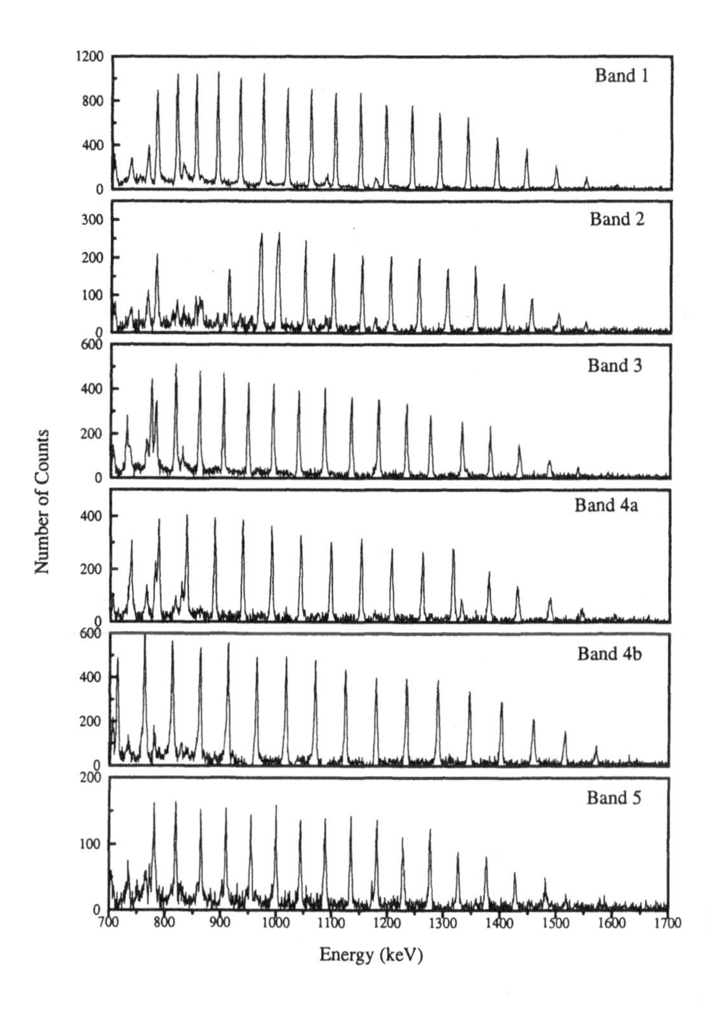

Figure 1 Coincidence spectra (quadruple gated) for SD bands 1-5 in ^{150}Gd. Gates were set on all band members.

The orbitals close to the neutron Fermi surface which satisfy the requirements outlined above are the [521]3/2, [402]5/2 and [514]9/2 Nilsson states. Of these available orbitals, it is suggested that bands 4a and 4b are based on the excitation of a neutron from the $\nu 7_2$ orbital into both signatures of the [402]5/2 Nilsson state. Of the remaining orbitals, the [521]3/2 and [514]9/2 are predicted (see Figure 1), to undergo an interaction with the $\nu 7_3$ intruder orbital at high rotational frequencies, whilst the [521]3/2 is further predicted to exhibit signature splitting at high rotational frequencies. This suggests that a neutron is excited from $\nu 7^2$ orbital into [521]3/2.

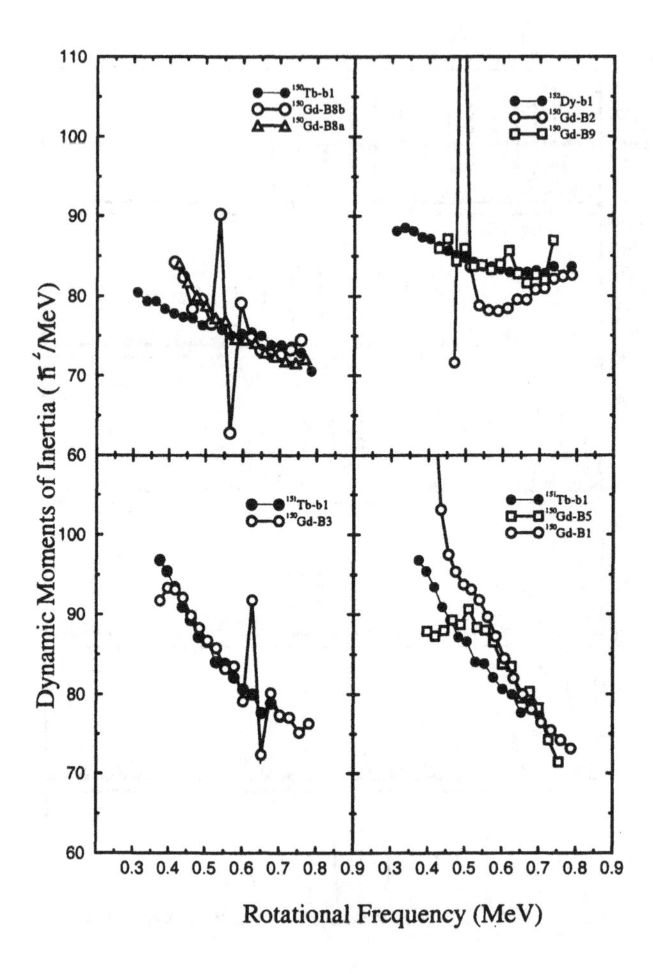

Figure 2 Dynamic Moments of Inertia for bands 2 , 3 , 5 , 8a, 8b and 9 in ^{150}Gd compared with the yrast SD bands of neighbouring nuclei.

3.4 Band 7

Band 7 carries 18% of the intensity of the yrast SD band in ^{150}Gd, suggesting that this band is either based on a single particle-hole excitation involving a large degree of excitation energy, or that there are two particle-hole excitations involved. From an observation of the $\mathcal{J}^{(2)}$ for this band, Figure 5, it is clear that band 7 has similar high-N intruder configuration to that of the yrast band in ^{148}Gd, i.e. $\pi 6^2 \nu 7^1 \otimes ([651]3/2)^{-1}$ [18]. In order to achieve this configuration a neutron must be excited from the 7_2 and the [651]3/2 orbitals into flat natural parity orbitals. The lack of signature partner to band 7 implies that the neutrons from the 7_2 orbital and the [651]3/2 orbital must occupy both signatures of the same Nilsson state. From Figure 4, there are two possible orbitals available to be occupied, namely the ν[402]5/2 and ν[514]9/2 Nilsson

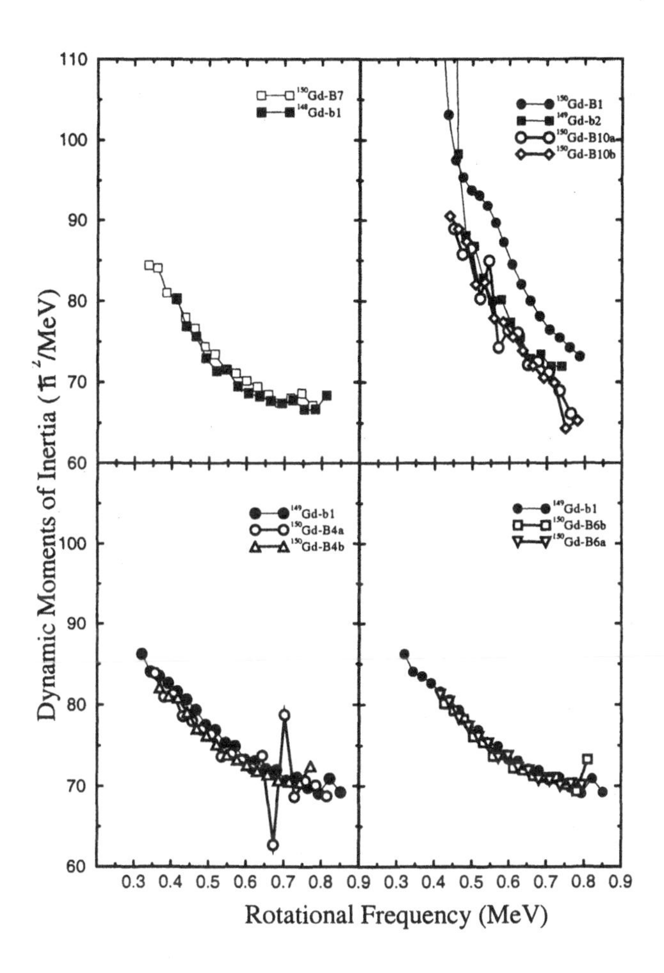

Figure 3 Dynamic Moments of Inertia for bands 1, 4a, 4b, 6a, 6b, 7, 10a and 10b in ^{150}Gd compared with the yrast SD bands of neighbouring nuclei.

states. As the specific ordering of the orbitals close to the N = 86 shell closure is deformation dependent, the actual configuration of band 7 is unclear, however, the configuration based on the occupation of the $\nu[402]5/2$ orbital is preferred based on the arguments presented in previous sections.

3.5 Bands 8a and 8b

The $\mathcal{J}^{(2)}$ moments of inertia for bands 8a and 8b are almost identical (see Figure 2), while the transition energies of band 8b appear to lie at the half-points for those of band 8a, indicating that these two bands are signature partner bands.

In order to determine the high-N configuration of these bands, the $\mathcal{J}^{(2)}$ moments of inertia have been compared with the $\mathcal{J}^{(2)}$ of the yrast SD bands

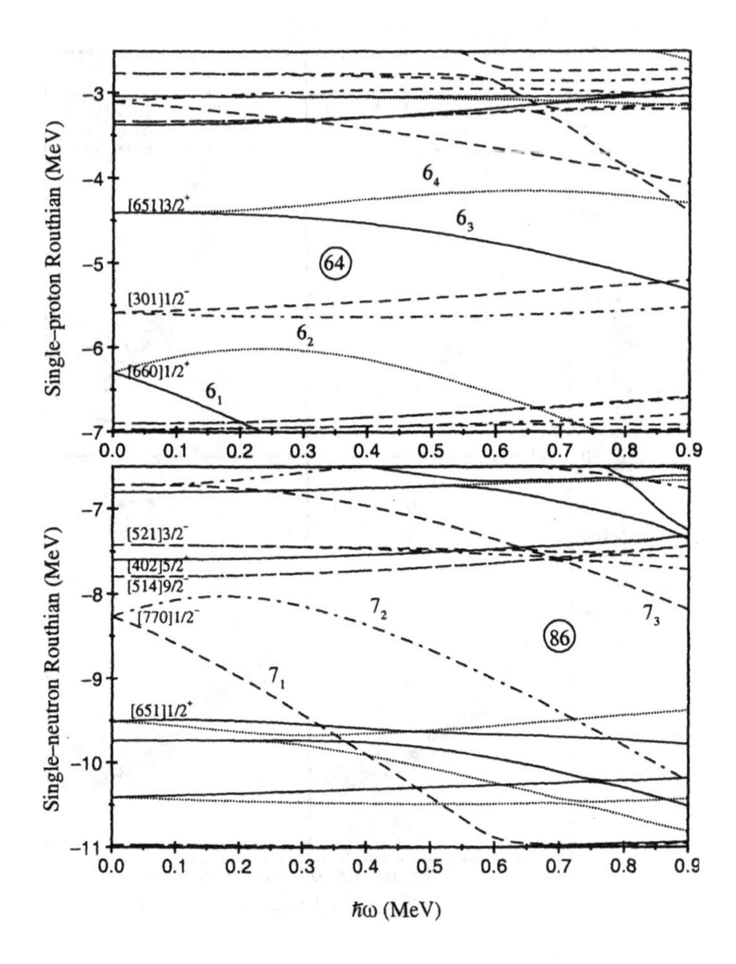

Figure 4 Cranked Woods-Saxon calculations for the single-proton (top) and single-neutron (bottom) orbitals close to the Fermi surface of ^{150}Gd, calculated for β_2 = 0.58, β_4 = 0.1 and γ = 0.0°. (Deformation parameters suitable for the yrast band in ^{150}Gd; dotted lines represent orbitals with $(\pi,\alpha)=(+,-1/2)$; Dashed for$(\pi,\alpha)=(-,-1/2)$; Solid for $(\pi,\alpha)=(+,+1/2)$; dot-dashed for $(\pi,\alpha)=(-,+1/2)$).

in the neighbouring nuclei (see Figure 2). As can be observed from Figure. 2, the $\mathcal{J}^{(2)}$ for these bands is similar to that for the yrast band in ^{150}Tb which has the high-N configuration, $\pi 6^3 \nu 7^1$ [19], suggesting that bands 8a and 8b are based on both a proton and neutron excitation out of the ^{150}Gd SD core.

Due to the similarities in the $\mathcal{J}^{(2)}$ between bands 8a, 8b and the yrast band in ^{150}Tb, it is conjectured that the proton is excited from the unfavoured signature of the [301]1/2 orbital into the 6_3 orbital (see Figure 1), while a neutron is excited from the 7_2 orbital into one signature of either of the [402]5/2, [514]9/2 or [521]3/2 orbitals. As the ν[402]5/2 state is expected to lie lower in energy than the $\nu[514]9/2$ and the $\nu[521]3/2$ states, it is natural to expect that the neutron is excited from the 7_2 orbital into one signature of the [402]5/2 orbital for band 8a, while the opposite signature is occupied for band 8b.

3.6 Band 9

A quadruple gated spectrum for band 9 is presented in Figure 5 As with the previous bands, the $\mathcal{J}^{(2)}$ for band 9 has been plotted and compared to the corresponding $\mathcal{J}^{(2)}$ moments of inertia of the yrast SD bands in adjacent SD nuclei. From Figure 2 it can be observed that the $\mathcal{J}^{(2)}$ moments of inertia for band 9 is almost identical to that for the yrast band in ^{152}Dy, suggesting the intruder configuration $\pi 6^4 \nu 7^2$, indicating that this band is based on a two proton particle-hole excitation from both signatures of the $\pi[301]1/2$ orbital into the 6_3 and 6_4 orbitals (see Figure 4).

Furthermore, from Figure 2 it can be observed that the data points for band 9 lie at almost exactly the same frequency as those for the yrast band in ^{152}Dy.

This observed similarity can be explained in terms of the properties of the orbitals involved in the excitation. It has been observed previously that single-particle excitations involving the [301]1/2 orbital result in an identical band with the yrast band in the Z + 1 nucleus, i.e. ^{150}Gd SD 3 $\rightarrow$ ^{151}Tb yrast SD, and ^{151}Tb SD 2 $\rightarrow$ ^{152}Dy yrast [6]. The observed similarities between these bands were explained in terms of Pseudo SU(3) symmetry (Ref. [7]).

3.7 Bands 10a and 10b

The transition energies of band 10a lie at the half-points of transitions in band 10b, with an r.m.s difference 2.66 keV indicating that these bands are also signature partners. From Figure 3 it can be observed that the $\mathcal{J}^{(2)}$ of these two bands is similar to the first excited band (band 2) in ^{149}Gd [21]. Due to the similarities in the $\mathcal{J}^{(2)}$ moments of inertia it is suggested that bands 10a and 10b have the same high-N configuration as band 2 in ^{149}Gd namely $\pi 6^2 \nu 7^2$ coupled to a hole in the positive signature of the $\nu[651]3/2$.

From Figure 4, the effect of the occupation of the [651]3/2 orbital is clear, as for the yrast band in ^{150}Gd it is occupied, leading to a greater average $\mathcal{J}^{(2)}$ moment of inertia when compared with the band 2 in ^{149}Gd. As the $\mathcal{J}^{(2)}$ moment of inertia for bands 10a and 10b lie close to that for band 2 in ^{149}Gd, it is conjectured that bands are based on the excitation of a neutron from the [651]3/2 into a flat natural parity orbital exhibiting little or no signature splitting (as these bands are assigned to be signature partners). Figure 4 in Ref. [20] shows the calculated single-particle levels for an appropriate deformation. From this figure it is clear that the $\nu[402]5/2$ orbital (while not closest to the Fermi surface) is the lowest lying orbital exhibiting the required behaviour. Thus, bands 10a and 10b are assigned to be based on the excitation of a neutron from the $\alpha = 1/2$ signature of the $\nu[651]3/2$ orbital into both signatures of the $\nu[402]5/2$ orbital.

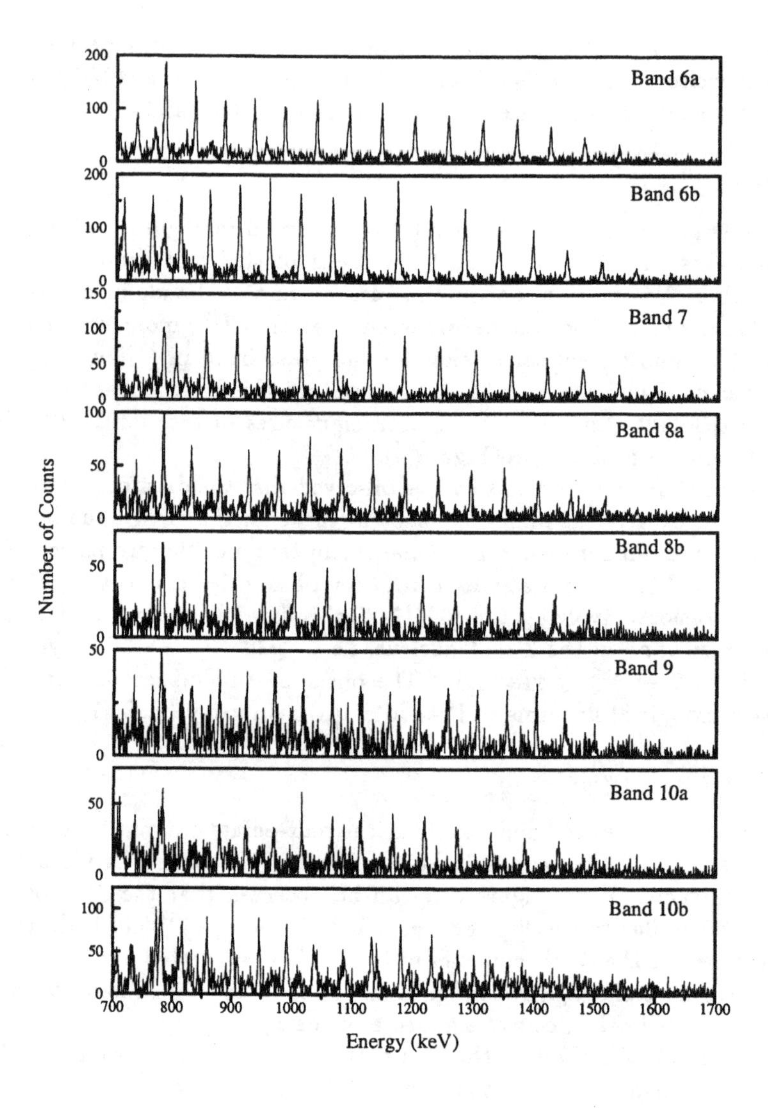

Figure 5 Coincidence spectra (quadruple gated) for SD bands 6-10 in ^{150}Gd. Gates were set on on all band members.

4 CONCLUSIONS

In summary, a total of eight new excited SD bands have been observed in the nucleus ^{150}Gd, with population intensities ranging from 6 to 19 % that of the yrast band. Six of these bands are assigned as signature partner bands based on single-particle excitation from neutron intruder states into the $\nu[521]3/2$ and $\nu[402]5/2$ natural parity states.

The observation of such a large number of excited SD bands in ^{150}Gd has allowed the confirmation and placement of orbitals lying close to the nuclear Fermi surface.

Acknowledgments

We wish to thank the crew of the 88" cyclotron for providing the beam and D.C. Radford for his analysis software [12]. One of us (DEA) acknowledges the support of EPSRC. Author, S. Ertürk would like to akcnowledge Nigde University Association, particularly the head of the Association, Mr. Ahmet Camci and the dean of the Science and Art Faculty, Prof. Dr. Suleyman Bozdemir for providing some financial help to the author to attend this conference. This work has been supported in parts by the U.S. Department of Energy under contracts No. DE-AC03-76SF00098 (LBL) and No. W-31-109-ENG-38 (ANL).

References

[1] P. Fallon *et al.*, Phys. Lett. **B218**, 137 (1989).
[2] S. Clarke, Ph.D Thesis, University of Liverpool, (1995), unpublished.
[3] C.W. Beausang *et al.*, Phys. Lett. B 417, 13 (1998).
[4] C. Baktash, Ann. Rev. Nucl. Part. Sci. 45, 485 (1995).
[5] W. Nazarewicz *et al.*, Nucl. Phys. A503, 3127 (1989).
[6] Byrski*et al.*, Phys. Rev. Lett. 64, 1650 (1990).
[7] W. Nazarewicz *et al.*, Phys. Rev. Lett. 64, 1664 (1990)
[8] C.W. Beausang *et al.*, Phys. Rev. Lett. 71, 1800 (1993).
[9] P. J. Nolan, F. A. Beck and David Fossan, Annu. Rev. Nucl. Part. Sci., 45, 561 (1994)
[10] P. Fallon *et al.*, Phys. Rev. Lett. 73, 782 (1994).
[11] P. J. Twin *et al.*, Phys. Rev. Lett. 57, 811 (1986).
[12] D.C. Radford, Nucl. Instr. Meth. A361 297 (1995).
[13] B. Haas *et al.*, Phys. Rev. Lett. 60, 503(1988).
[14] S. Ertürk *et al.*, to be published.
[15] I. Ragnarsson, Nucl. Phys. A557, 167c (1993).
[16] P.J. Twin *et al.*, in: Proceedings of the Conference on Nuclear Structure at the Limits, Argonne National Laboratory, July 22-26, (1996)
[17] B. Singh, R. B. Firestone and S. Y. F. Yu, Nucl. Inst. Meth. A78, 1 (1996).
[18] M. A. Deleplanque *et al.*, Phys. Rev. Lett. 60, 1626 (1988).
[19] M. A. Deleplanque *et al.*, Phys. Rev. C. 39 , 1651 (1989).
[20] Byrski*et al.*, Phys. Rev. C. 57, 1151 (1998).
[21] B. Haas *et al.*, Phys. Rev. C. 42, R1817 (1990). (1988).

DOUBLE-PHONON γ-VIBRATION OF DEFORMED NUCLEI

C. Fahlander

Division of Cosmic and Subatomic Physics
Lund University, Box 118, S-22100 Lund
Sweden

claes.fahlander@kosufy.lu.se

Abstract: The $K^\pi = 2^+$ rotational bands found at low energy in well-deformed nuclei are generally interpreted as arising from the excitation of a quadrupole collective single-phonon γ vibration. The presently known evidence for the expected $K^\pi=0^+$ and $K^\pi=4^+$ double-phonon γ vibrational states are discussed.

1 INTRODUCTION

In well-deformed nuclei two types of quadrupole shape oscillations are expected around the axially symmetric equilibrium state, one that breaks the axial symmetry, the γ vibration, and one where the axial symmetry is kept in the vibration, the β vibration. A typical low-lying spectrum of a prolate deformed nucleus shows a rotational ground band and low-lying $K^\pi = 0^+$ and $K^\pi = 2^+$ intrinsic excitations. The 2^+ states generally occur at an excitation energy of about 1 MeV, and are believed to be the one-phonon γ-vibrational states with a rotational band superimposed on the vibration. In the harmonic vibrational limit two double-phonon γ vibrational states, with $K^\pi=0^+$ and $K^\pi=4^+$, are expected at about twice the single-phonon energy. However, in this energy regime, at about 2 MeV, we are above the pairing gap, and thus also two quasi-particle excitations are expected in this region. The question of the localization of the double-phonon γ-vibrational strength has therefore been raised because mixing with the two-quasi particle states may limit the collectivity and cause fragmentation of the double-phonon strength over many states.

The Nucleus: New Physics for the New Millennium
Edited by Smit et al., Kluwer Academic / Plenum Publishers, New York, 2000.

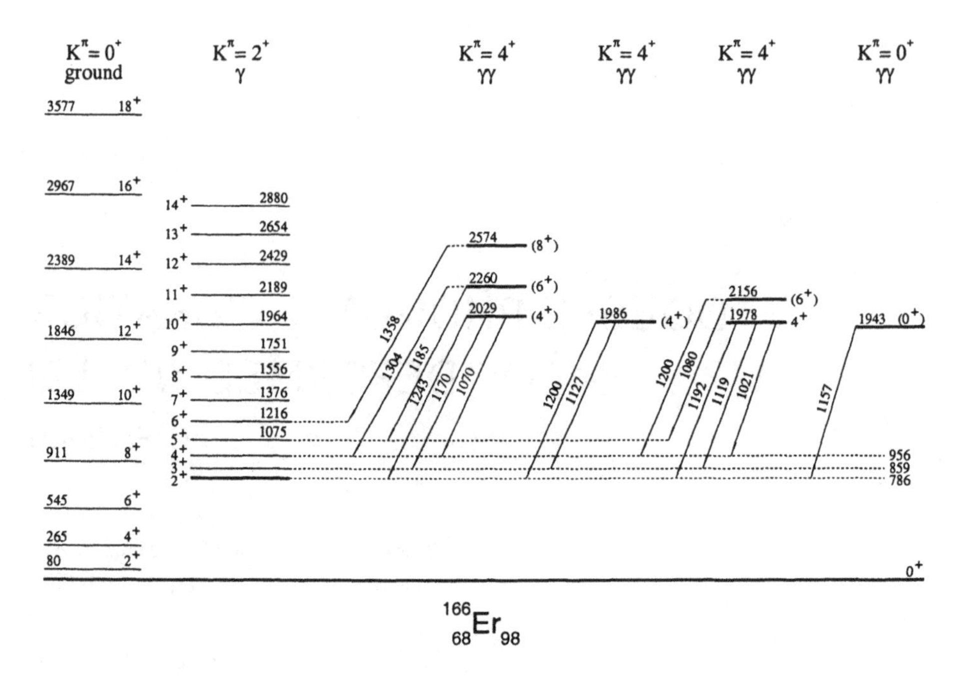

Figure 1 Partial level scheme of ^{166}Er

2 EVIDENCE FOR DOUBLE-PHONON STATES

The only unambiguous measure of double-phonon excitations are absolute E2 matrix elements connecting the double and single phonon states. They can be determined from lifetime measurements of states populated in for example light-ion reactions, such as inelastic neutron scattering or thermal neutron capture, or more directly in nuclear spectroscopy following Coulomb excitation.

Large components of the $K^{\pi} = 4^{+}$ double-phonon γ excitation were found in the transitional ^{186,188,190,192}Os nuclei by Wu et al. [1]. In well-deformed nuclei the strongest evidence for the existence of such two-phonon excitations is a $K^{\pi} = 4^{+}$ band in ^{232}Th populated in Coulomb excitation [2], and a $K^{\pi} = 4^{+}$ state at 2055 keV in ^{168}Er populated in the (n,γ) reaction [3], as well as in inelastic α scattering [4] and Coulomb excitation [5, 6, 7].

The Coulomb excitation experiment on ^{168}Er made use of the Heidelberg Crystal Ball spectrometer as a very efficient multiplicity filter to enhance low-multiplicity events such as those from the weakly populated double-phonon states [5, 6]. The same experimental setup was used to study also ^{166}Er, where two 4^{+} states, which both decay into the γ band, were observed, and a candidate to the 0^{+} two-phonon state was suggested [8] at an excitation energy of 1943 keV. In a more recent Coulomb excitation experiment on ^{166}Er involving the γ-ray detector system GASP of the Legnaro National Laboratory in Italy, the partial level scheme of Fig. 1 was determined [9]. A third 4^{+} state at 1986 keV

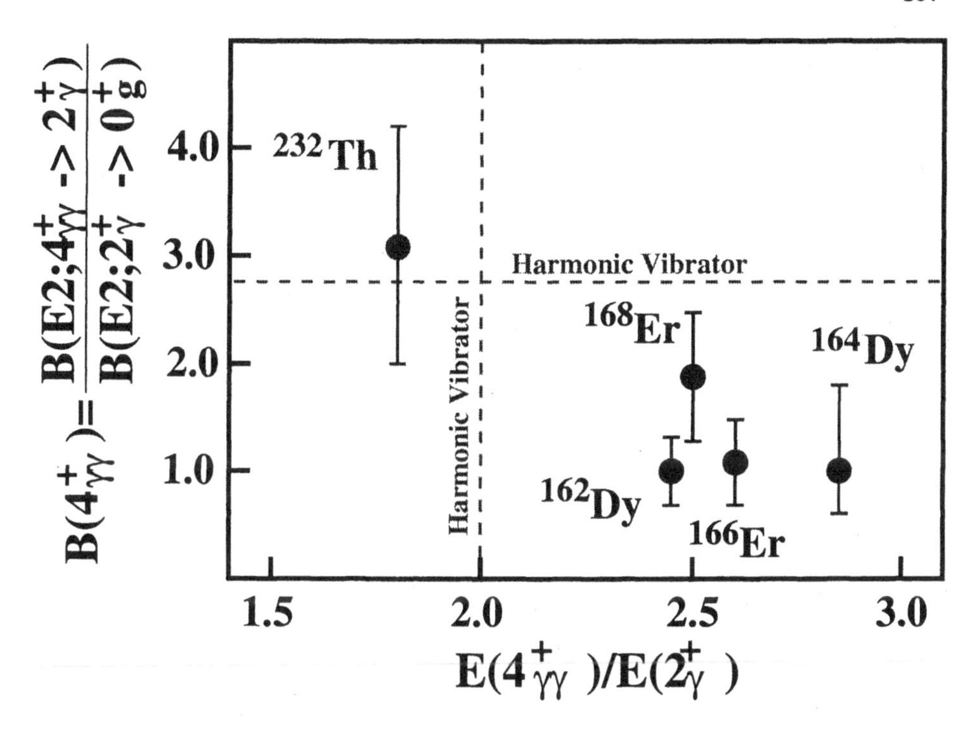

Figure 2 Energy ratios of the double-phonon to single-phonon states, and of the B(E2) ratios of the corresponding transitions

was suggested, and possible rotational states which couple to the vibration are observed on top of the 4^+ states at 1978 and 2029 keV. The angular distribution of the 1157 keV transition is isotropic, supporting the 0^+ assignment of the 1943 keV state [9]. This assignment is also supported by recent data from a neutron scattering experiment [10].

In the well-deformed rare-earth region 4^+ double-phonon γ vibrational states are known, besides in ^{166}Er and ^{168}Er discussed above, only in two more nuclei, namely in ^{162}Dy studied in a recent, yet unpublished, Coulomb excitation experiment [11], and in ^{164}Dy recently measured in neutron scattering [12]. The lifetime of a 4^+ state is also measured in ^{156}Gd [13], but there a significant fragmentation of the double-phonon strength is manifest.

3 DISCUSSION

The measured $B(E2,4^+_{\gamma\gamma} \to 2^+_\gamma)/B(E2,2^+_\gamma \to 0^+_g)$ ratios and the energy ratios $E(4^+_{\gamma\gamma})/E(2^+_\gamma)$ are plotted in Fig. 2 for the deformed nuclei discussed above. The harmonic vibrator values, 2.0 for the energy ratio and 2.78 for the B(E2) ratio, are indicated in the figure.

In the rare-earth nuclei the energy ratio $E(4^+_{\gamma\gamma})/E(2^+_\gamma) \approx 2.5$, i.e., larger than expected from the harmonic vibrator. There is a large positive anharmonicity, which result in an increase of the energy ratio and a reduction of the B(E2) value. It is manifested, however, that a rather significant part of the double-phonon strength is concentrated into one 4^+ state in these nuclei ($\approx$ 50%). The rest of the strength is likely to be fragmented over several other 4^+ states, three of which are observed in ^{166}Er. In contrast, the results in ^{232}Th show evidence for a rather pure harmonic double-phonon excitation, but with a slightly negative anharmonicity in the energies. A more elaborate discussion of the phonon character of these states can be found in ref. [14], where the intrinsic E2 matrix elements between the K=4 and K=2 bands are extracted after correction for band-mixing effects [14].

Already in 1982, before any double-phonon states were known experimentally, Dumitrescu and Hamamoto made some macroscopic and microscopic model calculations [15]. They showed that it was possible to calculate a large energy anharmonicity ($E(4^+_{\gamma\gamma})/E(2^+_\gamma) \approx 2.5$) in the Bohr hamiltonian by assuming that the potential energy surface of the nucleus has a minimum in γ slightly different from zero, i.e. if the γ vibration (dynamic triaxiality) takes place around a prolate deformed ground state, which is slightly triaxial.

It is also possible to understand the energy anharmonicities within the interacting boson model. The parameter χ of the IBM-1 hamiltonian varies from -1.3 in the SU(3) limit to 0 in the O(6) limit, and if one plots the calculated energy of the two-phonon state as a function of that parameter, it is found that most deformed nuclei have positive anharmonicity, and a few have negative anharmonicity. However, to understand the large observed anharmonicities in the IBM model one has to include a three-body term in the boson operators of the IBM hamiltonian [16]. With such a a cubic term the γ-band structure in ^{166}Er was well reproduced as well as the E2 transition rates [16]. To some extent such a cubic term is equivalent to the triaxiallity suggested by Dumitrescu and Hamamoto [15].

4 SUMMARY

In summary, candidates for the 4^+ double-phonon γ vibration have now been observed in five well-deformed nuclei. The only candidate for a 0^+ two-phonon γ vibrational state is found in ^{166}Er. In ^{232}Th a negative anharmonicity is observed in the energy of the double-phonon state. In the rare-earth nuclei positive anharmonicity effects result in an increase of the energy ratio, and in a reduction of the B(E2) value as compared to the harmonic vibrator prediction. This can be understood in the geometric model as being due to a γ vibration around a slightly γ deformed equilibrium shape of these otherwise strongly β deformed nuclei or by introducing a cubic term in the IBM hamiltonian.

References

[1] C.Y. Wu et al., Nucl. Phys. A607, 178 (1996)

[2] W. Korten et al., Phys. Lett. B317, 19 (1993)

[3] H.G. Börner et al., Phys. Rev. Lett. 66, 691 (1991)

[4] R. Neu, F. Hoyler, Phys. Rev. C46, 208 (1992)

[5] M. Heinebrodt, Diploma Thesis, University of Heidelberg (1994)

[6] T.Härtlein et al., European Physical Journal A **A2** 253 (1998)

[7] M. Oshima et al., Phys. Rev. C52, 3492 (1995)

[8] C. Fahlander et al., Phys. Lett. **B388**, 475 (1996)

[9] C. Fahlander et al., Proceedings of the International Symposium on Exotic Nuclear Shapes, Debrecen, Hungary, 1997

[10] P.E. Garret, et al., Phys. Rev. Lett. **78**, 4545 (1997)

[11] H. Bauer, T.Härtlein et al., private communication

[12] F. Corminboeuf et al., Phys. Rev. C56, R1201 (1997)

[13] A. Bäcklin et al., Nucl. Phys. **A380**, 189 (1982)

[14] C.Y. Wu and D. Cline, Phys. Lett. **B382**, 214 (1996)

[15] T.S. Dumitrescu and I. Hamamoto, Nucl. Phys. **A383**, 205 (1982)

[16] J.E. Garcia-Ramos et al., Phys. Lett. B, submitted

N-PARITY PROJECTED BCS THEORY

R. Balian[1], H. Flocard[2] and M. Vénéroni[2]

[1]CEA/Saclay, Service de Physique Théorique
91191 Gif-sur-Yvette Cedex, France
[2]Groupe de Physique Théorique, Institut de Physique Nucléaire
91406 Orsay Cedex, France

flocard@ipno.in2p3.fr

Abstract: Odd-even effects are manifest in many properties of small superconducting systems such as nuclei and nanometric aluminium grains. The BCS equations provide a natural tool to investigate these effects at zero temperature. Here we give and discuss the equations which generalize the BCS formalism at finite temperatures while ensuring (contrary to standard finite temperature BCS) an exact conservation of the particle number parity. Special attention is given to the low temperature limits in both the nuclear physics and the condensed matter contexts.

In nuclear physics, the qualitative differences of properties between even and odd systems have long been known, as well as their explanation by pairing correlations[1]. Indeed, the ground and lowest excited states of nuclei are generally well described by the BCS theory and its Hartree-Fock-Bogoliubov (HFB) generalization[2]. At zero temperature, the BCS state is a superposition of configurations with even particle number only, which allows the model to discriminate between odd and even systems. However, for small superconducting systems at non-zero temperature, the thermal BCS Ansatz mixes up configurations with even and odd particle numbers that have dissimilar physical properties. We have in mind metallic islands[3], ultra-small Aluminium grains[4] as well as hot nuclei which will soon be investigated intensively by means of the next generation of γ-ray arrays. The additional spurious mixing introduced by the temperature requires projecting the thermal BCS state onto subspaces having, at least, a well-defined number-parity (N-parity). Of course, a more faithful description would be furnished by the (more difficult) projection on states with a well-defined particle number. Nevertheless, the similarity of the properties of

The Nucleus: New Physics for the New Millennium
Edited by Smit et al., Kluwer Academic / Plenum Publishers, New York, 2000.

systems with the same N-parity suggests that the N-parity projection should already produce significant improvements.

The problem then is to devise an extension of the BCS theory which includes properly the N-parity projection. Several studies have been recently devoted to this question[5, 6] and formulas have been given whose interpretation and accuracy seem difficult to ascertain. In Ref.[7] , starting from a general variational principle, a method has been worked out for the construction of extended mean-field approximations at finite temperature which *consistently* incorporate projections, whether an invariance symmetry is broken or not. It seems worthwhile to derive correct equations for these projected mean-field approximations since (apart from exact calculations) they are most often the starting point for more sophisticated (and sometimes hypothetical) improvements. Here we discuss the relatively simple case of the N-parity projection

$$\hat{P}_\eta = \tfrac{1}{2}\left(\hat{1} + \eta\, e^{i\pi\hat{N}}\right) , \qquad \eta = \pm 1 \tag{1}$$

on the subspaces with even ($\eta = +1$) or odd ($\eta = -1$) particle number and we compare the properties of the grand-canonical BCS formalism when N-parity projection is enforced and when it is not. For the Hamiltonian we take the usual form

$$\hat{H} = \sum_p \epsilon_p (a_p^\dagger a_p + a_{\bar{p}}^\dagger a_{\bar{p}}) - \sum_{pq} G_{pq} a_p^\dagger a_{\bar{p}}^\dagger a_{\bar{q}} a_q \quad . \tag{2}$$

which corresponds to a set of doubly degenerate single-particle states (the labels p and $\bar{p}$ are associated with the same single-particle energy ϵ_p) and a pairing interaction coupling the p and $\bar{p}$ states with a strength characterized by the matrix elements G_{pq}. The (unnormalized) grand canonical density matrix is taken of the form $\hat{\mathcal{D}}_{\rm BCS} = e^{-\beta\hat{W}}$ where the operator $\hat{W}$ is taken of the independent quasi-particle form :

$$\hat{W} = (h - \sum_p e_p) + \sum_p e_p (\, b_p^\dagger b_p + b_{\bar{p}}^\dagger b_{\bar{p}}\,) \quad . \tag{3}$$

In Eq. (3) h is a c-number and $b_p^\dagger$ and $b_{\bar{p}}^\dagger$ denote the quasi-particle creation operators obtained from the particle operators by means of a BCS transformation:

$$\left\{ \begin{array}{rl} b_p = & u_p a_p + v_p a_{\bar{p}}^\dagger \\ b_{\bar{p}} = & u_p a_{\bar{p}} - v_p a_p^\dagger \end{array} \right. , \tag{4}$$

where the quantities u_p and v_p are constrained by the relation ${v_p}^2 + {u_p}^2 = 1$. For the N-parity projected density matrix we take $\hat{\mathcal{D}}_\eta(\beta) = \hat{P}_\eta\, e^{-\beta\hat{W}}$ with $\eta = \pm 1$. In each of the three cases (BCS, $\eta = +$, $\eta = -$) the optimal operator $\hat{W}$ is determined by independent variations with respect to the set of variational parameters : h, v_p and e_p. It must be kept in mind that self-consistency leads to three different sets of variational quasi-particle operators and energies.

Before a discussion of the results at finite temperature let us recall some properties of the $T = 0$ BCS approximation. At zero temperature, as said above, the BCS approximation takes exactly into account N-parity. The following formula gives the odd and even BCS ground state wave-function and the corresponding gap equation.

$$\begin{array}{cc} \text{Even, } \eta = + & \text{Odd, } \eta = - \\ |\Phi_+\rangle = \prod_p (u_p + v_p a_p^\dagger a_{\bar{p}}^\dagger)|0\rangle & |\Phi_-\rangle = a_0^\dagger \prod_{p\neq 0}(u_p + v_p a_p^\dagger a_{\bar{p}}^\dagger)|0\rangle \\ \Delta_{+p} = \frac{1}{2}\sum_q G_{pq}\frac{\Delta_{+q}}{\tilde{e}_q} & \Delta_{-p} = \frac{1}{2}\sum_{q\neq 0} G_{pq}\frac{\Delta_{-q}}{\tilde{e}_q} \end{array} \tag{5}$$

The odd wave-function is obtained by removing the pair associated with the level (hereafter labeled $p = 0$) whose single-particle energy ϵ_0 is closest to the fermi energy μ and by creating exactly one particle on this level. The odd and even gap equations differ only in that in the the former the summation does not involve the $p = 0$ index because this level is blocked by the unpaired particle created by $a_0^\dagger$. Depending on the level density at the Fermi energy, this blocking effect can induce a significant reduction of the gap. The gap formula involves the BCS quasi-particle energies $\tilde{e}_q = \sqrt{(\epsilon_q - \mu)^2 + \Delta^2_{\eta,q}}$. (In anticipation of the N-parity projected results, we have introduced a new symbol $\tilde{e}_q$ which until we consider the N-parity projected cases turns out to be identical to the variational parameter e_q in $\hat{W}$).

In the grand canonical BCS approximation the density matrix $\hat{\mathcal{D}}_{\rm BCS} = {\rm e}^{-\beta\hat{W}}$ is a statistical mixture of even and odd states. The relation of quasi-particle energies to the single-particle energies and the pairing gap is the same as at $T = 0$ while the gap equation becomes :

$$\Delta_{{\rm BCS}\,p} = \frac{1}{2}\sum_q G_{pq}\frac{\Delta_{{\rm BCS}\,q}}{\tilde{e}_q}\,t_q \tag{6}$$

It is therefore very close formally to the $T = 0$ even-N BCS equation from which it differs only via the multiplication of each term in the right-hand side by the thermal factor $t_q = \tanh\beta\, e_q/2$ where $\beta = 1/T$. As T grows, the thermal factor decreases leading to smaller values for the gaps until a transition temperature is reached beyond which pairing is no more active.

We discuss now results obtained by means of the formalism of ref.[7] for the N-parity projected grand canonical BCS approximation. In this case the variational density matrix is $\hat{\mathcal{D}}_\eta = \hat{P}_\eta\,{\rm e}^{-\beta\hat{W}}$. Then, for the gap equation one obtains

$$\Delta_{\eta\,p} = \frac{1}{2}\sum_q G_{pq}\frac{\Delta_{\eta\,q}}{\tilde{e}_q}\,t_q\,\frac{1 + \eta\,r_0\,{t_p}^{-2}\,{t_q}^{-2}}{1 + \eta\,r_0\,{t_p}^{-2}} \tag{7}$$

Compared to the unprojected case the sole formal change is the appearance on the R.H.S. of a factor which depends on the thermal factors defined above directly and indirectly through the quantity r_0

$$r_0 = \frac{\mathrm{Tr}\, e^{i\pi\hat{N}} e^{-\beta\hat{W}}}{\mathrm{Tr}\, e^{-\beta\hat{W}}} = \prod_p t_p{}^2 \tag{8}$$

The quasi-particle energies are not exactly of the standard BCS form. They can be written $e_q = \sqrt{(\epsilon_q - \mu)^2 + \Delta^2_{\eta,q}} + \eta\,\mathrm{Corr.Terms}$. The explicit form of the correction terms is given in ref.[7]. Since numerical calculations show that their magnitude is generally small we shall not consider them any further. The coefficients u_p and v_p are then given explicitly by the normalization condition $u_p{}^2 + v_p{}^2 = 1$ and by the relation $2u_pv_p/(u_p{}^2 - v_p{}^2) = \Delta_p/(\epsilon_p - \mu)$. All the quantities Δ_p, e_p, r_0, u_p, v_p depend on the N-parity index $\eta = \pm 1$. The standard BCS equations are recovered for $\eta = 0$.

An analysis of the last fraction in the R.H.S. of Eq.(7) leads to the following conclusions : (i) in the sum of Eq. (7), the modifications due to projection affect predominantly the terms corresponding to the single-particle states near the Fermi surface. (ii) the even-number projected gap is larger than the BCS one while the odd-number gap is smaller. (iii) larger differences with BCS are expected for the projection on an odd number of particles.

Once the self-consistent equations are solved, the *projected* average particle number, energy, entropy can be obtained from the formulae

$$\langle\hat{N}\rangle_\eta = \sum_p [1 - \frac{\epsilon_p - \mu}{e_p} t_p \frac{1 + \eta\, r_0 t_p{}^{-2}}{1 + \eta\, r_0}] \tag{9}$$

$$\langle\hat{H}\rangle_\eta = \sum_p [\epsilon_p - \frac{\epsilon_p(\epsilon_p - \mu) + \frac{1}{2}\Delta^2_{\eta p}}{e_p} t_p \frac{1 + \eta\, r_0 t_p{}^{-2}}{1 + \eta\, r_0}] \tag{10}$$

$$\tilde{S}_\eta = S_{\mathrm{BCS}} - \frac{\beta\eta\, r_0}{1 + \eta\, r_0} \sum_p e_p(t_p{}^{-1} - t_p) + \ln[\frac{1}{2}(1 + \eta\, r_0)] \tag{11}$$

In the formulae for $\langle\hat{N}\rangle$ and $\langle\hat{H}\rangle$ the projection (as compared to the grand canonical expressions of the same quantities) manifests itself only via the fraction $(1 + \eta\, r_0 t_p{}^{-2})/(1 + \eta\, r_0)$. Other thermodynamic quantities such as the Gibbs free energy $F_{\mathrm{G}\eta}$ and the partition function $\tilde{Z}_\eta$ can then be deduced by means of standard thermodynamics relations :

$$F_{\mathrm{G}\eta} = \langle\hat{H}\rangle_\eta - \mu\langle\hat{N}\rangle_\eta - \frac{1}{\beta}\tilde{S}_\eta \quad , \qquad \ln\tilde{Z}_\eta = -\beta\, F_{\mathrm{G}\eta} \tag{12}$$

One interesting property of the self-consistent formalism which leads to the above equations is that it preserves thermodynamic identities among the different approximate quantities. For instance one has

$$\langle\hat{H}\rangle_\eta - \mu\langle\hat{N}\rangle_\eta = -\frac{\partial \ln\tilde{Z}_\eta}{\partial\beta} \quad , \qquad \tilde{S}_\eta = -\frac{\partial F_{\mathrm{G}\eta}}{\partial T} \tag{13}$$

In order to establish a link with the BCS description of ground states, we study now the limit of Eqs. (6), (7) and (11) at small temperature. We first consider the situation where the single-particle level spacing at the Fermi surface is at the same time larger than the temperature and smaller than the $T = 0$ pairing gap. Nuclei and small metallic grains with pairing provide physical illustrations of such a situation.

For large values of β, the usual BCS gap equation becomes

$$\Delta_{\mathrm{BCS}\,p} = \frac{1}{2} \sum_{q} G_{pq} \frac{\Delta_{\mathrm{BCS}\,q}}{\tilde{e}_q} \left(1 - 2\mathrm{e}^{-\beta e_q}\right) \quad . \tag{14}$$

The terms proportional to $\exp(-\beta e_q)$ lead to a decrease of the gaps versus the temperature $T = 1/\beta$. In the same limit, the BCS entropy is approximated by

$$S_{\mathrm{BCS}} = 2 \sum_{p} \beta e_p \, \mathrm{e}^{-\beta e_p} \tag{15}$$

For the projection on even particle number the expansion of gap equation for $\eta = +1$ yields :

$$\Delta_{+\,p} = \frac{1}{2} \sum_{q} G_{pq} \frac{\Delta_{+\,q}}{\tilde{e}_q} \left(1 - 2\mathrm{e}^{-2\beta e_q} - 4 \sum_{r\,(r \neq p,q)} \mathrm{e}^{-\beta(e_q + e_r)}\right) \quad . \tag{16}$$

One thus retrieves, at $T = 0$, the BCS equation for even systems. The terms which control the low-temperature corrections to the gaps are also negative. However, they are now of the order of, or smaller than, $\exp(-2\beta\Delta_{+\,0})$ while they were of the order of $\exp(-\beta\Delta_{\mathrm{BCS}\,0})$ for BCS (near $T = 0$, $\Delta_{+\,0} \simeq \Delta_{\mathrm{BCS}\,0}$). This reflects the fact that any excitation with fixed N-parity requires the creation of at least two quasi-particles and therefore an energy larger than $2\Delta_{+\,0}$. Therefore the decrease of the pairing gaps with increasing temperature will be slower than for the BCS case. The low temperature behaviour of the entropy (11) is given by

$$\check{S}_{+} = 2 \sum_{p} \beta e_p \, \mathrm{e}^{-2\beta e_p} + 2 \sum_{p\,q\,(p \neq q)} \beta(e_p + e_q) \, \mathrm{e}^{-\beta(e_p + e_q)} \tag{17}$$

Like for all other thermodynamic quantities, the T-dependence is governed by terms smaller than $\exp(-2\beta\Delta_{+\,0})$.

For N odd ($\eta = -1$) and T small, we consider the limit of the projected gap equation for $p \neq 0$ (for a discussion of the case $p = 0$ see ref.[7]). At low temperature both the numerator and the denominator in the last fraction in Eq.(7) vanish as $\exp(-\beta e_0)$. After dividing out by this common factor, we obtain the well-conditioned equation:

$$\Delta_{-\,p} = \frac{1}{2} \sum_{q\,(q \neq 0)} \left\{ G_{pq} \frac{\Delta_{-\,q}}{\tilde{e}_q} + \left(G_{p0} \frac{\Delta_{-\,0}}{\tilde{e}_0} - G_{pq} \frac{\Delta_{-\,q}}{\tilde{e}_q} \right) \mathrm{e}^{-\beta(e_q - e_0)} \right\} \tag{18}$$

As expected, at $T = 0$, this equation is identical with that for the odd-N BCS ground state. As $\beta \to \infty$, the projected gap equation automatically singles out the level with the lowest quasi-particle energy and generates the odd particle-number BCS ground state. The blocking approximation thus emerges naturally in the zero-temperature limit. The low-temperature correction to the gap Δ_{-p} is positive and depends on terms which are proportional to the exponential of the difference $e_q - e_0$. Therefore, starting from the value provided by Eq. (5) (right column), one expects the gap to first *increase* with temperature as $\exp(-\beta(e_1 - e_0))$.

At low-temperature the entropy (11) behaves as

$$\check{S}_- = \ln 2 + \sum_{p\,(p\neq 0)} \beta\,(e_p - e_0)\,\mathrm{e}^{-\beta(e_p - e_0)} \tag{19}$$

When $\beta \to \infty$, the limit of $\check{S}_-$ is equal to ln2. This reflects the twofold degeneracy of the ground state.

In the physics of mesoscopic superconductors, one often considers another type of low temperature regime. It corresponds to a situation in which the twofold degenerate unperturbed energies ϵ_p are sufficiently dense near the Fermi surface to allow the replacement of a sum over p by an integral over $\epsilon \equiv \epsilon_p - \mu$ with a weight w_{F} equal to the level density. One also assumes that the pairing matrix elements are constant and equal to $\tilde{G}$ over an interval Λ around the Fermi energy μ, and that they vanish outside. The width of this interval, taken of the order of the Debye energy, is assumed to be much larger than the pairing gap. It must be finite, however, to ensure the convergence of the integrals.

The ordinary BCS gap equation at zero temperature

$$\frac{2}{\tilde{G}} = \int_{-\Lambda/2}^{\Lambda/2} \frac{w_{\mathrm{F}}\,d\epsilon}{\sqrt{\epsilon^2 + \Delta^2}} \quad , \tag{20}$$

then defines a p-independent quantity Δ that we shall take in this Subsection as a reference energy. In this field of physics, the "low-temperature" regime actually refers to an intermediate range of temperatures such that the conditions

$$\frac{1}{\Lambda} \ll \frac{1}{\Delta} \ll \beta \ll w_{\mathrm{F}} \tag{21}$$

are valid. Under these conditions the gap equation gives analytical expressions for the low temperature dependence of the BCS pairing gap :

$$\Delta_{\mathrm{BCS}}(\beta) = \Delta \left[1 - \sqrt{\frac{2\pi}{\beta\Delta}}\,\mathrm{e}^{-\beta\Delta}\right] \quad , \tag{22}$$

and of the BCS entropy :

$$S_{\mathrm{BCS}} = 2\,w_{\mathrm{F}}\Delta\,\sqrt{2\pi\beta\Delta}\,\mathrm{e}^{-\beta\Delta} \quad . \tag{23}$$

For the even particle number case, Eq. (7) provides an analytical expression for the low-temperature dependence of the even-N gap $\Delta_+(\beta)$:

$$\Delta_+ = \Delta \left[1 - 4\pi \frac{w_F}{\beta} e^{-2\beta\Delta} \right] \quad . \tag{24}$$

The extrapolation to zero temperature of Δ_+ is equal to the BCS gap Δ. Under the same conditions, one can can rewrite the expression for the entropy as

$$\check{S}_+ = 8\pi \left(w_F \Delta\right)^2 e^{-2\beta\Delta} \quad . \tag{25}$$

The BCS entropy is thus larger than $\check{S}_+$. Indeed, the disorder decreases when odd-particle components are removed.

In the odd-N case, we cannot use the equations obtained in the extreme low-T limit as a starting point since they are only valid when $w_F \leq \beta$ rests on the condition $1/\Delta \ll \beta$). We must work directly on the general projected equations (7) which, in the condensed matter low-T regime defined by (21), yields

$$\Delta_-(\beta) = \Delta \left[1 - \frac{1}{2w_F\Delta} + \frac{1}{4\beta w_F \Delta^2} \right] \quad . \tag{26}$$

At zero temperature the odd-N gap Δ_- is therefore smaller than the BCS gap by the quantity $1/2w_F$. This property is related with the blocking of the $p = 0$ level located at the Fermi surface, as described by the $T = 0$ odd-N gap equations (5). In the temperature range considered here, the gap grows linearly with temperature. In contrast, in the very low temperature (18) regime investigated above ($\beta \geq w_F$), the gap grows exponentially as $\exp(-\beta(e_1 - e_0)) \simeq \exp(-\beta/2{w_F}^2\Delta)$.

The odd-N projected thermodynamic quantities have a larger temperature variation than the corresponding BCS ones. This results from the fact that the excitation energies are then of the order of the inverse of ${w_F}^2\Delta$ and therefore much smaller than w_F^{-1} and therefore Δ. For the entropy (11) we obtain :

$$\check{S}_- = \ln 2 + \frac{1}{2} + \frac{1}{2}\ln\left(\frac{2\pi\, {w_F}^2 \Delta}{\beta}\right) \quad , \tag{27}$$

which is the expression of the *Sakur-Tetrode entropy for one classical particle* [8] of mass Δ, with an internal twofold degeneracy, enclosed in a one-dimensional box of length $2\pi\hbar\, w_F$. While elementary excitations of even-N systems require energies of at least 2Δ, the effective spectrum for odd-N systems is nearly continuous, with the form $m^2/2w_F^2\Delta$ where m is an integer. This occurence of *gapless elementary excitations* is consistent with recent experiments on the conductance of odd-N ultrasmall Aluminium grains[4]. The analogy with the Sakur-Tetrode entropy shows also that in the variational projected solution, the unpaired particle does not remain on the single-particle level of energy $\epsilon_0 = \mu$. The projected solution is a coherent superposition of configurations in which this particle explores all single-particle levels such that $\beta|\epsilon_p - \mu| \leq 1$. The fact that we recover a formula of classical thermodynamics is another indication that, strictly speaking, the regime considered here is not a low temperature regime.

The inequalities $\check{S}_- > S_{\rm BCS} > \check{S}_+$ implied by Eqs. (15,17,19) on the one hand and (23,25,27) on the other, reflect the following fact. When the system is constrained to have an even particle number, we have seen that its lowest excited states are obtained from the ground state by the creation of two quasi-particles, which requires at least an energy 2Δ. For an odd system, the excitation of the (gapless) unpaired last particle is much easier.

References

[1] A.Bohr and B.R.Mottelson, *Nuclear Structure*, Vol.II (W.A. Benjamin, 1975).

[2] P.Ring and P.Schuck, *The Nuclear Many-Body Problem* (Springer-Verlag, New-York/Berlin, 1980).

[3] M.T.Tuominen, J.M.Hergenrother, T.S.Tighe, and M.Tinkham, Phys. Rev. Lett. 69, 1997 (1992); P.Lafarge, P.Joyez, D.Esteve, C.Urbina, and M.H.Devoret, Phys. Rev. Lett. 70, 994 (1993).

[4] D.C.Ralph, C.T.Black, and M.Tinkham, Phys. Rev. Lett. 74, 3241 (1995); 76, 688 (1996); 78, 4087 (1997).

[5] D.V.Averin and Yu.V.Nazarov, Phys. Rev. Lett. 69, 1993 (1992); B.Jankó, A.Smith, and V.Ambegaokar, Phys. Rev. B 50, 1152 (1994); D.S.Golubev and A.D.Zaikin, Phys. Lett. A 195, 380 (1994); R.A.Smith and D.V.Ambegaokar, Phys. Rev. Lett. 77, 4962 (1996); K.A.Matveev and A.I.Larkin, Phys.Rev. Lett. 78, 3749 (1997).

[6] J. von Delft, A.D.Zaikin, D.S.Golubev, and W.Tichy, Phys. Rev. Lett. 77, 3189 (1996), F.Braun, J. von Delft, D.C.Ralph, and M.Tinkham, Phys. Rev. Lett. 79, 921 (1997).

[7] R.Balian, H. Flocard, and M. Vénéroni, Variational Extensions of BCS Theory, nuc-th/9706041, to be published in Physics Reports.

[8] R. Balian, "From Microphysics to Macrophysics", Vol. I, Chapt.4, Springer-Verlag, Berlin/Heidelberg, 1991.

ACCESS TO GAMMA-RAY SPECTROSCOPY OF NEUTRON-RICH *sdfp* SHELL NUCLEI

B.Fornal[1], R.Broda[1], W.Królas[1], T. Pawłat[1], J.Wrzesiński[1], D.Bazzacco[2], S.Lunardi[2], C.Rossi-Alvarez[2], G.Viesti[2], G. de Angelis[3], M.Cinausero[3], D.Napoli[3], K.Helariutta[4], P.M.Jones[4] R.Julin[4], S.Juutinen[4], A.Savelius[4], P.A.Butler[5], J.F.C.Cocks[5] J. Gerl[6]

[1]H. Niewodniczański Institute of Nuclear Physics, Cracow, Poland;
[2]l'Universita' and INFN, Padova, Italy;
[3]INFN Laboratori Nazionali di Legnaro, Italy;
[4]University of Jyväskylä, Finland;
[5]Oliver Lodge Laboratory, University of Liverpool, U.K.
[6]GSI, Darmstadt, Germany

Abstract: γ-rays in neutron-rich *sdfp* shell nuclei, produced in deep-inelastic processes during collisions of ^{37}Cl and ^{40}Ar ions on ^{208}Pb and of ^{48}Ca ions on ^{48}Ca, have been studied using large Ge multidetector arrays. Candidates for new yrast states in heavy argon and sulfur isotopes have been identified.

1 INTRODUCTION

In nuclear structure of poorly known neutron-rich nuclei around N=20 the excitations of neutrons across the N=20 gap play a crucial role. In particular, the Al, Mg and Na nuclei with N=20, according to experimental evidence and theoretical calculations, belong to the so called "island of inversion" [1] - their ground states are supposed to be dominated by the $(\nu f_{7/2})^2$ configuration, giving rise to large deformations. Also, neutron-rich nuclei close to the N=28 shell closure have recently attracted a particular interest. Werner et al. [2], on the basis of self-consistent mean-field calculations, suggested, that the major N=28 shell gap disappears when approaching Z=16.

Some of these nuclei have been investigated using a technique of intermediate-energy Coulomb excitation of radioactive beams. Energies of the 2_1^+ states and $B(E2;0_{g.s}^+ \rightarrow 2_1^+)$ values in $^{40,42}S$ and in $^{44,46}Ar$ [3] have been measured. The studies, however, could not locate higher excitations.

In a series of recent experiments we have shown that the yrast spectroscopy of hard-to-reach neutron-rich nuclei, populated in heavy-ion multinucleon transfer reactions (~15% above Coulomb barrier), can be studied very successfully in γ-γ thick target coincidence measurements (e.g. [4, 5]). In such deep-inelastic collisions the population of yrast states at moderate spins is strongly favoured.

2 EXPERIMENTS AND RESULTS

We have performed several experiments aiming at elucidating the structure of neutron-rich *sdfp* shell nuclei.

2.1 $^{37}Cl + {}^{208}Pb$ reaction

By using the beam of 230 MeV ^{37}Cl ions from the ALPI accelerator on a thick ^{208}Pb target and the GASP array, we identified candidates for two new negative parity states associated with the promotion of neutrons across the N=20 gap in neutron excessive ^{32}Si and ^{32}Al nuclei [6]. Figure 1 shows the decay scheme of ^{32}Al with a candidate for the 4^- excitation at 1178 keV identified in this work. Yrast states in ^{34}P N=19 isotone are displayed for comparison. In these N=19 isotones a significant lowering of the excitation energies of the 4^- levels, corresponding to $\nu d_{3/2} f_{7/2}$ configurations, occurs when going from Z=15 to Z=13. It is very likely that these configurations will become very low lying if not dominating ground states in nuclei with N=19 and Z=11,12. Such a scenario is indeed predicted by large-scale shell model calculations involving excitations of neutrons from the *sd* to the *fp* shell.

2.2 $^{40}Ar + {}^{208}Pb$ reaction

Another possibility to access new nuclei in the N=20-28 neutron-rich region is offered by an ^{40}Ar beam. We performed an experiment with that beam at the Accelerator Laboratory of the Jyväskylä University, using again a thick ^{208}Pb target. The γ-γ coincidences were collected with the DORIS array.

Among several interesting findings, we were able to gain new information on the structure of the ^{40}S nucleus. In this nucleus the first excited state 2^+ was identified at 891(13) keV using the method of intermediate-energy Coulomb

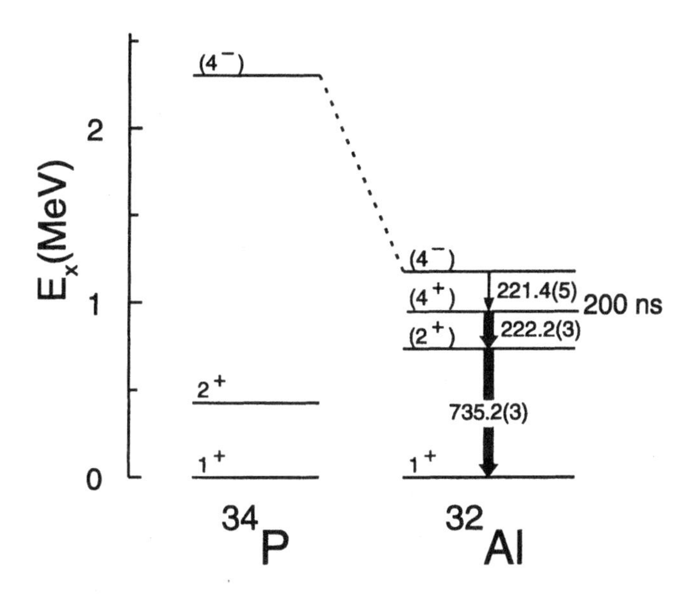

Figure 1 Decay scheme for ^{32}Al. Yrast states in ^{34}P isotone are shown for comparison

excitation of a radioactive beam. Another source of information on excited states in ^{40}S was the study of ^{40}P β-decay [7], in which ten γ-rays have been observed and placed in the level scheme. The 2^+ state, located in that work with higher precision at 903.5 keV, was found to be fed by several γ-rays, including 1013.1 and 1351.0 keV transitions.

By using our prompt gamma-gamma matrix and setting a gate on the 903.5 keV line we were able to display a coincident 1351 keV transition. There was no trace of any other line feeding the 2^+ state. The yield of 903 - 1351 keV coincidences was found to be consistent with the production yield of ^{40}S calculated using known γ-ray cascades emitted in the β-decay of ^{40}S itself. Taking into account that the multinucleon transfer processes, which are responsible for production of ^{40}S, populate selectively yrast states, we conclude that a level at 2254 keV, which decays by a 1351 keV γ-ray to the 2^+ state, is a perfect candidate for a 4^+ excitation in ^{40}S. Figure 2 shows the systematics of yrast states in heavy sulphur isotopes. It is worth noting that, firstly, the energy of the 2_1^+ in ^{40}S has dropped by $\sim$40% as compared to the ^{38}S isotope and, secondly, the 4^+ state in ^{38}S lies at 2.19 times the energy of the 2^+ level while the corresponding ratio in ^{40}S is 2.50. This might indicate a tendency towards deformation.

2.3 $^{48}Ca + {}^{48}Ca$ reaction

The studies of neutron-rich *sdfp* shell nuclei can be done also with a ^{48}Ca beam. We carried out an experiment at the Tandem accelerator in Laboratori

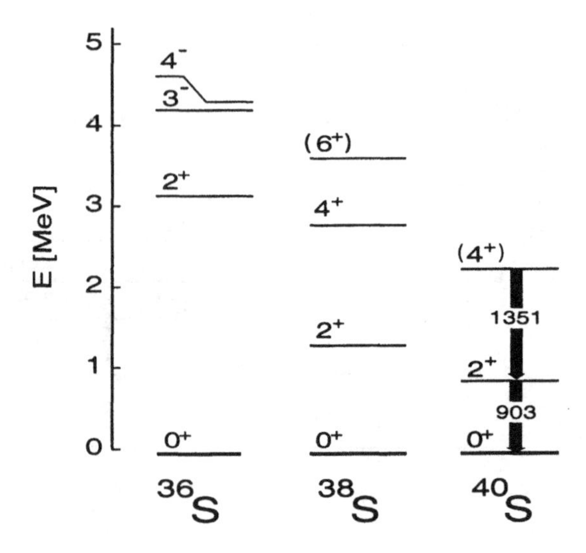

Figure 2 Systematics of yrast states in neutron-rich S isotopes. Arrows indicate the transitions observed in this work.

Nazionali di Legnaro bombarding a target of 0.74 mg/cm^2 ^{48}Ca (backed by 40 mg/cm^2 of evaporated ^{208}Pb) with a beam of 140 MeV ^{48}Ca ions. The γ-γ coincidences were collected with the Euroball array. Fusion-evaporation was a main reaction channel whereas multinucleon transfer processes, leading to nuclei in the vicinity of ^{48}Ca, accounted for less than 1% of the total reaction cross section. Despite this very low production yield, investigation of excited states in some nuclei around ^{48}Ca was possible due to the high resolving power of the Euroball array.

Scheit et al. [3] observed states in ^{44,46}Ar using the intermediate-energy Coulomb excitation method and proposed the 2_1^+ states of ^{44}Ar and ^{46}Ar at 1144(17) keV and 1554(26) keV, respectively.

We searched for these transitions in the $\gamma\gamma\gamma$ prompt coincidence cube from the ^{48}Ca + ^{48}Ca reaction by examining the cross coincidence relationship between complementary Ar and Ti reaction products. Double gates set on prompt transitions in ^{52}Ti displayed two γ-rays of 1159 keV and 1589 keV energy, which were not known previously. The reversed gates showed that both lines are in mutual coincidence and the intensity pattern of prompt yrast transitions from the ^{52}Ti isotope observed in coincidence with the 1159 keV and 1589 keV lines indicated clearly that these γ-rays occur in the partner nucleus ^{44}Ar. In addition, by setting a double gate on the 1159 keV and 1589 keV lines another 695 keV γ-transition in ^{44}Ar was found. Detailed examination of intensity relations between 1159, 1589 and 695 keV γ-rays pointed out that these transitions form a cascade thus connecting states at 1159, 2748 and 3443 keV excitation energy.

There is almost no doubt that a level at 1159 is the 2^+ excitation in ^{44}Ar observed at 1144(17) keV by Scheit et al. The higher states at 2748 and 3443 keV are perfect candidates for 4^+ and 6^+ excitations, respectively.

Similar analysis, by employing double gates set on known prompt γ-rays in ^{50}Ti, displayed a 1577 keV γ-ray from a partner nucleus ^{46}Ar. It is almost certain that this γ-ray is the sought $2^+ \rightarrow 0^+_{g.s.}$ transition observed by Scheit et al. at 1554(26) keV.

The systematics of known yrast levels in heavy Ar isotopes, including newly placed excitations is shown in Fig. 3.

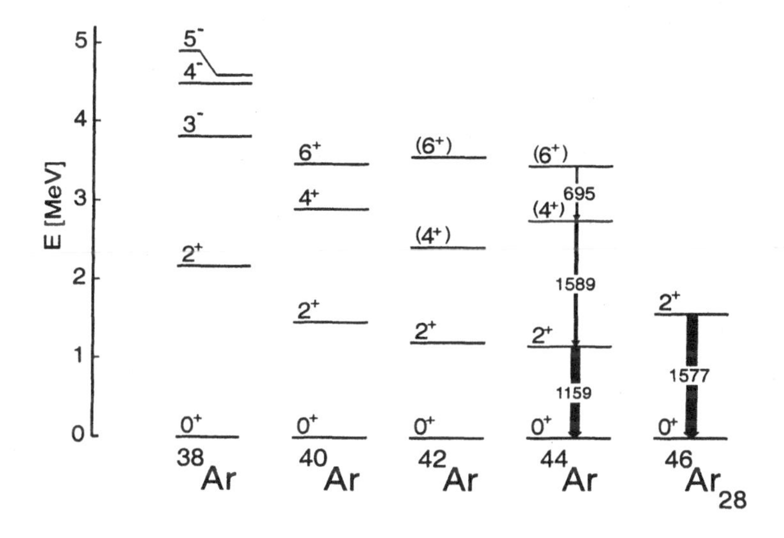

Figure 3 Systematics of yrast states in neutron-rich Ar isotopes. Arrows indicate the transitions observed in this work.

3 CONCLUSIONS

Theoretical description of the properties of neutron-rich *sdfp* shell nuclei was undertaken by Retamosa et al. [8] within the shell model framework. Very good agreement between predictions and experimental results on the 2^+ excitation energies and B(E2;$0^+_{g.s} \rightarrow 2^+_1$) values for ^{40,42}S and ^{44,46}Ar was achieved. A conclusion has been drawn that, at N=28 in the very neutron-rich regime the magicity persists. Our new data on higher excitations in ^{40}S and ^{44}Ar offer a more stringent test of these large-scale shell model calculations and interactions used. They also are in line with the development of collectivity predicted in heavy sulfur nuclei.

Acknowledgments

This work has been supported by the Polish Scientific Committee under grant no. 2P03B-150-10, and by the Italian–Polish collaboration.

References

[1] E.K.Warburton et al., Phys. Rev. C41 (1990) 1147.

[2] T. R. Werner et al., Phys. Lett. 335B, 259 (1994).

[3] H. Scheit et al., Phys. Rev. Lett. 77, 3967 (1996).

[4] R. Broda et al., Phys. Rev. Lett. 68, 1671 (1992).

[5] B. Fornal et al., Acta Phys. Pol. B26, 357 (1995).

[6] B. Fornal et al., Phys. Rev. C55, 762 (1997).

[7] J.A. Winger et al., in Proceedings of the International Conference *"Fission and Properties of Neutron-Rich Nuclei"*, Sanibel Island, Florida, USA, 10-15 November 1997, ed. J.H. Hamilton, A.V. Ramayya (World Scientific) (1998), p. 311.

[8] J. Retamosa et al., Phys. Rev. C55, 1266 (1997).

APPLICATIONS IN NUCLEAR STRUCTURE OF A FAMILY OF Q-DEFORMED ALGEBRAS

A.I. Georgieva[1], H.B. Geyer[2], M.I. Ivanov[1] and K.D. Sviracheva[1]

[1]Institute of Nuclear Research and Nuclear Energy,
Bulgarian Academy of Sciences, Sofia 1784, Bulgaria
[2]Institute of Theoretical Physics, University of Stellenbosch,
7599 Stellenbosch, South Africa

Abstract: We show how the introduction of a single parameter, which defines a family of q-deformed algebras, may be exploited to yield a much richer $su_q(2)$ like analytical spectrum than in the standard analysis. Generalizations to a q-deformed $su_q(3)$ algebra, which retains links to the standard Elliott model, are also briefly discussed.

Developments in physics of this century are intimately linked to the development and maturation of the corner-stone fields of relativity and quantum mechanics. While physics of the new millennium, together with the correlation and interpretation of its data may very well prove to require new mathematical structures, it is at the same time clear that the most obvious candidates are those that represent extensions and/or generalizations of present successful mathematical frameworks.

In the case of quantum physics the natural language is that of operators, associated with observables, and their corresponding algebras. A new develop-

The Nucleus: New Physics for the New Millennium
Edited by Smit et al., Kluwer Academic / Plenum Publishers, New York, 2000.

ment of this language towards the end of this century is the notion of quantum algebras which, from a mathematical viewpoint, are q-deformations of the universal enveloping algebras of the corresponding classical Lie algebras. In this framework one relates physical observables to a family of algebraic operators, each member of which is characterized by the value of the deformation parameter q. This naturally yields greater flexibility and richer structures to be exploited within the framework of algebraic description, and has indeed been used in a number of applications[1], although one has to acknowledge that a real "killer" physical interpretation of q in some application is still wanting.

Some applications, aiming to explore the possible role of the q-deformation parameter in the theory of nuclear collective structure, have for example been reported at the level of $su_q(2)$ [1, 3, 4]. Nevertheless further exploration of q-deformed group theoretical structures and methods, in particular linked to the q-deformed extensions of algebraic models associated with nuclear collective motion [5, 6], almost suggests itself.

A case in point concerns the reduction of quantum enveloping algebras to the $so_q(3)$ algebra of the angular momentum. This is a basic and crucial element, common to nuclear collective models based on the idea of a dynamical symmetry. However, the reduction of q-deformed algebras is a complicated problem, into which much effort has recently been put [12, 13].

One very natural way to obtain the embedding of the angular momentum algebra is to consider tensor operators [1], [6] with respect to it and then generate the higher rank algebras in terms of these operators. This is a procedure used in most of the well known algebraic models.

For the standard quantum enveloping algebra $su_q(2)$ it was shown that its generators could be realized in terms of annihilation a_i, creation $a_i^\dagger$ (with $(a_i^\dagger)^\dagger = a_i$) and number operators N_i of q-bosons ($i = \pm 1$) which satisfy

$$a_i a_i^\dagger - q^m a_i^\dagger a_i = q^{-mN_i}, \tag{1}$$

$$a_i a_i^\dagger - q^{-m} a_i^\dagger a_i = q^{mN_i}, \tag{2}$$

$$[N_i, a_i] = -a_i \ , \quad [N_i, a_i^\dagger] = a_i^\dagger, \tag{3}$$

where $m \geq 0$ is a new additional parameter.

The operators $a_i, a_i^\dagger$ and N_i for $i = \pm 1$ do not form an irreducible tensor operator [1] with respect to the boson oscillator representation of the $su_q(2)$ algebra

$$J_+ = a_1^\dagger a_2, \quad J_- = a_2^\dagger a_1, \quad J_0 = \tfrac{N_1 - N_2}{2}, \tag{4}$$

defined through the commutation relations of its generators $J_\pm$ and J_0,

$$[J_0, J_\pm] = \pm J_\pm, \quad [J_+, J_-] = [2J_0]\ , \tag{5}$$

where $[X] \equiv \frac{q^X - q^{-X}}{q - q^{-1}}$ and $[X]_m \equiv \frac{q^{mX} - q^{-mX}}{q^m - q^{-m}}$.

However, the following nontrivial modification of the creation and annihilation operators for $k = \pm 1$ introduced in eqs. (1) and (2),

$$b_k^\dagger = q^{\frac{k\sqrt{m}}{4}} a_k^\dagger q^{\frac{kmN_{-k}}{2}}, \tag{6}$$

$$b_k = q^{\frac{k\sqrt{m}}{4}} a_{-k} q^{\frac{kmN_{-k}}{2}}, \tag{7}$$

transforms them into a family of spinorlike tensors with respect to the standard $su_q(2)$ in (4):

$$\begin{array}{ll} \left[J_0, b_k^\dagger\right] = \frac{k}{2} b_k^\dagger, & [J_0, b_k] = \frac{k}{2} b, \\ J_+ b_1^\dagger = q^{\frac{m}{2}} b_1^\dagger J_+, & J_+ b_{-1}^\dagger = q^{-m} b_{-1}^\dagger J_+ + q^{\frac{-\sqrt{m}}{2}} b_1^\dagger q^{-mJ_0}, \\ J_- b_1^\dagger = q^m b_1^\dagger J_- + q^{\frac{\sqrt{m}}{2}} b_{-1}^\dagger q^{-mJ_0}, & J_- b_{-1}^\dagger = q^{-\frac{m}{2}} b_{-1}^\dagger J_-. \end{array} \tag{8}$$

The two components of the tensors $b_{\pm 1}, b_{\pm 1}^\dagger$ satisfy the commutation relations

$$[b_k, b_l^\dagger]_{q^\mu} = q^{\frac{-\sqrt{m}}{2} k} \delta_{k,-l} q^{-2K_m}, \qquad K_m = mJ_0, \tag{9}$$

$$[b_k, b_l]_{q^\lambda} = [b_k^\dagger, b_l^\dagger]_{q^\lambda} = 0, \tag{10}$$

$$\mu = \frac{-m(k-l)}{2}, \quad \lambda = -\mu \quad k, l = \pm 1, \tag{11}$$

where $[A, B]_{q^x} = AB - q^x BA$. Using spinors(6,7) with $m = \pm 1$, which correspond to a tensor of rank $j = |m|$, a q-deformed $sp(4, R)$-algebra is generated in refs. [10] and [11] by the possible tensor products of the two fundamental $SU_q(2)$-spinors

$$\{(b^\dagger)^1 \otimes (b^\dagger)^1\}_k^l := T_k^l, \qquad l = 1; k = 0, \pm 1; \tag{12}$$

$$\{b^1 \otimes b^1\}_k^l := \tilde{T}_k^l, \qquad l = 1; k = 0, \pm 1; \tag{13}$$

$$\{b^1 \otimes (b^\dagger)^1\}_k^l := I_k^l, \qquad l = 0, 1; k = -l, -l+1, .., l. \tag{14}$$

It is shown there, by calculating the commutation relations of the operators obtained, that they close a q-deformation of the "classical" $sp(4, R)$ algebra, as in the limit $q \longrightarrow 1$, its commutation relations are reproduced.

In the classical case this algebra [8] is of physical interest by itself and also is easily generalized to the higher rank cases – $sp(2n, R), n = 3, 4, \ldots$. In general the group $Sp(2n)$ can be used to describe pairing correlations in systems containing different kinds of particles [11], while the non-compact version of the symplectic group is applied in the description of collective vibrational excitations of a system of particles moving in an n-dimensional harmonic oscillator potential. With similar applications in mind in the q-deformed case we emphasise that a natural procedure entails embedding into a q-deformed $sp(4, R)$ algebra, by construction, a q-deformed $so(3)$-subalgebra, the latter being generated by the components of a first rank tensor $I^1_m(m = 0, \pm 1)$ in (14) which can be interpreted as an angular momentum operator.

We give here in more detail the results for the family of compact subalgebras generated by (14), obtained with the generalized spinors (6) and (7). The two sets of $b^\dagger_k, b_k$ $(k = \pm 1)$ define for the different values of $m \geq 0$ a family of q-deformed subalgebras $U_{\tilde{q}}(2)$ with explicit representation of the four generators:

$$I_{+1} = b^\dagger_{+1} b_{+1} \,, \quad I_{-1} = b^\dagger_{-1} b_{-1} \,, \quad I_0 = \frac{1}{[2]_m} \left(q^{m - \frac{\sqrt{m}}{2}} b^\dagger_{+1} b_{-1} - q^{-m + \frac{\sqrt{m}}{2}} b^\dagger_{-1} b_{+1} \right) \tag{15}$$

$$[N]_m = \left(q^{-\frac{\sqrt{m}}{2}} b^\dagger_{+1} b_{-1} + q^{\frac{\sqrt{m}}{2}} b^\dagger_{-1} b_{+1} \right) = [N_1 + N_{-1}]_m . \tag{16}$$

It is important to observe here that when working with the tensor representation of the algebra we can introduce boson expressions for the number operators. The operators (15) commute as a simple Lie algebra,

$$[I_{+1}, I_{-1}] = [2]_m \, I_0 q^{-2K_m} \quad , \quad [I_0, I_k] = k q^{-km} I_k q^{-2K_m} \quad , \quad [[N], I_{0,k}] = 0 \,,$$

and satisfy the conjugation relations

$$(I_k)^\dagger = I_{-k} \quad , \quad k = 0, \pm 1 \,.$$

The scalar $[N]$ generates the q-deformed subgroup $U_{\tilde{q}}(1)$ and is the Casimir of first rank for $U_{\tilde{q}}(2)$ which implies $U_{\tilde{q}}(2) = SU_{\tilde{q}}(2) \otimes U_{\tilde{q}}(1)$.

The second order invariant is defined to commute with the three generators $I_{\pm 1, 0}$, hence

$$C_2 = \frac{1}{[2]_m} \left(q^{-4} I_{+1} I_{-1} + q^4 I_{-1} I_{+1} \right) + I_0 I_0 . \tag{17}$$

After transformation the invariant is expressed through the self- conjugate generators of $U_{\tilde{q}}(2)$:

$$C_2^m = \frac{[N]_m}{[2]_m} \frac{[N+2]_m}{[2]_m} \qquad N = N_1 + N_{-1} . \tag{18}$$

We have thus obtained a family of Casimir invariants for each value of the parameter m, which can describe a family of collective bands of the same type whenever a phenomenological Hamiltonian

$$H_{coll} = \sum_m \kappa_m C_2^m \tag{19}$$

is expressed in terms of this second order Casimir operators. (Note that in the limit $q \to 1$ the q-generator relations coincide with the corresponding relations for the Lie algebra $su(2)$.) Furthermore the most general invariant operator with respect to the $SU_{\tilde{q}}(2)$, constructed as second order polinomial from the generators (13),(12) and(14) of the q-deformed $sp(4,R)$ algebra, is

$$S_2 = s_1 T^1_k . \tilde{T}^1_{-k} + s_2 \tilde{T}^1_k . T^1_{-k} + s_3 I_k . I_{-k} + s_4 [N]^2 = \tag{20}$$

$$s_1 [N][N-1] + s_2 [N+2][N+3] + s_3 [N][N+2] + s_4 [N]^2 \tag{21}$$

From this expression it is quite clear that four additional phenomenological parameters, together with the deformation parameter, are introduced in the invariant, thus allowing a larger variety of interactions for the corresponding Hamiltonian.

The states of a two-dimensional oscillator are constructed by acting with the operators (6) $b^\dagger_{\pm 1}$ onto the vacuum state- $\mid 0 \rangle$, defined as usual by the relations $b_\pm \mid 0 \rangle = 0$, $N_i \mid 0 \rangle = \mid 0 \rangle$, $\langle 0 \mid 0 \rangle = 1$. The orthonormal states in each fixed representation for $n = n_+ + n_-$ are:

$$\mid n_+, n_- \rangle = c(n_+, n_-)(b^\dagger_+)^{n_+}(b^\dagger_-)^{n_-} \mid 0 \rangle \tag{22}$$

where obviously the operators $N_\pm$ yield

$$N_{\pm 1} \mid n_+, n_- \rangle = n_\pm c(n_+, n_-)(b^\dagger_+)^{n_+}(b^\dagger_-)^{n_-} \mid 0 \rangle ,$$

from which it clearly follows that $J_0 \mid n_+, n_- \rangle = \frac{1}{2}(n_+ - n_-) \mid n_+, n_- \rangle$.

In order to find the action of the generators (15) on the basis states we need the following generalizations of the commutation relations (11):

$$b_k (b^\dagger_l)^n = q^{\frac{-m(k-l)n}{2}} (b^\dagger_l)^n b_k + \delta_{k,-l} q^{-\frac{k}{2}\sqrt{m}} [n]_m (b^\dagger_l)^{n-1} q^{-2K_m} . \tag{23}$$

From these relations and the explicit representation of the generators we obtain for the diagonal ones the following eigenvalues:

$$I_0 \mid n_+, n_- \rangle = \frac{1}{[2]_m} \{ q^{m(n_-+1)} [n_+]_m - q^{-m(n_++1)} [n_-]_m \} \mid n_+, n_- \rangle \tag{24}$$

and

$$\begin{aligned} [N]_m \mid n_+, n_- \rangle &= \{ q^{mN_{-1}} [N_{+1}]_m - q^{-mN_{+1}} [N_{-1}]_m \} \mid n_+, n_- \rangle \\ &= [n_+ + n_-]_m \mid n_+, n_- \rangle . \end{aligned} \tag{25}$$

In order to evaluate the action of the raising and lowering generators $I_\pm$ we first need the normalization factor $c(n_+, n_-)$ from (22). Assuming this to be real and using (23) together with mathematical induction gives

$$c(n_+, n_-) = \left\{ q^{\frac{m}{2} n_+ n_-} q^{\frac{\sqrt{m}(n_+ - n_-)}{4}} \sqrt{[n_+]_m ! [n_-]_m !} \right\}^{-1/2} . \tag{26}$$

The action of the operators $I_\pm$ on the normalized states of the basis (22) is obtained as

$$I_- \mid n_+, n_- \rangle = q^{\frac{m}{2}(n_- - n_+ + 1)} \sqrt{[n_+]_m [n_- + 1]_m} \mid n_+ - 1, n_- + 1 \rangle \tag{27}$$

$$I_+ \mid n_+, n_-\rangle \;=\; q^{\frac{m}{2}(n_- - n_+ - 1)}\sqrt{[n_+ + 1]_m [n_-]_m} \mid n_+ + 1, n_- - 1\rangle \quad (28)$$

The set of states$\{\mid n_+, n_-\rangle\}$ constitutes an orthonormal basis in the space of states H. This space is reduced with respect to N of expression (16) into subspaces H=$\sum \oplus$ H_n, with each component determined by a fixed value $n = n_+ + n_-$ through $N \mid n_+, n_-\rangle = n \mid n_+, n_-\rangle$. Each of these subspaces is therefore characterized by the action of an irreducible representation of the q-deformed $su(2)$ algebra. In this way the q-deformed oscillator $su(2)$-representation is reduced to a direct sum of irreducible representations, realizing a complete description of all the possible $su(2)$ representation for a given deformation.

We conclude by briefly referring to some recent work [14] which constructs a deformed $su_q(3)$ algebra in such a way that the reduction to $so(3)$ is preserved, while at the same time it allows a deformed quadrupole operator to be identified. This work exploits the fact that the symplectic algebras are particularly convenient for boson mapping and allows a direct generalization to q-deformation for the standard $SU(3)$ Elliott model [15].

AIG is supported through contract Φ-621 from the Bulgarian Ministry of Education and Science and HBG by grant 2034459 from the South African Foundation for Research Development.

References

[1] Biedenharn L C and Lohe M A, 1995 *Quantum Group Symmetry and q-Tensor Algebras* (Singapore: World Scientific)

[2] Raychev P, Roussev R and Smirnov Yu F 1990 *J. Phys. G: Nucl. Part. Phys.* 16 L137

[3] Gupta R K, Cseh J, Ludu A, Greiner W and Scheid W, 1992 *J. Phys. G: Nucl. Part. Phys.* 18 L73

[4] Bonatsos D, Faessler A, Raychev P P, Roussev R P and Smirnov Yu F, 1992 *J. Phys. A: Math. Gen* 25 L267

[5] Del Sol Mesa A, Loyola A, Moshinsky M and Velazquez V 1993 *J. Phys. A: Math, Gen.* 26 1147

[6] Gupta R K and Ludu A 1993 *Phys. Rev.* C 48 593

[7] Van der Jeugt J 1993 *J. Math. Phys.* 34 1799

[8] Quesne C 1993 *Phys. Lett. B* 298 344; *ibid* 304 81

[9] Smirnov Yu F, Tolstoi V and Kharitonov Yu 1991 *Sov. J. Nucl. Phys* 53 593

[10] Georgieva A 1993 *Tensorial structure of a q-deformed sp(4, R) superalgebra* Preprint IC/93/50 ICTP Trieste

[11] Georgieva A and Dankova Ts 1994 *J. Phys. A: Math. Gen.* 27 1251

[12] Castanos O, Chacon E, Moshinsky M and Quesne C 1985 *J. Math. Phys.* 26 2107

[13] Goshen S and Lipkin H J 1968 *On the application of the group Sp(4) or R(5) to nuclear structure* in *Spectroscopic and Group Theoretical Methods in Physics* eds. W A Friedman and H Feshbach (Amsterdam: North-Holland)

[14] Georgieva A I, Goleminov J D, Ivanov M I and Geyer H B 1999 *J. Phys. A: Math. Gen.* (in press)

[15] Elliott J P 1958 *Proc. Roy. Soc. A* 245 128; 526

MAGNETIC– ROTATIONAL BANDS IN Pb ISOTOPES

H. Hübel

Institut für Strahlen- und Kernphysik, Univ. Bonn,
Nussallee 14-16, D-53115 Bonn, Germany

Abstract: Regular sequences of strongly enhanced M1 γ–ray transitions have been reported in recent years in many near–spherical neutron–deficient Pb isotopes. They are not caused by quantal rotation of deformed charge distributions but by rotation of anisotropic arrangements of nucleonic currents. This new mode of nuclear excitation has been called *Magnetic Rotation.* The experimental properties of the magnetic–rotational bands are discussed. Evidence for their configuration and the angular momentum coupling scheme with its development along the bands due to the *Shears Effect* is presented.

1 INTRODUCTION

The low–energy excitations in nuclei can be divided into two groups: they may be of single–particle or of collective type. High angular momentum is most efficiently generated by collective rotation, but this mode of excitation is only observed in deformed nuclei. In this case the nuclear deformation specifies an orientation with respect to the angular momentum and quantal rotation may occur. On the other hand, in spherical nuclei high–spin states are formed by successive alignments of single–particle angular momenta. While collective rotation leads to the well known bands with energy spacings that follow an I(I+1) dependence, the single–particle excitations give rise to an irregular level pattern. The most regular rotational bands are observed in superdeformed nuclei, whereas spherical or weakly deformed nuclei do not show regular bands. Therefore, the discovery of long regular sequences of magnetic dipole transitions in near–spherical nuclei, like in the Pb and Bi isotopes [1, 2], was unexpected. As an example the γ–ray coincidence spectrum of such a band in ^{199}Pb is given in Fig. 1. The properties of these bands show that they represent a new mode of nuclear excitation, for which the name *Magnetic Rotation* was suggested [3].

The Nucleus: New Physics for the New Millennium
Edited by Smit et al., Kluwer Academic / Plenum Publishers, New York, 2000.

2 PROPERTIES OF THE BANDS

In many Pb and Bi isotopes regular sequences of M1 transitions that roughly follow an I^2 dependence of the energy levels have been found in recent years [1, 2]. As an example, Fig. 2 shows the partial level scheme of ^{199}Pb. The strong peaks in the spectrum of band 2 given in Fig. 1 are the $\Delta I=1$ M1 transitions. Some E2 crossover transitions are also observed, but they are very weak and hardly visible in the spectrum. The B(M1)/B(E2) ratios derived from the intensity ratios in the spectra are very large, they lie around 20 $(\mu_n/eb)^2$. Lifetime measurements as those described below show that the M1 transitions are indeed strongly enhanced and that the E2 collectivity is as small as expected for the near-spherical Pb isotopes. Clearly, in these cases one would not expect regular rotational-like bands. Only the irregular single-particle type level spectrum should occur, as seen at lower excitation energy and spin in Fig. 2.

The moments of inertia of the M1 bands are small ($\mathcal{J}^{(2)} \approx$ 15–25 $\hbar^2$/MeV). On the other hand, the ratios $\mathcal{J}^{(2)}$/B(E2) are exceptionally large; they are a factor of 10 or more larger than in well-deformed nuclei. Therefore, the moment of inertia cannot be related to the rotation of a deformed charge distribution as in normal rotational bands.

Another difference to rotational bands in deformed nuclei is the absence of signature splitting even at high angular momenta. This shows that the systems discussed here do not reach the rotational symmetry that gives rise to signature splitting.

Two interesting properties that are known in some deformed nuclei, band termination and identical bands, also occur for the M1 bands [4, 6]. Termi-

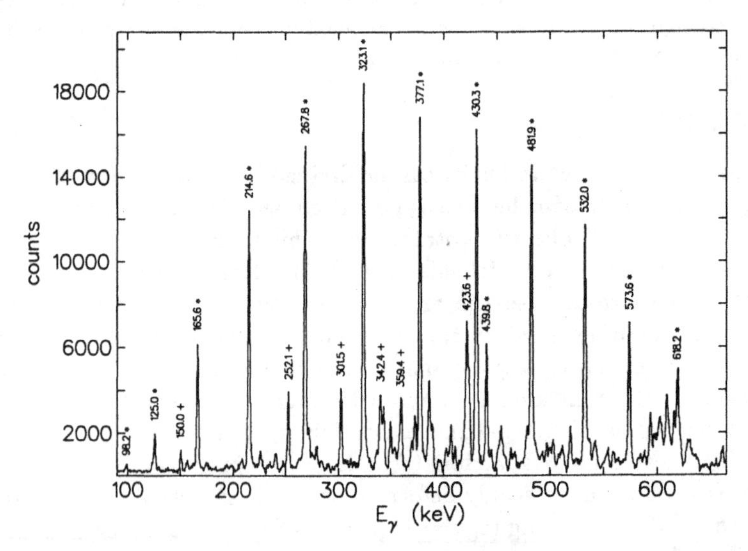

Figure 1 Gamma-ray coincidence spectrum of band 2 in ^{199}Pb (see level scheme in Fig. 2). The peaks marked by asterisks are the $\Delta I=1$ M1 transitions; the ones marked by plus signs belong to the normal irregular part of the level scheme into which the band decays.

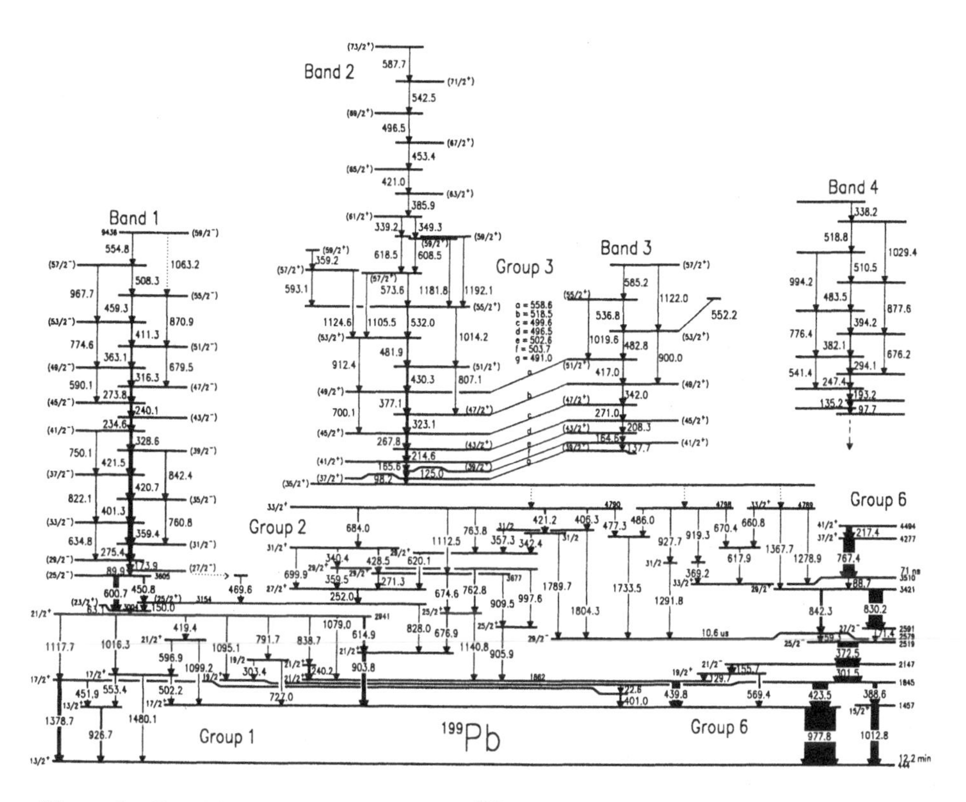

Figure 2 Partial energy-level scheme of ^{199}Pb [4, 5]. Only four of the five M1 bands known in this nucleus are shown.

nation of bands in deformed nuclei is reached when the angular momenta of the valence nucleons are aligned. In the Pb isotopes the bands terminate when the nucleons of a certain few-particle configuration are aligned, but then new bands may start at higher spin which will again terminate at higher maximum angular momenta.

The origin of identical bands in neighboring nuclei and the associated quantized alignment [7] is not yet understood. It has been suggested [8] that the pseudo-spin symmetry may offer a possible explanation. The clearest case involves the [301]1/2 orbital, which is a $\tilde{\Lambda} = 0$ pseudo-spin singlet, where pseudo-spin alignment can explain the identical bands in superdeformed nuclei in the A=150 region [9]. In the Pb isotopes the $\tilde{\Lambda} = 0$ neutron level [521]1/2 may play the same role to explain the identical M1 bands [4].

A final and very important observation is that the magnetic-rotational bands are only built on selective high-spin states. As we shall see in the following section, they are high-spin proton-particle excitations coupled to high-spin neutron-hole states. The bands are never built on any other states, in particular not on the individual proton-particle or neutron-hole states.

3 STRUCTURE AT THE BAND HEAD

High–spin proton states in the neutron–deficient Pb isotopes can be formed by excitations from the $s_{1/2}$ orbital just below the Z=82 shell gap into the $h_{9/2}$ and $i_{13/2}$ orbitals above that gap. A state of this type is the $\pi(h_{9/2}i_{13/2})_{11^-}$ excitation which is known in even Pb isotopes [2]. Such excitations have particle character because there are only a couple of protons above the shell gap. High–spin neutron states can be formed by hole excitations in the $\nu i_{13/2}$ subshell. Examples are the $i_{13/2}^{-n}$ $13/2^+$, 12^+ and $33/2^+$ states in the odd and even Pb isotopes, respectively [2].

The proton–particle and neutron–hole states discussed here are probably slightly oblate deformed. However, this deformation is not sufficient to explain the long regular bands. It is important to note that the bands are not built on these states directly but they are observed at higher spins and excitation energies. It was therefore suggested that the band heads are formed by a coupling of the proton–particle states, e.g. the $\pi(h_{9/2}i_{13/2})_{11^-}$ excitation mentioned above, to the neutron $i_{13/2}^{-n}$ hole states [4]. Optimal overlap of the density distributions of particles and holes is achieved for a perpendicular orientation of their orbitals. Therefore, the lowest energy in a particle–hole coupling is obtained for a perpendicular orientation of the particle and hole angular momenta. The coupling scheme is shown in the left-hand part of Fig. 3 for the example of the $\pi(h_{9/2}i_{13/2})_{11^-} \otimes \nu i_{13/3}^{-1}$ structure. The associated magnetic moments can be seen on the right–hand side of Fig 3. The perpendicular coupling of the proton–particle and neutron–hole spins results in a large component of the total magnetic moment perpendicular to the total nuclear spin. This breaks the symmetry of the quantal system in a similar way as the charge distribution does in deformed nuclei [3, 10, 11]. Such an arrangement may rotate about the axis I_{tot}. Using the concept of spontaneous symmetry breaking, it has been shown [11] that this leads to regular energy–level sequences if the proton and neutron spins are large and the configuration does not change much from state to state. Since it is the magnetic moment that rotates – and not the electric charge distribution – Frauendorf [3] suggested to name this new mode of excitation *Magnetic Rotation*, in contrast to the well–known bands in deformed nuclei which would then be called *Electric Rotation.*

In an alternative approach, Macchiavelli et al. [12, 13] analyzed the effective proton–neutron interaction in the magnetic–rotational bands and found that the $P_2(\Theta)$ term can give rise to a rotational spectrum. These authors suggest that such an interaction may be mediated through the core by particle–vibration coupling.

The configuration and the perpendicular coupling of the spins near the band head has recently been tested by a measurement of the g factor of the $29/2^-$ state at 2584 keV in ^{193}Pb [14] on which a magnetic–rotational band is built. The partial level scheme of ^{193}Pb is shown in Fig. 4. The half life of the state, $t_{1/2}$=9 ns, is just long enough to allow to measure the time–differential perturbed angular distribution in an external magnetic field. High–spin states

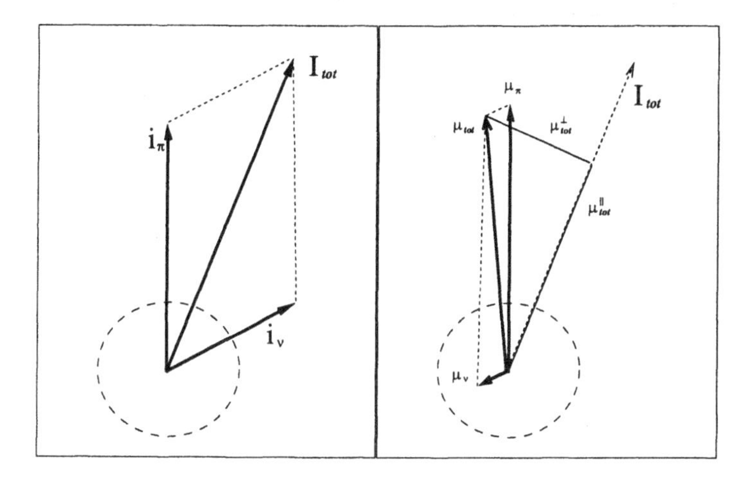

Figure 3 Angular momentum coupling scheme of a proton–particle/neutron–hole state (left) and the coupling of the associated magnetic moments (right).

in ^{193}Pb were populated in the ^{170}Er(^{28}Si,5n) reaction at a beam energy of 143 MeV. The pulsed beam was provided by the tandem accelerator at Legnaro. An external field of 3.14(5) T, which was periodically switched in direction, was used. The 158 keV transition in the decay of the isomer was observed as a function of delay time with respect to the narrow beam pulses in two planar Ge detectors placed at $\pm 135^o$ to the beam. The ratios R(t) of the intensities as a function of time for the two detectors are displayed on the right–hand side of Fig. 4. A fit to these curves gives a g factor of g = 0.68(3).

For the band head of band 1 in ^{193}Pb the configuration $(\pi(h_{9/2}i_{13/2})_{11^-} \otimes \nu i^{-1}_{13/2})_{29/2^-}$ was suggested [15, 16]. The g factor of this configuration can be calculated using experimental values for the g factors of its constituents. The values of the proton $(h_{9/2}i_{13/2})_{11^-}$ and the neutron $i^{-1}_{13/2}$ states, which were adopted [14], are $g(\pi) = 1.11(2)$ and $g(\nu) = -0.150(6)$, respectively. With these values one calculates g=0.71(4) for the $29/2^-$ band head, where the uncertainty was chosen to include the spread obtained when different experimental g factors are used.

The good agreement between the experimental and calculated g factors confirms the configuration as a proton $(h_{9/2}i_{13/2})_{11^-}$ excitation coupled to a neutron $i_{13/2}$ hole state. The spin of $29/2^-$ results from the near-perpendicular orientation of the proton and neutron spins, as shown in Fig. 3. Other conceivable configurations for the $29/2^-$ state are ruled out because they would result in g factors that are a factor of two or more smaller than the experimental value [14]. The result of this measurement is a direct confirmation of the approximately perpendicular coupling of the proton–particle and neutron–hole spins at the band head of a magnetic–rotational band. This special coupling gives rise to a large transverse component of the magnetic moment, which is required for the occurrence of magnetic rotation.

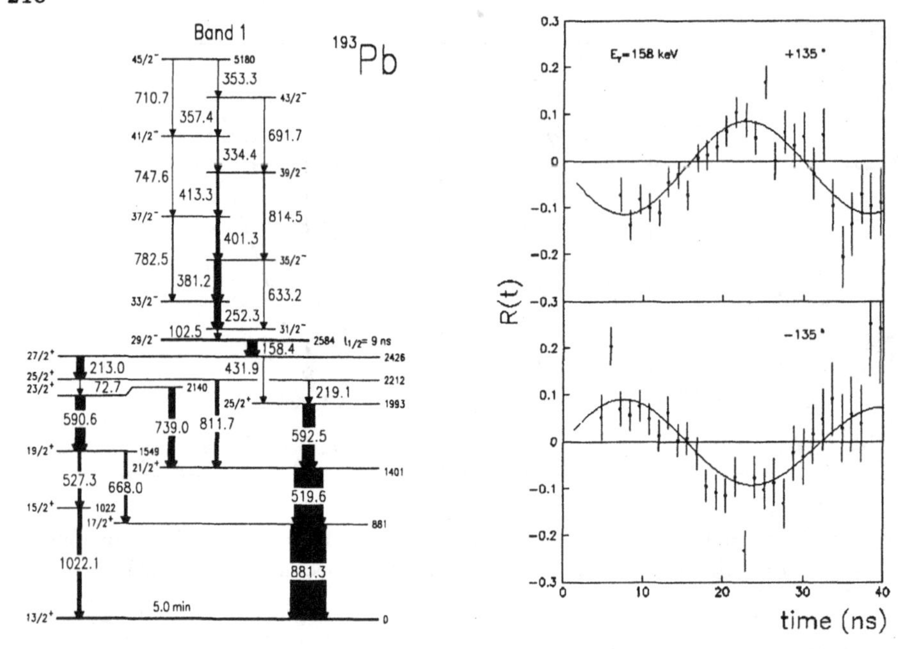

Figure 4 Partial energy-level scheme of ^{193}Pb (left) [15, 16] and the ratio $R(t)=(N_{up}-N_{down})/(N_{up}+N_{down})$ for the 158-keV transition measured in the two detectors at $\pm 135^o$ (right) [14].

4 SHEARS EFFECT

Near the bottom of the bands the directions of the proton–particle and neutron–hole spins do not coincide with the total angular momentum, see Fig. 3. As the system rotates about the axis of I_{tot}, Coriolis forces act on i_π and i_ν and cause them to align with I_{tot}. This costs energy because, as discussed above, the lowest-energy state is the one with the perpendicular coupling of the proton and neutron spins. This process increases the total nuclear spin I_{tot} until i_π and i_ν are fully aligned and the band terminates. Since this resembles the closing of the blades of a pair of shears, the name *Shears Bands* was suggested [4] for the magnetic–rotational bands.

Coriolis forces also act on the nucleons in rotating deformed nuclei. However, in deformed nuclei a large number of nucleons participates in the collective rotation. These nucleons are paired and the Coriolis force tries to break the pairs and to align their spins into the direction of the rotational angular momentum. Here, in the case of magnetic rotation, only a few unpaired high–spin particles and holes form the configuration. They do not fully align in one step but rather build a sequence of states with increasing energies and spins until the 'shears is closed'.

The shears effect has been investigated theoretically within the tilted-axis cranking (TAC) model [3, 4, 10]. In this approach uniform rotation about an axis that is not one of the principal axes of a deformed nucleus is considered. A slightly oblate deformed mean field was used in which the protons are initially aligned along the symmetry axis and the neutrons are aligned perpendicular to that axis. Solutions of the Hamiltonian are then calculated for cranking around the total angular momentum axis. The calculations show [4, 10] that the tilting angle between the total spin and the symmetry axis remains almost constant along the bands, as expected for the shears effect.

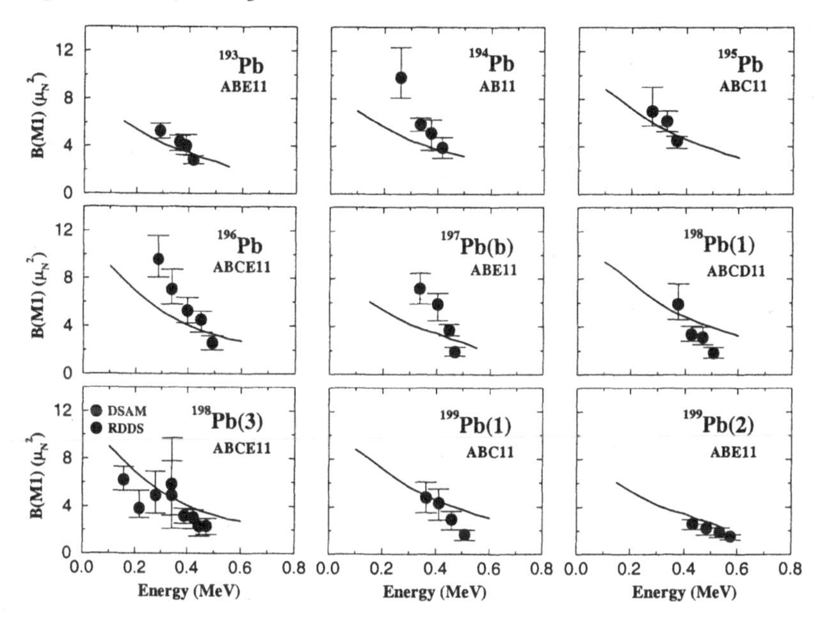

Figure 5 B(M1) values as a function of transition energy in magnetic–rotational bands in Pb isotopes [18, 19, 20]. The curves are the results of TAC calculations [17] for the adopted roton-particle/neutron-hole configuration.

The shears effect has also been investigated experimentally. A consequence of the alignment of the proton and neutron spins i_π and i_ν, respectively, along the direction of I_{tot} is a decrease of the transverse component of the magnetic moment $\mu_{tot}^{\perp}$, see right-hand side of Fig. 3. Since this component is responsible for the strength of the M1 radiation, the B(M1) rates should decrease with increasing spin along the bands. Lifetimes of states within shears bands have been measured by many groups (for refs. see [17]), but only recently the precision has been reached that is required to confirm the expected decrease of the B(M1) values [18, 19, 20]. Lifetimes were measured in Pb isotopes from ^{193}Pb to ^{199}Pb using the Doppler–shift attenuation method and, in the case of ^{198}Pb, also by the recoil–distance method. High–spin states in these nuclei were populated in the ^{172,174,176}Yb(^{26}Mg,xn), ^{186}W(^{18}O,xn) and ^{154}Sm(^{48}Ca,4n) reactions. Different backing materials (Al, W, Pb) were used to slow down the recoiling nuclei. Gamma–ray coincidences were measured with the Gammasphere spectrometer array at Berkeley. The B(M1) values extracted from the lifetimes are displayed in Fig. 5. The experimental points show the decrease predicted by the shears effect. For comparison, results of TAC calculations [17] are also included.

The consistency of the reduced transition probabilities with the shears effect may be tested in a rather simple way [12]. As pointed out above, the angle Θ between the proton and neutron spins decreases along the bands. This angle can be deduced from the measured spins of the bands. The B(M1) values then show the expected dependence on Θ: they decrease as Θ deceases with the closing of the shears.

In summary, the long regular sequences of strongly enhanced M1 transitions in neutron–deficient Pb isotopes can be explained in terms of magnetic rotation. Magnetic rotation is a new mode of excitation that can give rise to rotational–like bands in spherical nuclei. The configuration of the band head of a magnetic–rotational band in ^{193}Pb and the near–perpendicular coupling of the spins have been tested by a measurement of the g factor. Experimental B(M1) values show the decrease with increasing spin within the bands as expected for the shears effect.

References

[1] G. Baldsiefen et al. Proc. X. Intern. School of Nucl. Phys., Neutron Phys. and Nucl. Energy (1991), Varna, eds. W. Andreitscheff and D. Elenkov

[2] R.B. Firestone, Table of Isotopes, eds. V.S. Shirley et al., (Wiley 1996)

[3] S. Frauendorf, Z. Phys. A 358 (1997) 163

[4] G. Baldsiefen et al., Nucl. Phys. A 574 (1994) 521

[5] H. Hübel, Prog. Part. Nucl. Phys. 38 (1997) 89

[6] H. Hübel et al., Z. Phys. A 358 (1997) 237

[7] F.S. Stephens et al., Phys. Rev. Lett. 65 (1990) 301

[8] W. Nazarewicz et al., Phys. Rev. Lett. 64 (1990) 1654

[9] T. Byrski et al., Phys. Rev. Lett. 64 (1990) 1650

[10] S. Frauendorf, Nucl. Phys. A 557 (1993) 259c

[11] S. Frauendorf preprint (1998)

[12] A.O. Macchiavelli et al., Phys. Rev. C 57 (1998) R1073

[13] A.O. Macchiavelli et al., Phys. Rev. C 58 (1998) R621

[14] S. Chmel et al., Phys. Rev. Lett. 79 (1997) 2002

[15] G. Baldsiefen et al., Phys. Rev. C 54 (1996) 1106

[16] L. Ducroux et al., Z. Phys. A 356 (1996) 241

[17] M. Neffgen et al., Nucl. Phys. A 595 (1995) 499

[18] R.M. Clark et al., Phys. Rev. Lett. 78 (1997) 1868

[19] R.M. Clark et al., Phys. Lett. B 440 (1998) 251

[20] R. Krücken et al., Phys. Rev. C 58 (1998) R1876

MESON PRODUCTION NEAR THRESHOLD

Tord Johansson[1], H. Calén[1], J. Dyring[1], K. Fransson[1], L. Gustafsson[1], S. Häggström[1], B. Höistad[1], J. Johanson[1], A. Johansson[1], S. Kullander[1], R. J. M. Y. Ruber[1], J. Zlomanczuk[1], C. Ekström[2], K. Kilian[3], W. Oelert[3], T. Sefsick[3], R. Bilger[4], W. Brodowski[4], H. Clement[4], G. J. Wagner[4], A. Bondar[5], B. Shwartz[5], V. Sidorov[5], A. Sukhanov[5], A. Kupsc[6], P. Marciniewski[6], J. Stepaniak[6], J. Zabierowski[6], V. Dunin[7], B. Morosov[7], A. Povtorejko[7], A. Sukhanov[7], A. Zernov[7], A. Turowiecki[8], Z. Wilhelmi[8], V. Sopov[9], T. Tchernychev[9]

[1]Department of Radiation Sciences Uppsala University,
Box 535, S-75121 Uppsala Sweden
[2]The Svedberg Laboratory, Uppsala, Sweden
[3]IKP, Forschungszentrum Jülich GmbH, Jülich, Germany
[4]Physikalishes Institut, Tübingen University, Germany
[5]Institute of Nuclear Physics, Novosibirsk, Russia
[6]Institute for Nuclear Studies, Lôdz and, Warsaw, Poland
[7]Joint Institute for Nuclear Research Dubna, Moscow, Russia
[8]Institute of Experimental Physics, Warsaw University, Warsaw, Poland
[9]Institute of Theoretical and Experimental Physics, Moscow, Russia

tord.johansson@tsl.uu.se

Abstract: One of the main topics in the experimental programme at the CELSIUS cooler/storage ring of the The Svedberg Laboratory, Uppsala is meson production in light ion collisions near threshold. The PROMICE/WASA collaboration is studying π^0 and 2π production in pp collisions and η production in pN collisions. Some of the reaction channels have been measured for the first

The Nucleus: New Physics for the New Millennium
Edited by Smit et al., Kluwer Academic / Plenum Publishers, New York, 2000.

time in any detail. The experimental method is reviewed and some results are presented, together with an outlook.

1 INTRODUCTION

Cooler/storage rings offer high quality beams and can be equipped with thin windowless internal targets. These provide clean experimental conditions which are important for reaction studies close to the kinematic threshold. In fact, very little experimental information was available on meson production in this area before the advent of cooler/storage rings. Today, meson production close to threshold is one of the main topics of research at the CELSIUS, COSY and IUCF storage rings. The fact that only a few partial waves are involved near threshold should facilitate the theoretical interpretation of the data. Furthermore, relatively high and well-defined momentum transfers are involved, which makes such processes good candidates to probe short-range physics.

The CELSIUS ring at the The Svedberg Laboratory, Uppsala, provides protons up a maximum kinetic energy of 1.36 GeV. This makes pions and eta mesons energetically accessible for experimental studies in proton-nucleon collisions. A feature of reactions near threshold is that, due to kinematics, the reaction products are confined to a narrow cone around the forward direction in the laboratory system. A detection system covering the forward region has therefore complete acceptance for non-decaying reaction products near threshold. One complication at storage rings is that the beam pipe limits the access in the very forward direction. This drawback can be circumvented by using a magnetic system that bends charged particles out of the beam trajectory for subsequent detection. Alternatively, for short-lived mesons, one may detect their decay products. Both these techniques have been employed by the PROMICE/WASA collaboration. The experiments were performed using a cluster jet target giving a typical target density of 10^{14} atoms/cm^2. With a circulating beam of a few times 10^{10} protons, this provides luminosities of the order of 5×10^{30} cm^{2}s^{-1}.

2 THE PROMICE/WASA EXPERIMENT

2.1 The Experimental Set-up

The PROMICE/WASA (PW) experiment was designed to study meson production in the threshold region. It has the capability of measuring both photons and charged particles; a cross section of the set-up in the horizontal plane is shown in Fig. 1. The facility has two main constituents, *viz.* a Forward Detector (FD) and a Central Detector (CD). The forward detector has essentially full acceptance for charged particles emitted in the angular range between 4 and 21°. Its main components are a tracker (FPC), used for precise particle track reconstruction, a scintillator hodoscope for triggering and fast pixel determination (FHD), and a scintillator range hodoscope for energy measurements (FRH). The FD is complemented by a hodoscope (FVH) at the rear, to register penetrating particles, and four scintillators near the scattering chamber (FWC) for trigger purposes. The CD is made from two arrays of CsI(Na) crystals arranged in 7×8 matrices (CEC). In front of each array there are scintillators for charged particle identification/rejection (CDE). The characteristic performances of the PW experimental set-up are given in Table 1 and more details can be found in Ref. [4].

Table 1 Parameters and performances of the PROMICE/WASA detector

	Forward detector	Central detector
Scattering angle resolution	$< 1^\circ$ *	$< 5^\circ$ *
Energy resolution (relative)	$\approx 3\%$ *	$\approx 3\%$ * (charged) ≈ 8 MeV * (γ's) **
T_{max} charged pions (stopped)	150 MeV	190 MeV
T_{max} protons (stopped)	300 MeV	400 MeV
T_{max} deuterons (stopped)	400 MeV	450 MeV
Amount of material	$\approx 1\ X_0$	$\approx 16\ X_0$

* FWHM, ** @ 100 MeV

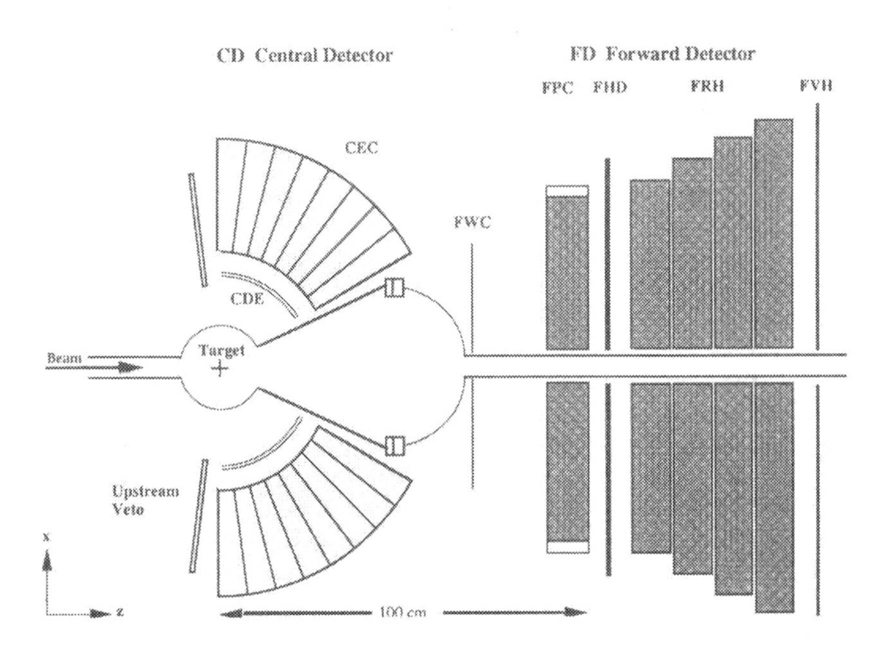

Figure 1 The PROMICE/WASA experimental set-up.

In addition, the CELSIUS magnets in the quadrant following the experimental set-up have been used as a spectrometer to detect charged reaction products emitted near zero degrees. This is illustrated in Fig. 2, which shows the location of the telescope for the threshold $pn \rightarrow d\eta$ measurement (see 2.5).

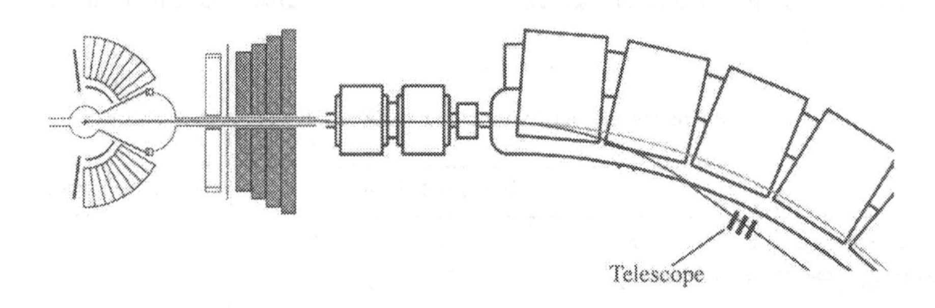

Figure 2 Zero-degree detection using the dipole magnets of the CELSIUS ring as a spectrometer. The position of a scintillator telescope is indicated.

2.2 The $pp \rightarrow pp\pi^0$ reaction

Neutral pion production in proton-proton collisions near threshold was for a long time believed to be well understood. When the first precision measurements were made at IUCF [21], it turned out that theory underestimated the cross section by a factor of five. Two different mechanisms were subsequently proposed to make up for the deficiency, *viz.* the inclusion of heavy meson exchanges *i.e.* short-range physics [19], and off-shell pion rescattering [17]. Both of them are likely important and the understanding of this reaction is still far from complete today.

The PROMICE/WASA collaboration has measured the $pp \rightarrow pp\pi^0$ cross section at several CM excess energies, Q_{cm} ($Q_{cm} = \sqrt{s} - \sum m_{final}$), between 0.5 and 14 MeV [3]. The results are shown in Fig. 3 together with results from IUCF [22] and a curve calculated using phase space and pp final state interaction (FSI), as given by the effective range approximation including Coulomb interaction (arbitrarily normalised) [28]. Such a curve describes the shape of the data well and there is excellent agreement between the IUCF and CELSIUS cross sections in the region of overlap. The PW measurement could be extended down to about 400 keV above threshold, where the protons escape undetected inside the beam pipe, by measuring the two γ's from the $\pi^0 \rightarrow 2\gamma$ decay in the CsI arrays. At the higher energies the two final state protons are measured in the FD and this allows the reconstruction of kinematically complete events to study differential cross sections.

The only surviving production amplitude at threshold corresponds to the transition $^3P_0 \rightarrow {}^1S_0\, s$, where the standard $^{2S+1}L_J$ notation is used for the

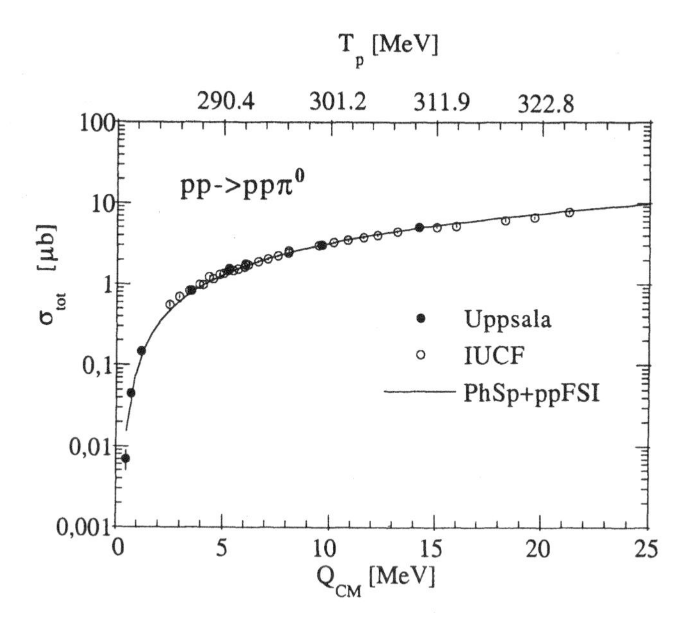

Figure 3 The $pp \to pp\pi^0$ cross section as a function of centre-of-mass excess energy Q_{cm} [3, 22]. The equivalent proton beam energy is shown at the top. The solid line represents a calculation using phase space and pp FSI (arbitrarily normalised).

pp system, with the lower case letter denoting the angular momentum of the meson. Such a transition leads to an isotropic angular distribution but an analysis of a high statistics experiment [29] undertaken at $Q_{cm} = 14$ MeV ($T_p = 310$ MeV) shows a clear anisotropy in the π^0-angular distribution as illustrated in Figs. 4a and 4b.

We have used a phenomenological description of the differential distributions for the acceptance corrections, taking the unpolarised matrix element to be of the form

$$|M|^2 = |A|^2 + \left[2\Re(A^*B)\frac{k^2}{\mu^2} + |B|^2\frac{k^4}{\mu^4}\right]\cos^2\theta_\pi + |C|^2\frac{q^2}{\mu^2} + |D|^2\frac{k^2q^2}{\mu^4}\sin^2\theta_{pp}. \tag{1}$$

The momentum of pion in the overall CM system is denoted by k, $2q$ is the relative momentum in the final two-proton system, and μ is the reduced mass in the final state. θ_π is the angle between the π^0 and the incident beam direction in the overall CM system, and θ_{pp} that of the pp relative momentum. The amplitude A corresponds to an admixture of the dominant ${}^3P_0 \to {}^1S_0\, s$ with a small amount of the ${}^3P_2 \to {}^1S_0\, d$ transitions, and B to the transition ${}^3P_2 \to {}^1S_0\, d$. The effect of the pp final state interaction has been taken into account by a multiplicative factor $F_{fsi}(q)$ calculated from the Paris wave function [18]. Agreement with the observed pion angular distribution is only found with $B = (1.2 \pm 0.2)A$ showing that pion d-waves cannot be neglected even this close to threshold. The k^2 dependence of the s-d interference term suggests that the angular asymmetry should increase with decreasing relative proton momenta. This is illustrated in Fig. 4b, where the pion angular distribution is plotted in three intervals of relative proton momenta. The data clearly follow this trend and there is good agreement with the parametrisation of the amplitudes (solid lines). These findings are also in agreement with data from RNCP where the same reaction has been measured as a function of beam energy for pp excitation energies below 3 MeV [20].

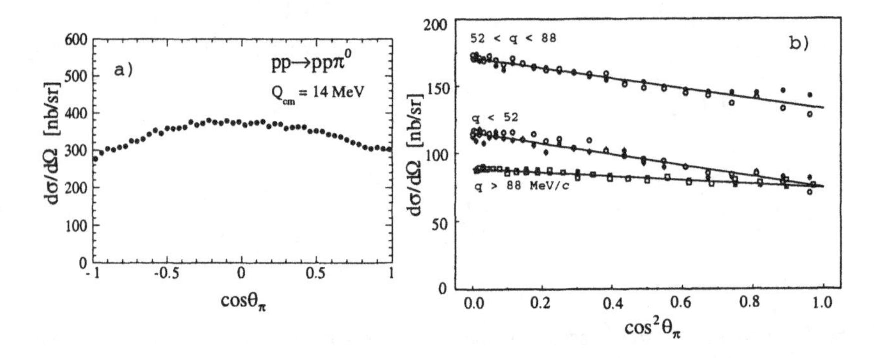

Figure 4 (a) Differential cross section for the reaction $pp \rightarrow pp\pi^0$ at 14 MeV excess energy as a function of the cosine of the pion production angle in the overall CM system. (b) The same data as in (a) as a function of $\cos^2\theta_\pi$ in three bins of excitation energy of the final state pp pair, $Q_{pp} < 2.88$ MeV, $2.88 < Q_{pp} < 8.25$ MeV, and $Q_{pp} > 8.25$ MeV respectively [29]. The data in the two hemispheres are shown as open and closed symbols respectively.

2.3 2π production in pp collisions

No experimental data exist for 2π production in pp collisions at excess energies below 55 MeV and data are also scarce for somewhat higher energies [24, 13, 12]. Theoretically it is generally assumed that these processes are dominated by the excitation of nucleon resonances, in particular the N*(1440) P_{11} resonance [1], which subsequently decays into a nucleon and two pions. We have collected data on the $pp \rightarrow pp\pi^+\pi^-$, $pp\pi^0\pi^0$, and $pn\pi^+\pi^-$ reactions in the region of $20 < Q_{cm} < 80$ MeV and the analysis is still in progress. Fig. 5 shows the existing data on the $pp \rightarrow pp\pi^+\pi^-$ reaction, which is the best measured reaction channel, together with a theoretical calculation using the N*(1440) as the main ingredient [1]. The arrows indicate the energies at which PW data have been taken and a first preliminary data point at $Q_{cm} = 21$ MeV is shown. Once the final cross sections are available, they will provide guidance for a better theoretical understanding of these processes.

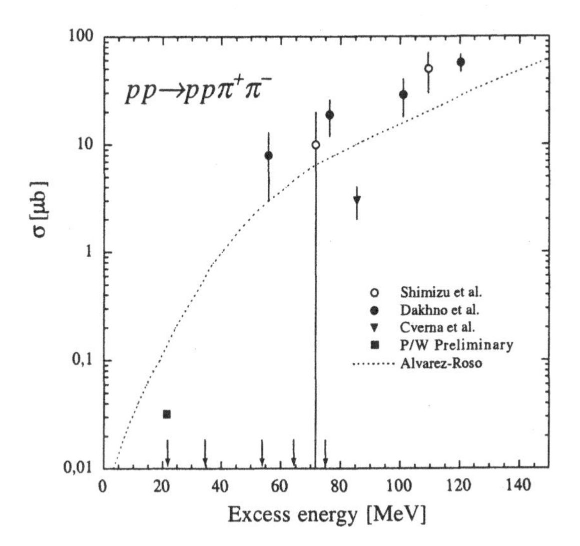

Figure 5 Total cross sections for the $pp \rightarrow pp\pi^+\pi^-$ reaction [24, 13, 12]. The arrows indicate the energies at which the PROMICE/WASA collaboration has taken data and the filled square shows the first preliminary data point. The dashed line is a theoretical calculation from Ref. [1].

2.4 η production in pp collisions

In contrast to the π^0 case, η production in nucleon-nucleon collisions near threshold is strongly influenced by the presence of a nucleon resonance. It is generally assumed that η production mainly proceeds *via* the excitation of the N*(1535) S_{11} resonance, which has a large decay branching ratio into the $N\eta$ channel. Theoretical calculations of η production are generally based on the one-boson-exchange model [16] and, although similar in spirit, they differ significantly in their assumptions of the relative importance of different meson exchanges. Therefore good data from different channels are needed in order to clarify the situation. Furthermore, the presence of the N*(1535) should influence the observables near threshold, where S-waves dominate, and give rise to a strongly attractive ηN FSI. It has in fact been speculated that the ηN interaction might be strong enough for quasi-bound systems to be formed [27], possibly even for the two-nucleon system [25].

The total cross section of the $pp \rightarrow pp\eta$ reaction has been measured at six different energies in the threshold region by the PW collaboration [5] using the same procedure as in the π^0 experiment. Our data are shown in Fig. 9 together with points from SATURNE [2, 10]. It is worth mentioning that, unlike the π^0 case, these data do not follow completely the shape of phase space modified by the pp FSI, suggesting the importance of the ηp FSI.

Differential cross sections have been extracted for the $pp \rightarrow pp\eta$ reactions at excess energies of $Q_{cm} = 16$ MeV and 36 MeV [7]. To carry out the acceptance calculations, we have parametrised the differential cross sections in an analogous way to Eq. 1, keeping terms upto power k^2 or q^2. Figs. 6a and 6b show the resulting η polar angular distributions in the overall CM system. The data at

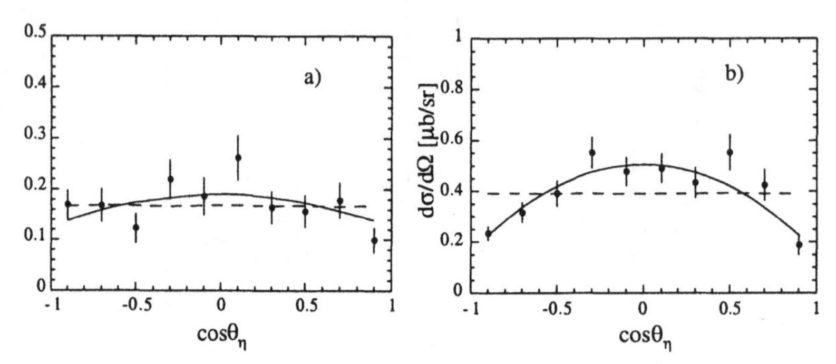

Figure 6 Differential cross sections in the η-production angle for the $pp \to pp\eta$ reaction at (a) $Q_{cm} = 16$ MeV ($T_p = 1295$ MeV) and (b) $Q_{cm} = 37$ MeV ($T_p = 1350$ MeV). The dashed curves represent isotropic emission, whereas the solid curves include the angular dependence of Eq. 1.

$Q_{cm} = 36$ MeV show an asymmetry indicating the presence of η d-waves at this energy. The fit using the amplitudes is shown as a solid line, whereas the phase space is given by a dashed line. Although the angular distribution at 14 MeV excess energy is marginally better reproduced with the inclusion of the $\cos^2\theta_\eta$ term (solid line) as compared to isotropic emission, no conclusions can be made at this energy about the presence higher partial waves. The importance of pp FSI is clearly manifested in the η kinetic energy distributions in Figs. 7a and 7b. These are strongly shifted towards higher energies with respect to phase space (dashed lines), with a consequent depletion at lower energies. This is a direct consequence of the strongly attractive pp FSI, which enhances events where the η recoils against two protons with low relative momentum.

2.5 *Quasi-free η production in pN collisions*

The $pp \to pp\eta$ reaction is a pure isospin-one channel and to learn more about the η production process in nucleon-nucleon collisions one should study also the isospin-zero channel. The pn channel is a mixture of $I = 0$ and 1, but previous data in this area have been very limited. Cross sections on η production have been obtained by unfolding spectra from the upper energy tail of a neutron beam [23], or deduced from inclusive measurements using a deuterium target [11]. With the PW set-up it was possible to isolate the exclusive pp and pn quasi-free reaction channels using a deuterium gas-jet target [6, 8, 9].

$$pd \to pp\eta + n_s \tag{2}$$

$$\to pn\eta + p_s \tag{3}$$

$$\to d\eta + p_s \tag{4}$$

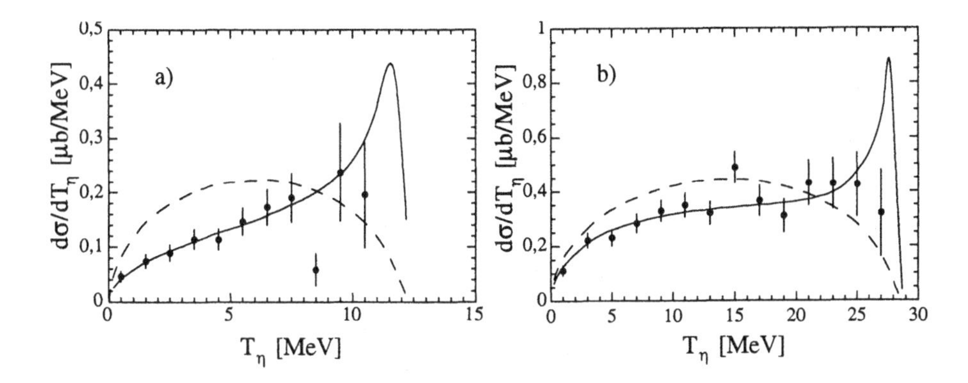

Figure 7 η kinetic energy distributions from the $pp \rightarrow pp\eta$ reaction at (a) Q_{cm} = 16 MeV and (b) Q_{cm} = 37 MeV. The dashed curves represent phase space distributions whereas the solid curves include the angular dependence of Eq. 1. It should be noted that the difference betweenthe solid line and phase space + pp FSI is marginal at Q_{cm} = 16 MeV and that, even at Q_{cm} = 37 MeV, phase space + pp FSI still dominates the shape.

where the s subscript denotes a slow "spectator" nucleon. The different reaction channels are identified by measuring the 2γ decay of the η in the CD together with charged particles in the FD. The Fermi motion of the target nucleon affects the CM energy of the system, consisting of the beam proton and the target nucleon, on an event-by-event basis. Despite staying at a fixed beam energy one can therefore obtain the excitation functions in Q_{cm}. Figure 8 shows the range of excess energies available at an incident proton kinetic energy of 1.36 GeV, taking the nucleon Fermi motion as given by the Paris potential [18].

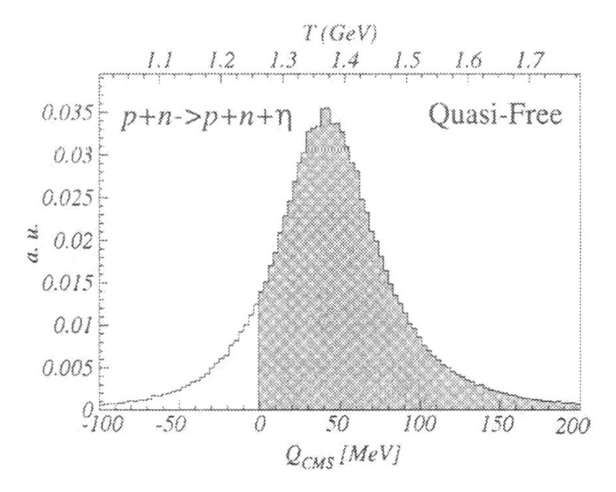

Figure 8 Distribution of CM excess energies, Q_{cm}, for the quasi-free $pn \rightarrow pn\eta$ process using a deuterium target. The upper scale gives the corresponding beam energy for a target nucleon at rest.

The measured data allow for a reconstruction of Q_{cm} for each event to a precision of about 5 MeV in the $d\eta$ and $pp\eta$ cases and $\approx$ 8 MeV for the $pn\eta$ case. Since the acceptances can be calculated for all three reaction channels, it is possible to extract the excitation functions for these quasi-free channels. The normalisation is made by evaluating the quasi-elastic pp scattering cross section from the same event sample. The measured excitation functions for the quasi-free reactions (1.2-4) are shown in Fig. 9, together with data on the free $pp \to pp\eta$ cross sections. The satisfactory agreement between the free

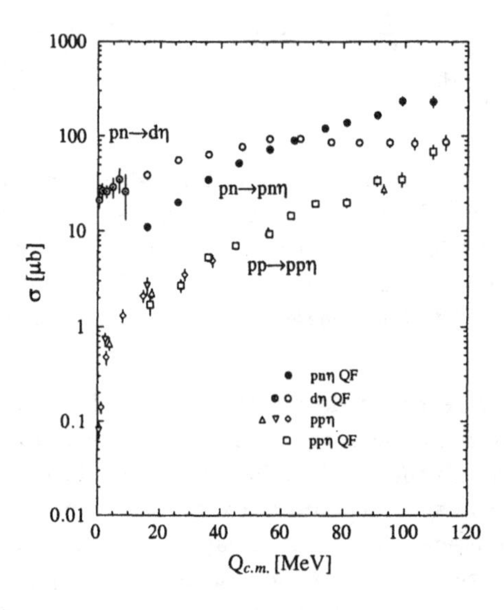

Figure 9 η production cross sections in the quasi-free pp and pn channels [6, 8, 9] together with cross sections for the free pp channel [5, 2, 10].

and the quasi-free pp data gives us confidence in the methodology. Several comments can be made on the resulting physics. The pn cross section is larger than the pp over the range of Q_{cm} covered. The $d\eta$ channel is dominant for $Q_{cm} < 60$ MeV, whereas the $pn\eta$ cross section is larger at the higher excess energies. The shapes of the $pn \to pn\eta$ and $pp \to pp\eta$ excitation functions are very similar, but with the $pn\eta$ being approximately 6.5 times larger. The ratio between these two cross sections, illustrated in Fig. 10a, shows that the $I = 0$ channel is substantially larger than the $I = 1$, and points to the importance of isovector exchange in these processes. This information will help to constrain theoretical models.

The $pn \to d\eta$ cross section is compared in Fig. 10b, with three curves. The dotted one corresponds to phase space and describes the shape of the data well up to $Q_{cm} \approx 60$ MeV when it starts to overpredict. The dashed curve includes, in addition to phase space, a Breit-Wigner function describing the shape of the $N^*(1535)$ resonance [6] and this reproduces the full energy dependence of the cross section quite well. This suggests that the rise of the cross section at low Q_{cm} is given by phase space, whereas the decrease above 60 MeV is governed by the tail of the N*(1535) resonance. This is the first explicit experimental evidence of the importance of this resonance in nucleon-induced η production. The solid curve in Fig. 10b corresponds to the phenomenological parametrisation of experimental data given in Ref. [23]. A direct comparison between these results is not possible since the parameterisation is valid only for $Q_{cm} < 10$ MeV. This is a region to which the first PW experiment was insensitive because the deuteron escapes undetected down the beam pipe.

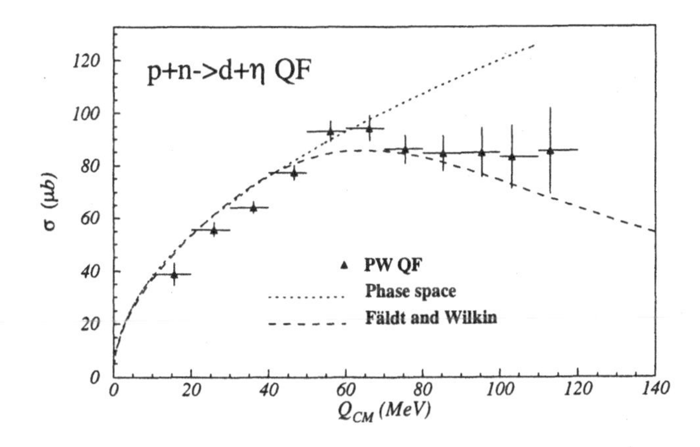

Figure 10 (a) The ratio between the quasi-free $pn \to pn\eta$ and $pp \to pp\eta$ cross sections [9]. (b) The quasi-free $pn \to d\eta$ cross section [6]. The dotted curve represents phase space and the dashed line is phase space with a Breit-Wigner curve describing the N*(1535) [15]. The solid curve is the parametrisation of the cross section from Ref. [23].

A dedicated measurement of the very near threshold region for this process was therefore made [8]. Deuterons emitted at zero degrees from the quasi-free $pn \to d\eta$ process were detected using the CELSIUS magnets as a spectrometer, as shown in Fig. 2. Taken together with the detection of two γ's from the η decay, this made it possible to make a measurement in the region covered by the Saclay experiment [23]. As can be see in Fig. 11a, these data do not show the spectacular increase seen in the Saclay data. On the other hand the cross sections do not behave like phase space when approaching threshold, but are fairly constant in the region $0 < Q_{cm} < 10$ MeV at a level of $20 - 30$ μb. This is further emphasised in Fig. 11b, where the ratio between the measured cross section and phase space expectation is shown. A relative enhancement of the cross section when approaching the threshold is characteristic of a strong final state interaction and could be consistent with a quasi-bound ηd system. The observed signal agrees in magnitude and shape to such an effect [25], which would correspond to an ηd scattering length of a few fermi.

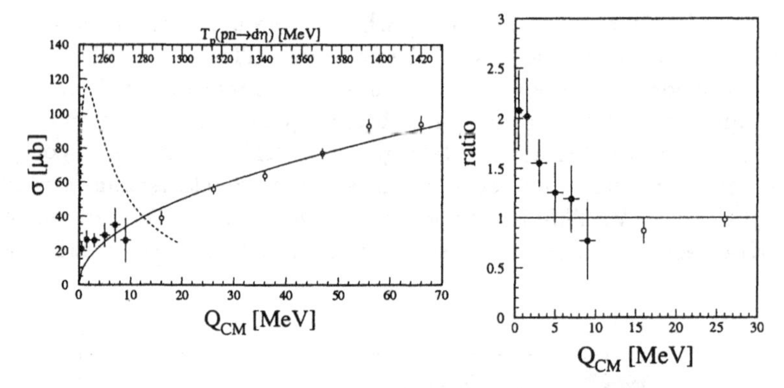

Figure 11 (a) Energy dependence for the quasifree $pn \rightarrow d\eta$ reaction near threshold (solid symbols) together with earlier data (open symbols). Also shown is a phase space curve, arbitrary normalised (solid line) and the parametrisation of the cross section from Ref. [23] (dashed line). The scale on top of the figure gives the corresponding beam energy of a fixed target experiment. (b) The ratio of the measured cross section and the two-body phase space. The phase space curve from (a) is shown as a solid line.

3 OUTLOOK

The total cross sections for light meson production are becoming reasonably well measured near threshold. To learn more one should make more complete studies by measuring the differential cross sections in detail. The first steps in this have been taken for the $pp \rightarrow pp\pi^0$ and $pp \rightarrow pp\eta$ reactions. A high acceptance over all of phase space is important for such experiments and the PW set-up has its limitations is in this respect. A new detector system with very high acceptance, WASA 4π, has been designed [26] and is presently being mounted at the CELSIUS ring. The equipment, which is shown in Fig. 12a, will be a very competitive facility. Other types of studies that are planned for this detector include meson decays and rare processes in general.

The forward part of the WASA detector is identical to the one used by the PW experiment, whereas the central part is new. A cross section of the central part is shown in Fig. 12b. The outer region (SEC) is made from a large array of CsI(Na) crystals with a geometrical coverage of 96% of 4π to measure photons and charged particles. This can be compared to the 7% solid angle coverage of the PW CsI(Na) arrays. The inner part is surrounded by a thin-walled (0.16 X_0) superconducting solenoid (SCS) and comprises a vertex detector, MDC, made from 16 layers of cylindrical drift chambers, and a plastic scintillator barrel (PSB). The central magnetic field is 1.3 T and the arrangement will allow for precise tracking and momentum measurements of charged particles emitted at larger angles. Access near the interaction region has been achieved through the development of a new internal target system which produces a train of frozen pellets of hydrogen or deuterium which are brought to intersect the circulating beam [14].

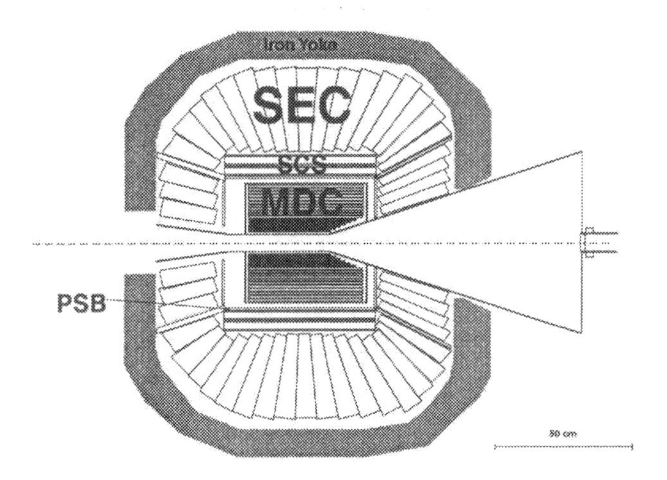

Figure 12 (a) The WASA 4π experimental set-up. (b) A cross section of the WASA central detector.

There are also plans for an energy upgrade of the CELSIUS-ring to increase the maximum proton energy from 1.36 to 1.85 GeV. This would greatly enhance the possibilities for detailed studies of η production and make the $pp \rightarrow pp\eta\pi^0$ reaction kinematically accessible. Heavier mesons, such as the ω and η', could then be produced in pd collisions.

4 CONCLUSIONS

Light meson production has been studied in light ion collisions at the CELSIUS cooler/storage ring in Uppsala by the PROMICE/WASA collaboration. Several new results have been found:

- An extension of the $pp \rightarrow pp\pi^0$ reaction data closer to threshold.
- Non-isotropic meson angular distributions are seen for both the $pp \rightarrow pp\pi^0$ and the $pp \rightarrow pp\eta$ reactions, showing an early onset of d-waves.
- The excitation functions for the $pn \rightarrow pn\eta/d\eta$ have been measured for the first time. The $pn \rightarrow pn\eta$ cross section is approximately 6.5 times higher than that for $pp \rightarrow pp\eta$, showing the dominance of the isospin zero channel. The shape of the $d\eta$ excitation function shows the explicit influence of the $N^*(1535)$ resonance.
- The very near threshold data on the quasi-free $pn \rightarrow d\eta$ reaction shows experimental support for a strong and attractive ηN interaction.

A second generation of experiments will use the WASA 4πdetector and allow for unprecedented measurements of differential cross sections as well as rare reactions and meson decays.

References

[1] L. Alvarez-Ruso, E. Oset and E. Hernandez. *Nucl. Phys.*. A 633:519, 1998.
L. Alvarez-Ruso. *Private communication.*

[2] A.M. Bergdolt et al. *Phys. Rev.*. D 48:R2969, 1993.
F. Hibou et al. *Phys. Lett.*. B 438:41, 1998.

[3] A. Bondar et al. *Phys Lett.*. B 356:8, 1995.

[4] H. Calén et al. *Nucl. Instr. and Meth.*. A 379:57, 1996.

[5] H. Calén et al. *Phys. Lett.*. B 366:39, 1996.

[6] H. Calén et al. *Phys. Rev. Lett.*. 79:2642, 1997.

[7] H. Calén et al. *Internal Report.* TSL/ISV-98-0198.

[8] H. Calén et al. *Phys. Rev. Lett.*. 80:2069, 1998.

[9] H. Calén et al. *Phys. Rev.*. C 58:2667, 1998.

[10] E. Chiavassa et al. *Phys. Lett.*. B 322:270, 1994.

[11] E. Chiavassa et al. *Phys. Lett.*. B 337:192, 1994.

[12] F.H. Cverna et al. *Phys. Rev.*. C 23:1698, 1981.

[13] L.G. Dakhno et al. *Sov. J. Nucl. Phys.*. 37:540, 1983

[14] C. Ekström et al. *Nucl. Instr. and Meth.*. A 371:572, 1996

[15] G. Fäldt and C. Wilkin. *Private communication.*

[16] J.F. Germond and C. Wilkin. *Nucl. Phys.*. A518:308, 1990.
J.M. Laget, F. Wellers and J.F. Lecolley. *Phys. Lett.*. B 257:258, 1991.
T. Vetter et al. *Phys. Lett.*. B 263:153, 1991.
A. Moalem et al. *Nucl. Phys.*. A589:649, 1995; A. Moalem et al., ibid. A600:445, 1996; E. Gedalin A. Moalem and L. Razdolskaja, *ibid* A634:368, 1998.
M. Batinic, I. Slaus and A. Svarc. *Phys. Scr.*. 56:321, 1997.

[17] E. Hernández and E. Oset. *Phys. Lett.*. B 350:158, 1995.

[18] M. Lacombe et al. Phys. Lett.. B 101:139, 1981.

[19] T.S.H. Lee and D.O. Riska. *Phys. Rev. Lett.*. 70:2237, 1993.

[20] Y. Maeda et al. πN Newsletter. 13:326, 1997.

[21] H.O. Meyer et al. *Phys. Rev. Lett.*. 65:2845, 1990.

[22] H.O. Meyer et al. *Nucl. Phys.*. A 539:633, 1992.

[23] F. Plouin, P. Fleury and C. Wilkin. *Phys. Rev. Lett.* 65:690, 1990.

[24] F. Shimizu et al. *Nucl. Phys.* A 386:571, 1982.

[25] T. Ueda. *Phys. Rev. Lett.*. 66:297, 1991.

S.A. Rakityansky et al. *Phys. Rev.*. C53:2043, 1996.

A.M. Green, J.A. Niskanen and S. Wycech. *Phys. Lett.*. B 394:253, 1996.

[26] *The Svedberg Laboratory Progress Report* 1996-1997.

http://www.tsl.uu/wasa.

[27] C. Wilkin, *Phys. Rev.* C47:R938, 1993.

[28] T. Wu and T. Ohmura. *Quantum Theory of Scattering*. Prentice-Hall, Cliffs, NJ, 1962.

[29] J. Zlomanczuk et al. Phys. Lett.. B 436:251, 1998.

OBJECT-ORIENTED PROGRAMMING FOR NUCLEAR PHYSICS MEASUREMENTS

K. Juhász[1], G.E. Perez[1,2], M. Emri[3], B.M. Nyakó[2], J. Szádai[2] and J. Molnár[2]

[1] Institute of Mathematics and Informatics, L. Kossuth University, Debrecen, Hungary
[2]Institute of Nuclear Research, Hungarian Academy of Sciences, Debrecen, Hungary
[3]PET Centrum, Medical University, Debrecen, Hungary

Abstract: A Windows-based software package, working with a dual-port incrementing matrix memory PC board in acquisition mode, has been developed for acquiring, storing, visualizing and analysing 2-dimensional spectra. It has been used for testing the performance of the electronics of a CsI(Tl) ancillary detector system of EUROBALL.

INTRODUCTION

The Institute of Nuclear Research has been participating in the development of a CsI(Tl) charged-particle detector system, intended as an ancillary detector of the EUROBALL γ-spectrometer. Using such detectors, the coincidence events, related to the emission of light charged particles (protons and alphas) in a nuclear reaction, can be enhanced (or vetoed). The discrimination of these particles at all energies is of vital importance. An improved detector wrapping [1] and a novel particle discrimination technique [2] have been developed for this purpose. The final version of the electronics of this ancillary system, to ensure compatibility with the electronics of EUROBALL, is being realised in VXI standard. The measurements for testing the performance of the prototypes of this dedicated electronics have been performed using a PC-based data acquisition system having an incrementing matrix-memory PC-board [3]. The present contribution gives details of a Data Acquisition and VISualization software package (DAVIS), which has been developed using the C++ object-oriented language, the Application Framework for Windows and the ObjectWindows Class Library of Borland C++ 3.1.

DESCRIPTION OF DAVIS

The program maintains the acquisition, visualization, storage and analysis of 2-dimensional spectra under the given hardware configuration of a specific experiment. The object class hierarchy of the program is shown in the flowchart. The

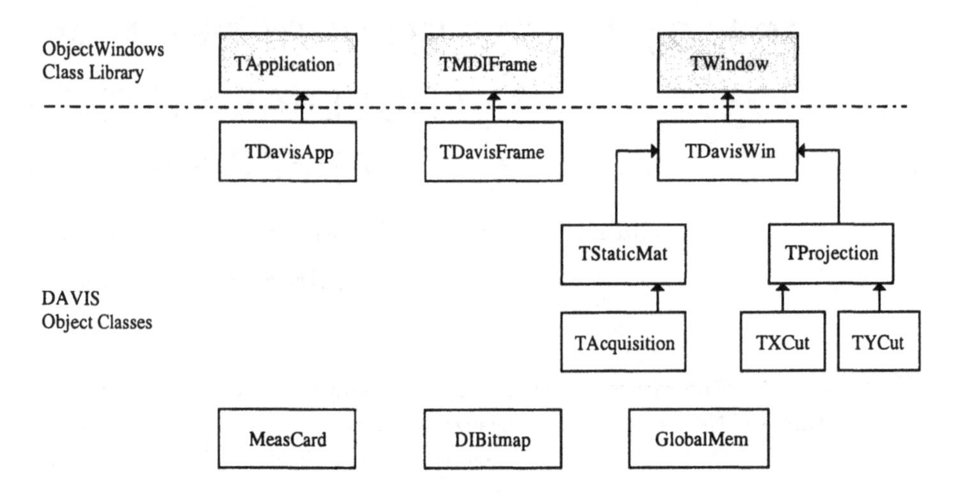

Figure 1 Flowchart of the object class hierachy.

main object classes, e.g. the application class (*TDavisApp*) and the base window object class (*TDavisWin*), are derived from ObjectWindows library classes. Multiple windows from within a single application frame are managed by the *TDavisFrame* MDI frame window class. *TAcquisition* inherits the *TStaticMat* class representing the 2D matrix objects, and contains the logic for measurement control. *TProjection* is an abstract class for manipulating any projections (cuts). *TXCut* and *TYCut* are derived classes for manipulating X and Y cuts, i.e. Y and X-projections of the matrix for the given X and Y regions, respectively.

The *MeasCard* class, the interface to the acquisition card, stores the multi-dimensional experimental data and maintains the full operation of the card: sets the hardware memory address of the card, initializes, reads (in real time) and resets the card, enables/disables (start/stop) acquisition. The *DIBitmap* class object is responsible for the manipulation and the on-screen graphical visualization of two-dimensional matrices. These are represented as 2D depth-sensitive colour bitmaps showing the actual memory contents of the card in real time. Global Windows memory allocation is managed by the *GlobalMem* class. In Measurement Control Mode the program can define 'new' measurement, set the 'number' of channels/detectors, preset measurement time, 'start', 'stop' and 'pause' the acquisition. Data are visualized as 2-dimensional (ParticleType-Energy) matrices and 1-dimensional (projection/slice) spectra both in acquisition and off-line analysis modes. Visualized 2D spectra can be saved to disk, printed/saved (as colour bitmaps), and X- and Y-cuts/full-projections of them can also be taken. Manipulations (expand/smooth, adjust Y scale, Gaussian fit) of 1D spectra are also possible.

The DAVIS program works under Windows 3.1 on any IBM PC compatible machine having a 486 or higher CPU, a minimum of 4 MB RAM and sufficient hard disk space (1 MB for the program; 256 KB/matrix/detector). In acquisition mode it needs a dual-port incrementing matrix memory PC card, which receives and stores two-dimensional data from the physical measurement, and is capable of managing simultaneous data collection and non-destructive memory read-out. Some simple data evaluation routines are also implemented.

APPLICATIONS IN NUCLEAR PHYSICS EXPERIMENTS

The program has been successfully applied in test measurements aming at the optimization of (a) the particle discrimination capability of the dedicated electronics [4] of a 25-element CsI detector modul (ChessBoard), and (b) the performance of the whole system with light charged particles produced in heavy-ion induced nuclear reaction. An example spectrum taken at SARA, Grenoble, using the ^{94}Mo+^{36}S reaction at E(^{36}S)=170 MeV is shown in the figure below.

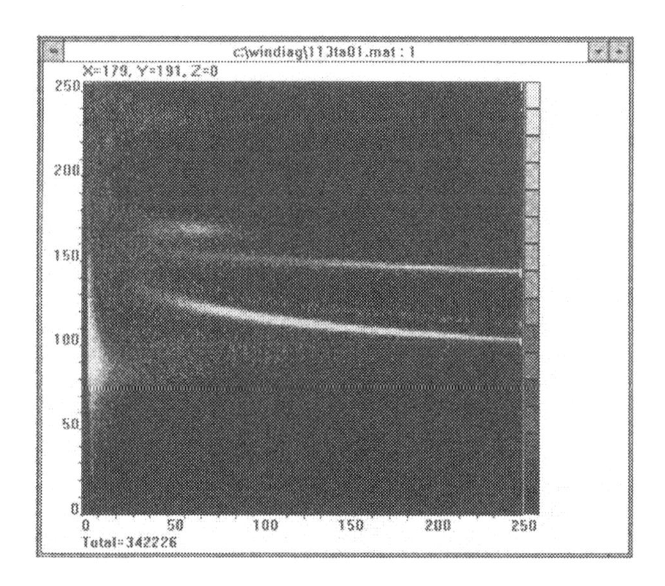

Figure 2 An example spectrum taken at SARA.

This detector module and its VXI electronics will be an integral part of the DIAMANT ancillary detector of the EUROBALL γ-spectrometer, the resolving power of which can be greately improved, using the type and energy information of the emitted charged particles [5]. The present software enabled most of the R&D work to be performed without having the expensive VXI infrastructure.

References

[1] G. Kalinka *et al.*, Nucl. Instr. Meth. submitted for publication.

[2] J. Gál *et al.*, Nucl. Instr. Meth. A366 (1995) 120.

[3] J. Molnár *et al.*, ATOMKI Annual Report 1995, p103.

[4] J. Gál *et al.*, Nucl. Instr. Meth. A399 (1997) 407

[5] J.N. Scheurer *et al.*, Nucl. Instr. Meth. A385 (1997) 501

RECENT RESULTS FROM IN-BEAM STUDIES OF VERY NEUTRON-DEFICIENT NUCLEI

Rauno Julin

Department of Physics, University of Jyväskylä (JYFL)
P. Box 35, FIN-40351 Jyvškylä
Finland*

Rauno.Julin@phys.jyu.fi

Abstract: Nuclear structure of very neutron deficient Z $\approx$ 82 nuclei and very heavy nuclei has been studied by employing in-beam γ-ray spectroscopic methods involving recoil-gating and recoil-decay tagging (RDT) methods.

1 INTRODUCTION

The very neutron deficient nuclei with Z > 75 are produced in fusion evaporation reactions and identified by determining their characteristic decay (mostly α decay) properties. Due to the small fusion cross-sections the conventional in-beam γ-ray spectroscopic methods even with the biggest Ge-detector arrays are insufficient for more detailed structure studies of these nuclei. The γ-rays of interest are obscured by the dominant background from fission and Coulomb excitation.

In the present work, we have utilised detection of forward-flying fusion residues in coincidence with prompt γ-rays to perform γ-γ coincidence measurements down to a production level of 100 μb. We have also shown that the short lifetimes of the very neutron deficient heavy and very heavy nuclei render it possible to utilise the characteristic decay (in these cases α decay) properties of these nuclei to collect clean γ-ray spectra down to a production level of 100 nb. Moreover, these decay properties provide us with an unambiguous identification of γ-rays with a certain nuclide. In these so-called recoil-gating and

*Jurosphere and SARI collaborations

The Nucleus: New Physics for the New Millennium
Edited by Smit et al., Kluwer Academic / Plenum Publishers, New York, 2000.

recoil-decay- tagging (RDT) experiments we have employed the JYFL gas-filled RITU separator [1] in combination with various Ge-detector arrays.

RITU is a charge and velocity focusing device designed for separating heavy fusion evaporation products with high (up to about 40%) transmission. Recoils are implanted into a 80 × 35 mm Si strip detector covering typically about 70% of the recoil distribution at the focal plane. It is divided into 16 strips each of which has a position resolution of about 300 μm in vertical direction, providing the granularity needed in correlating the recoils with their subsequent α decay.

2 THE JUROSPHERE AND SARI COLLABORATIONS

During the years 1997 and 1998, a series of in-beam γ-ray spectroscopy experiments using the Jurosphere and SARI Ge detector arrays at the RITU target area were carried out at JYFL. Both Ge detector arrays were provided by the French/UK Loan Pool. In 1997, during a 7 months campaign, about 3500 h of beam time were devoted to 25 experiments using the Jurosphere array. The SARI array was used for a period of about 5 months in 1998. During this time about 1840 h of beam time were allocated to 10 experiments. Most of these experiments were RDT measurements. In all experiments, beams provided by the JYFL-ECR ion source were accelerated by the K = 130 MeV cyclotron.

The Jurosphere array consisted of 10 TESSA-type and 15 EUROGAM Phase I Compton-suppressed Ge detectors. The absolute photo peak efficiency of the array for 1.3 MeV γ-rays was $\approx$ 1.5 %. Additionally, one TESSA type Ge detector was used to observe γ rays at the RITU focal plane, about 3 cm behind the Si strip detector.

The SARI array consisted of three or four segmented clover detectors positioned around the target position. The absolute efficiency of the array was 2-4 % at 1.3 MeV γ-ray energy depending on the number of the detectors and the target to detector distance. The SARI set-up also involved Ge detectors at the RITU focal plane to detect γ rays from long-lived isomers and the γ rays emitted following the α decays. A close geometry set-up of four TESSA detectors was used for this purpose in several experiments. In some of the experiments these detectors were replaced by a large volume clover detector (from GSI).

In the following sections new data for light Hg, Po and Rn isotopes and for a Z = 102 isotope ^{254}No will be discussed. Other highlights from the RDT experiments performed at JYFL include first observation of excited states in:

- ^{226}U from the reaction ^{208}Pb(^{22}Ne,4n) [2]
- ^{168,170}Pt from the reaction ^{112}Sn(^{58,60}Ni,2n) [3]
- ^{171}Ir from the reaction ^{116}Sn(^{58}Ni,p2n) [4]
- ^{184}Pb from the reaction ^{148}Sm(^{40}Ca,4n) [5]
- ^{206,208}Ra from the reaction ^{170,172}Yb(^{40}Ar,4n) [6]
- ^{164}Os from the reaction ^{106}Cd(^{60}Ni,2n) [7]

3 INTRUDER STRUCTURES IN Z > 82 AND Z < 82 NUCLEI

An improved RDT study of ^{176}Hg was carried out [8] to confirm the tentative assignments of ref. [9] and to look for prolate intruder structures earlier identified in the heavier Hg isotopes. Excited states of ^{176}Hg were populated via the ^{144}Sm(^{36}Ar,4n) fusion-evaporation channel. Altogether, 90000 ^{176}Hg α-decay events were recorded giving a production cross-section of about 5μb.

The energy spectrum of γ rays in coincidence with recoils is shown in Fig. 1(a). It is dominated by γ rays from ^{176}Pt produced via the (^{36}Ar,2p2n) fusion-evaporation channel. These γ rays are absent in Fig. 1(b), which shows a recoil-gated γ-ray spectrum obtained by correlating with the ^{176}Hg α decay. All the lines in this spectrum are assigned to originate from ^{176}Hg. The three strongest lines were seen by Carpenter et al. [9]. Recoil-gated α-tagged γ-γ coincidence data were also obtained. Analysis of such data (Fig. 1(c) and 1(d)) together with angular distributions reveals that the 613.3, 756.4, 551.0, 453.2 and 500.5 keV γ rays form an yrast E2 cascade.

The lowering of the transition energies in the yrast band above the 6^+ state is interpreted to be due to crossing of a prolate intruder band. In Fig. 2

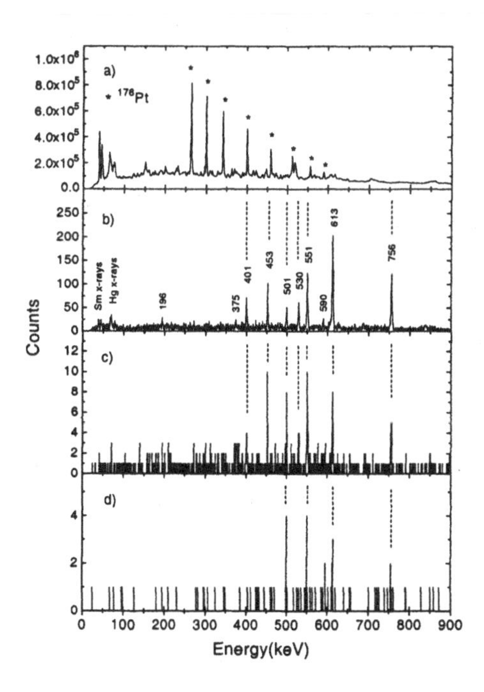

Figure 1 a) Energy spectrum of γ rays in coincidence with fusion-evaporation recidues detected in the RITU focal plane detector. b) γ- ray spectrum in coincidence with fusion-evaporation recidues and tagged with the ^{176}Hg α decays. c) Sum of recoil-gated and α-tagged γ-γ coincidence spectra gated on the seven strongest transitions in the spectrum of b). d) Recoil-gated and α-tagged concidence spectrum gated on the 453 keV transition.

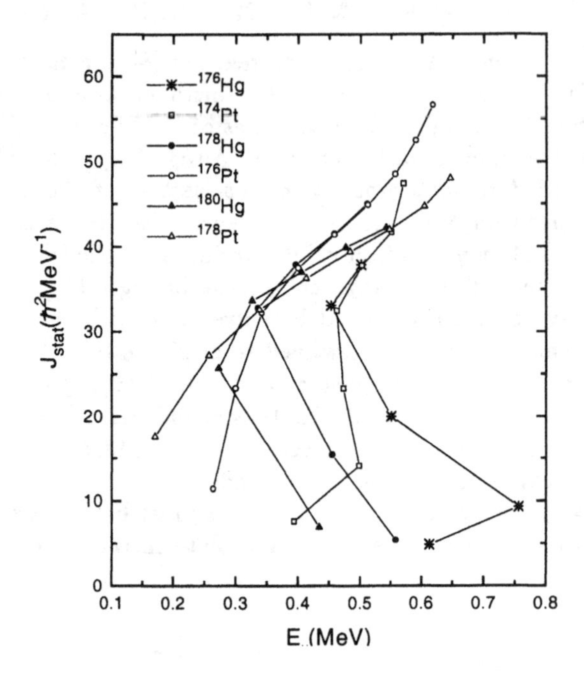

Figure 2 Static moments of inertia (J_{stat}) as a function of γ-ray energy for the Hg and Pt isotones with N=96, N=98, and N=100.

the static moments of inertia (J_{stat}) derived from the experimental yrast-level energies for the Hg and Pt isotones with N = 96, 98, and 100 are plotted as a function of γ-ray energy. The prolate band is manifested by the change towards a slightly increasing and smoothly behaving J_{stat}. Similarities in the J_{stat} values are especially striking between the pairs of isotones.

The level-energy systematics of odd- and even-mass Hg isotopes down to ^{176}Hg is displayed in Fig. 3. The energies of the first excited 2^+ and 4^+ states in ^{176}Hg lie higher than in any other Hg isotope revealing a transition towards a spherical ground state. A rapid increase in the excitation energy of the prolate intruder band with decreasing neutron number is also observed. The systematics of yrast levels for the odd-mass Hg isotopes, including our tentative candidates for ^{177}Hg from the same experiment, reveal interesting changes in the coupling of the $i_{13/2}$ neutron to the even-even Hg core. Similarly to the Pt isotopes [10], decoupling of the $i_{13/2}$ neutron hole from the oblate core in $N > 108$ is observed. In the $100 < N < 108$ isotopes the appearance of the yrast prolate intruder structure is seen as a change to a strong coupling of the $i_{13/2}$ neutron hole. A change towards a weak coupling scheme is indicated by our preliminary data for ^{177}Hg. As in the Pt isotopes it may reflect a coupling of the $i_{13/2}$ neutron to a nearly spherical core.

In a simple intruder picture [11] similar intruder structures as in the Hg isotopes (Z = 82-2) should be observed also in the Po isotopes (Z = 82+2). By

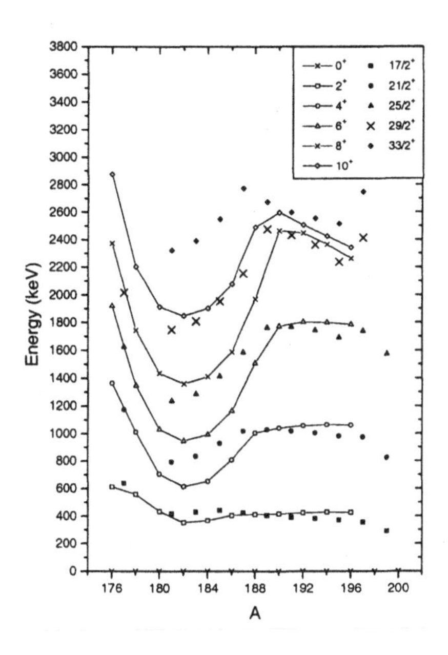

Figure 3 Energy level systematics in odd- and even-mass Hg nuclei.

employing the RDT and recoil-gating techniques we have obtained new in-beam γ-ray data for odd- and even-mass Po isotopes with $108 \leq N \leq 112$. A drop in level energies (Fig. 4) is observed with decreasing neutron number, revealing an appearance of intruding collective structures. Moreover, our new data for ^{192}Po shows that the intruder configurations have reached the ground state [12]. A coupling of the $i_{13/2}$ neutron to the even-even Po core different from that in Hg isotopes is observed. As pointed out in ref. [13] the near-degeneracy between the level energies for the odd and even neutron numbers represents a decoupling of the $i_{13/2}$ neutron hole, thus also revealing the oblate shape of the intruder band.

The study of possible intruder structures near the $N = 104$ neutron midshell in heavier nuclei above $Z = 82$ becomes increasingly difficult. A Jurosphere experiment for.a study of the $Z = 84$ nucleus ^{198}Rn was carried out [14] resulting as an α-tagged γ-ray spectrum shown in Fig. 5. The cross-section for the ^{166}Er(^{36}Ar,4n)^{198}Rn reaction was about 100 nb and therefore the rate in the RDT spectrum of Fig. 5 was only about 3 events per hour. Nevertheless, a clear pattern of peaks obviously forming an E2 cascade in ^{198}Rn is observed.

In Fig. 6 static moments of inertia extracted from the yrast line of ^{192}Po, ^{194}Po and ^{198}Rn are plotted as a function of γ-ray energy. Similarly to the intruder structures in the Hg isotopes (Fig. 2) a smooth behaviour of J_{stat} is observed. However, the values are much smaller than those for the prolate intruder bands in the Hg isotopes. Similarities of the J_{stat} values between ^{194}Po

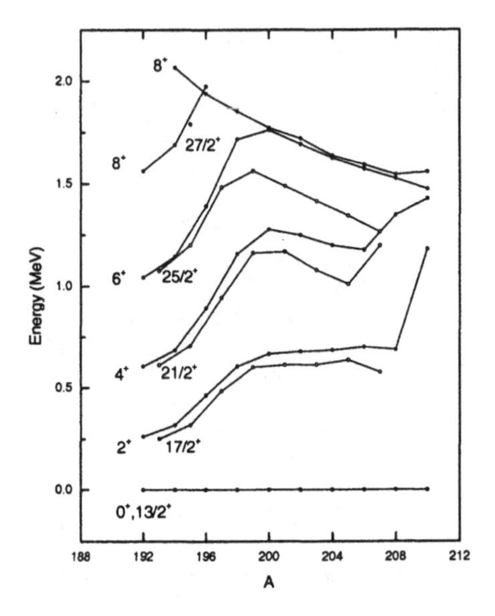

Figure 4 Level systematics for odd- and even-mass Po nuclei.

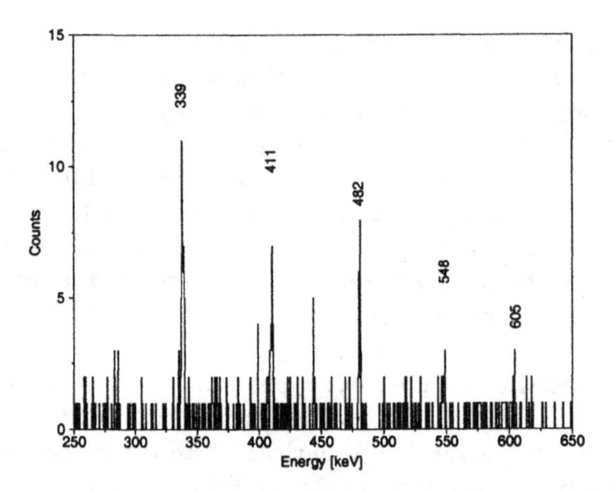

Figure 5 Prompt γ-ray spectrum gated with fusion-evaporation residues and tagged with the ^{198}Rn α decay.

and ^{198}Rn are striking especially at higher spin, indicating that the similar intruder configuration as in the Po isotopes are also appearing in the light Rn isotopes.

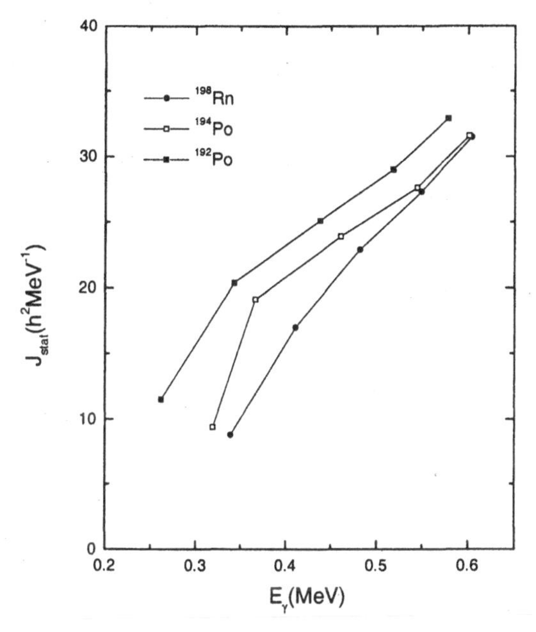

Figure 6 Static moments of inertia (J_{stat}) as a function of γ-ray energy for ^{192}Po, ^{194}Po and ^{198}Rn.

4 YRAST STATES OF ^{254}No

It is intriquing to apply the recoil-gating and RDT techniques for the stucture studies of heavy elements. In theoretical works the shell stabilization of trans-actinides has been found to originate from nuclear deformation with quadrupole and higher multipole moments. A direct test of these predictions is the determination of the level energies of the ground-state band in these nuclei. Population of this band would also provide valuable information on the spin-dependence of the fission barrier.

Fermium and nobelium nuclei are produced in fusion-evaporation reactions with cross-sections of the order of ten to few hundred nanobarn with an interesting exception of ^{254}No, which is found to be populated in a cold fusion of ^{48}Ca and ^{208}Pb with cross-section of 2 to 3 μb. Experiments to study excited states of ^{254}No have recently been carried out in two laboratories. In the Argonne National Laboratory, the Gammasphere array coupled to the Fragment Mass Analyser was used in a successful experiment in July, 1998 [15]. In August, 1998, the SARI array combined with RITU was employed for a similar study at JYFL. An about 0.5 mg/cm^2 thick ^{208}Pb target was bombarded with a beam of about 10 pnA of 219 MeV ^{48}Ca ions. The counting rate of the ^{254}No α particles was 20 - 40/h, and altogether 12000 full energy alpha particles were observed in the experiment.

The γ-ray spectra suffered from the lack of Compton-suppression for the Ge- clover detectors but still ^{254}No γ-rays were easely identified in the singles

γ- ray spectrum gated with the recoils and tagged with the 51 s α decay of ^{254}No. As the ^{254}No events fully dominate the recoil rate at the focal plane, the recoil gating was enough to generate a clean spectrum (Fig. 7). These data are in agreement with the Argonne experiment and in addition, a new 414 keV transition, tentatively assigned to the $16^+ \rightarrow 14^+$ transition, was observed. These data reveal that the fission barrier exists at least up to spin $16\hbar$ and that the moment of inertia derived from the transition energies continues its smooth increase. There is no indication of a possible backbend.

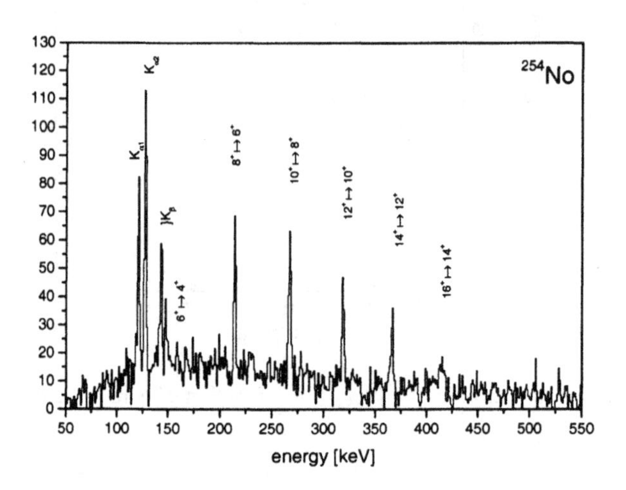

Figure 7 Recoil-gated singles γ-ray spectrum from the ^{48}Ca + ^{208}Pb reaction. The lines in the spectrum originate from ^{254}No.

Acknowledgments

This work was supported by the Academy of Finland and by the Access to Large Scale Facility program under the Training and Mobility of Researchers program of the European Union.

References

[1] M. Leino *et al.*. Nucl. Instr. Meth. B 99, 653, 1995.
[2] P.T. Greenlees *et al.*. J. Phys. G: Nucl. Phys. 24, L63, 1998.
[3] S.L. King *et al.*. Phys. Lett. B 443, 82, 1998.
[4] R.A. Bark *et al.*. to be published.
[5] J.F.C. Cocks *et al.*. Eur. Phys. J. A 3, 17, 1998.
[6] J.F.C. Cocks *et al.*. to be published.
[7] J. Simpson*et al.*. to be published.
[8] M. Muikku *et al.*. Phys. Rev. C 58, R3033, 1998.
[9] M.P. Carpenter *et al.*. Phys. Rev. Lett. B 78, 3650, 1997.

[10] B. Cederwall *et al.*. Phys. Lett. **B 443**, 69, 1998.

[11] J.L. Wood *et al.*. Phys. Rep. **215**, 101, 1992.

[12] K. Helariutta *et al.*. Phys. Rev. C **54**, R2799, 1996.

[13] N. Fotiades *et al.*. Phys. Rev. C **56**, 72, 1997.

[14] R.B.E. Taylor *et al.*. submitted to Phys. Rev. 1999.

[15] P. Reiter *et al.*. Phys. Rev. Lett. **82**, 509, 1999.

EXTRAPOLATION OF ASTROPHYSICAL S FACTORS TO ZERO ENERGY

S. Karataglidis, B. K. Jennings, and T. D. Shoppa

TRIUMF
4004 Wesbrook Mall,
Vancouver, B.C.
Canada, V6T 2A3

Abstract: We investigate the energy dependence of the astrophysical S factor for the reactions $^7\mathrm{Be}(p,\gamma)^8\mathrm{B}$, the primary source of high-energy solar neutrinos in the solar *pp* chain, and $^{16}\mathrm{O}(p,\gamma)^{17}\mathrm{F}$. Below 400 keV the energy dependence is well understood in terms of a subthreshold pole arising from the binding of the valence proton to the ^{8}B and ^{17}F ground states, respectively.

1 INTRODUCTION

The $^7\mathrm{Be}(p,\gamma)^8\mathrm{B}$ reaction, at energies of approximately 20 keV, plays an important role in the production of solar neutrinos [1]. The subsequent decay of the ^{8}B is the source of the high energy neutrinos to which many solar neutrino detectors are sensitive. The cross section for this reaction is conventionally expressed in terms of the S factor which is defined in terms of the cross section, σ, by:

$$S(E) = \sigma(E) E \exp\left[2\pi\eta(E)\right] , \tag{1}$$

The Nucleus: New Physics for the New Millennium
Edited by Smit et al., Kluwer Academic / Plenum Publishers, New York, 2000.

where $\eta(E) = Z_1 Z_2 \alpha\sqrt{\mu c^2/2E}$ is the Sommerfeld parameter, α is the fine structure constant, and μ is the reduced mass. The definition of the S factor eliminates from it most of the energy dependence due to Coulomb repulsion by factoring out the penetrability of a Coulomb potential. The reaction rate, obtained by folding the thermal distribution of nuclei in the stellar core with the cross section, peaks at approximately 20 keV, although this is temperature dependent and the peak energy varies by up to 3 keV. Because the cross section diminishes exponentially at low energies, the only accurate method of obtaining information about the S factor at energies of astrophysical interest is to extrapolate data taken at experimentally accessible energies ($E > 100$ keV). To do the extrapolation reliably we must understand the physics associated with the S factor.

Most calculations of the S factor follow the work of Christy and Duck [2]. (We refer the reader to that article for more details.) Here we present a brief overview of the model. The S factor, for the $^7Be(p,\gamma)^8B$ reaction, may be written as

$$S = C(I_0^2 + 2I_2^2)E_\gamma^3\left(J_{11}\beta_{11}^2 + J_{12}\beta_{12}^2\right)\frac{1}{1-e^{-2\pi\eta}}\,, \tag{2}$$

where

$$I_L = \int_0^\infty r^2 dr\; r\; \psi_{iL}(r)\psi_f(r) \tag{3}$$

$$C = \frac{5\pi}{9}\frac{1}{(\hbar c)^3}(2\pi\eta k)e^2\mu^2\left(\frac{Z_1}{M_1}-\frac{Z_2}{M_2}\right)^2 . \tag{4}$$

In Eq. (2), J_{LS} is the spectroscopic factor for a given angular momentum, L, and channel spin, S, β_{LS} is the asymptotic normalization of the bound state wave function, E_γ is the photon energy, and k is the momentum of the incident proton. The final bound state wave function $\psi_f(r)$ is normalized in the asymptotic region to $\psi_f(r) = W_{\alpha,l}(\kappa r)/r$ while the initial wave function reduces to the regular Coulomb wave function divided by $kr\sqrt{2\pi\eta}/(e^{2\pi\eta}-1)$. The unusual choice of normalizations is just to simplify the mathematics and generate integrals that are well-behaved at threshold. The initial state has both Coulomb and nuclear distortions. The Coulomb distortions are large and give the penetration factor included in the definition of the S factor, Eq. (1). They are included in all calculations. The nuclear distortions are much smaller but they are important and introduce a significant model dependence into the calculations.

The absolute magnitude of the S factor is determined primarily by the spectroscopic factor and the asymptotic normalization (see also Ref. [3]). The spectroscopic factor contains the many-body aspects of the problem and is calculable from standard shell model theory. The asymptotic normalization also depends on the many-body wave function, but is far more difficult to estimate from first principles: it requires detailed knowledge of how the 8-body wave function extends beyond the nuclear potential and its mapping to the Whittaker function in this region. This may be estimated crudely by approximating that behavior by using a suitably chosen Woods-Saxon wave function for the weakly bound proton. Instead we treat the overall factor, $A_n = J_{11}\beta_{11}^2 + J_{12}\beta_{12}^2$, as a free parameter, which is independent of energy, and determined by the S-factor data. For simplicity we will refer to this combination of asymptotic normalization and spectroscopic factor as the asymptotic strength.

To investigate the behavior of the integrals in Eq. (2), we first consider $\psi_f(r) = W_{\alpha,l}(\kappa r)/r$ for all radii and take $\psi_{i0}(r) = F_0(kr)/\{kr\sqrt{2\pi\eta}/(e^{2\pi\eta}-1)\}$. The s-wave integral then becomes

$$I_0 = \int_0^\infty dr\, r \frac{W_{\alpha,l}(kr)F_0(kr)}{k\sqrt{2\pi\eta}} \left(e^{2\pi\eta} - 1\right) . \tag{5}$$

The integral is smooth as k passes through zero and diverges as $k \to i\kappa$ ($E \to -E_B$). The nature of the divergence is determined by the asymptotic forms of the Coulomb wave function and Whittaker function for large r. There the Whittaker function is proportional to $r^{-|\eta k|/\kappa}e^{-\kappa r}$ [2] (ηk is independent of k). Above threshold the Coulomb wave function oscillates at large radii, however below threshold it is exponentially growing and is proportional to $r^{|\eta|}e^{|k|r}$. Thus the behavior of the integrand at large radius is

$$r^{1-|\eta k|(1/\kappa - 1/|k|)}\exp[-(\kappa - |k|)r] \tag{6}$$

and the integral diverges as

$$I_0 \sim 1/(\kappa - |k|)^2 \sim 1/(E_B + E)^2 = 1/E_\gamma^2 . \tag{7}$$

The S factor is proportional to $I_0^2 E_\gamma^3$, and gives rise to a simple pole in S at $E_\gamma = 0$. Hence, the product $E_\gamma S$ should be a straight line. This was demonstrated in [4]. However, the first correction term is not simply $1/E_\gamma$ but rather of the form $(1 + c\ln E_\gamma)/E_\gamma$, the logarithmic term coming from the $r^{-|\eta k|(1/\kappa - 1/|k|)}$ factor. Both the leading and first correction terms are determined purely by the asymptotic behavior of the wave functions. The second correction term, of order E_γ^0, is not determined purely by the asymptotic value of wave function alone but also depends on the wave function at finite r.

The presence of the pole suggests the S factor may be parametrized as a Laurent series:

$$S = d_{-1}E_\gamma^{-1} + d_0 + d_1 E_\gamma + \ldots \tag{8}$$

The coefficients of the first two terms, d_{-1} and d_0, are determined purely by the asymptotic forms of the wave functions while the third coefficient, d_1, is also dependent on the short range properties of the wave functions.

To account for nuclear distortions, which only affect the s wave contribution to the S factor [5], we construct a simple model where the initial state wave function is zero inside some radius, r_c, and a pure Coulomb wave outside. We impose the boundary condition that the wave function be zero at r_c. This generates a phase shift and is equivalent to having an infinitely repuslive potential with a radius r_c. The d-wave scattering state is taken to be an undistorted

Coulomb wave function, and the bound state is assumed to be a pure Coulomb state, described by a Whittaker function, for all radii.

We use this model, which preserved the character of the pole, along with the others available, to analyse the data and extract an S factor. This has been done in Ref. [5], and the results of the fits is shown in Fig. 1. From fitting the energy

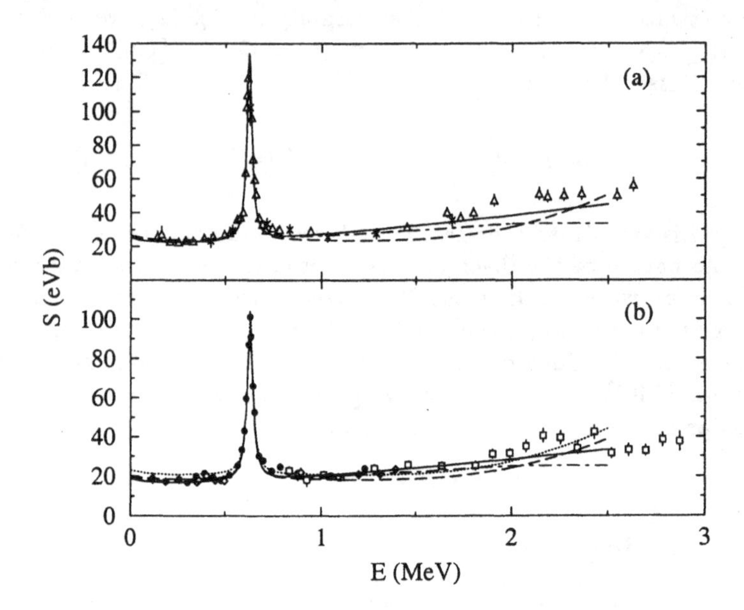

Figure 1 Fits to the data using various models of extrapolation [5]. The simple hard sphere model is indicated by the solid line in both cases.

range 0 to 400 keV we obtain $S(20) = 18.4 \pm 1.0 \pm 0.2$ eVb, or equivalently, $S(0) = 19.0 \pm 1.0 \pm 0.2$ eVb. The first error is experimental while the second is from model dependencies in the fit. Those model dependencies increase with energy, and so we also fit to 1.5 MeV, for which $S(20) = 17.6 \pm 0.7 \pm 0.4$ eVb and $S(0) = 18.1 \pm 0.7 \pm 0.4$ eVb.

In summary, we have determined the low energy behavior of the S factor for the $^7\mathrm{Be}(p,\gamma)^8\mathrm{B}$ to be dominated by a pole at $E_\gamma = 0$ which arises from the subthreshold ^{8}B ground state and the Coulomb interaction between the ^{7}Be and the proton and large radii. That pole induces the upturn at threshold in the S factor which is observed in the calculations, and is also seen in the data for the S factor for the $^{16}\mathrm{O}(p,\gamma)^{17}\mathrm{F}$ (0.498 MeV) reaction [6].

References

[1] J. N. Bahcall and M. Pinsonneault, Rev. Mod. Phys. 67, 885 (1992); J. N. Bahcall, S. Basu, and M. H. Pinsonneault, Phys. Lett. 433B, 1 (1998).

[2] R. F. Christy and I. Duck, Nucl. Phys. 24, 89 (1961)

[3] H. M. Xu, C. A. Gagliardi, R. E. Tribble, A. M. Mukhamedzhanov and N. K. Timofeyuk, Phys. Rev. Lett. 73, 2027 (1994).

[4] B. K. Jennings, S. Karataglidis, and T. D. Shoppa, Phys. Rev. C 58, 579 (1998).

[5] B. K. Jennings, S. Karataglidis, and T. D. Shoppa, Phys. Rev. C 58, 3711 (1998), and references therein.

[6] R. Morlock, *et al.*, Phys. Rev. Lett. 79, 3837 (1997).

SHAPE-COEXISTENCE IN ^{74}Kr

W. Korten

DAPNIA/SPhN, CEA Saclay
F-91191 Gif-sur-Yvette Cedex
France

wkorten@cea.fr

Abstract: Combined conversion electron (CE) and γ ray spectroscopy has been performed for ^{74}Kr. The possibility of electric-monopole (E0) decays was examined in order to confirm the evidence for a low-lying isomeric 0_2^+ state and to give insight into the mixing between the predicted coexisting prolate and oblate shapes.

1 INTRODUCTION

The nuclei in the mass region around A = 70-80 exhibit a wide variety of shapes. Qualitatively, their features can be understood as a consequence of the different energy gaps in the Nilsson single-particle levels at Z, N = 34 - 40. The coexistence of different shapes at low excitation energies is predicted by several different model calculations [1], using either the Hartree-Fock + BCS [15], the Nilsson-Strutinsky [3] or the Hartree-Fock Boguliubov (HFB) [4] approach. Especially some of the low-lying 0_2^+ states, frequently observed in these nuclei, are predicted to have an intrinsic shape or deformation different from that of the ground state. The nucleus ^{74}Kr is a prime candidate for shape coexistence, since the proton number Z = 36 and the neutron number N = 38 drive the nucleus simultaneously towards a well-deformed oblate and an equally well-deformed prolate shape.

Recently, using the fragmentation of a ^{92}Mo beam at GANIL [6], evidence for a low-lying isomeric 0_2^+ state in ^{74}Kr has been deduced from the observation of a delayed component in the $2_1^+ \rightarrow 0_1^+$ transition, but the isomeric transition itself, e. g. $0_2^+ \rightarrow 2_1^+$, was not observed. In that experiment the observational limit for γ decays was below 100 keV, so that the new 0_2^+ state is expected to be very close in energy to the 2_1^+ state. Therefore we attempted a direct observation of the E0 decay connecting the 0^+ states in ^{74}Kr.

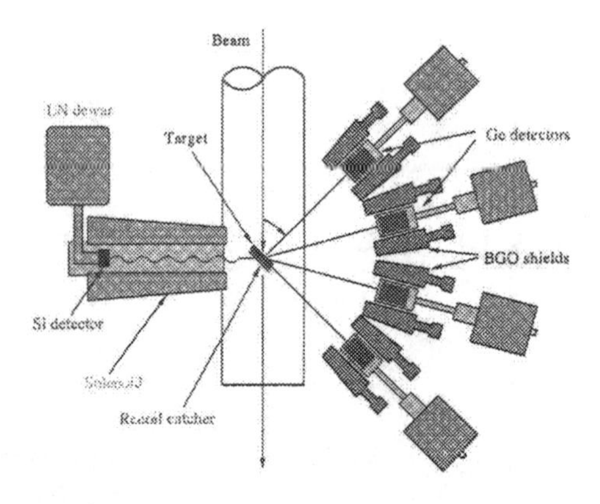

Figure 1 Experimental setup of GAREL+ (see text for details).

2 THE EXPERIMENT

The γ and conversion-electron (CE) decays of excited states in ^{74}Kr were studied in an experiment at the VIVITRON accelerator (IReS Strasbourg) using the GAREL+ setup, consisting of 14 large-volume Compton-suppressed Ge detectors and a magnetic solenoid spectrometer (β-TRONC [7]). With respect to the beam direction, the Ge detectors were positioned at angles of 46.4°, 75.0°, 105.0°, 133.6° and the CE spectrometer at 90.0°. The rather high efficiency amounted to 1.5 % at 1.3 MeV for the γ array and 1.5 % at 200 keV for the CE spectrometer. A ^{58}Ni target was bombarded with a pulsed beam of 60 MeV ^{19}F ions. In order to detect isomeric decays within the view of the detectors, a carbon catcher (770 μg/cm^2) was mounted behind the thin nickel target (520 μg/cm^2). Fig. 1 shows the experimental setup. Energy and timing information (with respect to the beam pulsing) have been measured for γ-ray and conversion-electron (CE) singles. $\gamma\gamma$ and CE-γ coincidences were also recorded.

3 RESULTS AND DISCUSSION

Since the cross section for producing ^{74}Kr in the reaction ^{19}F + ^{58}Ni amounts only to 20 mb, the singles spectra are dominated by other evaporation channels such as ^{74}Br and ^{74}Se. First of all, we searched for delayed E0 decays in the CE spectra, in accordance with the lifetime of the proposed 0^+ isomer [6]. Evidence for such an isomeric E0 transition was found at an electron energy of 495(1) keV (see Fig. 2), when applying a time gate of 5-40 ns. The E0 character of this transition is verified from a comparison with the γ-ray spectrum leading to the proposition of a new 0^+_2 level at 508(1) keV. In the inset of this figure the decay spectrum of the E0 transition is shown, together with a lifetime fit of 20(9) ns. A somewhat larger value of 42(8) ns was obtained by Chandler et al. [6] from the delayed $2^+_1 \rightarrow 0^+_1$ decay.

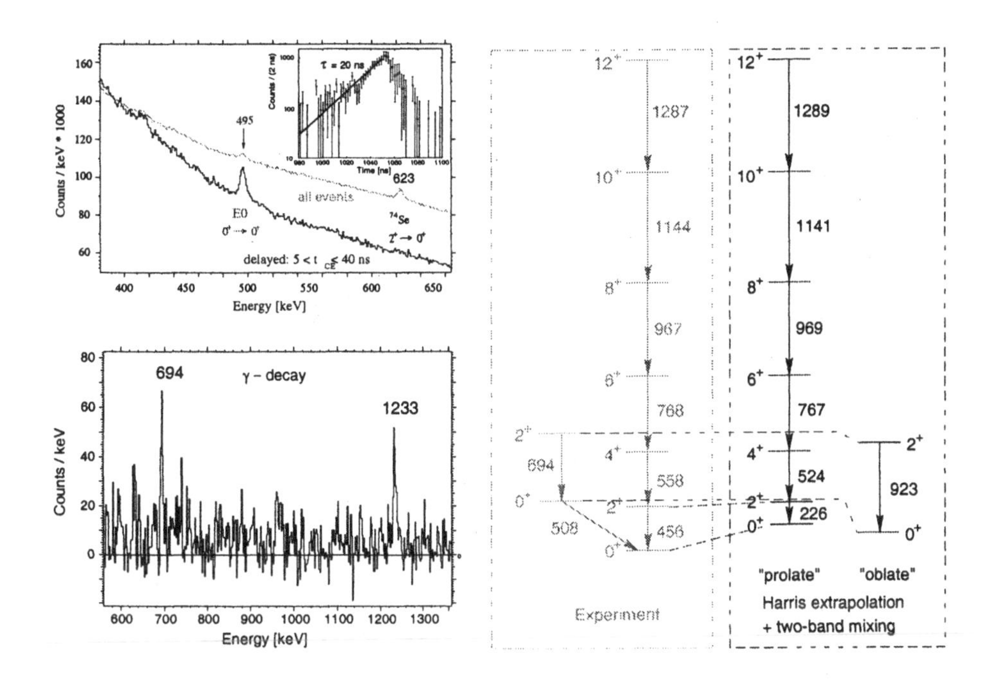

Figure 2 Left top : Conversion-electron singles spectrum obtained from the reaction $^{19}F + ^{58}Ni$ compared with the corresponding normalized time-gated spectrum ($5 < t_{CE} <$ 40 ns). In the inset the time spectrum for the 495 keV E0 transition is shown. Left bottom : The γ-ray spectrum gated by the 495 keV CE line shows two transitions feeding the 0_2^+ state. Right : New level scheme of ^{74}Kr as deduced from the experimental data and compared to a two-band mixing model.

An unambiguous assignment of the new 0^+ state, e.g. from an observation of the $0_2^+ \to 2_1^+$ transition was not possible. Two excited states decaying into the 0_2^+ state were, however, observed when tagging with the isomeric E0 decay (see Fig. 2). The 694 keV γ ray depopulates a state at 1202 keV which was previously observed and assigned as 2_2^+ level of ^{74}Kr [8]. A new level at 1738 keV is depopulated by the 1233.0(5) keV transition and fits energetically well with the band head of the γ-vibrational band from which so far only the odd-spin members have been observed experimentally. The 1233 keV transition was already identified to belong to ^{74}Kr [9], but could not be placed into the level scheme. Thus the assignment to ^{74}Kr is not only based on the delayed E0 decay, but also on the two γ transitions feeding the 0_2^+ level.

The partial level scheme of ^{74}Kr is also shown in Fig. 2, including the well-known rotational ground-state band as well as the new 0_2^+ and 2_2^+ states which are proposed to belong to the oblate well of ^{74}Kr. While the high spin states in ^{74}Kr show a rotational behaviour, the low spin states are perturbed, presumably due to the mutual interaction between the states of different intrinsic structure.

The position of the unperturbed states of prolate deformation can be obtained from a Harris-expansion of the high spin states. From the energies of the 0_2^+ and 2_2^+ states it is now possible to determine the unperturbed energies of the (presumably oblate) state as well as the mixing between the intrinsic states. The resulting unperturbed states are also shown in Fig. 2. From these data an interaction strength V = 254(1) keV and an energy difference of $|\Delta|$=19(6) keV for the unperturbed 0^+ states is deduced [10].

An important characteristics of the E0 decay is the possibility to determine the electric monopole strength. In order to calculate the E0 strength we used the formulas given in [11]. A complication arises from the fact that the 52 keV $0_2^+ \rightarrow 2_1^+$ transition in ^{74}Kr was not observed. An upper limit for the E0 transition strength can be obtained from the measured lifetime under the assumption that the 0_2^+ level decays only via the E0 transition. From the GANIL data [6] we can also estimate an upper limit of the partial E2 decay strength of $\lambda_{E2} \approx 80$ ns^{-1}. In this way we obtain limits for the E0 strength of $0.36(8) \leq |\rho(E0)| \leq 0.42(9)$. It is interesting to note that already the lower limit is larger than the prediction of $|\rho(E0)| = 0.17$ given in [6].

In conclusion, we have observed a new E0 transition from an isomeric 0_2^+ state in ^{74}Kr at 508 keV with a lifetime of 20(9) ns. The energetic position of the 0_2^+ state and the experimental limits for the E0 strength are evidence for a strong mixing of the two 0^+ states, supporting the picture of oblate-prolate shape coexistence. Since the branching ratio of the 0_2^+ decay in ^{74}Kr could only be estimated in this work, a future experiment is planned to determine this value more accurately. In addition, Coulomb excitation experiments with the EXOGAM γ-ray spectrometer using a radioactive ^{74}Kr beam from the SPIRAL facility at GANIL (France) are planned. A measurement of the intrinsic quadrupole moments of ^{74}Kr would give the final confirmation for the interpretation as coexisting prolate and oblate structures.

References

[1] J.L. Wood et al., Phys. Reports 215 (1992) 101

[2] P. Bonche et al., Nucl. Phys. A443 (1985) 39

[3] W. Nazarewicz et al., Nucl. Phys. A435 (1985) 397

[4] A. Petrovici et al., Nucl. Phys. A483 (1988) 317

[5] P.H. Regan et al., Acta Physica Polonica B28 (1997) 431

[6] C. Chandler et al., Phys. Rev. C56 (1997) R2924

[7] P. Paris et al., NIM A 357 (1995) 398

[8] D. Rudolph et al., Phys. Rev. C56 (1997) 98

[9] D. Rudolph, priv. comm.

[10] F. Becker et al., Eur. Phys. J. A4 (1989) 103

[11] A. Makishima et al., Nucl. Phys. A425 (1984) 1

HIGH-SPIN STATES IN THE TRANSITIONAL PALLADIUM NUCLEI OBSERVED IN HEAVY-ION INDUCED FISSION

T. Kutsarova[a,b], M.-G. Porquet[a], I. Deloncle[a], A. Minkova[a,c], E. Gueorguieva[c], P. Petkow[b], F. Azaiez[d], S. Bouneau[d], C. Bourgeois[d], J. Duprat[d], B.J.P. Gall[e], C. Gautherin[f], F. Hoellinger[e], R. Lucas[f], N. Schulz[e], H. Sergolle[d], Ts. Venkova[b], A. Wilson[a*]

[a] CSNSM, CNRS-IN2P3, F-91405 Orsay, France
[b] INRNE, BAN, 1784 Sofia, Bulgaria
[c] University of Sofia, Faculty of Physics, 1126 Sofia, Bulgaria
[d] IPN, CNRS-IN2P3, F-91406 Orsay, France
[e] IReS, F-67037 Strasbourg Cedex, France
[f] DAPNIA, CEA, F-91191 Gif sur Yvette, France

The low-spin collective excitations of even-even Pd nuclei in the A $\approx$ 110 mass region indicate a structural transition from an anharmonic vibrator to a gamma-soft rotor. This observation is supported by recent calculations of

*Present address: Dept. of Phys., Univ. of York, York Y01 5DD, UK

The Nucleus: New Physics for the New Millennium
Edited by Smit et al., Kluwer Academic / Plenum Publishers, New York, 2000.

the potential energy surfaces using different types of mean field approaches. In particular, prolate–to–oblate shape transition has been predicted at ^{111}Pd [1]. Experimental evidence for a shape change or shape coexistence can be obtained by high–spin studies of these transitional nuclei. Band crossings in the odd–N Pd nuclei are especially important to understand the origin of the alignments. No band structures built on the low–lying states were known in heavier odd Pd nuclei (A$\geq$109). High–spin states in these nuclei cannot be populated using fusion–evaporation reactions. The heavier neutron–rich ^{112,114,116}Pd nuclei have been studied from spontaneous fission of ^{252}Cf [2]. In this work the less neutron–rich Pd nuclei have been produced as fission fragments following the fusion reactions ^{28}Si + ^{176}Yb at 145 MeV. The ^{28}Si beam was provided by the Vivitron accelerator in Strasbourg. Gamma rays were recorded with the EUROGAM2 array [3] . The level schemes were constructed by examining γ– spectra extracted from two and three dimensional matrices.

The yrast band of ^{110}Pd has been extended to I^π=14$^+$ and the γ – band to I^π=8$^+$. In ^{108}Pd two new side bands of negative parity have been observed. The features of the γ – band of ^{110}Pd suggest significant γ softness in this nucleus. This observation is corroborated by the microscopic mesh Hartree-Fock-Bogoliubov calculations of potential energy surfaces [4]. The band structures in the odd–A ^{109,111}Pd nuclei have been identified using the method of cross–coincidences with γ–rays of the complementary fission fragments as developed for heavy–ion induced fission [5]. Two signature branches of a new band built on the $\nu h_{11/2}$ orbital have been identified in ^{109}Pd and the favoured signature of the $\nu h_{11/2}$ band has been observed for the first time in ^{111}Pd [6]. The spectroscopic properties as well as the rotational behaviour of these bands are characteristic of bands expected for a prolate nuclear shape in this mass region: (i) the large signature splitting observed in the $h_{11/2}$ bands is typical for a band built on an orbital lying close to the middle of the $h_{11/2}$ subshell on the prolate side. (ii) The favoured signature bands undergo delayed crossings due to blocking of $h_{11/2}$ orbital supporting a quasineutron character of the backbendings in the even–even ^{110,112}Pd nuclei expected for prolate ground–state shapes.

References

[1] P. Möller et al. *At.Data Nucl. Data Tables*, 59: 255, 1995.

[2] R. Aryaeinejad et al. *Phys. Rev.*, C48: 566, 1993.

[3] P.J. Nolan et al. *Ann. Rev. Nucl. Part. Sci.*, 45:561, 1994. *Nucl. Instr. Meth.*, A357:150, 1995.

[4] B. Gall, to be published

[5] M.G. Porquet et al. *Acta Phys. Polonica*, B27:179, 1996.

[6] T. Kutsarova et al. *Phys. Rev.*, C58: 1966, 1998.

FIRST IDENTIFICATION OF HIGH SPIN STATES IN ^{164}Ta

B.R.S.Babu[1], J.J.Lawrie, D.G.Roux[2], D.G.Aschman[2], R.Beetge[2], M.Fetea, G.K.Mabala[2], S.Naguleswaran, R. Nazmitdinov[3], R.T.Newman, C.Rigollet, J.F.Sharpey-Schafer, F.D.Smit, W.Whittaker[2].

National Accelerator Centre, P.O. Box 72, Faure 7131, South Africa
[1]On leave from University of Calicut, Calicut 673 635, India
[2] Department of Physics, University of Cape Town, Rondebosch 7700, South Africa
[3] Joint Institute for Nuclear Research, Dubna, Russia

Abstract: High spin states of ^{164}Ta were investigated for the first time. The residual nuclei were produced in the reaction ^{142}Nd(^{27}Al,5n)^{164}Ta at an incident energy of 150 MeV. The new AFRODITE array at the National Accelerator Centre, South Africa, was used to detect the resulting gamma radiation. A brief description of AFRODITE is given. Spectra of both the favoured and unfavoured sequences of one pair of signature partner bands as well as a tentative level scheme is presented. Further analysis is in progress.

The phenomenon of low-spin signature inversion of the $\pi h_{11/2} \otimes \nu i_{13/2}$ band has been studied systematically in the doubly-odd nuclei of the A = 160 region [1]. This effect is also known as anomalous signature splitting, wherein, below the inversion point, the excitation energies of the unfavoured band (with signature $\alpha = 1$, odd spins) are lower than those of the favoured band.

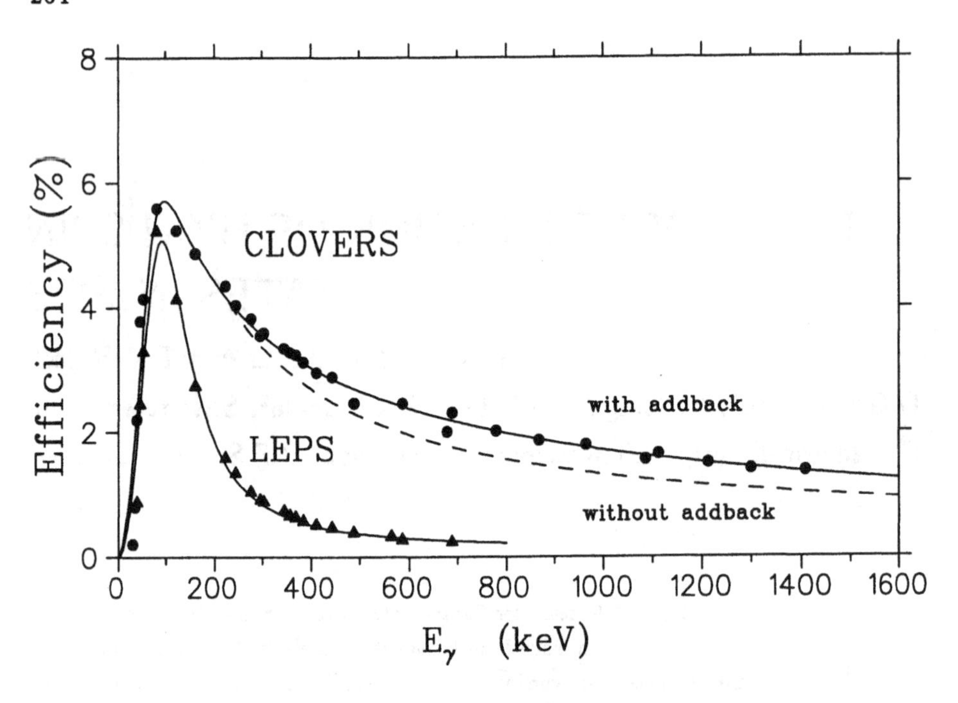

Figure 1 Total photopeak efficiency for LEPS and clovers.

This phenomenon has been extensively studied in different theoretical frameworks such as the Cranked Shell Model [2, 3], the Particle Rotor Model [4, 5, 6], the angular momentum projection method [7, 8] and the interacting boson model [9]. It has been predicted that low-spin signature inversion of doubly-odd nuclei would only be observed in the region of $62 < Z < 70$, and further, that the phenomenon may be a consequence of triaxiality [3]. However, these predictions are not consistent with subsequent observations of low-spin signature inversion in the nuclei with $Z = 71$ and 73 [1], and controversy also surrounds the different proposed mechanisms for the effect. The above indicates the need for an extension of the investigation of the systematic trends presented in [1] to different combinations of Z and N. We therefore decided to study high spin structure in ^{164}Ta, where none had hitherto been reported.

High spin states in ^{164}Ta were populated in the reaction $^{142}(^{27}\mathrm{Al},5\mathrm{n})^{164}$Ta at an incident energy of 150 MeV, using the separated sector cyclotron facility of the National Accelerator Centre, South Africa. A self-supporting ^{142}Nd target foil of thickness 700 μgcm^{-2}, enriched to about 99.7 %, was used. The γ-ray cascades resulting from the de-excitation of the residual nuclei were detected using the AFRODITE (African Omnipurpose Detector Array for Innovative Techniques and Experiments) array consisting of 8 clover detectors with BGO suppression shields and 7 LEPS detectors. Each clover consists of four n-type high-purity germanium detectors arranged in a square, similar to those used in EUROGAM II [10]. Each LEPS is a fourfold segmented planar germanium detector, and is equipped with a very thin beryllium entrance window. A detailed description of AFRODITE and its performance characteristics has been reported elsewhere [11]. A total of 350 million 3 or higher fold coincidence events were recorded. The data were sorted into clover-clover, LEPS-clover and LEPS-LEPS events to construct three matrices, viz. γ-γ, x-γ, and x-x.

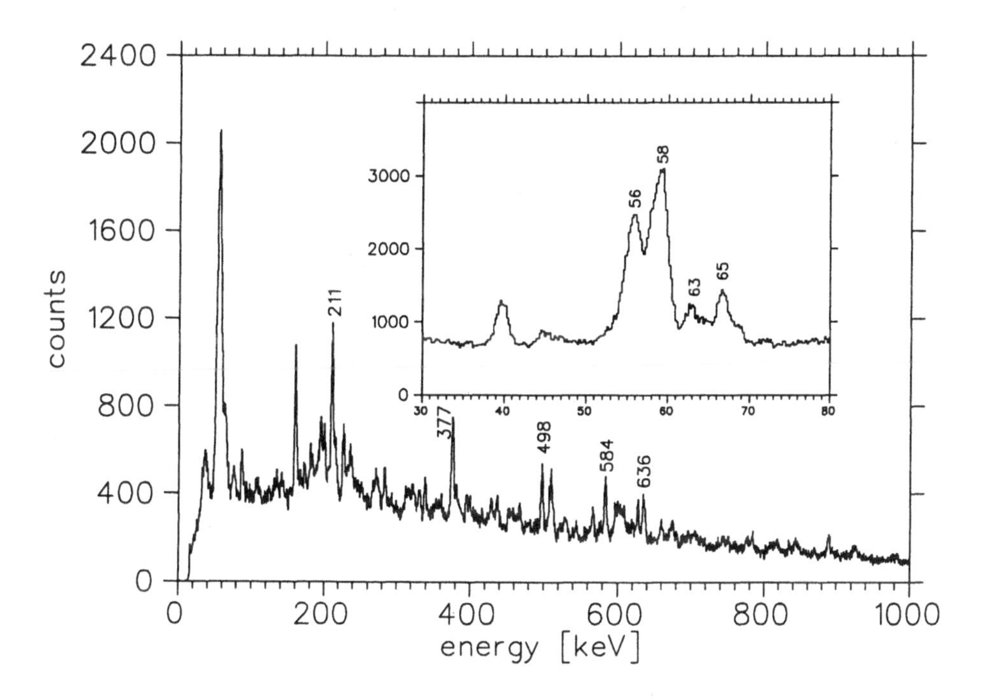

Figure 2 Total projection of clover-clover matrix showing Hf and Ta x-rays (insert).

The total photopeak efficiency for LEPS and clovers is shown in Figure 1. Figure 2 shows the total projection of the γ-γ matrix. No background has been subtracted. The inset clearly shows the separation of Hf (56 keV) and Ta (58 keV) x-rays, which were used to separate the ^{164}Ta from ^{164}Hf. ^{164}Ta is the second strongest channel produced at this beam energy, the first being ^{164}Hf. The tentative level scheme for ^{164}Ta, established from the coincidence relationships between gamma rays, is given in Figure 3. The tentative spin assignments are based on systematics. Summed gated spectra of the favoured ($\alpha = 0$) and the unfavoured band ($\alpha = 1$) are shown in Figures 4 and 5, respectively.

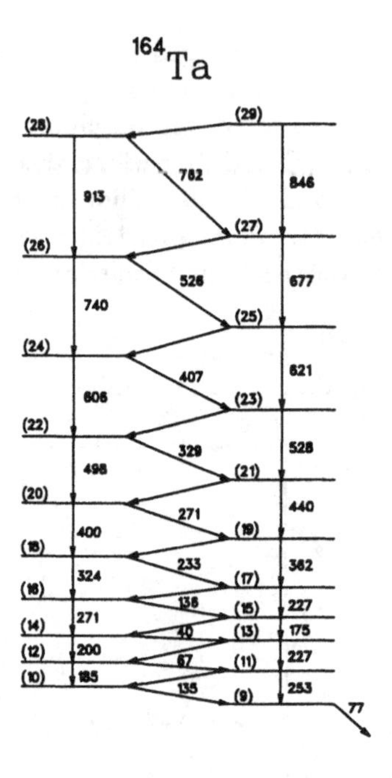

Figure 3 Tentative level scheme for ^{164}Ta.

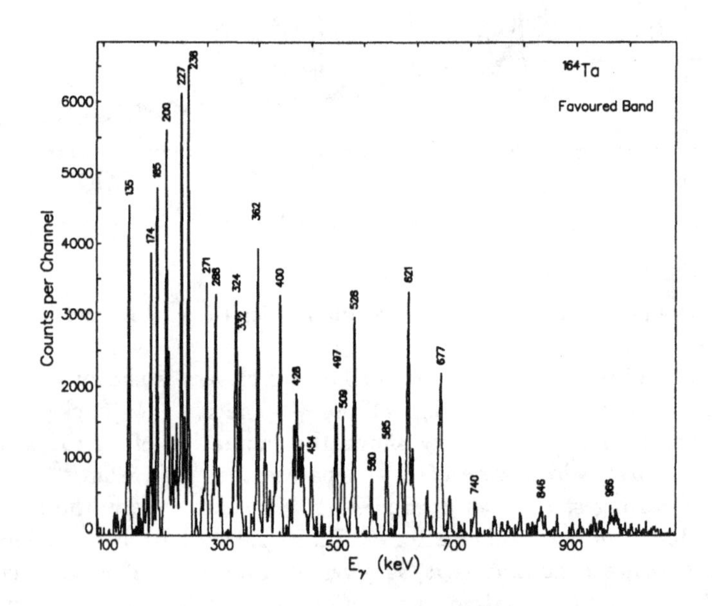

Figure 4 ^{164}Ta favoured band (α=0).

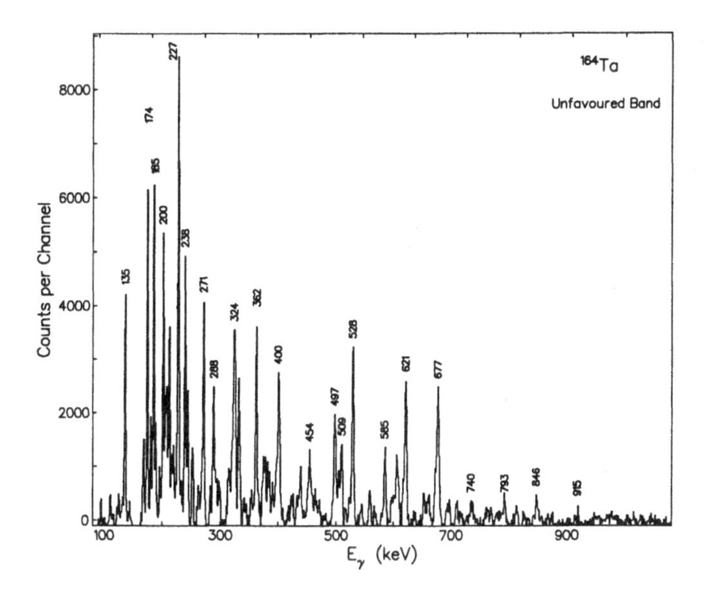

Figure 5 ^{164}Ta unfavoured band (α=1).

References

[1] Y.Liu et al., Phys Rev. C52 2514 (1995).

[2] M.Matsuzaki, Phys. Lett **B269**, 23 (1991).

[3] R.Bengtsson et al., Nucl. Phys. **A415**, 189 (1984).

[4] I.Hamamoto, Phys.Lett. **B235**, 221 (1990).

[5] P.B.Semmes and I.Ragnarsson, Proc. Int Conf. on High Spin Physics and Gamma Soft Nuclei, Pittsburgh, 1990 (World Scientific, Singapore, 1990) p500.

[6] A.K.Jain and A.Goel, Phys Lett **B277**, 233 (1992).

[7] K.Hara and Y.Sun, Nucl. Phys **A531**, 221 (1991).

[8] K.Hara, Nucl. Phys. **A557**, 449c (1993).

[9] N.Yoshida et al., Nucl. Phys. **A567**, 17 (1994).

[10] F.A.Beck, Prog. Part. Nucl. Phys. **28**, 443 (1992).

[11] B.R.S.Babu *et al.*, Proc. Nucl. Phys. Symp., Dept. of Atomic Energy, Government of India, **40B**, 416 (1997)

THE USE OF A REALISTIC EFFECTIVE NUCLEON-NUCLEON INTERACTION IN PRE-EQUILIBRIUM SCATTERING

R. Lindsay[1], Z. Karriem[1,2], W.A. Richter[3]
[1]Department of Physics,
University of the Western Cape,
Bellville, 7535, South Africa
[2]National Accelerator Centre, Faure 7131, South Africa
[3]Department of Physics,
University of Stellenbosch,
Stellenbosch, 7600, South Africa

Abstract: Calculations during the last 20 years have shown that the Feshbach-Kerman-Koonin model gives successful predictions of continuum scattering for (p,p') and (p,n) scattering. However, a very simple effective nucleon- nucleon interaction with an adjustable strength has usually been used in these studies. The use of a more realistic effective interaction is described in this contribution. Results of preliminary calculations indicate that cross-sections using such an interaction over-predict the measured cross sections indicating that aspects of the theory need to be reconsidered.

1 INTRODUCTION

The quantum mechanical description of pre-equilibrium nucleon scattering by Feshbach Kerman and Koonin(FKK)[1] has been used in various analyses since its publication in 1980. Several other formulations have followed[2][3], that agree with the FKK formulation for the first step, but differ in the multi-step formulation.

The Nucleus: New Physics for the New Millennium
Edited by Smit et al., Kluwer Academic / Plenum Publishers, New York, 2000.

Most subsequent theoretical work has concentrated on the validity of the various assumptions in the FKK formulation, while practical applications have been dominated by calculations using only a simplified central nucleon-nucleon interaction such as a delta function or a single Yukawa with a range of one fermi and an adjustable strength[4][5][6].

The FKK theory is a quantum mechanical description of the scattering which considers the energy transfer to the target as a serious of particle-hole excitations, similar in concept to the semi-classical exciton model due to Griffin[7]. These calculations have proved quite successful- see e.g. Richter et al. in these proceedings and [3]. There is, however, a need for calculations which can describe the data without the adjustable strength of the effective interaction which has been used in applications of the FKK theory. Pre-equilibrium cross-sections contribute the major part of the total reaction cross section in N-N scattering at incident energies above about 50 MeV. There is thus a need for predicting them for an arbitrary target at all energies. This is true in e.g. simulations used to study the proposal to transmute radio- active waste by accelerator driven nuclear fission, in simulations for proton and neutron therapy, as well as in radiation safety simulations. Calculations to describe other observables such as the recently measured analyzing powers [8], will also require an effective interaction which includes a fuller description of the central interaction as well as non-central parts of the NN interaction.

2 IMPORTANCE OF THE EFFECTIVE NN INTERACTION.

The developers of pre-equilibrium scattering codes based on the FKK theory used the nucleon-nucleon interactions which were well established at the time, namely the delta function and Yukawa interaction with one fermi range[3][9].

More realistic effective interactions such as the M3Y [10] and Love- Franey [11] interactions were being develped at the time, but were not yet well established. The present work attempts to apply the progress made in the realistic NN interaction field in the FKK model.

The importance of the interaction that is used can be seen from the expression for the double differential cross section as given by the first step in the FKK formalism:

$$\frac{d^2\sigma}{dEd\Omega} = \Sigma(2L+1)\omega(L,U) < \frac{d\sigma}{d\Omega} >_L$$

where ω is the spin and level density distribution at excitation energy U, and

$$< \frac{d\sigma}{d\Omega} >_L$$

is an L averaged DWBA cross section calculated as usual by

$$\frac{d\sigma}{d\Omega} \propto | \int \chi^{-*} < \psi_f | v_{N-N} | \psi_i > \chi^{+} |^2.$$

v_{NN} is the Nucleon-Nucleon interaction responsible for the particle-hole excitation. The first step cross section therefore depends on the *square* of the NN-interaction.

This dependence is even greater in the subsequent steps. The calculations in this presentation were calculated with the DWBA code DWBA91[12] which allows for the inclusion of a detailed NN-interaction and calculates the exchange part of the scattering.

The interaction used in this work is the one developed by the Melbourne group [13] The reason for this choice is the success that it has achieved in 200 MeV proton scattering, the energy used in this study.

3 RESULTS AND CONCLUSIONS

Preliminary results using the above description for *only the first step*, are shown in Figure 1 for proton scattering at 200 MeV.

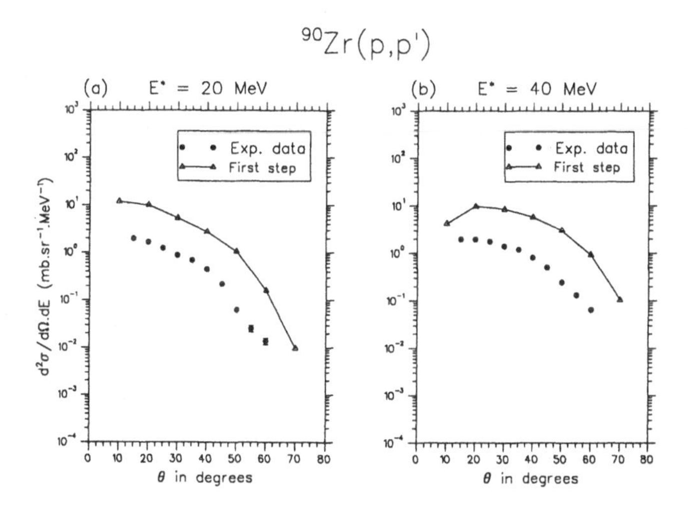

Figure 1 Predictions of a calculation for the first step cross section using the Melbourne interaction[13] compared to the data[5]

As can be seen, the results of only the first step already over-predicts the cross section. Earlier work using the M3Y interaction at a lower energy came to the same conclusion [14].This is not entirely unexpected, since the normalisation constants found in the prescription using a simple Yukawa interaction with adjustable strength, obtains V_0 values which are in general smaller that those found in particle-hole excitations. Calculcations have been performed to compare the transitions of the p-h states as a function of diferent excitation energies and L transfers as shown in Figure 2.

The extent of the over-prediction differs from state to state, but is found in all cases.

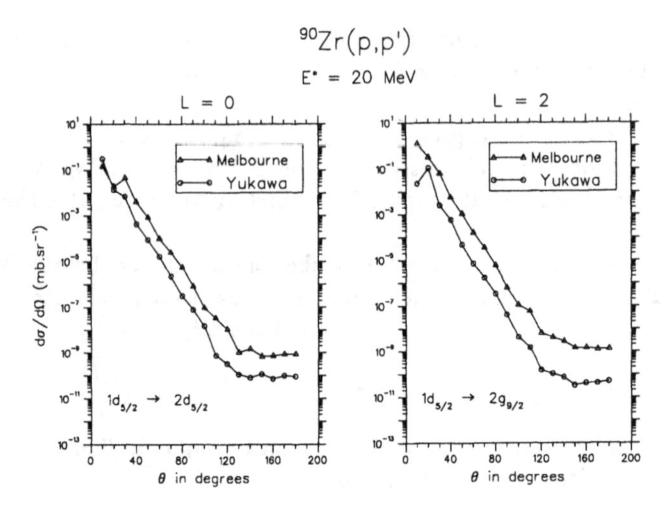

Figure 2 Comparison of predictions due to the Melbourne interaction and the phenomenological interaction with strength of 25 MeV.

These results imply that different aspects of the FKK formalism such as the nuclear model employed and the level densities used will have to be further investigated.

References

[1] H. Feshbach, A. Kerman, S. Koonin, Ann. Phys. 125, (1980) 477.

[2] R. Bonetti, A. J. Koning, J. M. Akkermans, P. E. Hodgson, Phys. Rep. 247, 1 (1994).

[3] Pre-equilibrium Nuclear Reactions, E Gadioli and P.E. Hodgson, Clarendon press, 1992.

[4] W. A. Richter, A. A. Cowley, R. Lindsay, J. J. Lawrie, J. V. Pilcher, S. V. Förtsch, R. Bonetti, and P. E. Hodgson, Phys. Rev. C46, 1030 (1992)

[5] W. A. Richter, A. A. Cowley, G. C. Hillhouse, J. A. Stander, J. W. Koen, S. W. Steyn, R. Lindsay, R. E. Julies, J. J. Lawrie, J. V. Pilcher, P. E. Hodgson. Phys. Rev. C49, 1001 (1994).

[6] A. J. Koning, M. B. Chadwick, Phys. Rev. C56,

[7] J. J. Griffin, Phys. Rev. Lett. 17, 478 (1996).

[8] S. W. Steyn, Ph. D. Thesis, University of Stellenbosch (1997)

[9] R. Bonetti, C. Chiesa, University of Milan, unpublished.

[10] G. F. Bertsch, J. Borysowicz, H. McManus and W. G. Love Nuclear Phys. A248, 399 (1977).

[11] W. G. Love and M. A. Franey, Phys. Rev. C24, 1073 (1981).

[12] J. Raynal, computer code DWBA91 (unpublished)

[13] Ph.D. thesis: S. Karataglidis, Univ. of Melbourne, Australia ; P. J. Dortmans and K. Amos, Phys. Rev. C49, 1309 (1994); S. Karataglidis, P. J. Dortmans, K. Amos and R. de Swiniarski, Phys. Rev. C52, 861 (1995).

[14] R Lindsay, Acta Phys. Slov. 45, 717 (1995).

GAMMA-RAY ARRAY PHYSICS

C.J. Lister

Physics Division, Argonne National Laboratory, Argonne, IL 60439-4843 USA
Lister@anlphy.phy.anl.gov

Abstract: The fusion of heavy ions at near-barrier energies offers the prospect of producing and studying a wide variety of exotic nuclear states with unusual angular momentum, isospin, and mass. However, a revolution in experimental technique has been needed to sort out the interesting states from the dross. The most successful path has been the development of large arrays of gamma detectors which are ideal for isolating the states of greatest interest and extracting their properties. I will review the need for large arrays, their history, characteristics and some contemporary problems which are being investigated. Finally, I will briefly visit the prospects for new technologies in the future which will allow us to reach almost all bound states.

1 INTRODUCTION

In this contribution I am going to discuss the development of large arrays of Compton Suppressed, High Purity Germanium (HpGe) detectors and the physics that has been, that is being, and that will be done with them. These arrays and their science have dominated low-energy nuclear structure research for the last twenty years and will continue to do so in the foreseeable future. John Sharpey-Schafer played a visionary role in convincing a skeptical world that the development of these arrays would lead to a path of enlightenment. The extent to which he succeeded can be seen both through the world-wide propagation of ever more sophisticated devices, and through the world-wide propagation of his students. I, personally, would not be working in research if

The Nucleus: New Physics for the New Millennium
Edited by Smit et al., Kluwer Academic / Plenum Publishers, New York, 2000.

it were not for John's inspirational leadership. I am eternally grateful to him. Many excellent reviews of array physics have been made in the past which can provide detailed background reading. The review by Paul Nolan[1], another ex-Sharpey Schafer student, is particularly comprehensive and clear.

2 THE NEED FOR LARGE GAMMA-RAY ARRAYS

In the beginning, in the '60's and '70's, well before I was a student, most nuclear structural information was gained from light-ion induced reactions or from beta-decay. The thrust of the research was aimed at PRECISION; precise measurements of the quantum numbers of SPECIFIC low-lying states. The experimental emphasis was CONTROL of what states were populated and how they were prepared. These measurements could then be used to test the (then) new models of nuclear structure, the spherical shell model, the deformed (Nilsson) shell model and the collective rotational and vibrational models. The tools matched the task. Light-ion reactions could be used to populate selected states, and careful choice of reaction and beam energy could allow rather few states of interest to be investigated selectively. Further, the choice of inelastic scattering, transfer- and fusion- reactions revealed diverse information about the states and their occupancy. Because rather few states were populated, and generally with reasonable cross-sections (>10's mb) the experimental equipment was, by today's standards, quite simple. In gamma-ray spectroscopy a single, small, Germanium detector could be used to measure lifetimes of states or angular distributions. Modest arrays of scintillating detectors could measure more complicated directional correlations. All this was good and great progress was made.

Then heavy-ions came along. They had great promise[2] in vastly widening the scope of nuclear spectroscopy to encompass nuclei at high excitation energy, high spin and far from stability. The possibility of populating EXOTIC and UNUSUAL states was opened up, and exotic and unusual nuclear models were suggested. However, heavy-ions carried a lot of baggage. Almost all control of the reactions was lost. The mutual Coulomb barriers were so high that following each reaction a continuum of many thousands of states were populated. There was initially no way to individually select and study them. It has taken more than twenty years to really come to terms with this fact and rebuild our experiments to take full advantage of the reactions we are forced to employ to prepare the exotic states. In these two decades gamma-ray spectroscopy has emerged as the most powerful experimental tool which can step into the heavy-ion chaos and extract useful physics.

Now, at the turn of the millennium, we are contemplating producing beams of unstable radionuclides[3, 4]. These beams will again broaden the scope of what is meant by nuclear structure research. In particular they will allow us to reach out towards the limits of nuclear binding and generalize nuclear structure models to encompass all bound systems. Already, theoretical models are in place which predict phenomena which do not occur along the valley of stability. In some ways life may become easier as these beams of exotic nuclei may allow us to go back to controlled reactions, in inverse kinematics, and return to using light-ion reaction selectivity. However, the lessons of heavy-ions have been well learned, and considerable thought is being dedicated to new experimental techniques in parallel with the development of new accelerators.

3 THE PROPERTIES DESIRED OF BIG ARRAYS

Following a heavy-ion reaction above the Coulomb barrier, particularly a heavy-ion fusion reaction, a hot blob of nuclear matter is produced which is internally excited to many 10's MeV and may have up to 100 units of angular momentum. Cooling starts by boiling off nucleons. The process is statistical, and involves protons, neutrons, alphas and lesser numbers of deuterons and heavier ions. Many final nuclides can be produced, frequently more than twenty of them, with production cross-sections varying from 100's mb to a few nb. The residues are still hot, and continue to cool to their ground states by emitting a series of gamma-rays, perhaps 30 in number and usually of different energy in each final nucleus. It is these gamma-rays which tell us about the quantum mechanics of the nucleus as it cools to its groundstate. Two distinct challenges emerge: identifying which nucleus emitted the gamma-rays, and then locating the emitting state in the nucleus. To reliably perform both tasks calls for a powerful and sophisticated experiment. Let us start in the gamma-ray sector and decide what characteristics are needed. I will then talk specifically about one experiment and the development of auxiliary trigger detectors to improve channel selection through isotope identification. The gamma-ray characteristics we need are:-

1. Energy Resolution: The overall gamma-ray spectrum may contain several hundred distinct transitions. Measuring the photopeaks accurately, and resolving close-spaced lines is a key to high-quality spectroscopy. High purity intrinsic germanium detectors, with efficiency approaching that of a $3'' \times 3''$ sodium iodide crystal are the detectors of choice, as they have energy resolution of 2×10^{-3}, compared to scintillator detectors which have 10^{-1} resolution.

2. Spectrum Quality (peak/total): Nuclear structure information is deduced through measuring the energy, intensity, and perhaps the polarization of gamma transitions between well defined quantum states. Only the photopeaks, when the full transition energy is measured, are of interest. However, nuclear transitions are frequently in excess of 1000 keV and are difficult to fully absorb, especially in a solid-state detector. To reduce the number of incompletely-absorbed events, each detector can be surrounded by an active anti-scattering veto shield. If designed carefully, the photopeak efficiency can be as high as 60% for 1.3 MeV gamma-rays, twice that of similarly-sized scintillators. The development of such "anti-Compton" shields was very important in developing the current generation of gamma-arrays.

3. High Efficiency: We have seen the problem with heavy-ion reactions is the fragmentation of the population over many states. Each state is modestly populated. To collect sufficient data on individual states requires an efficient detector, ideally covering 4π around the target and collecting 100% of the radiation.

4. Good Multi-Hit Capability: If several gamma-rays hit a single detector and "pile-up", the information is lost. Thus, the solid angle of each detector element needs to be less than 4π/(number of gamma rays), to keep the multi-hit probability low. In addition, for many reactions involving short-lived states, the gamma-rays are emitted while the source nucleus is moving. Consequently, Doppler corrections are needed which can be achieved if the angle of gamma emission is determined. More segmentation means better angle determination and better Doppler corrections and angular correlations.

5. Calorimetry: It has become clear that in many applications, catching the entire gamma-ray flux and measuring the total emitted gamma ray energy and the number of gamma-rays (the multiplicity) is very desirable for channel selection. This can be done by adding all the energies in each HpGe and BGO shield, or by having a separate "inner-ball" to measure the flux distribution.

6. High Countrate Capability: This simply allows MORE data to be collected, which is good. Typical contemporary detectors are limited to about 10,000 events/second.

A little History

1964 First Ge(Li) Detectors
Ewan and Taverdale Can. J. Phys. 42 (1964) 2286
1969 First Ge(Li) Detectors with Compton Shields
Ward, Hausser *et al.* (Chalk River)
1976 CSS g-g coincidence experiment
Bebehani *et al.* J. Phys. G2 (1977) 22
1980 Multiple CSS Array at N.B.I.
Twin, Herskind and Garrett
1984 CSS Array at Daresbury with BGO Calorimeter
Twin, Riley, Simpson, Nolen, Sharpey-Schafer *et al.*
1987 Essa 30 Array. First big international collaboration
(Daresbury, N.B.I., Milan)
1987 Gammasphere Concept
Stephens
1992 Big Arrays Arrive
GASP, Eurogam and Gammasphere
1990's The Era of Big Array Physics
Berkeley, Argonne, Legnaro, Strasbourg

4 GAMMASPHERE

To give some idea of contemporary research with big gamma-ray arrays I am going to present data from Gammasphere, the U.S. National Gamma Ray Facility. This is by no means meant to detract from the excellent work done at the other large arrays, either built (like GASP at Legnaro, or Euroball at Strasburg) or coming on line (like Afrodite at NAC, Jurosphere at Jyvaskyla, or the new Yale and Oak Ridge Arrays), nor to ignore the many new devices being developed in other laboratories. It is more a statement of what I am most familiar with, and perhaps a biased view of where we are headed.

Gammasphere has many of the desirable features which we have already discussed. For pioneering research in the high-spin regime, it was constructed with high photopeak efficiency (about 10% for 1.33 MeV gamma-rays), good energy resolution (<2.4 keV at 1.33 MeV), good photopeak-to-total response (> 55% of 1.33 MeV events are in the photopeak) and high granularity (>100 high-purity germanium (HpGe) detector channels, of which 65 are segmented, to allow precise Doppler correction and minimize the chance of "double-hits"). The device has a high degree of mechanical symmetry which is ideal for angular correlation studies. The whole spectrometer, both HpGe detectors and their BGO anti-Compton shields, can be used for photon calorimetry by adding the energy deposited in the nearly 900 active elements.

Gammasphere has been operating at the ATLAS accelerator at Argonne National Laboratory since January '98. It was built at Lawrence Berkeley Laboratory and used primarily as a powerful spectrometer for studying nuclei at the highest spins[5]. In the scope of this talk I am not going to discuss much about high-spin physics, mainly due to shortness of time, but also as there are many excellent talks at this conference which will emphasize this aspect of big-array physics.

When the device was moved to Argonne, it was proposed to harness its unique capabilities in a new way, by using its efficiency and granularity to tackle new physics problems, mainly concerning nuclei of astrophysical interest, very heavy nuclei, and nuclei far-from-stability along the proton dripline. In all these cases, the challenge is to study nuclei which are produced in heavy-ion reactions at the part-per-million level (sub-microbarn). To obtain a suitable degree of selectivity, Gammasphere has seldom been operated alone, and most usually with microball[6] and neutron detectors[7], forming an efficient light-particle detector, but most frequently with the Argonne Fragment Mass Analyzer (FMA) and its auxiliary detectors[8]. When a suitable triggering technique is available, and several are discussed later in this article, spectroscopic measurements have been made on states populated with cross-sections below 100nb. Such a level of sensitivity allows research on nuclei beyond the proton dripline in many cases, and studies of the heaviest nuclei beyond Z=100.

5 TRIGGERING

Three experimental directions have been followed in recent years in order to obtain the level of channel selection which is needed; using electromagnetic, light-particle, or final residue detection.

In electromagnetic selection the photons (gamma-rays or X-rays) are used to identify BOTH the nuclei AND the states of interest. Gammasphere is particularly good at this means of selection, as it is efficient for detecting many gamma-rays in coincidence. By selecting a series of gamma-rays of defined energy a unique pathway can sometimes be established which isolates the state of interest through its de-excitation pathway. This type of selection has dominated progress in "high-spin" physics for many years[5]. This method of channel selection is experimentally straightforward, as the gamma-ray detector is also the channel selection device, so Gammasphere can operate "standalone". However, it has shortcomings arising from the complexity of heavy-ion induced gamma-ray spectra and from the detector response, which are difficult to overcome and limit the overall sensitivity which can be achieved. Typically, selection at the 1mb level is possible using photons alone. A set of Gammasphere-compatible large-area LEPS detectors have recently been purchased to enhance selection of low-energy photons through improved resolution and timing. In particular, for heavy nuclei, triggering on atomic X-rays may become viable. The Gammasphere VXI electronics are also to be upgraded in order to improve the response to low energy photons. In some circumstances, all the elements in Gammasphere can be used for calorimetry, to determine the entry region which the final nucleus is in after particle evaporation. This calorimetric method is particularly useful when several different reaction processes are happening, and can be useful for suppressing (or enhancing) Coulomb excitation, fission, or transfer reactions, from fusion. When photon calorimetry is used with light particle evaporation, or residue detection, extra selectivity can be achieved.

In light particle channel selection, the light particles which are evaporated immediately following heavy-ion fusion are used to infer which states are reached in the reaction. For a known monoisotopic target and a pure beam, a measurement of the multiplicity of evaporated light particles (protons, deutrons, alphas, neutrons) can allow the final isotope to be inferred. Beyond that, if the particle energies and angles are measured, then the excitation energy in the nucleus can be extracted. The technique is efficient, flexible, and relatively inexpensive. In fact, it is frequently the case that all the evaporated particles need not be detected, but can be inferred, so there is great scope for creative data analysis to maximize the efficiency for any particular desired level of channel selection. Further, the velocity of the final residue can be extracted to improve Doppler correction of the subsequent gamma-radiation. The technique can be very sensitive and is only limited by target and beam contamination, count rates, and the overall efficiency of particle detection. With an efficient charged particle detector, like microball[6] and a neutron array[7], this method can be routinely effective down to the 10's of microbarn level.

Finally, channel selection can be achieved through direct detection of the residues produced in the heavy-ion reaction using a zero-degree spectrometer. The Fragment Mass Analyzer (FMA) at Argonne[8] is an electric-magnetic-electric dipole spectrometer which suppresses the non-interacting beam and mass-disperses the residues. The mass dispersion of the measurement can be as high as 1/450 FWHM. The ions are detected in transmission avalanche counters (PGAC) or Channel-Plate (CP) detectors. Beyond the focal plane several techniques can be used to provide further isotopic identification. For light (A<100) nuclei the characteristics of stopping in gas can give excellent Z-identification if E/A>1.5 MeV/A. For heavier nuclei the groundstate decay characteristics of alpha, proton, or beta-delayed protons can be used after the ions are embedded in large-area silicon strip detectors (DSSDs). The Recoil Decay Tagging method RDT [9] is then used to correlate prompt gamma-rays in Gammasphere with the subsequent decay. Finally, isomeric gamma-decays can be used for selection[10]. These techniques are all very sensitive, but inefficient (typically only a few percent) and are frequently limited only by the amount of data which is collected. Many experiments at the sub-microbarn level have been performed and some projects have reached the sub-100 nb level.

Naturally, these methods can all be used together to compliment each other in overdefining the channels of interest. Overdefinition is an important goal, as each technique has shortcomings which can lead to mis-identification. Determining the state of interest by independent means can remove these problems and lead to a new level of sensitivity. However, the challenge is making the experiments sufficiently efficient for useful physics. Even with perfect channel selection experiments at the 10's nanobarn level will be extremely difficult with Gammasphere, due to lack of raw detection efficiency. Consequently, considerable effort is being dedicated to a "next-generation" gamma-array with higher efficiency and faster countrate capability.

6 NEW DRIPLINE PHYSICS

More than 50 experiments have been performed since Gammasphere started operation at ANL in January 1998. There is obviously not enough space to discuss all these projects, so I will present just a few key topics in order to give a sense of some of the main thrust of research.

Light nuclei in the p- and sd- shell are still pivotal in our understanding of nuclear structure, as it is in these light nuclei that our theories are most developed, and our experimental data base is most complete. Every new model, be it relativistic mean field models, cluster models, monte carlo shell models or Hartree Fock models is tested here and compared to the existing excellent shell model calculations and to the data. Unfortunately, the location and gamma-decay of some of the most interesting high spin states are not experimentally known, as they are highly particle unbound by more than 10 MeV, so cannot be used to test the models. The states are expected to have very small radiative gamma-branches, perhaps of 10^{-3} of the population, so are very difficult to observe even with Gammasphere. However, by combining Gammasphere with the FMA, used as a time-of-flight spectrometer, great progress appears possible. The key is to use a two-body nuclear reaction so the kinematics of the reaction reveal which states were populated. Highly excited states mean there is little energy left for kinetics, so the residues recoil slowly. Thus, measuring the velocity of each ion marks its parentage, then the decay gamma rays can be sought. In principle this is a "singles" experiment as only the linking transition between the state of interest and the low lying states is sought. However, the branching ratios of the decay from each state to its low-lying daughters provides a critical test of the wavefunction. We have been using the $^{12}C(^{16}O,\alpha)^{24}Mg$ reaction to develop this concept. We have performed one successful experiment and have found candidates for J=10 states in the 18-21 MeV excitation range. However, we are having to learn special tricks to operate Gammasphere in 'singles' mode, as it's readout is optimized for low-rate, multiparameter data, but it appears considerable progress should be possible in seeking shell model states in ^{24}Mg, and also perhaps cluster states formed as resonances in $^{12}C+^{12}C$ reactions, and exotic high-spin shapes in ^{28}Si and ^{32}S.

Mirror nuclei, with conjugate proton and neutron numbers, have long been used as laboratories for precise measurements of the properties of the nuclear force, in particular its symmetries. At the level of Coulomb shifts arising from the different numbers of protons, a charge symmetric nuclear force would lead to identical spectra in mirror pairs. This is indeed the case for many low-lying states in light nuclei with "identicallity" at the level of 10's keV. For states which are not strongly bound this symmetry can break down as poorly bound protons (or neutrons) spill out of the central field and become even less bound. In intermediate mass nuclei this will always be the case when the N=Z line approaches the proton dripline. Here, comparison of T=+1/2 nuclear states (which are relatively well bound) to their T=-1/2 partner levels which may be technically unbound, will give us highly precise information from which we can extract information on modifications of surface diffuseness, single particle levels, pairing and residual interactions. Previous experiments[11] have reached the mid fp-shell, but using Gammasphere we have progressed to T=1/2 projects with A=53,65 and 79 so the approach to the proton unbound limit can be followed.

Groundstate proton radioactivity is a clean and unambiguous experimental signature that the proton dripline has been reached. Much progress has been made in understanding proton decay spectroscopy[12], which now extends from just above tin to above lead. This progress includes finding many new proton emitters, extraction of spectroscopic factors, studying deformed emission and observing fine structure. Studies of excited states in these nuclei is of great interest, as detailed spectroscopy can yield information on subtle modifications of the mean field and pairing in dripline and "post-dripline" nuclei. At some level all states in these nuclei should exhibit competition between gamma and particle decay, but the nature of that competition is extremely sensitive to the individual wavefunctions. Before the current round of Gammasphere experiments, only preliminary "singles" investigations using RDT on proton emitters had been reported[9], due to the low cross-sections for dripline isotope production and the inefficient experiments. Now, the technique has become routine and detailed gamma-ray correlation measurements can be made. Investigations of "in-beam" spectroscopy of groundstate proton emitters ^{109}I[13], ^{113}Cs, ^{141}Ho, ^{147}Hf, ^{155}Ta have all been made and are undergoing analysis. Several more experiments are awaiting beam time. Somewhat nearer to stability, investigations of groundstate alpha-emitters have begun to fill out the landscape between the dripline and the lightest nuclei previously studied and clarify systematic trends as the dripline is approached. Considerable progress has been made in the Os-Pt-Hg region[14].

In principle, radioactive beams can enhance the production of dripline nuclei. To date, the beams have not been of suitable quality or sufficiently intense to supplant careful stable beam research. In order to investigate where the break-even occurs for gamma-ray spectroscopy, and to identify when radioactive beams come into their own, we used a modest 56Ni beam which was produced for an astrophysical investigation[15]. The beam was not pure, containing ^{56}Co and ^{56}Fe, and was low in intensity, having about 10^4 ^{56}Ni ions/sec, more than a million times less intense than normal stable beams. However, by inducing fusion with ^{92}Mo and using the selectivity of the Gammasphere/FMA, gamma-rays associated with the decays of low lying states in $^{142,144}Dy$ were observed. These nuclei are already known, but lie at the periphery of stable-beam research. It was clear from this first investigation that only a modest improvement in the beam, perhaps to 10^6 particles/sec, would lead to nuclei which cannot presently be reached.

Gammasphere is a near-perfect device for Coulomb Excitation studies. Its efficiency, granularity and symmetry allow precise angular correlation measurements to be made which can be compared to theory. Two types of study are fruitful, either using thick targets for high yield, or using thin targets and correlating gammas with Rutherford scattered ions. Both techniques are being actively pursued in the Gammasphere program. Several thick target measurements of plutonium isotopes have been made following the technique suggested by Ward[16], in which the quantitative aspect of electromagnetic excitation is sacrificed for enhanced yield. In these studies a lead beam at an energy about 10% above the classical Coulomb barrier is used to maximize the probability of multi-step excitation. The nuclei are stopped in the target, so gamma-gamma correlations can be made free of Doppler shifts. Many new bands of states have been found and followed to high spin. One- and two- neutron transfer channels have been observed, found at the few % level, so neighboring odd-A nuclei can be simultaneously studied. To date, $^{238-44}Pu$ have been investigated[17]. It is clear that octupole correlations play an important role in all structural aspects of these nuclei. Strongly populated octupole bands have been found in all cases, which, with rotation, mix and perturb the groundstate band probably causing alignment and backbending anomalies.

Beyond plutonium, new elements have been synthesized to Z=112 by the meticulous work of the GSI and Dubna groups who inferred the relative stability of these elements through observing their groundstate alpha decays. These very heavy nuclei are all prevented from undergoing spontaneous fission by shell effects. Theoretical calculations indicate that deformation plays a key role in maximizing the binding energy. However, many open questions remain: how are the residues are formed in fusion reactions?, how deformed are they?, what deformed shell-configurations are responsible for the fission barrier?, and how does their shell-stability change with rotation? To address these issues, "in-beam" spectroscopy using Gammasphere coupled to the FMA is ideal. The recoils are identified following the GSI method of correlated alphas, but the RDT method can be then used to identify prompt gamma rays emitted shortly after formation of the nuclei while it is still rotating. Several experiments, producing the Z=102 element ^{254}No in the $^{208}Pb(^{48}Ca, 2n)^{254}No$ reaction have been completed successfully[18, 19], both at Gammasphere and using the RITU separator at Jyvaskyla, culminating with the identification of gamma-rays in the groundstate rotational band to spin $J^{\pi} = 18^{+}$. The deformation was found to be β=0.27(2), close to some theoretical predictions, and the intensities of the band indicate the shell-stabilized fission barrier is robust against rotation to surprisingly high spin. Many possibilities for novel experiments in these heavy nuclei now seem open.

7 FUTURE DIRECTIONS OF BIG ARRAY PHYSICS

We appear to have tamed heavy-ion fusion reactions for gamma-ray spectroscopy and are enjoying the fruits of our labours. Of course, the reactions themselves can still frustrate us in limiting which states are populated. Some exotic nuclei, and some states (particularly non-yrast states) are difficult to reach. For a while we have been improving our channel selection devices to increase the sensitivity of our experiments, and have reached a point where once again we need to make considerable progress in our gamma-detection technology. Even state-of-the art detectors like Gammasphere are simply not efficient enough to reach the states of greatest interest. How then to proceed? Three trends are apparent.

Firstly, and most excitingly, the production of high-quality radioactive beams of intensity similar to that we have available now in the stable domain is becoming a reality. Not only will this open whole new domains of nuclei to be studied, but it will also allow a much more diverse collection of experimental techniques to be used. This will profoundly influence the history of our understanding of nuclear structure.

Secondly, the gamma-arrays are changing, in efficiency, in granularity and in countrate capability. The efficiency issue is clear. Even Gammasphere, at 10% is not very efficient on an absolute scale, and clearly advancing towards 100% would be very advantageous. That is, the anti-Compton shields would be discarded and replaced with a solid Germanium shell. However, words are cheap, and exactly how this is to be achieved is not clear. Intense R&D is underway to explore the possibilities. In the germanium shell, the track of gamma-rays needs to be followed. That is, the detectors will become position sensitive.

This offers many new advantages in Doppler reconstruction and in polarization measurement. Finally, the detectors will count at least 10 times faster than contemporary detectors. This capability comes from enhanced segmentation and new signal processing methods. It is interesting that we take it completely for granted that faster computers and data acquisition systems will be coming shortly, as this has been the case during the era of big arrays. However, perhaps one change of importance will be that more intelligent "front-ends" are used instead of spooling vast amounts of data to some storage medium.

Finally, I would point to sociological changes. When I started research, a scientist with a few students could build his own equipment and be a world leader. Now this is not the case and the best research groups are large close-knit teams of scientists with diverse, but overlapping skills. This will continue, and I think our field will be strengthened by it, though I know many researchers fear it. When Gammasphere moved to Argonne, I proposed selecting twelve key experiments and focusing on doing those projects as well as possible, allocating each one month of beam time and getting all the interested scientists to work together. Naturally, perhaps rightly, the community and the Program Advisory Committee found this highly amusing. It did not happen, and we have continued "business as usual" doing many three and four day projects. However, in fact, many of these experiments have great overlap, and if the scientists involved had worked together to build one or two superior experiments, I think the cause would have been further advanced.

It may be thought that building of big experiments and big teams of scientists would lead to the diminishing role of individual contributions. This is simply not so, and larger groups of people working harmoniously together involve more inspiring and wiser leadership. In South Africa today this can be seen everywhere around us and at every level, in the leadership of Nelson Mandela here on a national scale, in John Sharpey-Schafer's leadership of the NAC, and in the enthusiastic and committed contributions of the many new young people coming into the field. This is an exciting time both for South Africa and for nuclear physics.

Acknowledgments

I would like to thank my colleagues at ANL and LBNL for operating Gammasphere with passion and the upmost professionalism. Working with the many user groups has provided many new ideas and challenges. I apologize for not being able to discuss more of the excellent projects which are underway. This work was supported by many grants from the U.S. Department of Energy, NSF, and European Funding Agencies, including the ANL DOE contract No. W-31-109-ENG-38.

References

[1] P. J. Nolan, F. A. Beck, and D. Fossan, Ann. Rev. Nucl. Part. Sci. 45, 561 (1994).

[2] Proceedings of the Heavy-Ion Summer Study, Oak Ridge Tennessee, ed. S. T.Thornton (July 1972), unpublished ORNL CONF-720669.

[3] Concept for an Advanced Exotic Beam Facility, ANL 1995, unpublished.
[4] Proceedings of the Workshop on the Scientific Opportunities for an Advanced ISOL Facility, Columbus, OH 1997, unpublished.
[5] F. S. Stephens *et al.*, Gammasphere Proposal LBNL (1998), Gammasphere: The Beginning 1993-1997, ed. M. Riley http://www-gam.lbl.gov (1998).
[6] D. G. Sarantites *et al.*, Nucl. Inst. Meth. A381, 481 (1996).
[7] D. P. Balamuth *et al.*, Proposal for a Neutral Array for Gammasphere, unpublished (1998).
[8] C. N. Davids *et al.*, Nucl. Inst. Meth. B70, 358 (1992); B. B. Back *et al.*, Nucl. Inst. Meth. A379, 206 (1996).
[9] R. S. Simon *et al.*, Z. Phys. A325, 197 (1986); E. S. Paul *et al.*, Phys. Rev. C51, 78 (1995).
[10] D. Seweryniak *et al.*, Phys. Lett. B440, 246 (1998).
[11] C. D. O'Leary *et al.*, Phys. Rev. Lett. 79, 4349 (1997).
[12] P. J. Woods and C. N. Davids, Ann. Rev. Nucl. Part. Sci. 47, 541 (1997).
[13] Ch-H. Yu *et al.*, Phys. Rev. Lett. C59, R1834 (1999).
[14] M. P. Carpenter *et al.*, Phys. Rev. Lett. 78, 3650 (1997).
[15] K. E. Rehm *et al.*, Phys. Rev. Lett. 80, ;676 (1998).
[16] D. Ward *et al.*, Nucl. Phys. A600, 88 (1996).
[17] I. Wiedenhoever *et al.*, submitted to Phys. Rev. Lett. (1999).
[18] P. Reiter *et al.*, Phys. Rev. Lett. 82, 509 (1999).
[19] M. Leino *et al.*, submitted to J. Phys. G (1999).

STUDY OF THE STRENGTH DISTRIBUTION OF THE PRIMARY γ-RAYS IN THE DECAY FROM SUPERDEFORMED STATES IN ^{194}Hg

A. P. Lopez-Martens

Institut de Recherches Subatomiques UMR7500, 67300 Strasbourg, France

Abstract: One-step decay pathways between superdeformed (SD) and normally deformed (ND) states have been observed in ^{194}Hg and ^{194}Pb nuclei. In all the other studied cases of the mass 190 region, the intensities of the so called "one-step linking transitions" are too weak for any decay scheme to be firmly established. To understand the origin of such strong intensity fluctuations, the strength distribution of the primary γ rays in the decay from superdeformed states is investigated by applying the Maximum Likelihood Method. For the ^{194}Hg case, it is concluded that the observed primary transitions might consist of the strongest transitions selected stochastically from a Porter-Thomas distribution and that the presence of strong one-step links in ^{194}Hg represents a very lucky case.

1 INTRODUCTION

Since the discovery in 1986 of the SD rotational band in ^{152}Dy[1], many more such structures have been observed in other nuclei of the mass 150 region, and also in mass 130, 190, 80 and recently mass 60 nuclei. The common feature to all these bands is their gradual feeding at high angular momentum and their

The Nucleus: New Physics for the New Millennium
Edited by Smit et al., Kluwer Academic / Plenum Publishers, New York, 2000.

sudden depopulation at low spin, when the nucleus changes shape. Because it is difficult to determine how the nucleus then decays to its ground-state, most SD states have unknown excitation energy, spin and parity.

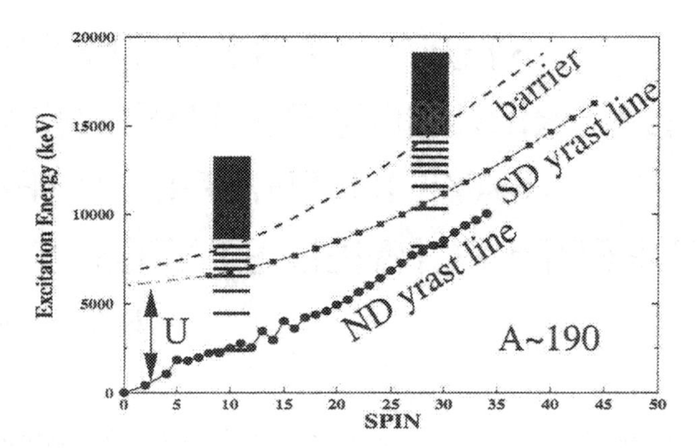

Figure 1 Excitation energy vs Spin plane for a typical mass 190 nucleus. The SD and ND yrast lines, as well as the height of the barrier seperating the 2 nuclear shapes are represented. The level density of ND states is schematically shown at spin 10 and 30 $\hbar$. The excitation energy of the SD states relative to the ND states is denoted by U.

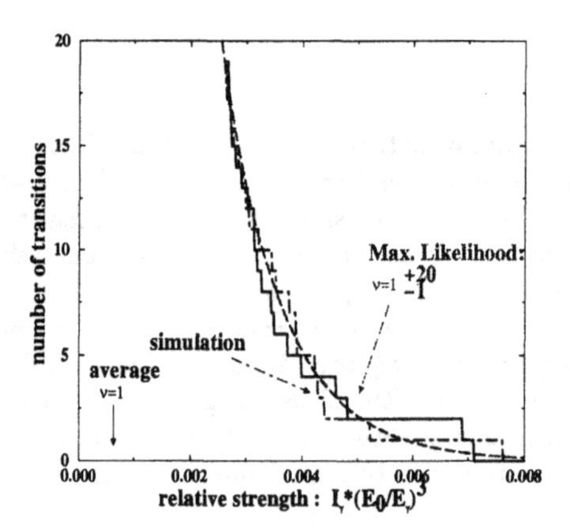

Figure 2 Experimental cumulative strength distribution (solid line), most likely cumulative distribution (long dashed line) and cumulative strength distribution of a set of 19 strengths which result from random sampling of a Porter-Thomas distribution (dashed-dotted line). The value of the most likely average strength is marked by an arrow and is clearly much smaller than the experimental strength threshold ω_{low}=0.0026

SD states decay to ND states because of the conjuncture of many effects, 4 of which are illustrated in Figure 1. As the nucleus decays along a SD band, the transition rate to decay to the next SD state decreases. At the same time, the excitation energy of the next SD state above the ND yrast line increases. This yields an increase in the level density of surrounding ND states, making it easier and easier for the SD state to mix with them. Another important factor is the decrease in the height of the potential barrier separating the 2 types of nuclear shapes. This again, together with the increase in pairing correlations, makes it easier for the SD state to couple to its neighbouring ND states through the barrier.

The tunneling occurs as a motion in collective coordinate space. If the excitation energy of the decaying SD state is such that the motion on the ND side stays within collective degrees of freedom, the SD state will couple to vibrational states in the first well [2]. Although the amplitude of the SD wavefunction in the first well is most probably small, it can precipate the decay because the decay rate is much larger there than in the SD well.

On the other hand, if the excitation energy of the decaying state is large enough, the collective states on the ND side will be thoroughly mixed into the dense background of multi-particle states. The coupling will manifest itself as an admixture into the SD state of one of these complex states. The decay is then due to the statistical coupling of the tail of the SD wavefunction in the first well and the wavefunctions of low energy ND states [3]. The question is "does the decay fom SD states occur in the ordered regime, in the chaotic regime or somewhere in between?"

2 DECAY SPECTRUM

In 1994, it was discovered that the decay spectrum is a quasicontinuum on which a few discrete lines are superimposed [4]. This was done by identifying all the components of the spectrum of γ-rays emitted by a SD nucleus, from its formation in a fusion-evaporation reaction until it reaches its quasi-spherical ground-state.

First come transitions from thermally excited states at rapid rotation. Then comes the long series of regularly spaced transitions which form the SD rotational band. Then follows the decay spectrum and finally the transitions emitted by the nucleus once it has reached the ND yrast line. From a quasicontinuum analysis, the excitation energy of the 10^+ SD state in ^{192}Hg is estimated to be 4.3$\pm$0.9 MeV above yrast. From a fluctuation analysis, the average number of transitions sampled in the decay is found to be of the order of 10^4 [7]. In the few cases where high energy transitions directly connecting the SD states to known ND states are observed, the absolute excitation energies, as well as the most probable spins and parities, can be extracted for SD states [6, 7, 8]. All this information tells us that the decay from SD states is very fragmented and occurs at high excitation energy (between 3 and 4 MeV above yrast). This supports a statistical decay scenario via the admixture of a hot compound state.

The remarkable similarity in shape and magnitude of the decay spectrum in ^{194}Hg and the spectrum following thermal neutron capture in ^{167}Er is further evidence that the decay from SD states might well be a statistical process. At the neutron resonances, it has been shown that the strength distribution of the primary γ-rays is a χ^2 distribution with one degree of freedom, often called Porter-Thomas distribution [9]. This is the hallmark of a statistical decay and is a direct consequence of random matrix theory. Compared to other χ^2 distributions, the Porter-Thomas distribution yields much larger strength fluctuations.

These stochastic fluctuations might explain the differences observed in the decay spectra of ^{194}Hg and ^{192}Hg. These 2 nuclei are very similar in many respects, yet strong one-step linking transitions have been observed in ^{194}Hg and not in ^{192}Hg. To find out whether this is due to an accidental enhancement and suppression of random matrix elements or to special structural effects, the strength distribution of the primary transitions in ^{194}Hg must be determined.

3 PRIMARY-TRANSITION STRENGTH DISTRIBUTION

Two high-statistics GAMMASPHERE experiments were added up and the spectrum of γ-rays emitted by the ^{194}Hg SD nucleus was sorted by setting coincidence gates on 2 transitions in the band. Since the excitation energy of the SD states at the point of decay is $\sim$ 4.2 MeV, any resolved line with transition energy well above half this value can be considered to be a primary line. In total, 41 such lines were observed, but only 31 stood out from the noise of the spectrum. The decay occurs at the SD 12^+ and 10^+ states at roughly the same excitation energy and with similar intensity. We therefore assume that the strengths of the transitions stemming from these 2 states are selected from a common χ^2 distribution.

Primary transitions following neutron capture have been shown to have a ratio of $E1/M1$ strengths of the order of 5-7. This is why it is assumed that the primary lines in ^{194}Hg are of $E1$ nature. From the measured intensities I_γ and transition energies E_γ of the observed primary lines, the relative strengths ω are computed. One then defines the strength threshold ω_{low} above which no line has been missed. In total, 19 strengths are found to lie above the threshold. Now we seek the χ^2 distribution which is most likely to describe the data above ω_{low}. We do this by means of the Maximum Likelihood Method.

4 RESULTS

It turns out that the most likely distribution is a Porter-Thomas distribution ($\nu = 1$) [10]. This is shown in Figure 2. However the uncertainty in the value of ν is very large. This is because the sample size is very small and because only the tail of the distribution is accessed experimentally, and it is at low strength that the different χ^2 distributions really differ.

From a fluctuation analysis, we know that there are $\sim$600 primary lines above 2.6 MeV. Another estimate for this number is given by dividing the total strength above 2.6 MeV by the average strength θ. This yields 550. Yet another estimate is given by the value of the most likely cumulative strength distribution at the origin: 400. All these independent numbers agree quite well and strongly support the value $\nu = 1$. Taking into account the uncertainty in θ, the most likely distribution predicts that there should be 2-6 lines with strengths larger than 8θ. Experimentally, there are 3 such lines, 2 of which are single-step links.

It seems very unlikely that such large strengths should have been selected for this particular type of transition. This may be the sign of some selection rules at play. However, faced with the question "what distinguishes the SD states in ^{194}Hg from those in ^{192}Hg ?", we have no answer, and must conclude that the presence of strong one-step links in ^{194}Hg is the luck of the draw.

5 CONCLUSION

It seems that the decay from SD states in ^{194}Hg is mediated by an admixed component of a compound ND state and that the occurrence of strong one-step links represents a lucky case. However, a definite conclusion cannot be drawn since the primary transitions are so weak that we can detect only the stronger transitions in the tail of the strength distribution. A more favourable case to study is ^{194}Pb since the excitation energy of the SD states at the point of decay is lower, and so in this case, the average strength will be larger than the experimental threshold.

If it is found that the decay has not reached the chaotic limit, it might be possible to address the recent question " is the decay from SD states an example of chaos assisted tunneling ?". One could also imagine a systematic search for primary γ-rays and by comparing different nuclei with different decay-out excitation energies, one could attempt to locate the energy region where the onset of chaos occurs.

References

[1] P. J. Twin *et al.* Phys. Rev. Lett. 57(1986) 811
[2] P. Bonche *et al.* Nucl. Phys A519 (1990) 509
[3] E. Vigezzi, R. A. Broglia and T. Døssing. Phys. Lett. B249 (1990) 163
[4] R.G. Henry *et al.* Phys. Rev. Lett. 73 (1994) 777
[5] A. Lopez-Martens *et al.* Phys. Rev. Lett 77 (1996) 1707
[6] T.L. Khoo *et al.* Phys. Rev. Lett. 76 (1996) 1583
[7] A. P. Lopez-Martens *et al.* Phys. Lett. B380 (1996) 18
[8] K. Hauschild *et al.* Phys. Rev. C55 (1997) 2819
[9] C.E. Porter and R.G. Thomas. Phys. Rev. 104 (1956) 483
[10] A. Lopez-Martens *et al.* to be published in Nucl. Phys. A

γ-RAY STUDIES OF INDUCED FISSION $^{12}C+^{238}U$

M. Houry[1], Ch. Theisen[1], R. Lucas[1],
F. Becker[1], W. Korten[1], Y. Le Coz[1]
G. Barreau[2], T.P. Doan[2]
J. Durell[3], D. Grimwood[3], A. Roach[3], A.G. Smith[3], B.J. Varley[3]
Th. Ethvignot[4]
I. Deloncle[5], M.G. Porquet[5]
A. Astier[6], S. Perries[6], N. Redon[6]

[1]DAPNIA/SPhN CEA Saclay 91191 Gif sur Yvette France
[2]CENBG Gradignan 33175 France
[3]Dept. of Physics and Astronomy, University of Manchester UK
[4]CEA/DIF/SPN Bruyères le Chatel 91190 France
[5]CSNSM IN2P3-CNRS Orsay 91405 France
[6]IPN Lyon F-69622 Villeurbanne France

rlucas@cea.fr

Abstract: A fission fragment detector, SAPhIR, using photovoltaic cells has been developed and used with the multidetector array EUROBALL to select fission fragments produced in the $^{12}C+^{238}U$ reaction. These fission fragments, belonging to different regions of the mass chart, are very neutron rich. Their structure and, in particular, isomers identified in the range of 20ns up to 2μs, have been studied.

1 INTRODUCTION

Fission is known to provide a large number of neutron-rich nuclei which are, for the most part, accessible in no other manner and which are released with various excitation energies, spins and deformations. The development of high resolution γ-ray arrays has led to important developments in the spectroscopy

The Nucleus: New Physics for the New Millennium
Edited by Smit et al., Kluwer Academic / Plenum Publishers, New York, 2000.

of fission fragments. To increase reaction channel selectivity and to identify in a better way the fission fragment isomers, we have developed a modular fission fragment detector based on photovoltaic cells. This detector, SAPhIR (Saclay Aquitaine Photovoltaic Isomer Research), already used in many experiments [1], is a collaboration between CEA Saclay, CENBG Bordeaux and CEA/DIF Bruyères le Châtel.

2 THE SAPHIR DETECTOR

The solar cells are 500μm layers of polycrystalline silicon. Their main characteristics are a thin depletion layer, a low resistivity and a large capacitance. The charge collection is made by funnelling [2]. They do not need any bias voltage and are very easy to handle. Typically we used 1.2 cm$\times$2.5 cm cells. For fission fragments from a ^{252}Cf source, with a miniaturized preamplifier (2 cm$\times$2 cm) developed at Saclay, the resolution is similar to that obtained with a standard surface barrier detector.
New VXI cards (including both time and energy channels) dedicated to SAPhIR (but usable for other Si ancillary detectors) have been constructed at CEA Saclay and have been successfully tested [3], [4], [5].

3 STUDY OF FISSION FRAGMENTS

Isomeric levels provide additional crucial information about nuclear softness, high order deformation parameters, pairing correlations etc. The reaction ^{12}C + ^{238}U was studied at 90 MeV incident energy to investigate nuclei populated around the $A \simeq 120$ mass region. The fission products were identified in a 32 photovoltaic cells version of the SAPhIR detector and the coincidence γ-ray cascades were detected with EUROBALL III installed in Legnaro. The delayed spectroscopy has been completed by the analysis of prompt structures built on these states. This was done with EUROBALL III alone and by using a thick target to avoid Doppler corrections. More than 300 nuclei were identified and more than 1 giga event with multiplicity $\geq$3 were registered.

New isomers in the range 20ns-2μs were found and level schemes of numerous nuclei have been constructed up to relatively high spins. With the resolving power of the array we used, detailed level schemes of isotopes ranging from near the stability line to as much as ten neutrons beyond the most neutron-rich stable isotope have been obtained. In this work our knowledge of Cadmium isotopes has been considerably enhanced completing the detailed investigation near the shell closure at Z=50 [6]. If the decay scheme of any particular nucleus is well understood, the angular momentum carried away by the fission fragment is deduced from the nuclear level scheme and the intensities of the γ-ray transitions. By combining the two experiments (thin and thick targets), levels constructed on top of new isomeric levels have been obtained, increasing the knowledge of the corresponding nuclear structure.

These experiments have shown that a multidetector like EUROBALL associated with ancillary detectors is a very powerful tool to perform spectroscopic studies and to clarify very low intensity phenomena.

Acknowledgments

The authors wish to thank B. Cahan, A. Le Coguie, G. Durand, N. Karkour, G. Linget for their help in the production of the VXI cards and mechanical staff in Bordeaux for the construction of the detector. Authors also wish to thank the Legnaro staff for providing good beams.

References

[1] C. Gautherin et al, Eur. Phys. J. A1, 391 (1998) and PHD (1997)

[2] F.B. Mc Lean IEEE Trans. on Nucl. Sci. N6, 2018 (1982)

[3] Ch. Theisen et al, AIP conference proceedings, Seyssins 1998 p 143

[4] M. Houry et al, AIP conference proceedings, Seyssins 1998 p 220 and to be published

[5] G. Barreau et al, to be published

[6] A. Roach et al, to be published

NEW MEASUREMENTS OF LEVEL DENSITIES

E. Melby, L. Bergholt,
M. Guttormsen, S. Messelt, J. Rekstad, A. Schiller, and S. Siem

Department of Physics, University of Oslo
P. Box 1048 Blindern, 0316 OSLO
Norway

elin.melby@fys.uio.no

Abstract: An iterative procedure for simultaneous extraction of the level density and the γ-ray strength function from a set of primary γ-ray spectra is used. This procedure opens new perspectives in the search for thermodynamic phase transitions and the order to chaos transition, and has so far been used to study nine different nuclei in the rare earth region.

Recently a new method of extracting the experimental level density and γ-ray strength function from measured γ-ray spectra was reported [1]. Both the level density and the γ-ray strength function for the two nuclei, ^{162}Dy and ^{172}Yb, were studied by means of the (^{3}He,α)-reaction. The results showed significant non-statistical behaviour [1]. The level density revealed step-like enhancements in the region below 5 MeV of excitation energy in both nuclei. Especially, there was found to be an enhancement of levels just above 2Δ in excitation energy, where Δ is the pairing gap energy. This is in agreement with theoretical calculations by Ref.[2] and is tentatively interpreted as the expected gradual breakdown of the pairing correlations, with an enhancement of two-quasiparticle levels just above 2Δ. In the γ-ray strength function, peaks are found at $E_\gamma \approx 3$ MeV in the two nuclei. These enhancements of γ-ray strength are probably due to favoured transition energies, and may then indicate remains of order at high excitation energy [1].

As a part of a systematic study of rare earth nuclei, experiments have been performed at the Oslo Cyclotron Laboratory, especially looking at the nuclear level density, the γ-ray strength function, and possible dependency of the K-quantum number in the γ-decay from highly excited nuclear states. The nuclei

The Nucleus: New Physics for the New Millennium
Edited by Smit et al., Kluwer Academic / Plenum Publishers, New York, 2000.

^{143}Nd, ^{148}Sm, ^{161}Dy, ^{166}Er, and ^{171}Yb were produced by the (^{3}He,α)-reaction, while ^{149}Sm, ^{162}Dy,^{167}Er and ^{172}Yb was produced by the (^{3}He,^{3}He')-reaction. The beam energy in all the experiments was $E_{^3\mathrm{He}} = 45$ MeV. The γ-rays were detected in 28 NaI detectors in coincidence with the charged ejectiles, which were detected with 8 particle telescopes.

Preliminary results show that the non-statistical features reported in Ref. [1] is not due to the reaction mechanism. Both the nuclei ^{162}Dy and ^{172}Yb have been studied with the (^{3}He,^{3}He')-reaction as well as with the (^{3}He,α)-reaction. These reactions represent two very different population mechanisms, and also different ground state spins of the targets are implied. The results for ^{162}Dy from the two reactions are shown in Fig. 1. Apart from the deviation at low excitation energy, the level densities are found to be very similar.

The experimental methods make it possible to study the dependency of the γ-ray strength function and the level density on the K-quantum number and nuclear deformation. The experimental level density also opens new possibilities for extracting thermodynamical quantities such as entropy, temperature, and heat capacity as a function of excitation energy in the nucleus.

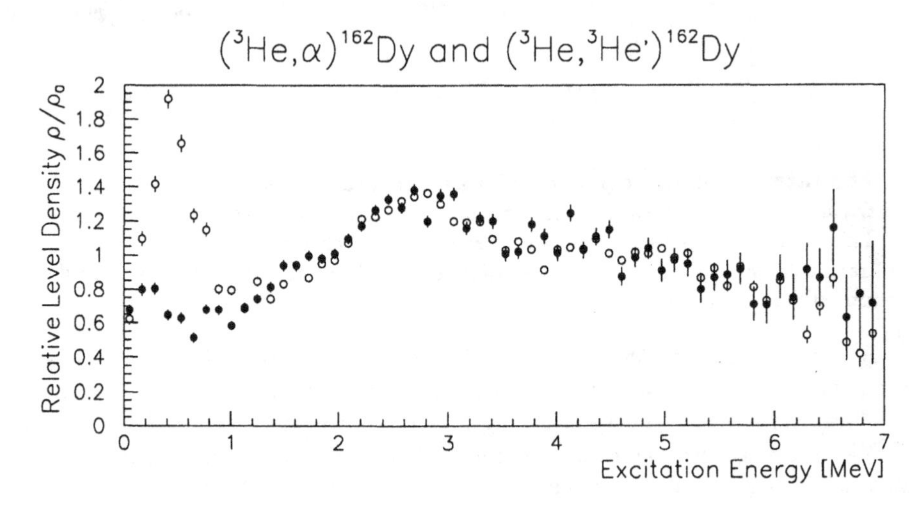

Figure 1 : Level density for ^{162}Dy extracted by means of the (^{3}He,α)-reaction (closed circles) and the (^{3}He,^{3}He')-reaction (open circles) are practically identical. To show the fine structure in the level densities more clearly, both level densities are divided by a best fit function of the form $\rho_0 = Ce^{E_x/T}$, where C and T are constants.

References

[1] T. S. Tveter et al., Phys. Rev. Lett. 77 (1996) 2404

[2] T. Døssing et al., Phys. Rev. Lett. 75 (1995) 1276

STUDY OF HIGH-SPIN STATES IN ^{190}Pt

E. Gueorguieva[1,2], P. Boutashkov[1], A. Minkova[1], C. Schück[2], Ch. Vieu[2], M. Kaci[2], J.S. Dionisio[2], Ts. Venkova[1]

[1]Faculty of Physics, University of Sofia, Sofia, 1164, Bulgaria
[2]C.S.N.S.M. CNRS-IN2P3, F-91405 Orsay, France

animin@rose.phys.uni-sofia.bg

Abstract: The present work reports on a study of high spin states in ^{190}Pt. On the basis of TRS calculation, an interpretation is proposed for the new structures as well as for the previously known 10^- isomer.

1 EXPERIMENT

The ^{190}Pt nucleus was a by-product in a reaction aimed to study the superdeformed states in ^{191}Au [1]. It was produced in the p-channel of ^{186}W(^{11}B,p6n) reaction obtained at beam energies of 84 and 86 MeV with a stack of two $280\mu g/cm^2$ ^{186}W self-supporting foils. The experiment was carried out with the Eurogam2 multidetector array operating at the Vivitron accelerator in Strasbourg. Events were recorded when at least five unsuppressed detectors fired in prompt coincidence. A total of 9×10^8 four and higher fold events were collected. The total yield of ^{190}Pt nucleus was only about 4% of the total reaction cross-section.

2 RESULTS AND DISCUSSION

The γ-γ coincidences have been studied from a cube and matrices constructed off-line. By using the RADWARE software the known ^{190}Pt level scheme [2, 3] has been extended with 40 new transitions establishing 33 new levels, placed up to spin 32 and 8.1 MeV excitation energy. Angular distributions of the observed γ-rays were analyzed at the 8 angles of Eurogam2 from special data sorts with energy gates. From this the multipole order of most of the transitions could be

The Nucleus: New Physics for the New Millennium
Edited by Smit et al., Kluwer Academic / Plenum Publishers, New York, 2000.

determined. The transition intensities have been obtained from the same data sort and summation of all angles.

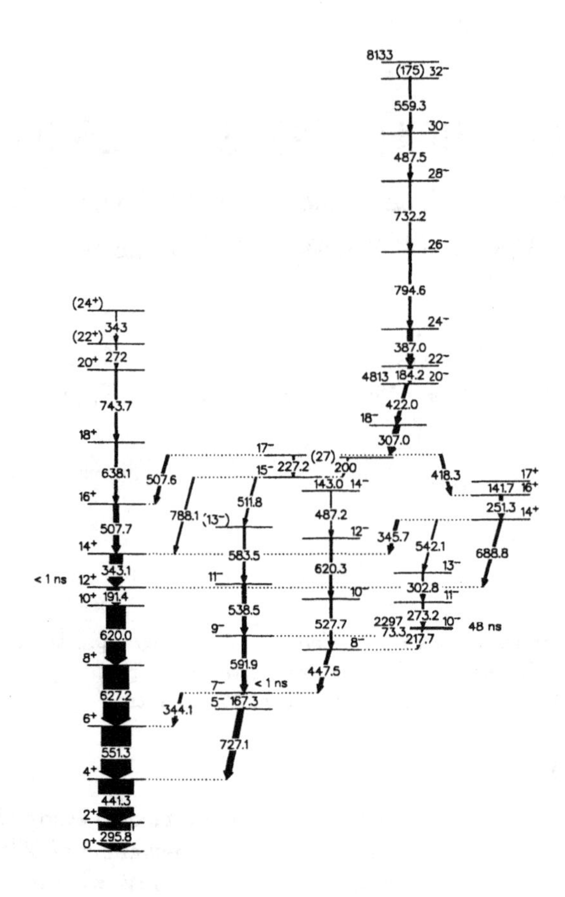

Figure 1 Partial decay scheme of ^{190}Pt showing the semi-decoupled band built on 5^-, 8^- levels, the new observed band built on the 20^- level and its decay to the 10^- isomer structure

A partial decay scheme of ^{190}Pt is shown in Fig. 1. The new high-spin (I=20^-) rotational band at 4813 keV feeds the new 15^-, 17^- and 18^- levels which have most probably a four quasiparticle structure, the negative parity band built on the 7^- isomer and the ground state band. Another decay path has been established through the second positive band to the levels belonging to the negative band built upon the previously known 10^- isomer at 2297 keV [2]. A recent measurement of the g_k-factor of this 10^- isomeric state [?] confirms its structure as resulting from the coupling of two neutrons from the $9/2^-[505]$ x $11/2^+[615]$ orbitals [2]. The high-spin band built on the 20^- level has probably the configuration ν $[i_{13/2}h_{9/2}] \otimes \nu\ i^2_{13/2}$ or ν $[i_{13/2}h_{9/2}] \otimes \pi\ h^2_{9/2}$.

The TRS calculations [4] show that the nucleus in its ground state is very γ-soft. It has a triaxial minimum at γ -90° for the isomer 10^- and at γ -70° for the 20^- high-spin band. It seems that the nucleus keeps its triaxial shape but a pronounced minimum appears. The hindrance in the isomeric decay might be due to a change in the γ-softness of the nucleus.

Acknowledgments

One of the authors (A.M.) is indebted to the French MRT for financial support.

References

[1] C.Schück *et al.*, Phys. Rev. C 56, R1667 (1997).
[2] S.Hjorth *et al.*, Nucl.Phys. A262, 328 (1976).
[3] J.C.Cunnane *et al.*, Phys. Rev. C, 13, 2197 (1976).
[4] R.Wyss *et al.*, Nucl. Phys., A511, 324 (1990)

PROJECTILE FRAGMENTATION STUDIES WITH THE MOST EXOTIC NUCLEI

David J. Morrissey

National Superconducting Cyclotron Laboratory
Michigan State University, East Lansing, Michigan 48824
United States*

morrissey@nscl.msu.edu

Abstract: The A1200 projectile fragment separator has been operated at the NSCL for more than seven years. This device has delivered exotic nuclei from the fragmentation of a large variety of projectiles that extends out to the limits of stability. The complexity of the measurements with the most exotic nuclei has evolved during this period from simple identification and decay studies to sophisticated studies of secondary nuclear reactions. Recent examples that will be presented include the coulomb excitation of neutron-rich nuclei in the sulfur-argon region with comparative studies of the (p,p') reaction with the same nuclei and a study of the (d,n) reaction of ^{7}Be to form the ground state of ^{8}B performed in inverse kinematics as a prototype for studying direct reactions. The results of all of these reactions were compared with a number of DWBA analyses and indicate the need for more theoretical work. The A1200 will be de-commissioned in July, 1999 and will be replaced with a new device with a much larger acceptance and higher resolution. The progress of the up-grade project is also very briefly summarized.

ionIntroduction

The present experimental program at the NSCL is based on ECR injection of the K1200 cyclotron for the production of a wide variety of radioactive nuclear beams (RNB). Radioactive ion beam experiments have made up more than 50% of the program since the initial operation of the A1200 separator and now are in excess of 70%. Such a large fraction has placed increased demands

*Funding provided by grant NSF PHY-95-28844.

on the accelerator for higher intensities of unusual, stable ions. For example, heavy-ion beams of Ag, Cd, Fe, Mo, and many other elements have been used. The sophistication of the RNB experiments has also changed dramatically over the last few years. The early experiments with beams from the A1200 concentrated on the nuclear structure information available from searches for the limits to stability and beta-decay properties of exotic nuclei. Whereas, present day experiments are, more commonly, detailed studies of nuclear structure by direct reactions in reverse kinematics.

The operating principles of in-flight separation of radioactive ions [1] and present generation fragment separators [2, 3] have been described in recent review articles. Briefly, the projectile fragmentation technique converts a primary, heavy-ion beam into secondary, exotic, heavy-ion beams with a transmission target that is thin enough that the ions retain most of their initial velocity. The mixture of unreacted primary and secondary ions are first filtered to select a single magnetic rigidity, $B\rho = mv/q$, by a dispersive beam line in conjunction with an aperture. Isotopic selection is completed by passing the ions through an energy degrading "wedge" from which ions entering with a single $B\rho$ but with different atomic numbers emerge with different momenta. A second dispersive beamline then provides, in most cases, isotopic separation. The nature and thickness of the production target and the energy degrader, as well as the sizes of momentum apertures, are parameters that are adjusted to control the secondary beam intensity and purity.

The most important features of secondary beams from projectile fragmentation facilities are that they are not limited by chemical selectivity and are provided within a few microseconds (or less) of their production. Thus, projectile fragments are routinely available, albeit some at low rates, out to the limits of stability. This technique generally provides 'very fast' ion beams but has been used to provide secondary beams with E/A as low as $\sim$25 MeV. The emittances of the beams are large, $\sim 20\pi$ mm-mr, and the present generation of experiments rely on tracking the incident beam particles onto the target or collimating the secondary beam if the intensity permits. Worldwide, five laboratories are presently operating fragmentation facilities: GANIL (France), GSI (Germany), RIKEN (Japan), the COMBAS separator at Dubna, and the A1200 at the NSCL. The A1200 at the NSCL and the FRS at GSI are placed at the beginning of the beam transport system giving them the ability to send secondary beams to any experimental area [4]. The importance of this special feature can be seen in the fact that radioactive ion beams were used in all of the experimental vaults at the NSCL within approximately eighteen months of initial operation.

The results of two very recent studies using exotic nuclei from the A1200 separator are described in the following sections. The present generation of experiments are characterized by kinematic coincidence measurements between the scattered exotic beams and the other reaction product. As examples, the coulomb excitation and inelastic proton scattering of neutron-rich sulfur and argon isotopes is described and the proton stripping reaction with ^{7}Be.

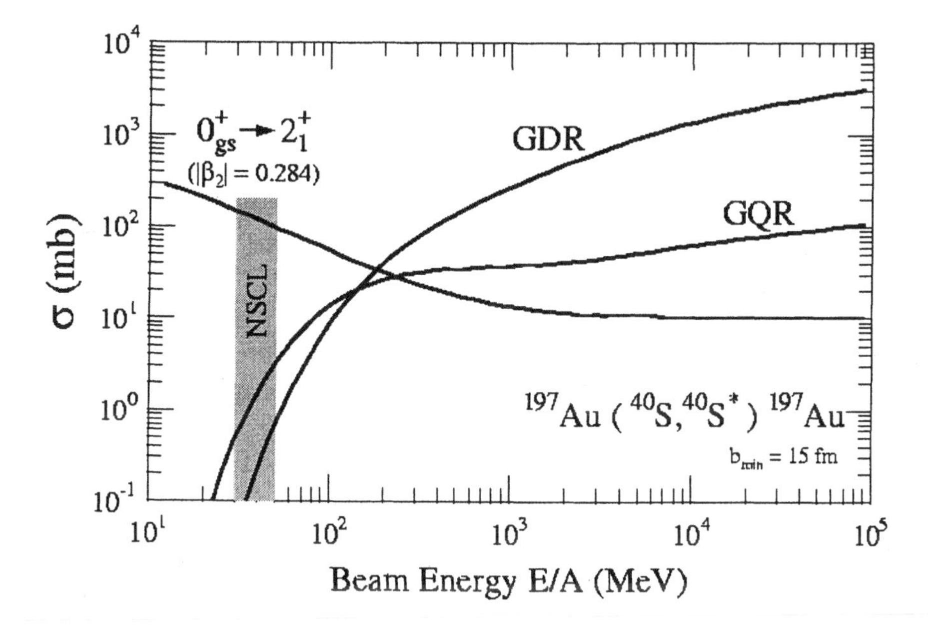

Figure 1 The cross sections for the Coulomb excitation of ^{40}S projectiles by a gold target are shown as function of bombarding energy per nucleon. The excitation of the first 2^+ state gradually falls with energy whereas the high-lying giant dipole and giant quadrupole resonances grow (GDR and GQR, respectively). The energy range of the NSCL is shown by the band.

1 NEUTRON-RICH SULFUR AND ARGON ISOTOPES

Coulomb excitation of intermediate energy (~50 MeV/A) heavy-ions has been shown to be a very powerful technique for studying the evolution of nuclear structure. Glasmacher[5] has recently reviewed the theoretical and experimental bases of the technique. It turns out that the high velocity of the projectiles limits the excitation to the first excited state, see figure1 from [6]. One of the strongest features of the technique is that the first excited states of a series of exotic nuclei can be readily measured at one time. The level structure and the transition matrix elements by which these levels are coupled can provide strong information on the shape and general properties of the nuclei. The energy of the transition between the first excited and ground states is obtained directly from the measurements and the transition matrix element can be extracted from the cross section.

One of the intriguing results from recent studies of the most exotic nuclei is the indication that the shell closures that are so evident among the stable nuclei seem to weaken for the most neutron-rich nuclei. This evidence comes from the (inferred) lack of stability of ^{28}O[7, 8] and studies of the first excited

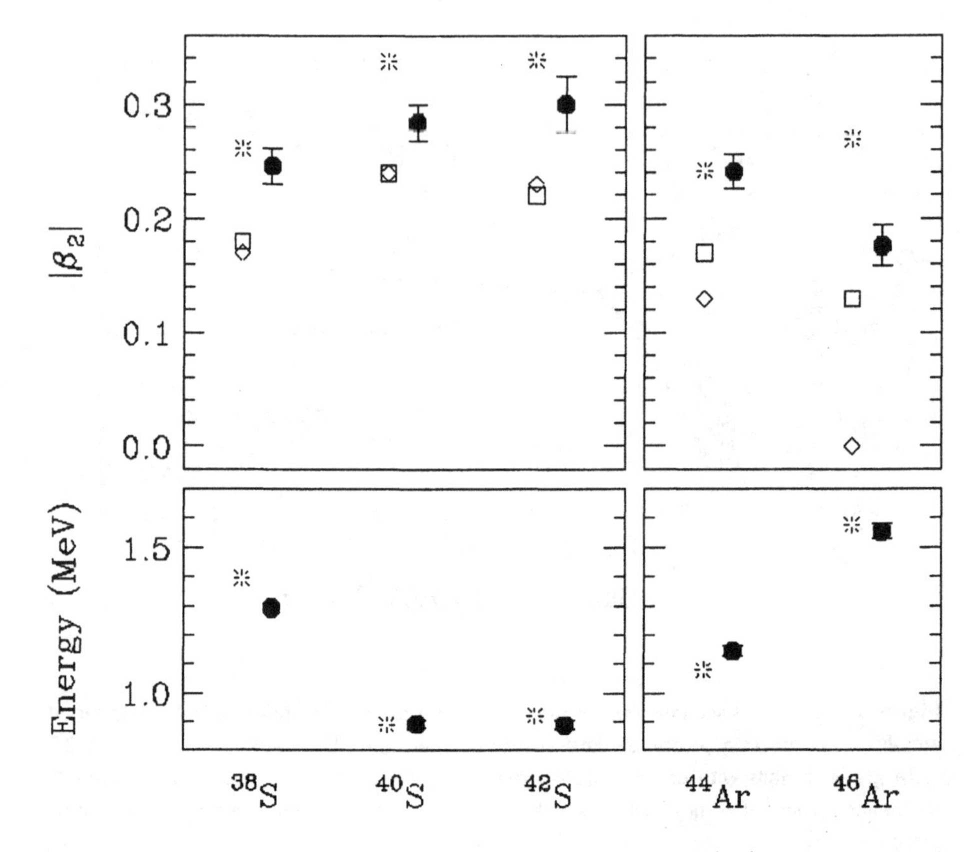

Figure 2 The experimental quadrupole deformation parameters $|\beta_2|$ (solid points) are compared in the top panels to the results of calculations with a shell model (stars), a relativistic mean field model (open diamonds), and a Hartree-Fock model (open squares). The experimental excitation energies are contained in the lower panels. All from the work of Scheit[6].

states by coulomb excitation[9, 10, 11]. A series of measurements of neutron-rich, even-even argon and sulfur nuclei was performed to systematically study the evolution of shell-strength in the Z$\leq$20 region by Glasmacher, et al. The same experimental setup based on a large array of position sensitive NaI(Tl) detectors was used for all the measurements in order to provide a uniform survey. Known transitions in reference nuclei, e.g., ^{40}Ar are used to provide a check on the new measurements. In this case, 38 NaI(Tl) detectors each 18 cm long and 5.75 cm in diameter were positioned around the coulomb excitation target in three concentric rings, parallel to the beam direction. The energies and interaction points of the photons were measured in coincidence with the scattered secondary ions. The ions themselves were identified, event-by-event, in a telescope at zero degrees to insure that only coulomb excitation had taken place in the target. The power of this technique can be appreciated from the results for heavy argon and sulfur isotopes shown in figure2.

An important goal of studies of exotic neutron-rich nuclei is to develop tools that are sensitive to the neutron density distribution. The coulomb excitation work just described is sensitive to the proton density distribution. If we consider nuclear excitations and nucleon scattering in particular, the Pauli principle makes the like-nucleon interaction about three times weaker than the interaction between unlike nucleons. Therefore, simple proton scattering processes, elastic and inelastic, should be more sensitive to the neutron distribution than the proton distribution. Moreover, comparisons of the excitation of low-lying states by proton scattering with coulomb excitation should provide some insight to differences in the proton and neutron distributions as pointed out by Suomijärvi[12].

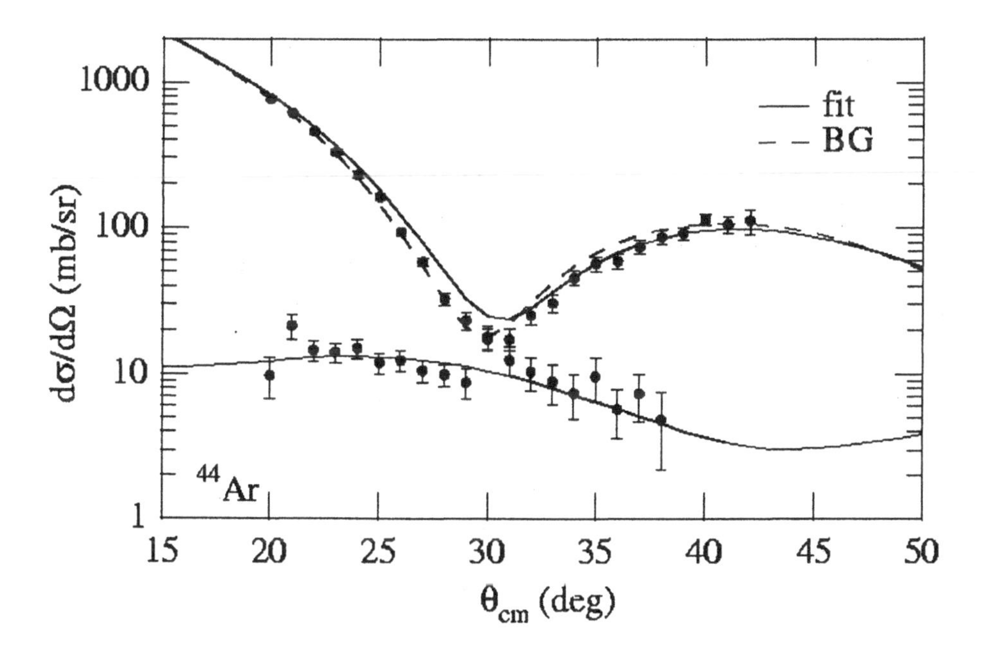

Figure 3 The measured cross section for elastic (upper curves and points) and inelastic (lower) proton scattering by ^{44}Ar. The curves are the results of DWBA calculations using the Becchetti and Greenlees global potentials (dashes) and a fit to the elastic data (solid).

A collaboration between IPN-Orsay, Florida State University and Michigan State University has carried out a series of proton scattering reactions in reverse kinematics with radioactive ion beams. The initial studies were performed with ^{38}S but have been extended to the heavy argon isotopes[6]. The experimental setup is similar to that used for coulomb excitation in that the projectile must be identified near zero degrees in coincidence with the scattered protons and the scattering angle of the proton must be accurately measured. A set of detector telescopes was placed near 90° in the laboratory system that used resistive silicon-strips to localize the scattered proton. Typical results for the elastic and inelastic scattering are shown in figure3. The data are well represented by the DWBA calculations made with the ECIS code using the global optical model potentials developed by Becchetti and Greenlees[13] for the reactions of stable nuclei. A somewhat better fit to the data can be obtained by adjusting the parameters as shown by the solid curve. The value of the $|\beta_2|$ that can be extracted from the inelastic scattering data is relatively insensitive to the optical model parameters. The results indicate that the deformation parameters obtained with the nuclear excitation are systematically larger than those obtained with the electromagnetic excitation. Detailed measurements such as these for a variety of nuclei extending away from the valley of stability are and will present a challenge to theoretical models.

2 THE d(^{7}Be,^{8}B)n REACTION

The measurement of the cross sections for single-nucleon transfer reactions is a technique that has been used to probe nuclear structure for many years. The ^{8}B nucleus has been the subject of much study recently, in particular, several groups have tried to determine whether or not ^{8}B exhibits a proton halo[14], and references therein. These studies have left the situation ambiguous and with the uncertainty they bring to the underlying structure of ^{8}B, we performed a different experimental study: the (d,n) single nucleon transfer reaction. The short half-life for ^{7}Be (53 days) makes the traditional approach of studying the valence proton state of ^{8}B via direct reactions with a ^{7}Be target very difficult. However, the availability of energetic beams of exotic nuclei allow this reaction to be studied in reverse kinematics. A similar study, but at an extremely low bombarding energy, has been reported[15].

It is interesting to note that the d(^{7}Be,^{8}B)n reaction is also important from a very different standpoint. Xu, et al. have shown that the asymptotic normalization coefficient (ANC) for the overlap of the ^{7}Be + p in ^{8}B extracted from an analysis of the transfer reaction can be used to extract the S factor for the proton capture reaction on beryllium[16, 17]. This ^{7}Be(p,γ) reaction has itself been the subject of intense study as a possible solution to the solar neutrino puzzle. However, it is difficult to obtain a very accurate determination of the overlap from studies of direct reactions such as this due to the large number of model parameters.

The A1200 fragment separator provided a secondary beam of ^{7}Be nuclei (25 MeV/nucleon) from the reaction products from the interaction of a ^{12}C beam (60 MeV/nucleon) in a 587 mg/cm^2 beryllium target using a 425 mg/cm^2 aluminum wedge. The intensity of the ^{7}Be beam was carefully monitored and was approximately 335,000 particles per second. The secondary target was an 18 mg/cm^2 deuterated polyethylene foil. The contribution from reactions on the carbon nuclei were measured with a 31 mg/cm^2 carbon foil and subtracted from the data.

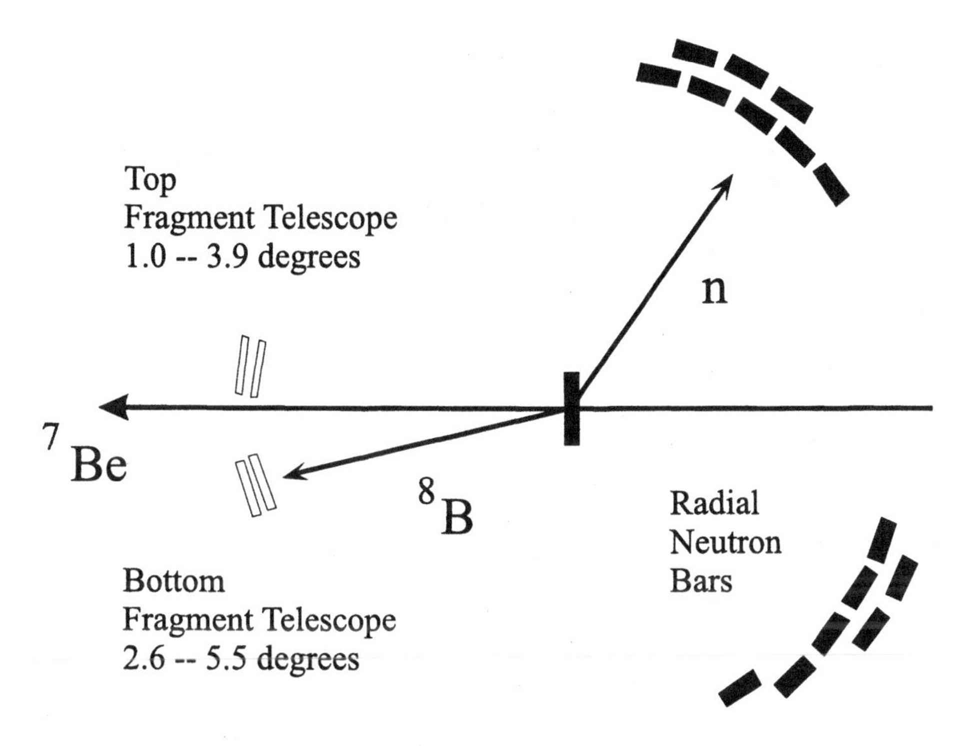

Figure 4 Schematic diagram of the detector system. The silicon telescopes were positioned above and below the plane of the beam and each neutron detector subtended an arc 90° by 4.4° at a 1 meter radius.

The scattered charged particles were identified and measured in a pair of silicon telescopes placed above and below the beam. The telescopes consisted of a double-sided silicon strip detector (5cm x 5cm x 300 or 500μm, 16x16 strips) and a silicon PIN diode detector (5cm x 5cm x 500μm) placed at lab angles that ranged from either 1.0° to 3.9° or 2.6° to 5.5°. Neutrons in kinematic coincidence with the scattered particles were detected in the NSCL neutron time-of-flight array[18]. The neutron array was calibrated by observing the beta-delayed neutrons from a ^{17}N source produced by the A1200 and implanted into a beta-detector at the target position[19]. A schematic layout of the two types of detectors is shown in Figure 4. The overall response of the detector system was checked by measuring the d(^{12}C,^{13}N)n reaction at 22 MeV/nucleon under identical, i.e., low intensity, conditions. These results are not presented here due to constraints on space.

The cross section for quasi-elastic scattering was obtained by grouping the charged particles identified in each of the 256 pixels in each silicon telescope according to laboratory angle. Note that the (only bound) excited state in this system could not be resolved from the ground state by the silicon detectors. The cross section for the transfer reaction was obtained by a similar grouping of the single-particle inclusive data from the silicon telescopes and by the neutron coincidence data. The resulting angular distributions for both reactions are shown in figure 5, the error bars indicate the combined uncertainties (ordinate) and the angular bin width (abscissa).

Note that the energy spread of the incident beam does not allow the separation of inelastic scattering of ^{7}Be from elastic scattering. Moreover, the deuteron was not detected in coincidence with the scattered ^{7}Be nuclei and its fate was unknown. That is, the deuteron might have been excited into the continuum and broken-up. Such undetected breakup processes are a potential difficulty in reverse kinematics studies with deuterium. As small number of coincidences between neutrons from the breakup of deuterium and quasielastic ^{7}Be nuclei were observed although the neutron array was not positioned optimally. The breakup cross section corresponded to 73±25 mb at an angle of 17° and thus makes some contribution to the largest angle data in figure 5.

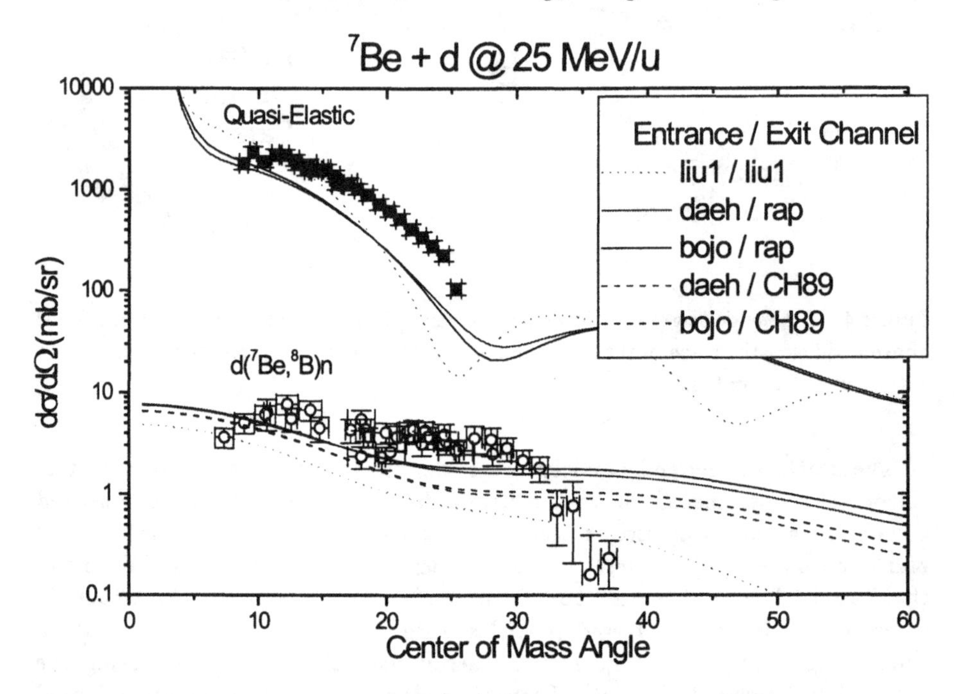

Figure 5 Results for the quasielastic scattering and transfer reaction are compared to DWBA calculations with several optical potentials from the literature, see the text for details.

The measured angular distributions are compared with the predictions of finite-range DWBA calculations performed with the FRESCO [20] coupled-channels computer code in figure 5. All calculations were performed with the deuteron bound by a gaussian potential, and ^{8}B bound by the potential of Tombrello [21], a Woods-Saxon with V_0= 32.62 MeV, r_0=1.25 fm, a=0.65 fm, giving an ANC value of 0.764. The calculations were performed using several different sets of optical parameters from the literature. Global deuteron parameters from Daehnick et al. [22] were used for the ^{7}Be + d entrance channel, while the global parameters from Rapaport [23] and the CH89 set [24] are used for the ^{8}B + n exit channel. The entrance channel parameter set labeled as CFP1 was constructed from the ^{8}B real potential described by Tombrello and the ^{9}Be imaginary potential described by Mani et al.[25]. For completeness, calculations using the parameter sets determined by Liu et al. [15, 17] for the same reaction at a much lower energy are also shown. The difficulty of performing a traditional DWBA analysis on reactions with radioactive beams should be apparent. For example, there is little hope to be able to measure the exit channel optical model potential for ^{8}B and a neutron. Although reasonable agreement was obtained with one parameter set, further analysis is necessary. In addition, the analysis was found to be more sensitive to the parameters of the optical model potentials than to the structure of ^{8}B.

3 FACILITY UPGRADE

The National Superconducting Cyclotron Laboratory (NSCL) at Michigan State University is implementing a plan to reconfigure the existing K500 and K1200 superconducting cyclotrons from independent operation into coupled operation in order to address the need for very heavy ion beams and very intense unusual beams for secondary ion production in a cost-effective way. The NSCL has refurbished the K500 cyclotron to inject intense beams of low-charge heavy ions radially into the K1200 cyclotron. These beams will be stripped and accelerated in the larger machine. The existing fragment separator at the beginning of the beamlines (the A1200) will be completely replaced with a substantially improved device (the A1900). A schematic diagram of the new floor-plan is shown in figure6, the refurbished cyclotrons and the new separator will match on to the existing experimental vaults.

The heart of any radioactive beam program is the device that separates and analyzes the reaction products and provides the secondary radioactive beams. The new fragment separator will have a collection efficiency approaching 100% for fragmentation products compared to the present 2-4% for the A1200 [26]. The improved magnet layout will allow for better radiation shielding, and improved optics will provide higher purity secondary beams. The parameters of the separator are indicated in Table 1

At the time of the conference the K500 has been fully restored and had been operated and tested with the existing ECR ion sources for approximately six months. The performance of the refurbished machine was substantially improved and comparable to that of the K1200 machine. The construction of the magnets for the A1900 fragment separator is well underway. Eight superconducting quadrupole triplets and one spare are in production. The first complete triplet magnet assembly and cryostat are complete, all three magnets have been operated and are being mapped. The construction of the four large 45°-dipoles is also underway. Five bobbin assemblies (one spare) have been completed and the cryostats are being welded. The experimental program will continue with the existing facility (K1200/A1200) until approximately July/1999. Following

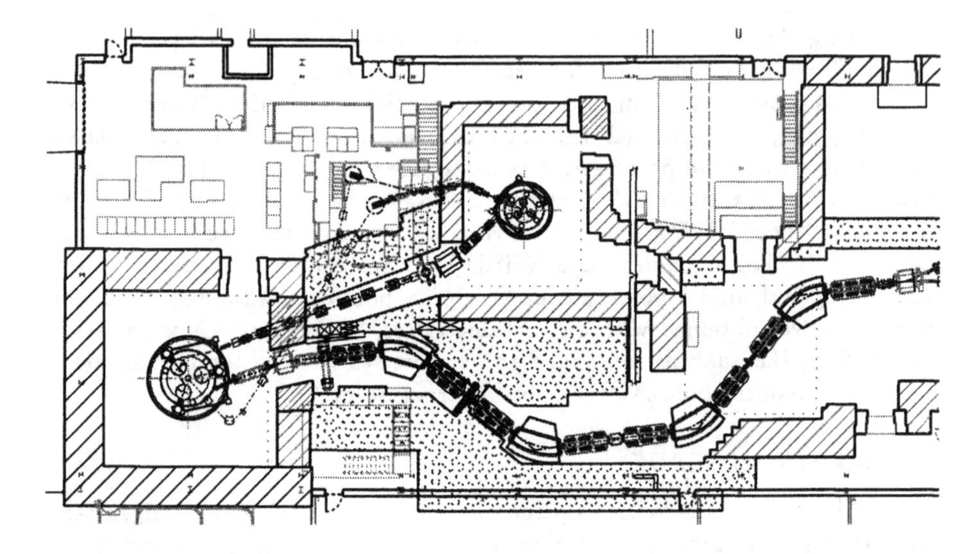

Figure 6 A schematic diagram of the coupled cyclotron floor-plan under construction at the NSCL. The beam will travel from the smaller cyclotron near the center of the figure into the larger cyclotron and then into the new separator.

Table 1 Ion-optical parameters of fragment separators.

Parameter	A1900	A1200†	RIPS‡
Solid Angle (msr)	8	0.8	5
Momentum Acceptance (%)	~5.5	3	6
Resolving Power*	2915	2400	1500
Intermediate Image§			
Magnification M_x	2.04	0.7	-1.6
Magnification M_y	0.75	3.0	-5.7
Dispersion (cm/%)	5.95	1.67	2.4
Length (m)	35	22	21
Maximum Gradient (kG/cm)	2.2	3.5	1.41
Dipole Full Vertical Gap (cm)	9	5	14
Relative Acceptance	18	1	12.5

†–Symmetric medium acceptance optical mode [4]
‡–As described in Ref. [27]
*–Resolving Power = Dispersion/X_0M_x, where X_0 is the beam spot size at the target (assumed to be 1 mm) and M_x is the magnification.
§–First order, at the wedge position.

the completion of the experimental work, the facility will be reconfigured during an eighteen month shutdown and a six month commissioning period.

Acknowledgments

The author wishes to acknowledge a long and productive collaboration with Dr. B.M. Sherrill that has supported most of this work. The author would especially like to acknowledge the work of Drs. T. Glasmacher and H. Scheit on Coulomb excitation that was presented here. The coupled cyclotron project and construction of the A1900 separator is the work of the entire NSCL staff and is supported by the National Science Foundation under the grant PHY-95-28844. Special recognition should be given to Dr. C.K. Gelbke, the laboratory director, and to Dr. R.C. York, the project leader.

References

[1] D.J. Morrissey and B.M. Sherrill, Phil. Trans. R. Soc. Lond. A356 (1998) 1985.

[2] B.M. Sherrill, Proc. 2nd Intl. Conf. Radioactive Nuclear Beams, Louvain-la-Neuve, 1991, Th. Delbar, Ed., (IOP Publishing, London, 1991) 3.

[3] G. Münzenberg, Nucl. Instrum. Meth. B70 (1992) 265.

[4] B.M. Sherrill, D.J. Morrissey, J.A. Nolen, Jr. and J.A. Winger, Nucl. Instrum. Meth. B56/57 (1993) 1106.

[5] T. Glasmacher, *Annu. Rev. Nucl. Part. Sci.* 48 (1998) 1.

[6] H. Scheit, Ph. D. Thesis, Michigan State University, (1998).

[7] M. Fauerbach, et al. *Phys. Rev.* C 53 (1996) 647.

[8] O. Tarasov, et al. *Phys. Lett.* B409 (1997) 64.

[9] H. Scheit, et al. *Phys. Rev. Lett.* 77 (1996) 3967.

[10] T.Glasmacher, et al. *Phys. Lett.* B395 (1997) 163.

[11] R.W. Ibbotson, et al. *Phys. Rev. Lett.* 80 (1998) 2081.

[12] T. Suomijärvi, et al., Nucl. Phys. A616 (1997) 295c, and J.H. Kelley, et al., Phys. Rev. C56, R1206 (1997).

[13] F.D. Becchetti, Jr. and G.W. Greenlees, Phys. Rev. 182 (1969) 1190.

[14] B.S. Davids, et al. *Phys. Rev. Lett.* 81 (1998) 2029.

[15] Weiping Liu, et al. *Phys. Rev. Lett.* 77 (1996) 611.

[16] H.M. Xu, et al. *Phys. Rev. Lett.* 73 (1994) 2027.

[17] C.A. Gagliardi, et al., *TAMU Prog. in Res.* (1997) I17.

[18] R. Harkewicz, et al., *Phys. Rev.* C44 (1991) 2365.

[19] K.W. Scheller, et al., *Phys. Rev.* C 49 (1994) 46.

[20] I.J. Thompson, *Comp. Phys. Reps.* 7, (1988) 167.

[21] T.A. Tombrello, *Nvcl. Phys.* 71 (1965) 459.

[22] W.W. Daehnick, et al., *Phys. Rev.* C 21 (1980) 2253.

[23] J. Rapaport, *Phys. Reps.* 87 (1982) 25.

[24] R.L. Varner, *Phys. Reps.* 201 (1991) 57.

[25] G.S. Mani, et al., *Nucl. Phys.* A 165 (1971) 145.

[26] Morrissey, DJ, and the NSCL Staff, 1997 *Nucl. Instrum. Meth.* **B126** 316.

[27] T. Kubo, et al., Proc. 1st Intl. Conf. Radioactive Nuclear Beams, Berkeley, Ca, 1989, W.D. Myers, J.M. Nitschke, and E.B. Norman, Eds., (World Scientific, Singapore, 1993) 563.

SYSTEMATICS OF LOW-LYING DIPOLE STRENGTH IN HEAVY NUCLEI FROM NUCLEAR RESONANCE FLUORESCENCE EXPERIMENTS

A. Nord[1], F. Bauwens[2], O. Beck[1], D. Belic[1], P. von Brentano[3], J. Bryssinck[2], D. De Frenne[2], T. Eckert[1], C. Fransen[3], K. Govaert[2], L. Govor[4], R.-D. Herzberg[3], E. Jacobs[2], U. Kneissl[1], H. Maser[1], N. Pietralla[3], H.H. Pitz[1], V. Yu. Ponomarev[5], V. Werner[3]

[1] Institut für Strahlenphysik, Universität Stuttgart, 70569 Stuttgart, Germany
[2] Vakgroep Subatomaire en Stralingsfysica, 9000 Gent, Belgium
[3] Institut für Kernphysik, Universität zu Köln, 50937 Köln, Germany
[4]Russian Scientific Centre "Kurchatov Institute", Moscow, Russia
[5]Bogoliubov Laboratory of Theor. Physics, Joint Inst. of Nucl. Research, Dubna, Russia

Abstract: Low-lying dipole excitations in heavy nuclei were systematically investigated in nuclear resonance fluorescence (NRF) experiments at the bremsstrahlung facility installed at the Stuttgart 4.3 MeV DYNAMITRON accelerator. The systematics of the M1 *Scissors Mode* in deformed odd-mass nuclei is presented. New results are reported on strong E1 excitations in spherical semi-magic Z=50 isotopes.

The Nucleus: New Physics for the New Millennium
Edited by Smit et al., Kluwer Academic / Plenum Publishers, New York, 2000.

1 INTRODUCTION

Low-lying dipole excitations in heavy nuclei met with an increased interest in recent years. The prediction [1] and the subsequent discovery [2] of the orbital $M1$ *Scissors Mode* in deformed nuclei in 1984 stimulated a large number of both experimental and theoretical work (for references see, e.g., recent reviews [3, 4, 5]). A topic of current interest is the study of the fragmentation of these dipole mode in the neighbouring odd-mass nuclei [6]. On the other hand, also enhanced electric dipole excitations ($E1$) were expected and observed in heavy nuclei. In semi-magic spherical nuclei like the $N = 82$ isotones (see [7, 8, 9, 10, 11] and refs. therein) or the even-even Sn-isotopes [12, 13] the corresponding 1^- states were interpreted as the spin 1 member of the $1^-, 2^-, \ldots, 5^-$ quintuplet due to a two-phonon coupling of the quadrupole and octupole phonons. NRF is an outstanding tool to investigate these dipole modes providing model independent information on excitation energies, lifetimes, reduced transition probabilities, spins and parities.

2 EXPERIMENTAL TECHNIQUES

The NRF measurements reported on have been performed at the bremsstrahlung facility of the Stuttgart 4 MV DYNAMITRON accelerator [3]. The high current DC electron beam (maximum current $4mA$; typical current $0.8\ mA$) is bent by $120°$ and focused on the bremsstrahlung radiator target. The excellent quality of the bremsstrahlung beam enables one to run NRF experiments at two different setups simultaneously. At the first NRF site the scattered photons are detected now by three carefully shielded Ge(HP)-γ-spectrometers (efficiencies of 100% relative to a 3"×3" NaI(Tl) detector) placed at scattering angles of $90°$, $127°$, and $150°$ with respect to the incident beam. At the second site two sectored single crystal Ge–Compton polarimeters [14] are installed at slightly backward angles of $\approx 95°$ with respect to the photon beam. These detectors measure the linear polarization of the resonantly scattered photons providing the parity information.

3 M1 EXCITATIONS

The *Scissors Mode* in odd-mass nuclei was detected first in the Dy isotopes ^{163,161}Dy [15, 16] where a concentration of dipole strength was observed fitting into the systematics of the *Scissors Mode* in the neighbouring even-even isotopes ^{164,162,160}Dy [16, 17]. However, first investigations of the odd-mass Gd nuclei ^{157,155}Gd surprisingly revealed an extreme fragmentation of the low-lying dipole strength into about 100 transitions in the energy range 2–4 MeV [6, 16]. Further NRF studies including the odd-proton nuclei ^{159}Tb and ^{153,151}Eu established a systematics showing an increasing fragmentation and strong reduction of the detected strength with decreasing mass number A. The Darmstadt group could recently demonstrate by a statistical fluctuation analysis of the NRF spectra [18, 19] that the missing strength in the experimentally observed strength distributions in these odd-mass nuclei is hidden in the continuous background,

due to an extreme fragmentation of the *Scissors Mode*.
The spectrum of our latest experiment on ^{165}Ho is shown in figure 1. Figure

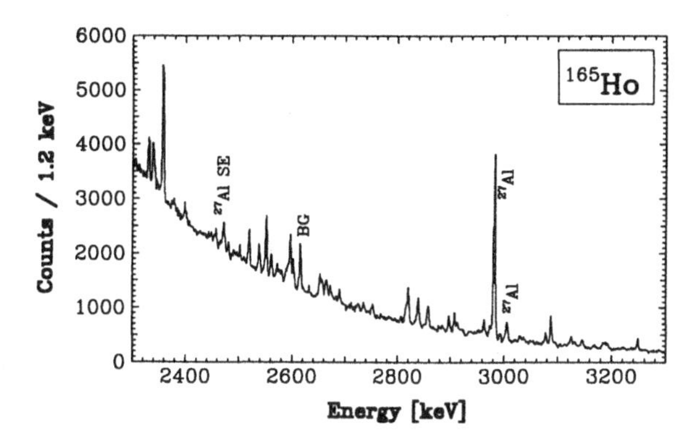

Figure 1 Photon scattering spectrum of ^{165}Ho. Peaks marked by "Al" correspond to photon flux calibration lines.

2 shows the resulting dipole strength distribution together with the complete systematics of odd-A isotopes. Plotted is the spin-weighted reduced ground state transition width $g\Gamma_0^{red} = (2J+1)/(2J_0+1)\cdot\Gamma_0/E_\gamma^3$ as a function of the excitation energy. This quantity is directly proportional to the reduced excitation

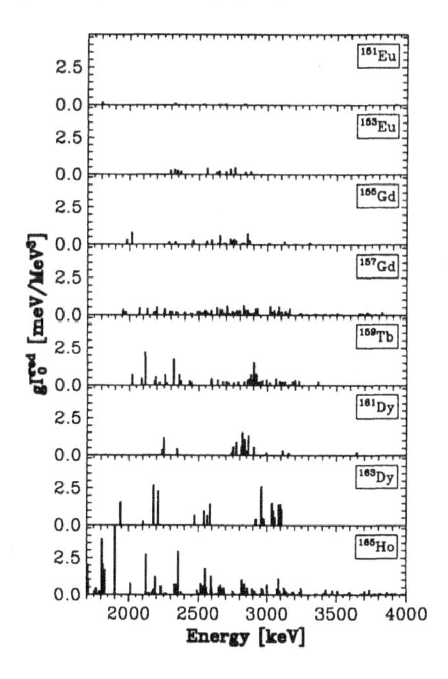

Figure 2 Dipole strength distributions in odd-mass rare earth nuclei as measured in NRF experiments at the Stuttgart DYNAMITRON (see text).

probabilities $B(M1)\uparrow$ and $B(E1)\uparrow$, respectively. The observed strong fragmentation without any concentration of strength in ^{165}Ho is not expected from the systematics and up to now is not understood. Moreover, the total strength in the energy range between 2-4 MeV amounts to about 36 meV/MeV^3 which is as much as observed in the neighboring even-even isotopes and hence much more then ever found in an odd-mass nucleus.

4 E1 EXCITATIONS

In semi-magic vibrational nuclei, coupling between the quadrupole and octupole phonons is expected to produce a quintuplet of negative-parity states (1^- to 5^-). The spin 1 member of this multiplet can ideally be investigated in NRF experiments. These two-phonon $E1$ excitations are well known for the even-even $N = 82$ nuclei [7]. They have also been observed recently in the Sn isotopes (Z=50) [12, 13]. Figure 3 shows the results obtained in recent NRF experiments performed at the Stuttgart NRF-facility. In the investigated isotopes

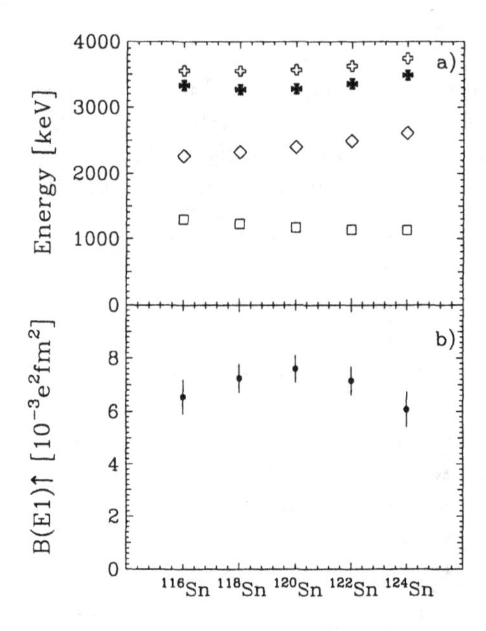

Figure 3 Upper part (a): Energies of the 2_1^+ (open squares), 3_1^- (open diamonds) one-phonon states and the 1^- (full crosses) two-phonon states in ^{116,118,120,122,124}Sn compared to the sum energies $E_x(2^+) + E_x(3^-)$ (open crosses). Lower part (b): Experimental $B(E1)\uparrow$ values for the $E1$ two-phonon excitations.

^{116,118,120,122,124}Sn all quantities are remarkably constant: The energies of the one-phonon excitations E_{2^+} and E_{3^-}, of the two-phonon excitations E_{1^-}, and the absolute excitation strengths $B(E1)\uparrow$. For two of the dipole excitations (in ^{118,122}Sn) the negative parity could be established examplarily by linear polarization measurements [20], for two others (in ^{116,124}Sn) it was determined by analyzing power measurements using linearly polarized bremssstrahlung [12]. The enhanced $E1$ strengths can be explained within the Quasiparticle Phonon Model [13] or by including 1p-1h admixtures, at the tail of the electric giant dipole resonance, into the collective two-phonon states [11, 12].

Acknowledgments

The support by the Deutsche Forschungsgemeinschaft under Contract Nos. Br 799/9-1, Kn 154/21,Kn 154/30 and the Fund for Scientific Research-Flanders is gratefully acknowledged.

References

[1] N. Lo Iudice *et al., Phys. Rev. Lett.* 41, 1532 (1978).
[2] D. Bohle *et al., Phys. Lett.* 137B, 27 (1984).
[3] U. Kneissl *et al., Prog. Part. Nucl. Phys.* 37, 349 (1996).
[4] A. Richter, *Prog. Part. Nucl. Phys.* 34, 261 (1995).
[5] D. Zawischa, *J. Phys. G: Nucl. Part. Phys.* 24, 683 (1998).
[6] A. Nord *et al., Phys. Rev.* C 54, 2287 (1996).
[7] R.-D. Herzberg *et al., Nucl. Phys.* A592, 211 (1995).
[8] T. Belgya *et al., Phys. Rev.* C 52, R2314 (1995).
[9] M. Wilhelm *et al., Phys. Rev.* C 54, R449 (1996).
[10] M. Grinberg *et al., Nucl. Phys.* A573, 231 (1994).
[11] K. Heyde *et al., Phys. Lett.* B 393, 7 (1997).
[12] K. Govaert *et al., Phys. Lett.* B 335, 113 (1994).
[13] J. Bryssinck *et al., submitted to Phys. Rev.* C.
[14] B. Schlitt *et al., Nucl. Instr. a. Meth. in Phys. Res.* A 337, 416 (1994).
[15] I. Bauske *et al., Phys. Rev. Lett.* 71, 975 (1993).
[16] J. Margraf *et al., Phys. Rev.* C 52, 2429 (1995).
[17] C. Wesselborg *et al., Phys. Lett.* 207B, 22 (1988).
[18] J. Enders *et al., Phys. Rev. Lett.* 79, 2010 (1997).
[19] J. Enders *et al., Phys. Rev.* C 57, 996 (1998).
[20] D. Belic *et al.*, Annual Report 1997, IfS Stuttgart, and to be published.

SYSTEMATICS OF TERMINATING CONFIGURATIONS IN A ≈100 Pd NUCLEI

B.M. Nyakó[1], J. Gizon[2], A. Gizon[2], J. Timár[1], L. Zolnai[1], G.E. Perez[1], Zs. Dombrádi[1], D. Sohler[1], A.J. Boston[3], D.T. Joss[3,a], E.S. Paul[3], A.T. Semple[3], N.J. O'Brien[4], C.M. Parry[4], A.V. Afanasjev[5,b], I. Ragnarsson[5]

[1] Institute of Nuclear Research of the Hungarian Academy of Sci., Debrecen, Hungary
[2] Institut des Sciences Nucléaires, IN2P3-CNRS/Université J. Fourier, Grenoble, France
[3] Oliver Lodge Laboratory, University of Liverpool, Liverpool, U.K.
[4] Department of Physics, University of York, Heslington, U.K.
[5] Department of Mathematical Physics, University of Lund, Lund, Sweden

Abstract: High-spin states of $^{100-103}$Pd have been studied using high-fold γ-ray coincidence data obtained with the EUROGAM-2 spectrometer and the ^{70}Zn(^{36}S,xn) reaction at E=130 MeV. New high-spin bands were found and the previously known ones were extended up to higher spins. Several of the observed bands are assigned as terminating configurations through comparison with Nilsson-Strutinsky cranking calculations. The terminating bands, many of which have been observed up to the predicted terminating state, are found to have systematically the same (N-shell)-(high-j) configurations in these nuclei.

1 INTRODUCTION

Experimental observations show that collective and single-particle excitation modes often compete in generating high-spin states of nuclei. Such a competition is manifested in the phenomenon of band termination: the collective rotation of a deformed nucleus is terminated, when increasing the spin along a band, due to a change in the shape of the nucleus from collective deformed to near-spherical or non-collective deformed. In this terminating state the nuclear spin is built up solely from the spin contributions of the individual nucleons and consequently it has a maximum value for a given configuration.

It was predicted [1] that in $A \approx 100$ Pd and Ru nuclei the valence-space configurations remain yrast from the low-spin (near-prolate) rotational states up to

The Nucleus: New Physics for the New Millennium
Edited by Smit et al., Kluwer Academic / Plenum Publishers, New York, 2000.

the terminating single-particle state. With the primary aim to search for such band terminations in the ^{102}Pd nucleus we carried out an experiment recently, from which ^{102}Pd was found to be the first nucleus in which both valence-space and core-excited terminating configurations were observed [2]. Subsequently, a systematic search for terminating bands was undertaken in the neighbouring nuclei produced in the same experiment. In this contribution we report on the new experimental results obtained for the collective structures in $^{100,101,103}Pd$, which are proposed to have terminating configurations on the basis of comparison with Nilsson-Strutinsky cranking calculations. The terminating bands are found to show systematically the same (N-shell)-(high-j) configurations.

2 EXPERIMENTAL TECHNIQUES AND DATA REDUCTION

High spin states in the nuclei $^{100-103}Pd$ were populated in the $^{70}Zn(^{36}S,xn\gamma)$ reaction at a bombarding energy of 130 MeV, using the beams of the Vivitron tandem accelerator at IReS, Strasbourg. Two stacked self-supporting foils of Zn, enriched to 70% in ^{70}Zn and each having a thickness of 440 $\mu g/cm^2$ were used as the target. Coincidence γ-rays were detected with the EUROGAM-2 spectrometer [3]. A total of 6×10^8 four-fold and higher-fold Compton-suppressed coincidence events were stored on magnetic tapes. These events were unpacked off-line into a *total* $E_{\gamma1} - E_{\gamma2} - E_{\gamma3}$ triples-coincidence cube, and, using the higher-fold events, into *gated* (quadruples) cubes by requiring coincidences with the strongest transitions in the yrast cascades of each nucleus. For data reduction the RADWARE spectrum analysis package [4] was used. The level schemes of the studied nuclei have been constructed on the basis of the triples and quadruples γ-ray coincidence relations, and the intensity and energy balances extracted from these coincidence cubes. In order to achieve firm spin and parity assignments for the levels, directional correlation (DCO) ratios and γ-ray linear polarizations of the more intense transitions were measured.

3 DISCUSSION OF LEVEL SCHEMES

Using the data of the above experiment, the level schemes of the studied nuclei have been considerably extended. In this section, however, only those parts of the level schemes for $^{100,101,103}Pd$ are discussed briefly which are relevant for the band termination systematics. For a detailed discussion of the ^{102}Pd level scheme the reader is referred to Ref. [2].

The partial level schemes for ^{100}Pd and ^{101}Pd, and for ^{103}Pd, as derived from the present work, are shown in Figs. 1 and 2, respectively. As for ^{100}Pd, our data confirm the level scheme obtained from a NORDBALL experiment [5], while the possibility of performing DCO as well as linear polarisation analysis (using the clover detectors of EUROGAM) enabled an unambiguous negative parity assignment for band 1. The upper part of this band has been rearranged, and extended up to spin 25 $\hbar$, compared to the previous level scheme [6].

In ^{101}Pd band 1 and band 4 were previously known [7] up to $35/2^-$ and $29/2^+$, respectively. From the present experiment they have been extended up

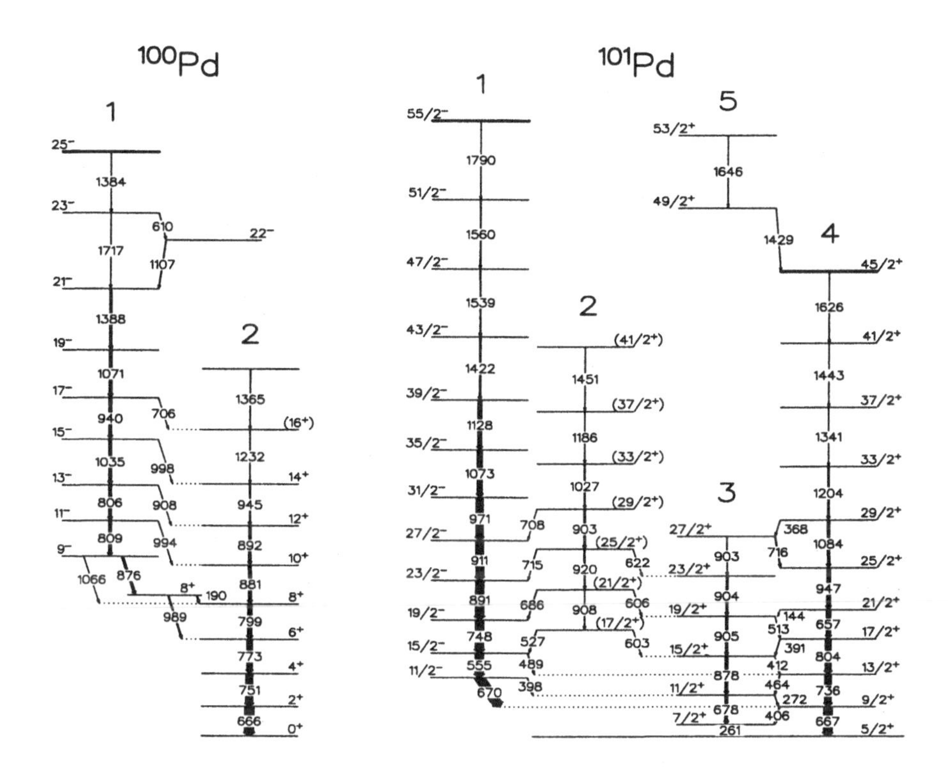

Figure 1 Partial level schemes of ^{100}Pd and ^{101}Pd obtained in the present work. Level and γ-ray energies are given in keV, the widths of the arrows are proportional to the γ-ray intensities.

to $55/2^-$ and $45/2^+$, and new bands (band 2 and band 5 which feeds to band 4 via a supposedly stretched E2 transition) have also been observed. We have found another, most probably terminating, high-spin structure, which feeds to bands 3 and 4, however, as its placement is still ambiguous, it will not be discussed here.

The level scheme of ^{103}Pd, compared to the most recent one proposed in Ref. [8], has been substantially modified. The newly established links between band 2 and the negative-parity yrast band (band 1) have resulted in new spin values for the states in band 2, which has been extended up to $I^\pi = 65/2^+$. Due to the strong inter-band M1 transitions between bands 2 and 3, new spin assignments are also given to the levels in band 3, (for details see Ref. [9]). This band extends up to $I^\pi = 59/2^+$. The negative-parity yrast band, built on the $11/2^-$ state at 785 keV, has been established up to the $I^\pi = 51/2^-$ level at E_x=11.638 MeV. Similarly to the case in ref. [8], no direct links have been found between band 7 and the lower energy bands. However, based on the changes of the level spins for bands 2 and 3, and on the characteristics of the

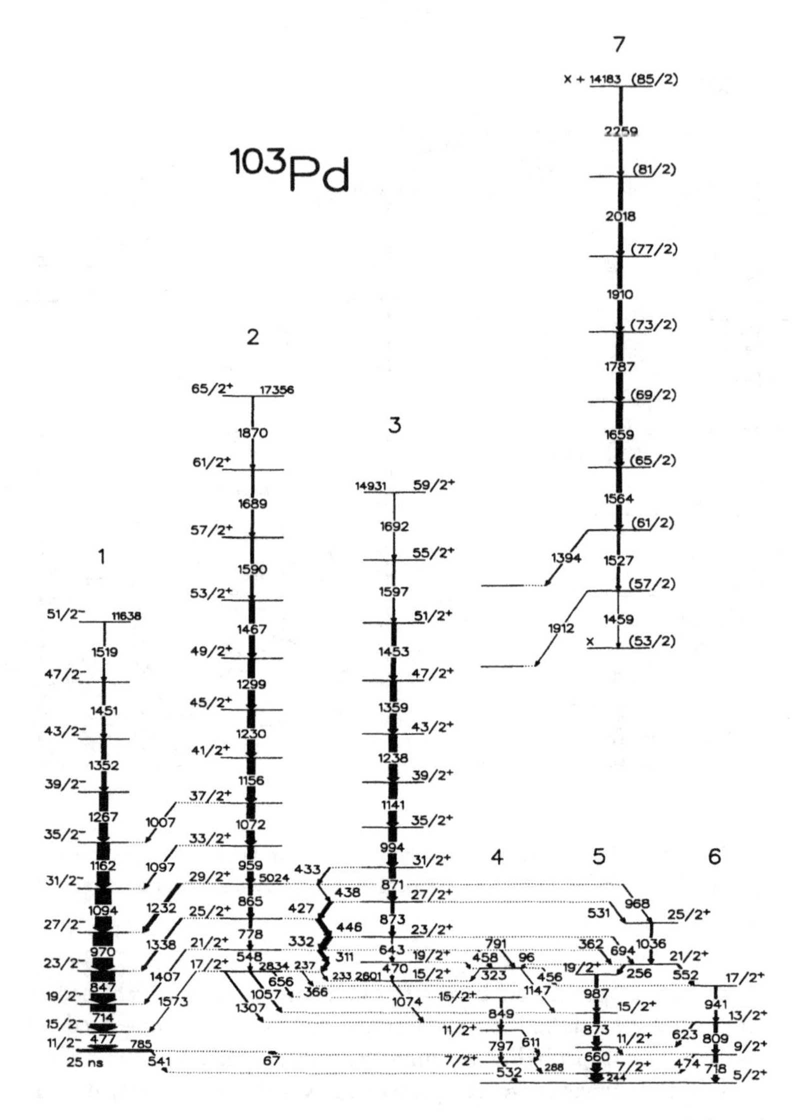

Figure 2 Partial level scheme of ^{103}Pd obtained in the present work.

decay from band 7 to these bands, we propose new, though tentative spins for band 7, as well.

4 THEORETICAL INTERPRETATION

The experimental results obtained for these Pd isotopes have been interpreted using Nilsson-Strutinsky cranking calculations. In our formalism [10] the configurations are determined by the number of particles in the N-shells of the rotating basis. An additional feature [11] is that after the diagonalisation, we identify the orbitals of dominant high-j character and thus we can also distinguish between particles in the intruder high-j shells and in the other (low-) j shells. The configurations can then be identified by the number of particles (or holes) in the different j-shells, or groups of j-shells, relative to a closed core. The energy of each configuration at each spin is minimized in the deformation space $(\varepsilon_2, \varepsilon_4, \gamma)$ which allows the development of collectivity within specific configurations to be traced as a function of spin. The pairing correlations are neglected in the calculations, which could thus be considered as realistic only at high spins, say above $\sim 15 - 20\hbar$. As the single-particle parameters are not very well known in the $A \approx 100$ region, we used the standard parameters of Ref. [10] for the Nilsson potential.

In the case of the $^{100-103}$Pd (Z=46) nuclei, the configurations are labelled, for simplicity, by the shorthand notation $[(p_0)p_1p_2, n]$, as they are defined by the number of proton holes (p_0) in the $N = 3$ shell, by the number of protons (p_1) and (p_2) in the $g_{9/2}$ and $h_{11/2}$ orbitals, respectively, and by the number of neutrons (n) in $h_{11/2}$ orbitals, relative to the ^{90}Zr core. In this notation, particles

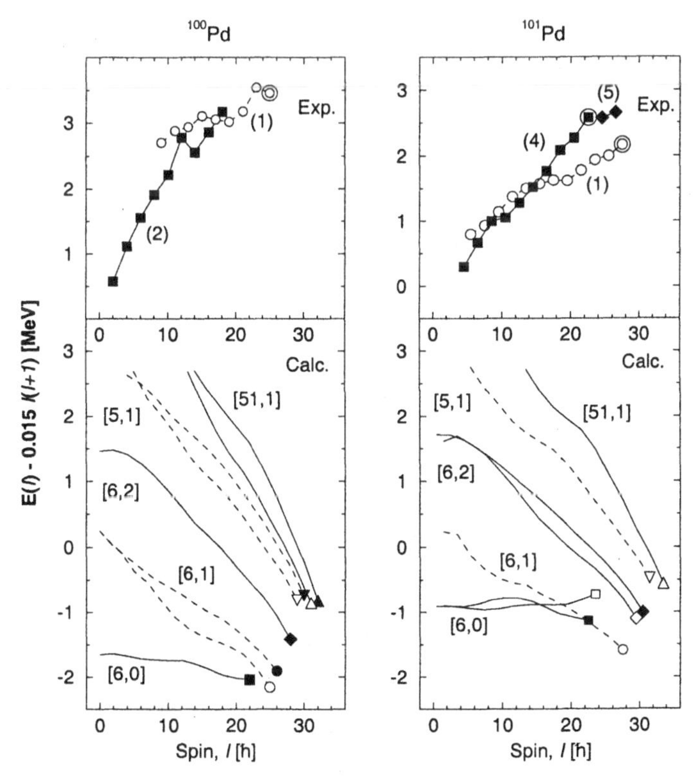

Figure 3 Comparison of experimental bands and lowest energy calculated valence-space and core-excited configurations for ^{100}Pd and ^{101}Pd. The observed bands are labelled with the numbers used in Fig. 1. Calculated curves for positive (solid) and negative (dashed) parity bands are labelled with their configurations. Terminating states are encircled (Exp.) or indicated by the symbol (Calc.) of the assigned band.

in low-j orbitals $(d_{5/2}, g_{7/2})$ are not indicated, while p_0 and p_2 are omitted, when their values are zero. In Fig. 3 the experimental energies of the observed high-spin bands for ^{100}Pd and ^{101}Pd are compared with the corresponding theoretical ones, relative, in each case, to the same rigid rotor reference. Similar comparisons for ^{102}Pd and ^{103}Pd are shown in Fig. 4 (see also Refs. [2] and [9]). The assignments of the calculated configurations to the experimental bands are based on similarities in their energies, relative positions, slopes, signatures and terminating spins with the calculated ones. Note, however, that especially for ^{100}Pd and ^{101}Pd, the agreement between calculations and experiment is only qualitative, which can be understood from the small collectivity and the low spin values which means that pairing is not negligible.

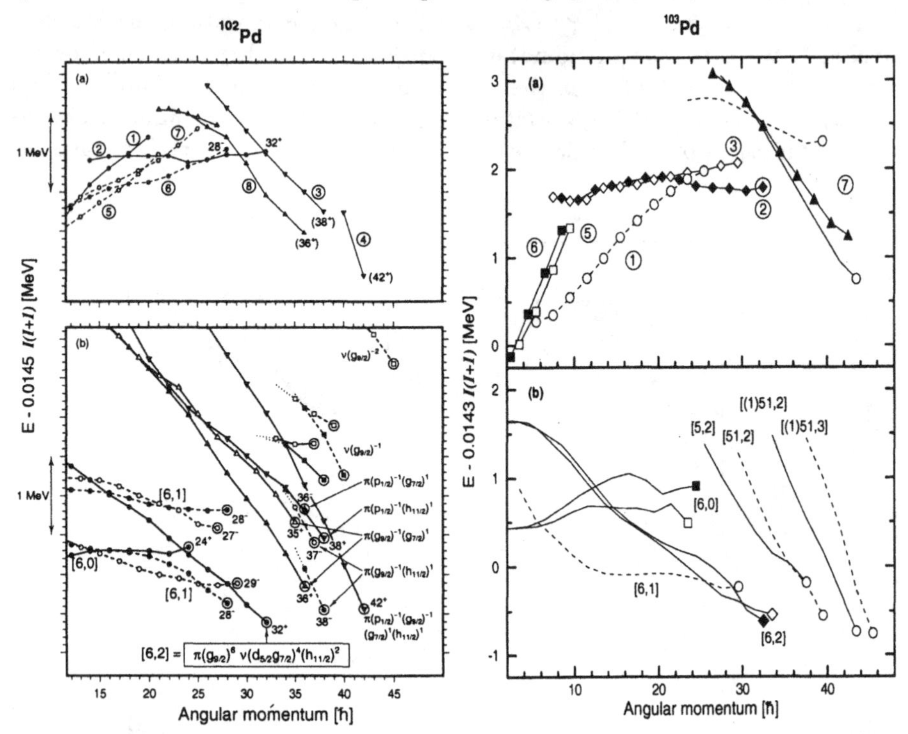

Figure 4 Comparison of the experimental bands and calculated configurations for ^{102}Pd and ^{103}Pd. See caption to Fig. 3 for the use of lines and symbols.

The following discussion of the systematics is based on the interpretation of bands 1, 2, 5 and 6 as valence-space terminating configurations, and bands 3 and 4 as core-excited configurations (see. Fig. 4), in ^{102}Pd [2].

5 SYSTEMATICS OF TERMINATING CONFIGURATIONS

Many terminating bands have been found in ^{100,101,103}Pd having systematically the same (N-shell)-(high-j) configurations as those found earlier in ^{102}Pd [2]. Only the number of $d_{5/2}g_{7/2}$ neutrons is different in the bands having the same configurations in the different nuclei. The configuration systematics is summarised in Fig. 5.

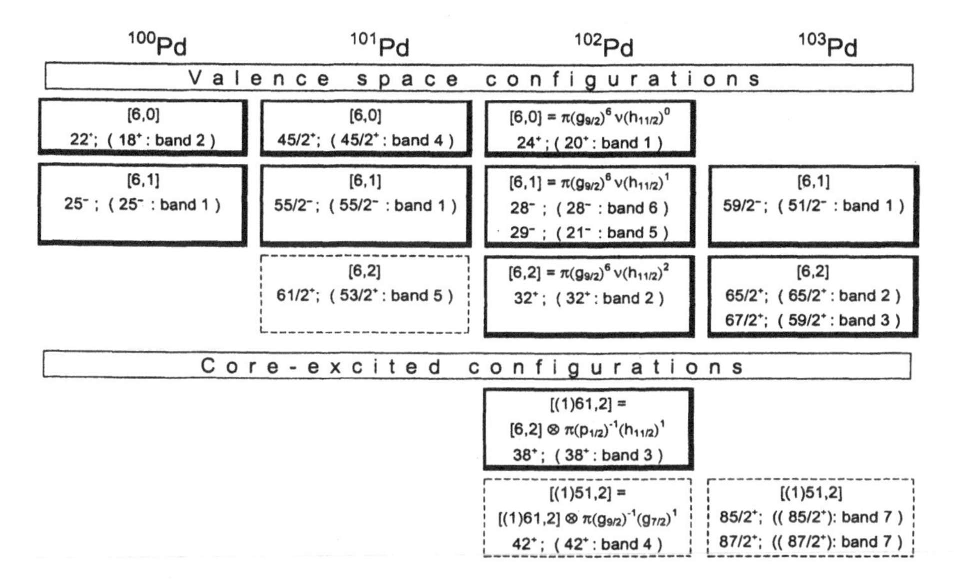

Figure 5 The systematics of terminating bands in $^{100-103}$Pd nuclei. The proposed configurations, the terminating and highest observed spins of the labelled bands are shown. Dashed blocks indicate tentative assignments

Valence-space configurations: In ^{102}Pd the yrast band (band 1) corresponds to the [6,0] configuration which contains 6 protons in the $g_{9/2}$ and 0 neutrons in the $h_{11/2}$ orbitals. The yrast bands in ^{100}Pd (band 2) and ^{101}Pd (band 4) are also proposed to have this configuration, which are the lowest ones in both nuclei at medium and low spins. They are observed almost up to termination.

The lowest negative-parity bands are calculated to correspond to the [6,1] configuration, with 6 protons in the $g_{9/2}$ and 1 neutron in the $h_{11/2}$ orbitals. At termination, it is non-yrast in ^{103}Pd, it becomes yrast in ^{102}Pd, ^{101}Pd and ^{100}Pd (see Figs. 3 and 4). In the latter two nuclei, the bands gradually become non-yrast at medium and lower spins. Accordingly, those negative-parity bands which are yrast at termination have been observed up to the predicted terminating states at 28^- (^{102}Pd), $55/2^-$ (^{101}Pd) and 25^- (^{100}Pd), while the other negative-parity bands, band 1 in ^{103}Pd and the α=1 signature band 5 in ^{102}Pd, are observed only close to termination (see. Fig. 5).

The next lowest energy positive-parity configuration, the [6,2] one, corresponding to 6 protons in the $g_{9/2}$ and 2 neutrons in the $h_{11/2}$ orbitals, is the yrast or the yrast positive-parity configuration at high spins. With decreasing neutron number and spin it becomes non-yrast. This explains why the observed bands with proposed [6,2] assignment are strongly populated and seen up to or close to termination in ^{103}Pd (bands 2 and 3) and ^{102}Pd (band 2), while only some high-spin members of it are observed in ^{101}Pd (tentative assignment to band 5), and none in ^{100}Pd. In ^{103}Pd band 2, the α=1/2 signature branch of this configuration, is observed up to $65/2^{+}$, while band 2 in ^{102}Pd is seen up to 32^{+}, the predicted terminating states.

Core-excited configurations: In ^{102}Pd core-excited terminating configurations have also been observed. To band 3 we assign the [(1)61,2] configuration, which is formed from the [6,2] configuration by raising one proton from the $p_{1/2}$ to the $h_{11/2}$ orbital. To band 4 we tentatively assign the [(1)51,2] configuration, which is formed by raising one more proton from the $g_{9/2}$ to the $g_{7/2}$ orbital. In ^{103}Pd, this latter configuration is assigned tentatively to band 7.

6 SUMMARY

In conclusion, the detailed study of $^{100-103}$Pd has resulted in the observation of several new high-spin bands and the extension of previously known ones to higher spins. Using Nilsson-Strutinsky cranking calculations, terminating configurations have been assigned to most of these bands, many of which were observed up to the predicted termination. For the terminating bands in the studied nuclei an (N-shell)-(high-j) configuration systematics is presented.

Acknowledgements: EUROGAM 2 was funded jointly by the French IN2P3 and the U.K. EPSRC. This work was supported in part by the exchange programmes between CNRS and the Hungarian Academy of Sciences, between the British Council and the Hungarian Ministry of Culture and Education, by the Hungarian Scientific Research Fund, OTKA, No. 20655 and by the Swedish Natural Science Research Council.

(a) present address: School of Sciences, Staffordshire University, Stoke on Trent, U.K.
(b) permanent address: Nuclear Research Center, Salaspils, Latvia.

References

[1] I. Ragnarsson, A.V. Afanasjev, J. Gizon, Z. Phys. A 355, (1996) 383.

[2] J. Gizon et al., Phys. Lett. B, 410 (1997) 95.

[3] P.J. Nolan, F.A. Beck and D.B. Fossan, Ann. Rev. Nucl. Part. Sci. 44 (1994) 561.

[4] D.C. Radford, Nucl. Instr. and Meth. A 361 (1995) 297.

[5] G.E. Perez et al., to be published.

[6] S.K. Tandel et al., Z. Phys. A 357, (1997) 3.

[7] R. Popli et al., Phys. Rev. C22 (1980) 1121.

[8] D. Jerrestam et al., Nucl. Phys. A 557 (1993) 411c.

[9] B.M. Nyakó et al., submitted to Phys. Rev. C.

[10] T. Bengtsson and I. Ragnarsson, Nucl. Phys. A 436 (1985) 14.

[11] A.V. Afanasjev and I. Ragnarsson, Nucl. Phys. A591 (1995) 387;
I. Ragnarsson, V. P. Janzen, D. B. Fossan, N. C. Schmeing, R. Wadsworth, Phys. Rev. Lett. 74 (1995) 3935.

ENHANCED E1-DECAY FROM TRIAXIAL SD BANDS IN ^{164}Lu

S.W. Ødegård[1,2], G.B. Hagemann[1], S. Törmänen[1], P.O. Tjøm[2], A. Harsmann[1], M. Bergström[1], R.A. Bark[1], B. Herskind[1], G. Sletten[1], A. Görgen[3], H. Hübel[3], B. Aengenvoort[3], U. van Severen[3], C. Ur[5], H.J. Jensen[2,6], D. Napoli[4], S. Lenzi[5], C. Petrache[5], C. Fahlander[4,7], H. Ryde[7], A. Bracco[8], S. Frattini[8], R. Chapman[9], D.M. Cullen[10], S.L. King[10]

[1] *NBI, Univ. of Copenhagen, Denmark,* [2] *Univ. of Oslo, Norway,* [3] *Univ. of Bonn, Germany,* [4] *LNL, INFN, Legnaro, Italy,* [5] *INFN, Padova, Italy,* [6] *KFA Jülich, Germany,* [7] *Univ. of Lund, Sweden,* [8] *Univ. of Milan, Italy,* [9] *Univ. of Paisley, Scotland,* [10] *Univ. of Liverpool, UK*

Abstract: In a search for exotic structures in odd-odd ^{164}Lu, eight new, presumably triaxial, superdeformed bands were found. Two of the bands are connected to normal-deformed structures and the E1 strength of the connecting transitions appear to be enhanced over single particle expectations, possibly due to octupole correlations.

Nuclei with N$\sim$94 and Z$\sim$71 constitute a new region of exotic shapes, coexisting with normal prolate deformation [1-4]. They provide a unique possibility of studying superdeformed (SD) shapes with a pronounced triaxiality. Two such cases, interpreted as most likely corresponding to the $\pi i_{13/2}$ configuration, have recently been found in ^{163,165}Lu [3, 4]. Large Q_t values, corresponding to $\varepsilon_2 \sim 0.4$ (SD) with $\gamma \sim +18^\circ$, were derived for the triaxial SD band in ^{163}Lu [4] from both Recoil Distance and Doppler Shift Attenuation Method measurements. Calculations [3] with the Ultimate Cranker (UC) code[1] have revealed that large deformation minima are actually expected for all elements of the symmetry group (π, α) in ^{163,165}Lu. That is, the large deformation is due not only to the deformation-driving effect of the $\pi i_{13/2}$ intruder orbital, but also is the result of a re-arrangement of the core. The neutron number N$\sim$94 is crucial, as a gap in single particle energy appears at large values of

[1]Extensive use of the program "NUSMA"[5] has been applied in the analysis of the UC calculations

The Nucleus: New Physics for the New Millennium
Edited by Smit et al., Kluwer Academic / Plenum Publishers, New York, 2000.

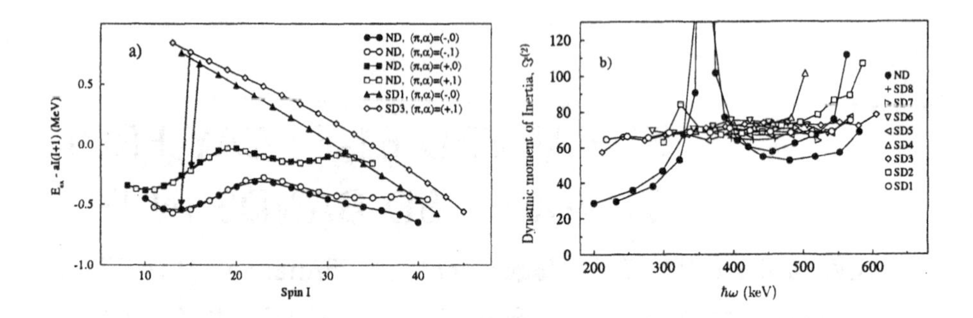

Figure 1 a): Exitation energy relative to a rigid rotor for SD1 and SD3 together with the yrast ND band. The connecting transitions are marked with arrows. b): Dynamic moment of inertia for all SD bands together with the yrast ND band.

γ ($\sim \pm 20^\circ$). Calculations of the potential energy surfaces for the lowest expected configurations in odd-odd ^{164}Lu for both positive and negative parity show local minima with large deformation and $\gamma \sim \pm 20^\circ$ in addition to minima at normal deformation (ND).

In one of the first Euroball experiments, a search for triaxial SD structures in ^{164}Lu, eight new presumably triaxial SD bands were observed[6]. The bands may belong to both of the calculated minima with $\gamma \sim \pm 20^\circ$. Figure 1 shows the dynamic moments of inertia for all eight SD bands, which behave in a similar way as those found in ^{163,165}Lu[3, 4], together with the level energies of SD1, SD3 relative to the yrast ND band.

Two of the strongest populated bands, labeled SD1 and SD3, have been connected to known [7, 8], ND bands in ^{164}Lu. These bands together with the connected ND states are shown in Fig. 2a. The band SD1 decays to several states of both positive and negative parity, which suggests the spin and parity assignments shown in fig. 2a. These assignments together with the measured DCO ratio of 1.0(2) imply that the strongest transition of 1128 keV is of stretched dipole character. The theoretical values for the Euroball geometry using summed gating on all detector angles [9] are about 0.8 and 1.4 for stretched dipole and quadrupole transitions, respectively. For band SD3, the single decay of 1532 keV to the yrast ND 14^- state with a DCO ratio of 0.9(2) is assigned as a stretched dipole transition, which determines the spin of band SD3. Since a pure M1 transition of such high energy is very unlikely, we assume it is of E1 character, and SD3 is therefore assigned positive parity.

With a detailed and complex decay-out of SD1 and a single decay of SD3, it is interesting to compare the E1 strength to statistical expectations. Using the expressions for the total transition probabilities T(E1) and T(E2) together with the expression for the reduced B(E2) transition probability, one may express

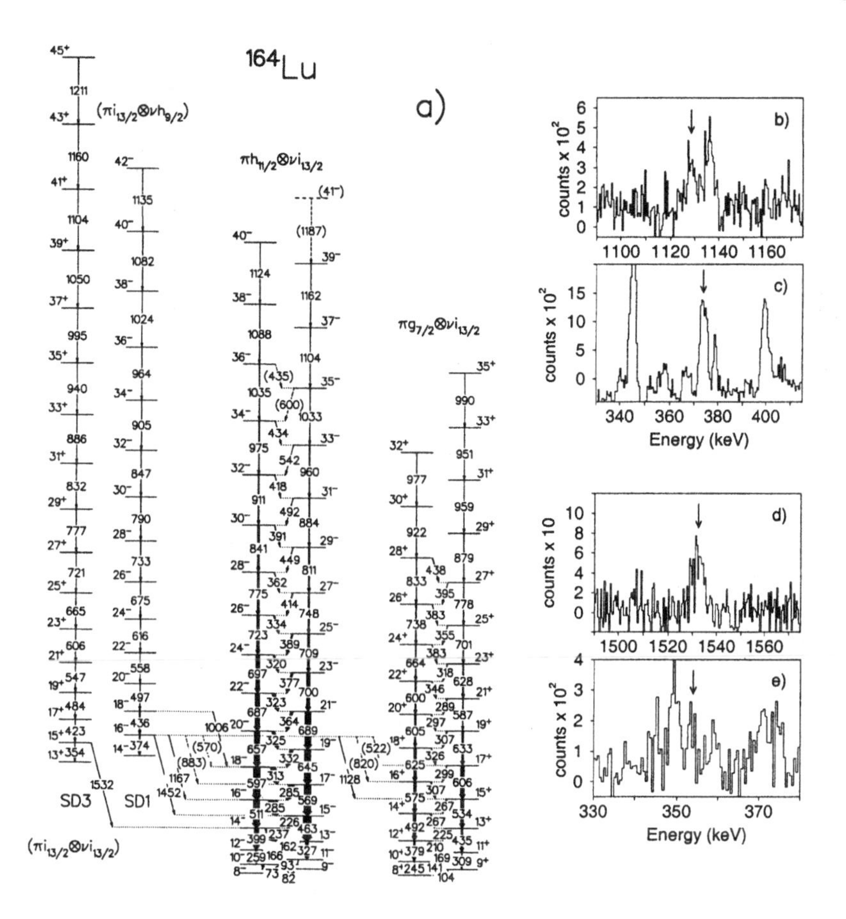

Figure 2 a): Partial levelscheme of ^{164}Lu showing the SD and ND bands which have been connected. b): 1128 keV transition connecting SD1 to ND. c): SD1 in-band transition used in the estimate of the B(E1) value of the transition in b). d): 1532 keV transition connecting SD3 to ND. e): SD3 in-band transition used in the estimate of the B(E1) value of the transition in d). Spectra in fig. b-e) have not been intensity corrected.

the reduced B(E1) transition probability as

$$B(E1) = 7.62 \times Q_t^2 \times \frac{T_\gamma(E1)}{T_\gamma(E2)} \times \langle IK20|I-2K\rangle^2 \times \frac{E_\gamma(E2)^5}{E_\gamma(E1)^3} \times 10^{-4}[e^2 fm^2].$$

The strength of the E1 decay is estimated assuming $Q_t(SD) = 11b$, comparable with the measured deformation in ^{163}Lu[4]. Also it is assumed that $\frac{K}{I} \ll 1$, giving the value 0.35 for the square of the Clebsch Gordon coefficient. From the intensities of the out of band E1 transition and the in band E2 transition the B(E1) strength was estimated. The intensities were obtained by applying single gates in SD gated $\gamma - \gamma$ matrices. Gates on all transitions in SD1 from the 497 keV to the 1082 keV, excluding the 617 keV, gave the spectra in Fig.

2b,c). The ratio T(E1)/T(E2) of the efficiency corrected intensities was found to be 0.49(17) for the decay out of SD1. In a same way using three gates, 423 keV to 546 keV, gave the spectra in Fig. 2d-e). In this case, for the decay out of SD3, the the ratio of the efficiency corrected intensities was found to be 1.7(6). Applying this expression, $B(E1) = 0.8(3) \cdot 10^{-4} e^2 fm^2$ and $0.85(30) \cdot 10^{-4} e^2 fm^2$ for the 1128 keV and 1532 keV transitions, respectively. Expressed in Weisskopf units this gives $\sim 4 \cdot 10^{-5}$WU for both transitions, which is around 400 times faster than the E1-decay found for the (axially symmetric) SD to ND states in ^{194}Hg [10], and only $\sim$ 6 times slower than octupole-enhanced E1 transitions between some of the ND bands in the same nucleus, ^{164}Lu. The E1 decay from SD1 is associated mainly with an $h_{9/2}$ SD to $i_{13/2}$ ND quasineutron transition, whereas the E1-decay from SD3 is associated with an $i_{13/2}$ SD to $h_{11/2}$ ND quasiproton transition. Octupole enhancement [11] is found between ND bands of similar structure in odd-N and odd-Z rare earth nuclei and may therefore be present in both of these different E1 transitions.

For SD1 the values of B(E2,SD→ND) are $10^3 - 10^4$ times reduced compared to the SD in-band B(E2) values. This is quite different from the decay of the triaxial SD band in ^{163}Lu. In that nucleus the decay to the normal deformed structures takes place through an isolated mixing at I=21/2 with an interaction strength of around 20 keV of the SD with the $[411]1/2^+$ configuration [12].

This project has been supported by the Danish Natural Science Foundation, the EU TMR project no ERBFMBICT961027, the Swedish Natural Science Research Council, the Research Council of Norway, BMBF Germany and the UK EPSRC. The dedicated help from staff and Euroball support groups at the INFN laboratory in Legnaro is highly appreciated.

References

[1] S. Åberg, Nucl. Phys. A520 (1990) 35c and refs. therein

[2] I. Ragnarsson, Phys. Rev, Lett. 62 (1989) 2084

[3] H. Schnack-Petersen et al., Nucl. Phys. A594 (1995) 175

[4] W. Schmitz et al., Nucl.Phys. A539 (1992) 112 and Phys. Lett. B303 (1993) 230

[5] see http://www.matfys.lth.se/~ ragnar/ultimate.html

[6] S. Törmänen et al., Il Nuovo Cimento 111A (1998) 685 and to be published in Phys. Lett. B

[7] X.-H. Wang et al., Nucl. Phys. A608 (1996) 77

[8] M.A. Cardona et al., Phys. Rev. C56 (1997) 707

[9] M. Palacz et al., Nucl. Phys. A625 (1997) 162

[10] G. Hackman et al., Phys. Rev. Lett. 79 (1997) 4100

[11] G.B. Hagemann and I. Hamamoto, Phys. Rev. C47 (1993) 2008

[12] J. Domsheit et al., to be published

VARIATIONAL APPROACH TO MEDIUM MASS N=Z NUCLEI

A. Petrovici

National Institute for Physics and Nuclear Engineering
R-76900 Bucharest, Romania

spetro@roifa.ifa.ro

Abstract: Results are presented concerning shape coexistence and neutron-proton correlations at low and high spins in the N=Z nuclei ^{72}Kr and ^{74}Rb obtained within the *complex* version of the Excited Vampir variational approach. The pair structure and the average pairing gaps of the realistic wave functions are investigated in order to evaluate the pairing correlations.

1 INTRODUCTION

The medium mass nuclei in the A=70 mass region display a remarkable diversity of shapes and rapid changes in structure with particle number, angular momentum and excitation energy. Qualitatively, this can be understood from the deformed independent particle shell model, which predicts the occurence of large gaps at different deformations in the single particle spectra of many nuclei in this mass region. Thus a strong competition between the corresponding many-particle configurations and consequently the coexistence of bands with differently deformed structure are to be expected. In addition, since in these nuclei protons and neutrons fill the same single particle orbits, their alignment with increasing angular momentum will be strongly competitive and may even occur simultaneously. This effect should be particularly enforced in the self-conjugate N=Z nuclei. For nuclei close to the N=Z line furthermore the neutron-proton pairing correlations can be expected to play a significant role in their structure and decay. Important questions in this context are the relative importance of isoscalar and isovector neutron-proton pairing, the possible coexistence of both modes in the same nucleus and the dependence of these correlations on angular momentum and deformation.
For a unified description of the interesting properties of these nuclei at low as

The Nucleus: New Physics for the New Millennium
Edited by Smit et al., Kluwer Academic / Plenum Publishers, New York, 2000.

well as at high angular momenta, obviously a model is needed, which can account for the delicate interplay between collective and single particle degrees of freedom. It should furthermore treat like–nucleon and neutron–proton pairing correlations on the same footing and, last but not least, the handling of realistic model spaces as well as general two–body forces should be numerically feasible. These requirements are fulfilled by the *complex* Excited Vampir approach. This model is based on chains of variational calculations using symmetry projected Hartree–Fock–Bogoliubov (HFB) vacua as test configurations. The use of essentially *complex* HFB transformations allows one to account for all kinds of two nucleon correlations even though time reversal and axial symmetry are still imposed on the underlying HFB transformations. In order to have direct access to the competition between like-nucleon isovector pairing and the isovector (T=1) and isoscalar (T=0) neutron-proton pairing, we investigated particular parts of the two-body densities which reflect these aspects.

2 THEORETICAL FRAMEWORK

We investigated the lowest 0^+ states and the lowest positive-parity even-spin states up to high angular momenta in ^{72}Kr and ^{74}Rb. First the Vampir solutions, representing the optimal mean-field description of the yrast states by single symmetry-projected HFB determinants were obtained. Then the Excited Vampir approach was used to construct additional excited states by independent variational calculations. Finally, the residual interaction between the lowest orthogonal configurations was diagonalized. We define the model space and the effective Hamiltonian as in earlier calculations [1] for nuclei in the $A \simeq 70$ mass region : a ^{40}Ca core is used and the valence space consists out of the $1p_{1/2}$, $1p_{3/2}$, $0f_{5/2}$, $0f_{7/2}$, $1d_{5/2}$ and $0g_{9/2}$ oscillator orbits for both protons and neutrons. As effective two-body interaction a renormalized nuclear matter G-matrix based on the Bonn One-Boson-Exchange potential is used. We analyze the wave functions in terms of nucleon pairs coupled to particular quantum numbers using the 'pair number operator' defined by

$$\rho_{(M)}^{JTT_z\pi} \equiv \frac{1}{2} \sum_{n_il_ij_in_kl_kj_k} \delta\Big((-)^{l_i+l_k},\pi\Big)(-)^{j_i+j_k-M}(-)^{1-T_z}$$

$$\times \sum_{m_im_k\tau_i\tau_k} \langle j_im_ij_km_k|JM\rangle\langle\frac{1}{2}\tau_i\frac{1}{2}\tau_k|TT_z\rangle c^\dagger_{j_im_i\tau_i}c^\dagger_{j_km_k\tau_k}$$

$$\times \sum_{m_rm_s} \langle j_k-m_rj_i-m_s|J-M\rangle\langle\frac{1}{2}-\tau_k\frac{1}{2}-\tau_i|T-T_z\rangle c_{j_km_r\tau_k}c_{j_im_s\tau_i}$$

which can be considered as the trace of the scalar part of the two-body density [2]. The expectation value of this operator within a particular state counts the number of pairs coupled to particular angular momentum, parity and isospin quantum numbers (J^π,T,T_z). Obviously it fulfils a trivial sum rule : summing over J, π, T and T_z one obtains the total number of distinct pairs which can be formed in the chosen nucleus: A(A-1)/2.

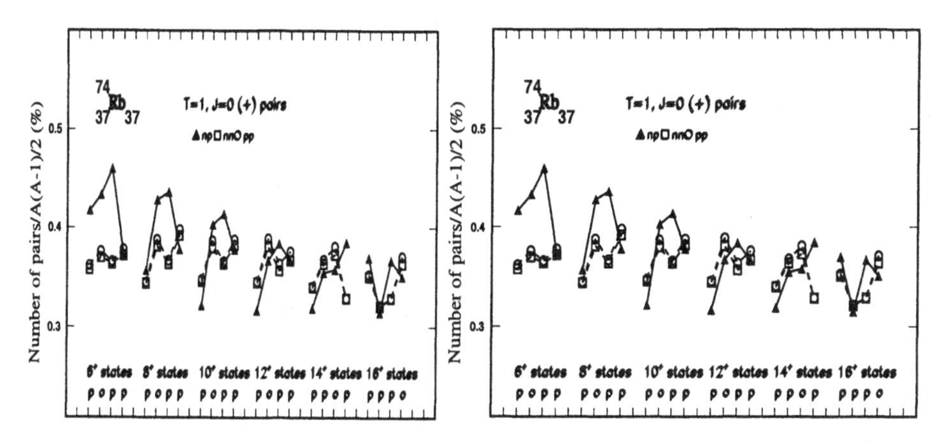

Figure 1 The number of nn, pp and np pairs with T=1, $J^{\pi}=0^{+}$ of the lowest few even-spin positive-parity states in ^{74}Rb, and ^{72}Kr nuclei in percent of A(A-1)/2.

3 RESULTS AND DISCUSSION

We investigated the lowest few even-spin positive-parity states up to spin 20^{+} in ^{72}Kr and 16^{+} in ^{74}Rb. In ^{72}Kr, for the 0^{+}, 2^{+} and 4^{+} states we found that the first minimum is oblate (o) and the second one is prolate (p) deformed in the intrinsic system. For all investigated states from spin 6^{+} to 20^{+} the main projected determinant for the yrast solution is prolate deformed. The first excited state for the spins 6^{+}, 8^{+}, 10^{+}, 12^{+} is based on an oblate deformed configuration. In ^{74}Rb the yrast band is prolate and the first excited one is oblate deformed up to spin 12^{+}. In both nuclei at higher angular momenta other prolate deformed states become energetically favoured with respect to the oblate band. Strong oblate-prolate mixing is present at low spin while at intermediate and higher spin values the mixing with other prolate deformed configurations starts to become dominant and larger differences are seen in the structure of the various states in the two nuclei. A 'pair structure' analysis has been performed, and the average pairing gaps of the dominant intrinsic configuration have been calculated. Fig. 1 displays the number of neutron-neutron (nn), proton-proton (pp) and neutron-proton (np) isovector $J^{\pi}=0^{+}$ pairs obtained for the lowest few 0^{+} states of the considered nuclei. The amount of T=1 np pairs clearly exceeds the nn and pp ones in the odd-odd nucleus, but becomes comparable with the latter in the even-even one. The amount of isovector $J^{\pi}=0^{+}$, as well as the isoscalar $J^{\pi}=1^{+}$ pairs reveals significant variations with excitation energy and changes in the intrinsic deformations of the underlying configurations.
In ^{74}Rb with increasing spin the amount of the isovector $J^{\pi}=0^{+}$ np pairs decreases strongly for the yrast states and moderately for the calculated first excited states as can be seen from Fig. 1. This behaviour is also reflected in the corresponding average pairing gap. It is decreasing drastically with increasing spin for the main configurations of the yrast states from 0.587 MeV for the 0^{+} state down to almost zero at spins 12^{+} and 14^{+}. On the other hand almost

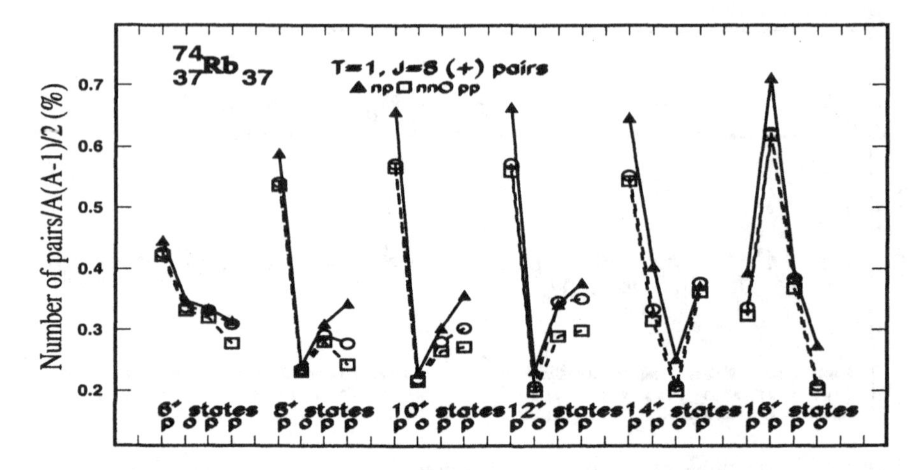

Figure 2 Same as in Fig. 1, but for isovector $J^{\pi}=8^{+}$ pairs in ^{74}Rb.

no decrease is observed for the first and the second excited states. The number of np pairs coupled to T=0, $J^{\pi}=1^{+}$ indicates such a strong dependence on the structure of each state, too. Important for the alignment properties are the isovector $J^{\pi}=8^{+}$ and isoscalar $J^{\pi}=9^{+}$ pairs. Fig. 2 displays a sharp increase in the aligned isovector $J^{\pi}=8^{+}$ correlations on the yrast line in between spin 6^{+} and 10^{+} with the np pairs showing a stronger contribution than the like-nucleon ones. A similar increase is seen in the isoscalar $J^{\pi}=9^{+}$ pairs. It is obvious that the pair structure analysis should be done state by state, since the behaviour of the yrast state can be completely different than that displayed by the first few excited states. Opposite to the results for ^{74}Rb, in ^{72}Kr all types of isovector $J^{\pi}=0^{+}$ pairs yield about the same contribution. Furthermore a slower variation of the number of np pairs with increasing spin is observed. The alignment is less pronounced than in the odd-odd nucleus. The number of the aligned isovector $J^{\pi}=8^{+}$ and isoscalar $J^{\pi}=9^{+}$ pairs display a slower increase with angular momentum than obtained in ^{74}Rb. The isovector 0^{+} and 8^{+} np pairs do not dominate over the corresponding like-nucleon pairs as it is the case in the odd-odd nucleus.

In summary, the faster alignment of the np pairs with respect to the like-nucleon ones displayed by the yrast states of the odd-odd ^{74}Rb is not found in the even-even ^{72}Kr. The strong dependence of the number of low spin pairs and aligned pairs on deformation, however, is found in both these N=Z nuclei.

References

[1] A. Petrovici, K.W. Schmid and A. Faessler. Neutron-proton pairing correlations in medium mass N≃Z nuclei. *Nucl. Phys.*, A, *accepted for publ.*

[2] A. Petrovici, K.W. Schmid and A. Faessler. Two-body densities of realistic nuclear wave functions. *Z. Phys.*, A437: 87-98, 1993.

GAMMA-RAY SPECTROSCOPY OF FISSION FRAGMENTS INDUCED BY HEAVY IONS: FROM NUCLEAR STRUCTURE OF NEUTRON-RICH NUCLEI TO MECHANISM STUDIES

M.-G. Porquet

CSNSM IN2P3-CNRS
Bât 104-108
91405 Orsay Campus,
France

porquet@csnsm.in2p3.fr

Abstract: About eighty nuclei in the A $\approx$ 100 mass region have been observed as fission fragments following the fusion evaporation ^{28,30}Si + ^{176}Yb at 145 MeV bombarding energy. These nuclei have been identified from their γ-ray cascades detected with the EUROGAM2 array. The level schemes of several stable or neutron-rich nuclei have been extended to higher spins. From coincidences between transitions in complementary fragments, γ-rays de-exciting new high-spin states of hitherto unexplored isotopes can be assigned and some aspects of the fission mechanism can be studied.

1 INTRODUCTION

Taking advantage of the new high-efficiency Germanium arrays, providing greatly improved sensitivity and selectivity in γ-γ coincidence spectrometry, the study of γ-rays emitted following spontaneous fission has recently been a precious source of information about the structure of very neutron-rich nuclei [1, 2]. Another technique, using fission induced by heavy ions, can be used in order to populate high spin states of some stable or neutron-rich nuclei which cannot be populated using the standard fusion-evaporation reactions [3]. A first experiment performed with the EUROGAM2 array at Strasbourg has been devoted

The Nucleus: New Physics for the New Millennium
Edited by Smit et al., Kluwer Academic / Plenum Publishers, New York, 2000.

to the study of high-spin states of nuclei with A = 80-120. About eighty nuclei have been obtained as fission fragments following the fusion reaction ^{28,30}Si + ^{176}Yb at 145 MeV bombarding energy. In this contribution we report on the method of identification of these fission fragments by means of their prompt γ-ray emissions. Level schemes of several stable and neutron-rich nuclei have been extended or newly established. Some of them are discussed. Moreover odd-even effects, for an odd proton number as well as an odd neutron number have been observed in the individual yields of fission fragments obtained from their emitted γ rays.

2 EXPERIMENTAL TECHNIQUES

The experiment has been performed at the Vivitron accelerator at Strasbourg, using beams of 145 MeV ^{28,30}Si. A 1.5 mg/cm^2 target of ^{176}Yb was used, onto which a backing of 15 mg/cm^2 Au had been evaporated in order to stop the recoiling nuclei. Gamma-rays were detected with the Eurogam2 array [4] . This spectrometer consisted of an array of 54 escape-suppressed germanium detectors, 30 of which were large-volume coaxial detectors positioned at backward and forward angles with respect to the beam. The remaining 24 detectors, arranged in two rings close to 90 degrees to the beam direction, were four-element "clover" detectors. The data were recorded in an event-by-event mode with the requirement that a minimum of five unsuppressed Ge detectors fired in prompt coincidence. In the reaction induced by ^{30}Si a total of 40 million events were recorded. Using the ^{28}Si beam, a total of 540 million coincidence events were collected, out of which 135 million were three-fold, 270 million four-fold, and 108 million five-fold. The off-line analysis consisted of both usual γ-γ or γ-γ-γ sorts [5] and multiple-gated spectra [6].

3 RESULTS

3.1 Mass, charge and spin distributions in fragments

We have identified γ-ray cascades belonging to about eighty A $\approx$ 100 nuclei produced as fission fragments and stopped in the target backing before emitting γ-rays. The mass region reached extends from Z = 28 to Z = 56 and from N = 36 to N = 78, along the valley of stability and beyond it towards the neutron-rich side. As expected for a fissioning system having excitation energy [7], the most produced nuclei (A $\approx$100, Z $\approx$40-44) are located around half the mass of the compound nuclei (^{206,204}Po) after neutron evaporation. Example of multi-gated spectra is shown in Fig. 1. From such spectra, new γ-rays can be identified and level schemes of several well-produced nuclei have been already extended to spin $\approx$16 $\hbar$. For less-produced fragments, such as ^{80,82}Se, the yrast states have been identified up to spin $\approx$10$\hbar$ [8] : only the first yrast states, 2^+ and 4^+, were already known in these two nuclei [9].

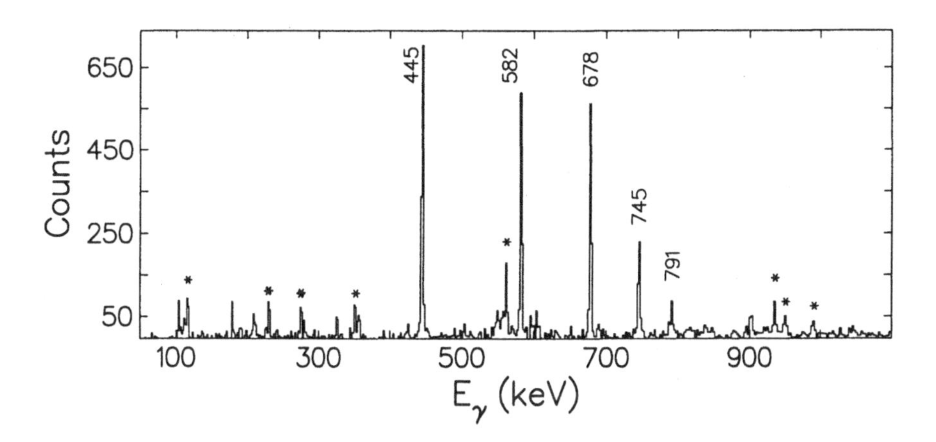

Figure 1 Spectrum of rays in double coincidence with the 270 keV ($2^+ \rightarrow 0^+$) and 730 keV ($10^+ \rightarrow 8^+$) transitions of ^{106}Ru. Band members are labelled with their energy in keV. The transitions marked with a star belong to the complementary fragments.

Prompt coincidences occur between transitions emitted by complementary fragments. For example the complementary fragments of ^{110}Pd produced in the two reactions used in this work can be observed in the spectra shown in Figure 2 : In the first case (^{28}Si) the predominant complementary fragment is ^{88}Sr, while in the second case (^{30}Si), it is ^{90}Sr. When summing the number of protons of complementary fragments, one obtains exclusively 84, the proton number of the Po compound nuclei, for both two data sets. The summed numbers of the neutrons are mainly 116 in the first reaction and 114 in the second case, that is six neutrons less than the neutron numbers of the compound nuclei. By means of the γ-ray identification of the complementary fragments, the total number of emitted neutrons is measured very precisely, although we do not know exactly how many neutrons have been emitted, before fission, by the compound nucleus and, after fission, by the fragments. The pre-fission and post-fission neutrons were directly measured for several compound nuclei at various excitation energies [10]: At 65 MeV excitation energy ^{210}Po emits $\approx$3 neutrons before fission and each fragment emits $\approx$1.7 neutrons, our results are consistent with these numbers.

The identification of even-even nuclei is straightforward as at least the energy of the first transition ($2^+ \rightarrow 0^+$) is always known from radioactivity studies. From coincidences occurring between transitions in even-Z complementary fragments, γ-rays de-exciting high-spin states of several odd-N isotopes have been identified in this work. For example, several γ-ray cascades associated to the $\nu h11/2$ sub-shell have been proposed in ^{109,111}Pd [11].

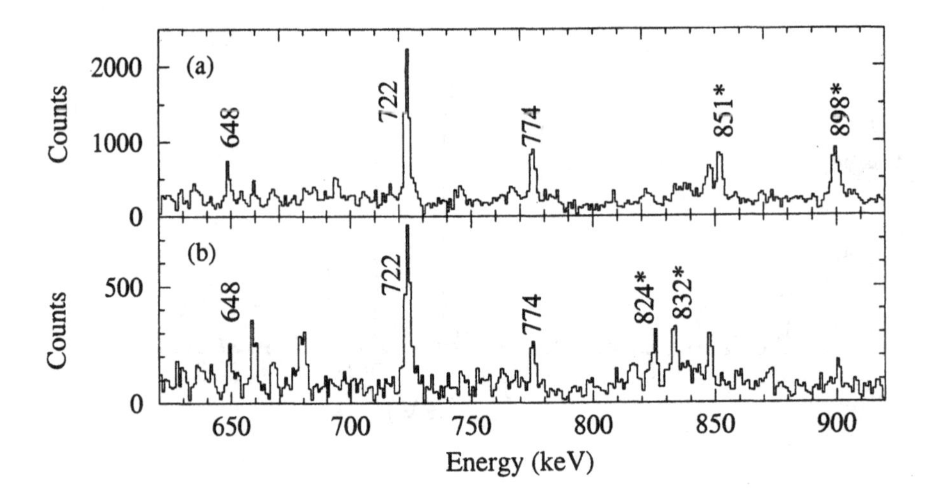

Figure 2 Single gated spectra of ^{110}Pd produced in the reactions
(a) ^{28}Si(145 MeV) + ^{176}Yb, the transitions marked with a star belong to ^{88}Sr.
(b) ^{30}Si(145 MeV) + ^{176}Yb, the transitions marked with a star belong to ^{90}Sr.

We have also observed odd-odd partitions of the even number of protons of the Po compound nuclei. In this mass region, very few level schemes of odd-Z nuclei were previously known at high spin. Indeed, only the lighter masses of each isotopic series have been studied from heavy-ion-induced reactions. Moreover radioactive decays of odd-A isotopes only populate low-spin states of the daughter. From γ-ray coincidences between odd-Z complementary fragments, new high-spin level schemes have been built. For example, from the transitions de-exciting the high-spin states of ^{89}Y [12], those corresponding to the neutron-rich ^{109}Rh have been identified [13]; conversely from the known transitions of ^{103}Rh [9], high-spin states of the neutron-rich ^{95}Y have been established [14].

3.2 *Shapes of nuclei around ^{110}Pd*

The high-spin states of several odd-N, odd-Z and odd-odd nuclei around ^{110}Pd have been obtained in this work : $^{109,111}_{46}$Pd [11], $^{113,115}_{48}$Cd [15], $^{107,109}_{45}$Rh [13], $^{106,108}_{45}$Rh and $^{110,112}_{47}$Ag [14]. Their rotational behaviours are those expected, in this mass region, for rather prolate nuclear shapes. It is worth noting that a sharp transition between prolate and oblate shapes has been predicted to occur between ^{108}Rh and ^{109}Rh, ^{110}Pd and ^{111}Pd, ^{111}Ag and ^{112}Ag, ^{114}Cd and ^{115}Cd [16], respectively. Our results are at variance with such a prediction, while they point out the importance of triaxiality, which have been already stated for the more neutron-rich nuclei from the properties of the γ-band observed in the even-even Ru isotopes [1]. The rotational band built on the ground state ($\pi g_{9/2}$ parentage) of $^{105,107,109}_{45}$Rh and $^{107,109}_{47}$Ag is characterized by a large signature splitting (see Figure 3). On the other hand, the signature splitting observed in the yrast band of the odd-odd $^{104,106,108}_{45}$Rh and $^{108,110,112}_{47}$Ag nuclei, built on the configuration $\pi g_{9/2}\ \nu h_{11/2}$, is much smaller. That means that the additionnal occupancy of the $\nu h_{11/2}$ drives the nuclear deformation towards a more symmetric shape.

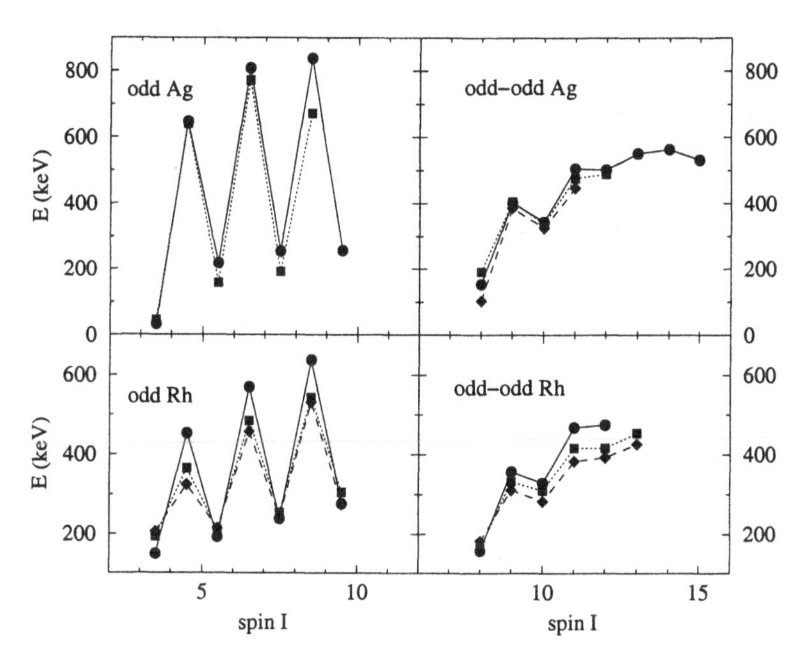

Figure 3 Energy of the γ-ray decaying the state of spin I+1 towards the state of spin I as a function of spin value I, for the yrast bands observed in several odd and odd-odd Rh and Ag.

3.3 *Individual yields of fragments*

Up to now, individual yields of fragments from heavy-ion induced fission have never been measured. In principle, the relative yield of each fragment can be measured from the number of γ-rays populating its ground state. To reach this goal the low-energy part of the level schemes have to be known very precisely. Moreover the γ transitions have to be disentangled as there are a lot of rays very close in energy in the spectra. For example, between 777 and 779 keV, we have identified eight transitions de-exciting states of ^{82}Kr, ^{89}Y, ^{108}Pd, ^{111}Pd, ^{86}Rb, ^{96}Mo ($2^+ \rightarrow 0^+$), ^{85}Rb, and ^{100}Mo. In that respect, while peak intensity measured in γ-spectra built without any condition is useless, the high-fold events as obtained when using the new germanium arrays can be used with benefit. The problem of obtaining correct relative intensities from high-fold events has been solved by creating multiple-gated spectra directly from data on tapes with the code Fantastic [6].

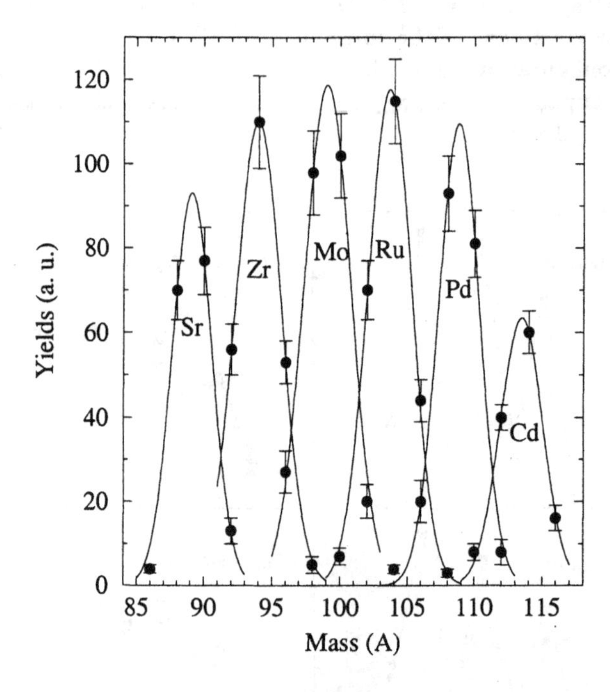

Figure 4 Yields of the even-even nuclei produced as secondary fission fragments from the fusion reaction $^{28}Si+^{176}Yb$ at 145 MeV bombarding energy. They have been normalized to the yields of $^{98,100}Mo$ and fitted using Gaussian functions.

The relative yields obtained in this experiment for the even-even nuclei belonging to six isotopic series, Sr, Zr, Mo, Ru, Pd, and Cd, are given in Figure 4. They have been normalized to the yields of $^{98,100}Mo$ and fitted using Gaussian functions. We have also measured the yields of several odd-Z and odd-N fragments. We have selected only the most produced odd-A fragments, in which even the very low-intensity cascades populating the ground state (and isomeric states, if any) can be fairly easily measured in our data set. Results are presented in Figure 5, for many groups of three nuclei with consecutive mass values A, A+1, A+2, either three isotones (even N) or three isotopes (even Z). Odd-even effects can be clearly observed, as well as on proton number as on neutron number: the yields of odd-mass nuclei are always smaller than those of the even-even neighbours, by a factor around 2. Such effects are observed for the first time in fission induced by heavy ions at a bombarding energy just above the Coulomb barrier.

Numerous measurements for spontaneous or low-energy induced fission of actinides have shown that the charge distributions exhibit an odd-even staggering (see for example [17]). In phenomenological models, it has been assumed that this effect originates from the proton pairing in the fissionning nucleus.

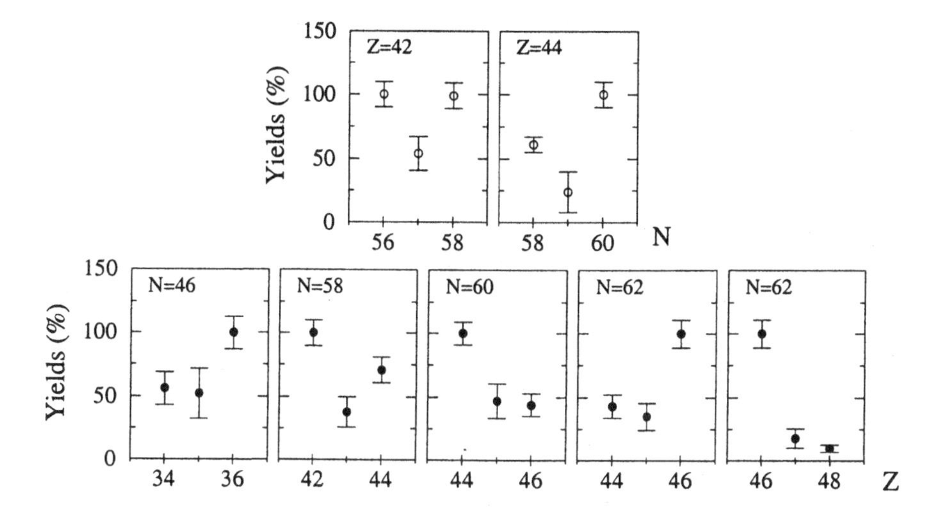

Figure 5 Yields of some odd-A (odd-Z or odd-N) fragments compared to the ones of neighbouring even-even nuclei, produced as secondary fission fragments from the fusion reaction ^{28}Si+^{176}Yb at 145 MeV bombarding energy. In each case, the maximum yield has been normalized to 100.

The survival of proton pairs would indicate that there is only a small amount of intrinsic excitation energy during the fission process. A model of pair-breaking [18] reproduces satisfactorily experimental data. Results obtained in this work could be interpreted along the same lines. Fission of ^{200}Po, which could be produced at very low excitation energy (after the emission of four pre-fission neutrons) and at high spin (around 35 $\hbar$), would involve very low intrinsic excitation energy, as a lot of energy is frozen as rotational energy (in this mass region the energy of the 35 $\hbar$ yrast state is around 12 MeV). It is also worth noting that in fissionning nuclei with high values of fissility parameter there is some heating during the descent from saddle to scission and a certain fraction of this thermal excitation is dissipated in breaking pairs. Such a heating process cannot occur in fissionning nuclei with low values of fissility parameter (such as Po isotopes), as the scission point is very close to the saddle point.

Conclusion

The reactions ^{176}Yb + ^{28,30}Si (145 MeV) have been used to produce the fissile compound nuclei ^{204,206}Po. The fission induced by heavy ions, when associated with a very sensitive gamma array, is a very powerful technique to study high-spin states of some stable or neutron-rich nuclei which, at this present time, cannot be populated in another way. Moreover the individual identification of each fragment by means of its emitted γ transitions allows us to have a new insight into the mechanism of fission induced by heavy ions.

Acknowledgments

I am very gratefull to Prof. J. Sharpey-Schafer for his precious support when this work was in its infancy. The experiment described here has been done by a collaboration of six laboratories, CSNSM (Orsay), IReS (Strasbourg), IPN (Orsay), INRNE (Sofia), University of Sofia, and DAPNIA/SPhN (Saclay). I would like to acknowledge all my colleagues which have been involved in the analyses of data, particularly Dr I. Deloncle. The Eurogam project was funded jointly by EPSRC (UK) and IN2P3-CNRS (France).

References

[1] I. Ahmad and W.R. Phillips, *Rep. Prog. Phys.* 58 , 1415 (1995)

[2] J. Hamilton et al., *Prog. Part. Nucl. Phys.* 35 , 635 (1995)

[3] M.-G. Porquet et al., Workshop on High Angular Momentum Phenomena, Piaski, Poland (August 23-26, 1995), *Acta Phys. Polonica* 27, 179 (1996)

[4] P.J. Nolan et al., *Ann. Rev. Nucl. Part. Sci.* 45, 561(1994)

[5] D Radford, *Nucl. Instr. and Meth.* A361, 297 (1995)

[6] I. Deloncle et al., *Nucl. Instr. and Meth.* A357, 150 (1995)

[7] R. Vandenbosch and J. Huizenga, Nuclear Fission, Acad. Press, N Y, 1973

[8] M.-G. Porquet et al., Second International Workshop on Nuclear Fission and Fission-product spectroscopy, Seyssens (France), April 22-25, 1998

[9] R. B. Firestone, Table of Isotopes, 8th edition, Wiley, New York, 1996

[10] D.J. Hinde et al., *Nucl. Phys.* A452, 550 (1986)

[11] T. Kutsarova et al., *Phys. Rev.* C58, 1966 (1998)

[12] L. Funke et al., *Nucl. Phys.* A541, 241 (1992)

[13] Ts. Venkova et al., to be published.

[14] M.-G. Porquet et al., to be published.

[15] N. Buforn et al., to be published.

[16] P. Möller et al., *At. Data Nucl. Data Tables* 59, 255 (1995)

[17] J.-P. Bocquet and R. Brissot, *Nucl. Phys.* A502, 213c (1989)

[18] H. Nifenecker et al., *Z. Phys.* A308, 39 (1982)

ANALOGIES BETWEEN TERMINATING BANDS IN Sn/Sb AND Zn NUCLEI AND MAGNETIC BANDS IN Pb NUCLEI

Ingemar Ragnarsson

Department of Mathematical Physics,
Lund Institute of Technology,
P.O. Box 118, S-221 00 Lund, Sweden

ingemar.ragnarsson@matfys.lth.se

Abstract: Terminating bands in ^{62}Zn and ^{109}Sb and magnetic bands in ^{199}Pb are described within a principal axis cranking approach. At high spin, the energy of these bands as a function of spin are described surprisingly well in all cases. It remains to be investigated if also the transition probabilities within magnetic bands can be understood within this approach.

1 INTRODUCTION

Two types of nuclear excitations which have attracted much interest recently are terminating bands and magnetic bands. Terminating bands [1] are formed in configurations with several valence particles (and holes) outside closed shells; so many that a collective band is formed but so few that the maximal spin which can be formed from them is within experimental reach. It is especially the properties of these bands close to termination in terms of energy cost to build angular momentum (often expressed as a $\mathcal{J}^{(2)}$ moment of inertia), collectivity etc. which have been considered. Magnetic bands (see [2, 3] and references therein) are also formed in nuclei with a few valence particles and/or valence holes outside closed shells. They are characterized by large $M1$ transition probabilities caused by close to perpendicular proton and neutron spin vectors. This is typical for high-K bands where for example the protons have a large spin component along the symmetry axis (a high value of K) but where the perpendicular collective spin component is largely built from the neutrons.

The Nucleus: New Physics for the New Millennium
Edited by Smit et al., Kluwer Academic / Plenum Publishers, New York, 2000.

One feature which has been specifically emphasized for magnetic bands is the decrease of the $M1$-values at high spin [2]. This decrease can be understood as a decrease of the spin component along the symmetry axis, i.e. the effective K-value goes down [4]. Indeed such a decrease is what is expected when the band comes close to termination [1] where, in the terminating state, all spin vectors are quantized along one axis, i.e. their expectation values of j along an axis perpendicular to this quantization axis must be zero. It is however only recently that the influence of this alignment process on the $M1$ strength has been realized [4] even though possible $M1$ transitions at high spin in terminating bands were discussed long ago [5].

If it should be possible to observe the $M1$-transitions and their decrease in rotational bands, the $B(E2)$ transitions should not dominate, i.e. the deformation should not be too large. This condition suggests that the number of valence particles outside closed shells should be limited. From that point of view, the bands which have recently been discovered [6, 3] in Pb nuclei are ideal. Furthermore, making a two-particle two-hole excitation across the $Z = 82$ gap, it is possible to take advantage of the two high-j shells above the gap, $i_{13/2}$ and $h_{9/2}$, placing the two protons in orbitals with highest possible spin components along the (oblate) symmetry axis so that a total K of $6.5 + 4.5 = 11\hbar$ is obtained. Then a rotational band might be built on this band head if the high-K proton spin is combined with the spin from the neutron holes in the $N = 126$ core. These neutron holes tend to make the nucleus slightly oblate as required to make a high-K proton state.

Terminating bands in the $A = 110$ region [7, 1] and in Zn nuclei [8, 9] have been successfully studied using the one-dimensional cranking approach, i.e. where only the spin component along one principal axis (the rotation axis) is accounted for. In the Pb bands, this is not a reasonable approximation at low spin values, where it seems that two perpendicular spin vectors of similar amplitude are active. Consequently, these bands have mainly been studied using the tilted axis cranking (TAC) [4] approach where the two perpendicular spin vectors are accounted for. Our main interest is however, what happens to these bands when they come close to termination. A possible termination of band 1 in ^{199}Pb was discussed in Ref. [10], where however the spin contribution from the low-j particles was not estimated from the contributing subshells.

An interesting observation is that the configurations of the $M1$ bands in Pb nuclei show large similarities with the configurations of the terminating bands in the $A = 110$ mass region [7] and in Zn nuclei [8, 9] if particle states and hole states are interchanged. Thus, Fig. 1 illustrates typical distributions of valence particles (filled circles) and holes (open circles) in terminating bands in the $A = 60$ region and $A = 110$ regions and in a magnetic band in ^{199}Pb. The figure shows for example that the high-j protons in the $i_{13/2}$ and $h_{9/2}$ subshells in Pb isotopes have an analogue of two $g_{9/2}$ proton holes in the $A = 110$ region and a $f_{7/2}$ proton hole in Zn nuclei. Similarly, the $i_{13/2}$ neutron holes in Pb nuclei correspond to $h_{11/2}$ and $g_{9/2}$ particles, respectively, in the other two regions.

Note that the bands in ^{62}Zn [9] do not only terminate in a well-defined way but another interesting feature is that they, in a similar way as the Pb bands, are formed from signature partner bands connected by strong $M1$-transitions.

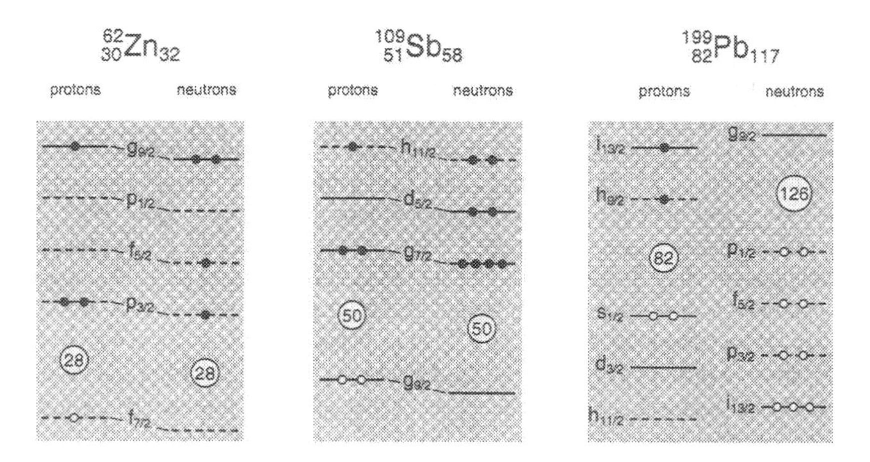

Figure 1 Distribution of valence particles and holes over subshells in typical terminating and magnetic bands. Particles relative to the closed core nuclei ^{56}Ni, ^{100}Sn and ^{208}Pb and shown by filled circles and holes by open circles. In the calculations, these subshells are mixed and the orbitals are then characterized according to their main components. Furthermore, no attempt is made to distinguish between the different low-j shells, e.g. $p_{3/2}$, $f_{5/2}$ and $p_{1/2}$ in ^{62}Zn or the $s_{1/2}$ and $d_{3/2}$ proton subshells for ^{199}Pb.

2 CONFIGURATION-CONSTRAINED NILSSON-STRUTINSKY CRANKING CALCULATIONS.

At the termination of a rotational band, all spin vectors are quantized along the same axis. Consequently at termination and, considering the continuous change of all properties, close to termination, the use of one-dimensional cranking can be well justified. Furthermore, at high rotational frequencies with several broken pairs, pairing correlations should be of minor importance and may therefore be neglected. With these approximations, we can use the formalism introduced in Ref. [11] and later somewhat refined in Refs. [7, 12] to describe the configurations shown in Fig. 1. Using this approach, we can fix different configurations and microscopically calculate how the spin is built from all occupied orbitals, i.e. with no need for any core, effective charges etc. Furthermore for the different configurations, the energy is minimized with respect to deformation at each spin value. Standard parameters [11, 1] are used in these calculations with no specific fits for the different nuclei.

Calculations and experiment are compared in the form of spin I vs transition energy E_γ in Fig. 2. The configurations illustrated in Fig. 1 are assigned to the highest spin band in ^{62}Zn, band 1 in ^{109}Sb and the upper branch of band 1 in ^{199}Pb. The configuration of the lower spin band in ^{62}Zn has only one $g_{9/2}$ neutron [9] while the configuration compared with band 2 in ^{199}Pb is different from what is generally used, see below. Also the transition energies resulting from *rigid rotation* are shown in Fig. 2. It is evident that the observed bands deviate strongly and in typical ways from rigid rotation and that these deviations are described with a good accuracy by the calculations. The Zn and Sb bands are observed and calculated up to the maximum spin values which can be built in the configurations according to the Pauli principle. In these bands, the energy cost to obtain the last spin units before termination is quite high. These features have been discussed extensively, see e.g. Refs. [7, 9, 1].

In the Pb bands, an even higher energy cost to obtain the highest spin values is observed and calculated. While calculations and experiment are only compared at high spin for the Pb bands, the full experimental bands are drawn in the lower left panel of Fig. 2. The 'standard interpretation', see e.g. Ref. [3], is that these bands correspond to a different number of aligned $i_{13/2}$ holes. If the number of aligned $i_{13/2}$ holes coincides with the total number of $i_{13/2}$ holes, these configurations can be labelled $(p_{1/2},f_{5/2},p_{3/2})^{n-9}(i_{13/2})^{-n}$ with $n = 1$ for the lower branch of band 1, $n = 2$ for the lower branch of band 2, $n = 3$ for the upper branch of band 1 and $n = 4$ for the upper branch of band 2. However, for one and possibly two aligned $i_{13/2}$ holes, the configuration might be better described as having also two $i_{13/2}$ holes which are paired off to $I \approx 0$. With these assignments, a smooth variation of the alignment shift between the ^{199}Pb bands would be expected. As seen in Fig. 2, this is not the case; instead the alignment in the lower branch of band 2 is close to that for the upper branch of band 1 but much larger than the alignment in the lower branch of band 1 (the spin values in band 2 have not been measured but comparisons with the corresponding band in ^{197}Pb [13] give strong support to the present assignments).

A different configuration assignment for band 2 is supported by the present calculations. Thus the agreement between calculations and experiment for band 2 is not so good for an $(i_{13/2})^{-4}$ assignment but it is much better if the upper branch of band 2 is assigned to a configuration with 3 $i_{13/2}$ neutron holes but then one proton hole in $h_{11/2}$ and one in $(s_{1/2},d_{3/2})$ instead of two holes in $(s_{1/2},d_{3/2})$. This is thus the calculated configuration which overlaps with the upper branch of band 2 of ^{199}Pb in Fig. 2. Note, however, that for our general conclusions, this difference in configuration assignment for band 2 is not important.

The calculated deformation trajectories in the (ε_2,γ)-plane for the configurations of Figs. 1 and 2 are shown Fig. 3. The bands in ^{62}Zn and ^{109}Sb, which are mainly built from *particles* relative to closed cores, start out close to prolate at low spin. With increasing spin, they gradually traverse the γ-plane and end up in non-collective terminating states at oblate shape ($\gamma = 60°$). These general predictions have been confirmed in recent life-time measurements [14, 9]. The Pb configuration on the other hand is dominated by holes relative to the core and consequently, the rotational band is built from collective rotation at oblate or near-oblate shape at low spin. With increasing spin, it approaches the $\gamma = -120^\circ$ axis, corresponding to non-collective rotation at prolate shape, but it never reaches this axis; instead it remains slightly collective even for spin values beyond I_{max}; the maximum spin value which can be formed from the valence particles and holes as defined from their distribution over high- and low-j subshells at low spin

Let us consider in some more detail the maximum spin which can be built from the particles in the active j-shells for the ^{199}Pb configuration assigned to band 1 (Fig. 1). A maximum value of $2\hbar$ is obtained for the $N = 4$ ($s_{1/2}$,$d_{3/2}$) proton holes while the $h_{9/2}$ and $i_{13/2}$ protons contribute with $11\hbar$. Combined with $6.5 + 5.5 + 4.5 = 16.5\hbar$ for the 3 $i_{13/2}$ holes and $2.5 + 1.5 + 1.5 + 0.5 + 0.5 - 0.5 = 6\hbar$ for the $N = 5$ neutron holes (if it is rquired that there are 3 such holes of each signature), this leads to a total $I_{max} = 35.5\hbar$, i.e. a few spin units higher than the highest observed spin in band 1, $31.5\hbar$. However, as mentioned above, the calculated bands do not terminate; instead the minimum in the energy surfaces is found at $\varepsilon = 0.10 - 0.15$ close to oblate shape, $\gamma \approx -65^\circ$. Indeed, it is possible to follow the bands in a continuous way to spin values well above I_{max}, see Fig. 3. This must mean that at high rotational frequencies, other j-shells than those indicated in the configuration labelling get important amplitudes in the wave-functions. The configuration assigned to band 2 shows similar properties but has somewhat larger tendencies to terminate in a non-collective prolate state ($\gamma = -120^\circ$).

The spin contribution from the different subshells have been analyzed previously for the terminating bands in the $A = 110$ region [12, 1]. It is then interesting to make a similar analysis for the ^{199}Pb configuration of Fig. 1. At very low spins, the spin is almost exclusively built from the $i_{13/2}$ holes coming very close to the 'maximal value' $16.5\hbar$ already at very low rotational frequencies. At intermediate spin values, the contribution from the $N = 5$ neutron holes is rather large, exceeding $6\hbar$ already for total spin values below $I = 30\hbar$. Note that if the active $N = 5$ neutron orbitals had no components outside the subshells $p_{1/2}$, $f_{5/2}$ and $p_{3/2}$, the maximum spin from these 6 holes is $6\hbar$. The fact that this value is exceeded already at a rather low frequency must mean that the active hole orbitals continuously get larger and larger components in the higher-j $N = 5$ subshells, $h_{9/2}$ and $f_{7/2}$.

At low rotational frequencies, the spin vectors of the $h_{9/2}$ and $i_{13/2}$ protons are expected to be approximately perpendicular to the rotation axis. Consequently, at low spin they have a small contributions to the total spin in the present approximation. However, their contribution increases continuously reaching $\sim 2.5\hbar$ and $4.5\ \hbar$, respectively, at a total spin of $30\hbar$. Consequently, the perpendicular spin must decrease continuously leading to decreasing $M1$ transition probabilities with increasing spin. Indeed, this should be a general effect for all bands with large $M1$ transition probabilities; with increasing spin values all spin vectors will gradually align along one axis and the perpendicular components will decrease leading to decreasing $B(M1)$-values.

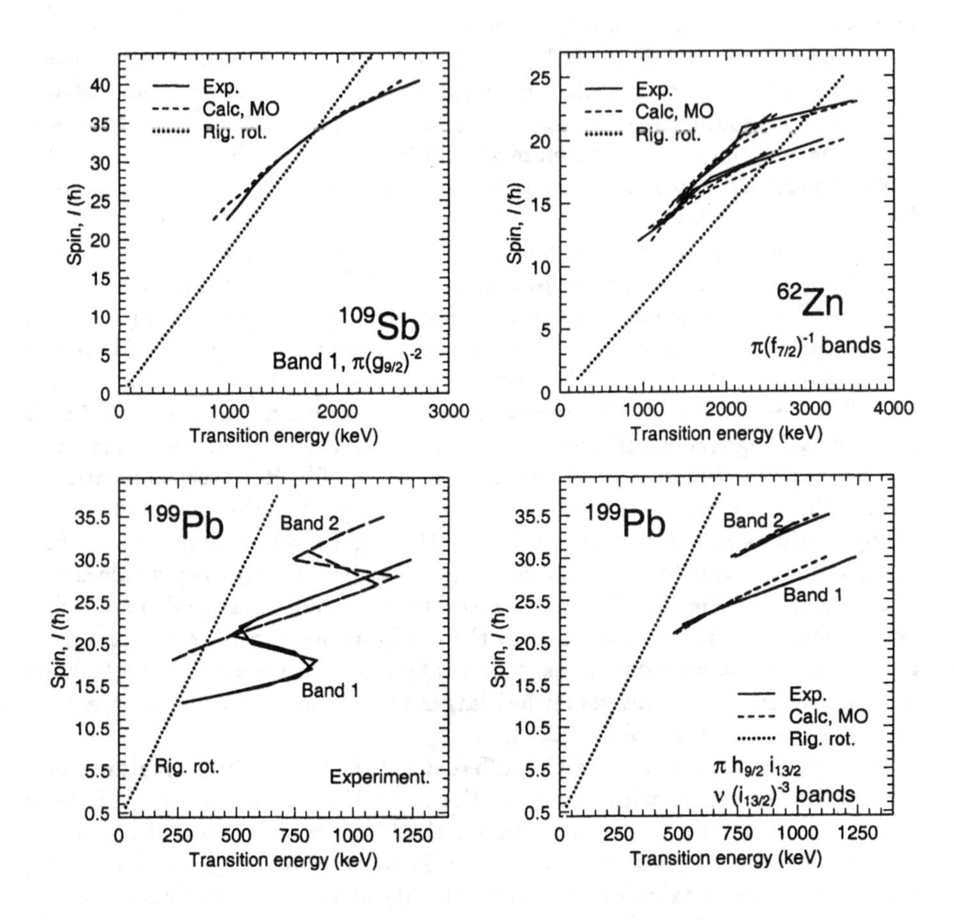

Figure 2 Spin I vs transition energy E_γ for terminating bands with experimentally measured spin values in ^{109}Sb and ^{62}Zn and for magnetic bands in ^{199}Pb. The lower left panel shows experimental data for the bands 1 and 2 in ^{199}Pb through their respective backbends. In the other three panels, experimental data are compared with calculations within the cranking Nilsson-Strutinsky approach with the modified oscillator. Typical features of the calculated configurations are indicated in each panel. In the comparison in the lower right panel, only the region above the backbends for the ^{199}Pb bands is shown. Both of the calculated configurations have three neutron holes in $i_{13/2}$ but differ from the orbitals of the proton holes. Thus, both these holes are in the $(s_{1/2},d_{3/2})$ orbitals for the configurations compared with band 1 while one hole is in the $(s_{1/2},d_{3/2})$ orbitals and one in $h_{11/2}$ for the calculated configuration compared with band 2. Note that both signatures are plotted for the Pb bands, but (except in band-crossing regions) they coincide so that only one line can be discerned both in experiment and in calculations. In all panels, the transition energies resulting from rigid rotation at a typical prolate deformation ($\varepsilon_2 \approx 0.25$) are shown by a dotted line.

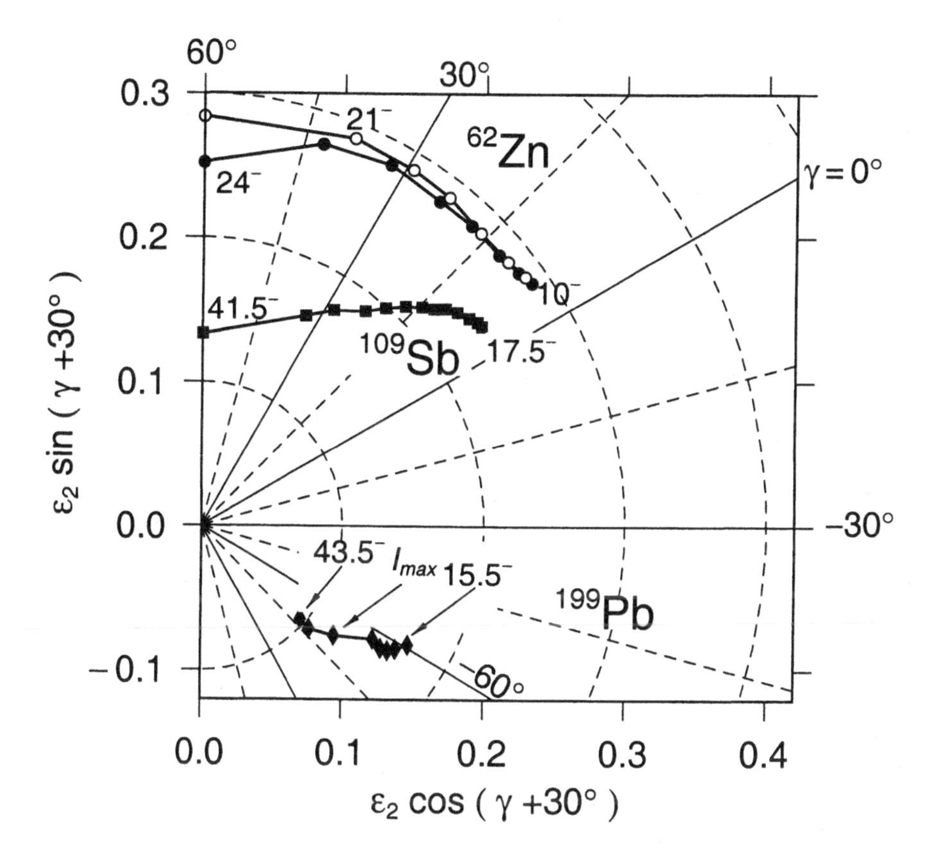

Figure 3 Calculated deformation trajectories for configurations considered in Fig. 2. The deformation is shown in steps of $2\hbar$ for the Zn and Sb bands and in steps of $4\hbar$ for the Pb band. Only the band with the proton holes in $(s_{1/2}, d_{3/2})$ is shown for ^{199}Pb. The discontinuities in this trajectory are not numerically significant. The configuration with one of the proton holes in $h_{11/2}$ instead shows a larger tendency to terminate in a non-collective state when reaching I_{max}; the maximum spin value which can be formed from the valence particles and holes as defined from their distribution over high- and low-j subshells (Fig. 1) at low spin. The calculated deformation for the I_{max}-state is indicated for the ^{199}Pb band.

3 DISCUSSION AND SUMMARY

The analogy between typical terminating bands and typical magnetic bands has been pointed out. Both these kinds of bands are characterized by a limited number of particles and holes outside closed shells, corresponding to small collectivity and a value of I_{max} which is not too high and within experimental reach.

An attempt was made to describe both terminating bands and magnetic bands within a principal axis cranking approach where thus only the spin contribution along the rotational axis is considered. This approach has previously been shown to be very successful for terminating bands. The present calculations show that also the high spin parts of magnetic bands can be described quite accurately within this approach what concerns energy vs. spin. An interesting feature is that the magnetic bands in Pb nuclei appear not to terminate at $I = I_{max}$.

Transition probabilities have not been considered in the present calculations. However, it is evident that with increasing spin, the spin vectors of all valence particles and valence holes will to a larger and larger extent become aligned along one axis. Consequently, the $B(M1)$ transition probabilities will decrease with spin in general agreement with experiment.

The question is then if also the $B(E2)$ transition probabilities can be understood. The small deformations we calculate for the Pb bands suggest small $B(E2)$ values. Even so, it seems that the observed values [2], or the upper limits put on the observed values, are even smaller than suggested from a rotational formula. However, for the Pb bands, a direct use of the rotational formula might be questionable. One specific feature is that for the ^{199}Pb configuration shown in Fig. 1, there is almost no spin contribution from the protons except from the 'high-K' orbitals, $h_{9/2}$ and $i_{13/2}$. Indeed, with a very small collective spin component from the protons, a very small $B(E2)$ is expected. Thus, we conclude that additional studies are needed before it can be determined if the very small observed $B(E2)$ transition probabilities in some magnetic bands can be understood from 'standard rotation' or if some very different features are present in these bands.

References

[1] A.V. Afanasjev, D.B. Fossan, G.J. Lane and I. Ragnarsson, Phys. Rep., to be publ.

[2] R.M. Clark *et al.*, Phys. Rev. Lett. 78, 1868 (1997).

[3] H. Hübel, Prog. Part. Nucl. Phys. 38, 89 (1997) and these proceedings.

[4] S. Frauendorf, Nucl. Phys. A 557, 259c (1993).

[5] T. Bengtsson and I. Ragnarsson, Physica Scripta T5, 165 (1983).

[6] G. Baldsiefen *et al.*, Nucl. Phys. A 574, 521 (1994).

[7] I. Ragnarsson, V.P. Janzen, D.B. Fossan, N.C. Schmeing and R. Wadsworth, Phys. Rev. Lett. 91, 3935 (1995).

[8] A. Galindo-Uribarri *et al.*, Phys. Lett. B422, 45 (1998).

[9] C.E. Svensson *et al.*, Phys. Rev. Lett. 80, 2558 (1998).

[10] H. Hübel *et al.*, Z. Phys. A358, 237 (1997).

[11] T. Bengtsson and I. Ragnarsson, Nucl. Phys. A 436, 14 (1985).

[12] A.V. Afanasjev and I. Ragnarsson, Nucl. Phys. A 591,387 (1995).

[13] G. Baldsiefen *et al.*, Nucl. Phys. A 587, 562 (1995).

[14] R. Wadsworth *et al.*, Phys. Rev. Lett. 80, 1174 (1998).

A DECADE OF PROTON CONTINUUM SCATTERING AT THE NAC – WHAT HAVE WE LEARNT?

W.A. Richter[1], S.W. Steyn[1], A.A. Cowley[1], G.C. Hillhouse[1], J.A. Stander[1], J.W. Koen[1], G.J. Arendse[1], S. Wyngaardt[1], R. Lindsay[2], G.F. Steyn[3], J.J. Lawrie[3], S.V. Förtsch[3], J.V. Pilcher[3], P.E. Hodgson[4], and M.B. Chadwick[5]

[1],Department of Physics, University of Stellenbosch, Stellenbosch, 7600, South Africa
[2]Department of Physics, University of the Western Cape, Bellville 7530, South Africa
[3]National Accelerator Centre, Faure 7131, South Africa
[4]Department of Physics, University of Oxford, OX1 3RH, United Kingdom
[5]University of California, Theoretical Division,
Los Alamos National Laboratory, Los Alamos, NM 87545, USA

Abstract: Double differential cross sections for inclusive (p,p') reactions have been measured at the National Accelerator Centre, Faure, South Africa for a range of targets with mass numbers between 12 and 197, and for incident energies from 80 to 200 MeV. Extensive comparisons have been made with multistep direct calculations based on the Feshbach, Kerman and Koonin theory. Good general agreement has been found using a short-range Yukawa two-body interaction, and some systematic discrepancies at the lowest and highest excitation energies have also been investigated. In the most recent experiments analyzing powers have been measured in addition to the cross-sections.

1 INTRODUCTION

The object of this talk is to give an overview of the work on proton continuum scattering at the National Accelerator Centre over the last decade, and to give some indication where an extension of this work may lead to in the near future. When the 200 MeV open-sector cyclotron became fully operational towards the end of 1986, it was imperative to find niches in the broad spectrum of experimental nuclear physics which could be investigated with the unique multipurpose facility available, viz. basic research in nuclear physics and chemistry, radiotherapy and medical research, and isotope production. In terms of

The Nucleus: New Physics for the New Millennium
Edited by Smit et al., Kluwer Academic / Plenum Publishers, New York, 2000.

output of papers and students, complementary theoretical activity stimulated, and international collaboration, the work on (p, p') turned out to be such a niche.

One of the main reasons for the interest in proton-nucleus scattering in the 100 MeV region leading to the continuum is that a significant part of the cross-section consists of pre-equilibrium reactions. Apart from fundamental studies of nuclear reactions, such cross-sections are of particular interest to proton radiotherapy, which forms an integral part of the applications of the separated-sector cyclotron at NAC. To explain the emission of particles from such reactions Griffin (1966) [1] proposed the exciton model. In this model the incident particle experiences a succession of energy losses due to collisions with nucleons in the nucleus, creating a particle-hole excitation in each collision by promoting the struck nucleon to an energy level above the Fermi surface. A sequence of particle-hole excitations of greater complexity arises as a result e.g. 1p-1h, 2p-2h, 3p-3h, ... A fully quantum-mechanical theory of the process was formulated by Feshbach (1973,1977) [2],[3] and Feshbach, Kerman and Koonin (1980) [4]. Two distinct types of processes play a role, depending on the degree of excitation:

1) The statistical multistep direct emission of the projectile (MSD), in which the projectile remains unbound at all times.

2) The statistical multistep compound emission of a nucleon (MSC), in which all particles are bound.

In the latter process a nucleon is typically emitted with energies up to a few tens of MeV, and can be largely excluded from a study of the MSD process by limiting the emission energy. We have generally restricted measured emission energies in our studies to at least 20 MeV lower than the incident energy.

The study of proton scattering to the continuum is important for a number of reasons:

- It sheds light on the nucleon-nucleus reaction mechanism.
- It allows a study of the nucleon-nucleon (N-N) interaction at energies typically encountered in a nucleus.
- Modifications in the N-N interaction due to the nuclear medium can be studied via quasielastic scattering (QES).
- Relativistic models, e.g. Dirac-based, can be tested provided polarization transfer observables can be measured.

  From the point of view of practical applications, there are also a number of issues of growing importance:

- Optimization of dose delivery in proton radiotherapy requires nuclear cross-section data.
- Accelerator-driven technology to transmute radioactive waste.
- Radiation protection studies.

2 SUMMARY OF (p,p′) WORK

The following is a summary of inclusive (p,p') reaction studies carried out at the National Accelerator Centre since 1989. Targets and incident energies are listed. The spokesperson for each experiment is indicated in brackets.

Table 1 : (p,p') EXPERIMENTS ON THE 200 MEV OPEN-SECTOR CYCLOTRON, NATIONAL ACCELERATOR CENTRE, FAURE

Experiment	Targets	Energies / Reference
• Sept. 1987 (Cowley)	^{12}C	$E_p = 90,\ 200$ MeV NPA 485 (1988) 258
• Jun. - July 1988 (Cowley)	^{58}Ni, ^{100}Mo, ^{197}Au	$E_p = 120,\ 150,\ 175,\ 200$ MeV Z.Phys A 336 (1990) 189 PRC 43 (1991) 691 PRC 46 (1992) 1030
• Nov. 1989 (Cowley)	^{90}Zr	$E_p = 80,\ 120$ MeV PRC 43 (1991) 678
• Feb. – Aug. 1991 (Richter)	^{90}Zr, ^{89}Y, ^{92}Mo, ^{94}Mo, ^{96}Mo, ^{98}Mo	$E_p = 120,\ 160,\ 200$ MeV PRC 49 (1994) 1001
• May. – Oct. 1992 (Richter)	^{115}In, ^{141}Pr, ^{167}Er, ^{173}Yb, ^{181}Ta	$E_p = 120,\ 150,\ 175,\ 200$ MeV PRC 54 (1996) 1756

• Feb. – March 1995. (Cowley) Targets of biological interest (also studied (p,d), (p,t), and at 200 MeV $(p,{}^3He)$ and (p,α) as well)

Targets	Energies / Reference
^{12}C, ^{14}N, ^{16}O	$E_p = 150,\ 200$ MeV To be published in Nucl. Phys. A

- June - July 1996; Dec. 1996 - Jan. 1997 (Richter)

POLARIZED BEAM: MEASURE DDX* AND A_y

$$\left.\begin{array}{l} ^{40}\mathrm{Ca} \\ ^{24}\mathrm{Mg} \\ ^{51}\mathrm{V} \end{array}\right\} \quad E_p = 150,\ 165\,(\mathrm{Ca}),\ 186\,\mathrm{MeV}$$

* Double differential cross-section

From the above table it is evident that the targets used covered a wide range of mass numbers. One of the reasons for this is that the mass dependence of the two-body effective interaction used was one of the important issues investigated. The incident proton energies employed ranged between 80 MeV to 200 MeV. Double differential cross-sections were measured in all cases and in the last two runs analysing power was measured in addition.

The basic experimental setup used in most of the experiments has been described in refs. [5, 6, 7]. Proton beams were delivered by the separated-sector cyclotron of the National Accelerator Centre to a 1.5-m-diameter scattering chamber which contained the targets and detector telescope(s). The detector arrangement usually consisted of a simple $\Delta E-E$ detector telescope comprising one or more Si surface barrier detectors and a NaI stopping detector. A plastic scintillator active collimator provided a veto signal for protons scattered off the brass collimators. For the ΔE Si detectors a ^{228}Th source was used for the calibration. The NaI detector energy calibration was based on elastic and inelastic scattering of protons from ^{12}C and H in a CH_2 target. Standard $\Delta E-E$ particle identification and timing techniques were used. Angular distributions were generally measured in steps of 5° or 10° for angle values up to about 160°.

For the analyzing power measurements two identical detector telescopes were placed at identical angles with respect to the incoming beam, but these measurements will be discussed in a separate section.

For the study of the targets of biological interest (C,N,O) the target gas was contained in a cylindrical cell of 10 cm in diameter, having entrance and exit windows of 6 μm Havar foil. The gas cell was connected via a pressure regulator to an inlet manifold through which it could be filled with any of the target gases. Isobutane was used as the carbon target since the experimental set-up was for a gas target, unsuitable for solid targets. Measurements with the cell filled with hydrogen enabled the subtraction of the p+^{1}H contribution in this case. The effective target length and solid angle were defined by means of a double-aperture collimator system having passive and active collimators at both the front and rear apertures. In this way the background caused by scattered particles generated by the Havar windows of the gas cell could be reduced to a level below 2%. Data were collected at scattering angles ranging from 20° to 150°, with all angles determined to better than 0.05°. The effective target length varied between 6 and 19 mm, depending on the scattering angle. The solid angle subtended was 1.5 msr and the angular resolution 2°.

3 THE STATISTICAL MULTISTEP DIRECT REACTION THEORY

Extensive comparisons for most of the measured double differential cross-sections for targets in Table 1 have been made with theoretical values calculated from the FKK theory [8],[9], [10], [11], [12], [13]. The calculations for the nuclei of biological interest are still in progress. The details of the fully quantum-mechanical FKK theory have been presented in several papers [4] and a review article [14].

In our calculations the multistep direct code of Bonetti and Chiesa [15] has been used. The DBWA cross-sections for a selection of single-particle transitions based on a spherical Nilsson shell model are first calculated using the programme DWUCK4. The global optical potentials of Schwandt et al. [16] and Madland [17] have generally been employed. These cross-sections are then used in a subroutine MUDIR to calculate the contributions to the double differential cross-section for a sequence of steps corresponding to different particle-hole excitations. In a typical calculation about 6 different steps are used. Many extensions and improvements to the code have been made subsequently [18]. In most multistep calculations a simple short-range Yukawa interaction with effective strength V_0, obtained from normalizing to the data, has been used for the two-body effective interaction.

4 COMPARISON WITH NAC DATA

A representative example of an angular distribution comparison is shown in Fig. 1.

The double differential cross-sections were measured for different emission energies of the outgoing proton, at 20 MeV intervals. This corresponds to excitation energies $U = E_p - E_p'$ of the residual nucleus in the case of one-proton emission. It is evident that good general agreement is obtained between theory and experiment over many orders of magnitude. At the highest and lowest emission energies there are systematic discrepancies which are found for all the cases investigated.

In order to explain the discrepancies, it was necessary to consider a variety of assumptions and approximations in the application of the theory, and the probable magnitude of their effects on the final calculated quantities.

5 REFINEMENTS IN THE APPLICATION OF THE THEORY

An extensive list of factors which have a bearing on the accuracy of the calculations have been given some consideration. Many of these effects have been discussed in the papers connected with this work (See in particular ref. [9], [11], [19], [12]), but some merit special mention here. The magnitudes of calculated

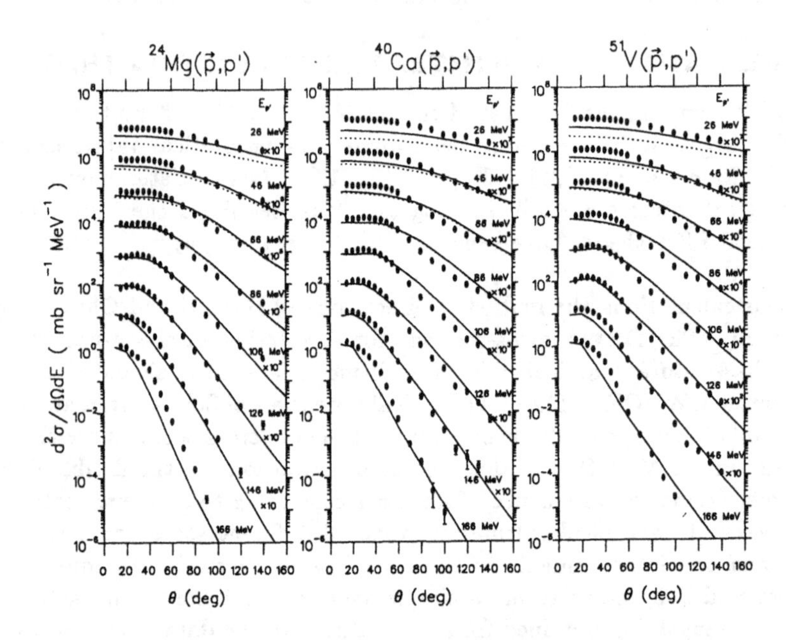

Figure 1 Angular distributions for ^{24}Mg($\vec{p}, p'$), ^{40}Ca($\vec{p}, p'$) and ^{51}V($\vec{p}, p'$) at 186 MeV incident energy and various emission energies $E_{p'}$. The symbols represent experimental results while the curves are results of MSD calculations. The dashed lines correspond to primary emission only while the solid lines take the emission of two protons into consideration. Statistical error bars are shown where these exceed the symbol size. Results are multiplied by the indicated factors for display and are given in the laboratory system.

MSD cross-sections show a significant dependence on the optical potential employed. For example, in ref. [9] it is shown that despite the similarity of the Schwandt and Madland optical potentials, there is more than a factor of two difference in the cross-sections for the same set of input parameters with the two optical potentials. Because the calculations are normalised to the experimental data by an appropriate choice of the effective interaction strength V_0, it means that the choice of optical potential affects the V_0 value extracted, a point to bear in mind when comparing different V_0 values.

Two factors turn out to be of particular importance, viz. the inclusion of multiple nucleon emission in the theory, and the use of a more realistic interaction which incorporates the spin dependence and the isospin dependence of the two-body N-N interaction. At the incident energies employed at NAC it becomes very likely that a second nucleon is emitted in the multistep chain. Two-nucleon emission has been addressed by Chadwick *et al.* [20] in an approximate way, in which a number of simplifying assumptions are made, e.g.

that the second nucleon is emitted from the same step (i.e. np-nh excitation) as the first, primary nucleon. This method has been used in conjunction with the Milan code and generally results in a significant reduction of the discrepancy between theory and experiment for forward angles and high excitation energies. An example of taking into account this effect is shown in Fig. 1.

As regards the two-body interaction, the simple central Yukawa interaction conventionally used cannot reproduce analyzing power measurements because it lacks a spin dependence. In addition, it is expected that the strength of the two-body interaction should increase along the multiple scattering chain as the projectile loses energy, and the precise form of the energy dependence of V_0 is only known approximately from phenomenology. A promising development is the formulation of a two-component multistep direct theory due to Koning and Chadwick [21]which distinguishes between protons and neutrons in the multistep cascade. This extension of the multistep direct theory is essential when considering more realistic interactions, which would necessarily have different interaction strengths for n-p and p-p interactions.

Recently a fully microscopic interaction has been developed from an N-N g-matrix approach by Amos and his group at the University of Melbourne, and it has yielded impressive results for the agreement of predicted analyzing powers with experimental measurements for numerous nuclei [22]. This would be an appropriate interaction to use in the multistep calculations, and its implementation for this purpose using the Milan code of Bonetti is in progress. In addition a nonlocal optical potential which is based on proper antisymmetrization of the projectile and target nucleons can be obtained by folding the N-N g-matrices with the appropriate density matrices. The program DWBA91 of Raynal [23] allows calculations with such a realistic interaction. Good agreement obtained with experiment for analyzing powers in particular makes the interaction a prime candidate for multistep direct calculations and for comparison of the results with polarized proton continuum scattering data.

6 ANALYZING POWER MEASUREMENTS

In the first run (June - July 1996; refer to table 1) the polarization of the incident protons varied for various segments of the run and fairly low values were obtained. In the next run stable polarization values between 70 - 80 percent were obtained during the experiment. Beam polarization was checked between measurements by using the scattering chamber as a polarimeter with a ^{12}C target. Double differential cross sections were measured as well as analyzing power $A_y = \frac{1}{P}\frac{\sigma_L - \sigma_R}{\sigma_L + \sigma_R}$ for angles from 15° to 80° in 5° steps, and thereafter in 10° steps to 140°.

7 CONCLUSIONS

The bulk of the world's (p, p') data has been provided by the proton continuum scattering programme at NAC. Extensive comparisons of the angular distribu-

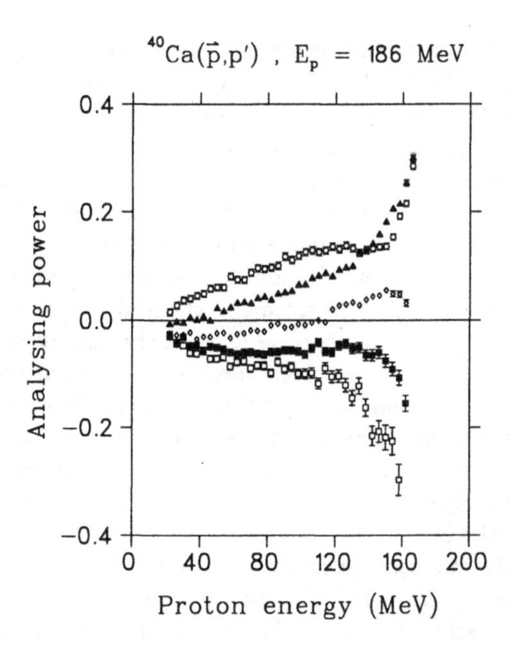

Figure 2 Analyzing power for ^{40}Ca($\vec{p}, p'$) at incident energy of 186 MeV as function of emission energy and various angles of emission. The open circles correspond to 15°, the solid triangles to 25°, the diamonds to 35°, the solid squares to 45° and the open squares to 55°. The symbol size correspond to the estimated statistical error in the data where error bars are not shown.

tions measured with theoretical analyses based on the Feshbach, Kerman and Koonin multistep direct theory have shown the theory to be generally applicable over a wide mass range, and also down to very low masses ($A = 12$), and a wide range of incident energies, up to the highest energy available (200 MeV). Cross-section data can be quite well reproduced as a rule using a short-range central Yukawa interaction. Deviations at the extremes of excitation can at least be qualitatively explained, but more complete quantitative work on, for example multinucleon emission, is required. Nonetheless, even approximate corrections for this effect gives much better agreement with the data.

The results of MSD calculations and the final fitted V_0 values obtained are quite sensitive to the set of input parameters used. Therefore comparisons of results obtained with different MSD codes should take into account the dependence on the input parameters.

In order to explain analyzing power data the full complexity of the nucleon-nucleon interaction must be considered. This includes the spin dependence as well as the isospin dependence, so that a two-component theory of the multistep chain must be used together with such realistic interactions.

8 THE FUTURE

The NAC polarized beam facility provides excellent analyzing power data and should be fully utilized in order to place constraints on MSD calculations. Because of the importance of multinucleon emission at typical NAC energies, and the uncertainty as to the relative contribution of such secondary emission, experiments using some coincident detector array to obtain more reliable estimates of the amount of secondary emission in MSD reactions would be very useful. Also experiments at other facilities to measure complete sets of polarization transfer observables (or at NAC with a focal plane polarimeter) need to be considered in the future. This will enable choices between relativistic models of the effective interaction to be made, or test the adequacy of non-relativistic models.

References

[1] Griffin, J. J. Phys. Rev. Lett. 17, 478.

[2] Feshbach, H., Proc. Int. Conf. on Nuclear Physics, Munich, eds., J. J. de Boer and H. J. Mang (North-Holland, Amsterdam) Vol II, p. 631.

[3] Proc. Int. Conf. on Nulcear Reaction Mechanisms, Varenna (CLUED), p. 1.

[4] H. Feshbach, A. Kerman, S. Koonin, Ann. Phys. 125, (1980) 477

[5] S. V. Förtsch, A. A. Cowley, J. J. Lawrie, D. M. Whittal, J. V. Pilcher and F. D. Smit, Phys. Rev. C43, 691 (1991)

[6] J. V. Pilcher, A. A. Cowley, D. M. Whittal, and J. J. Lawrie, Phys. Rev. C40, 1937 (1989)

[7] A. A. Cowley, S. V. Förtsch, J. J. Lawrie, D. M. Whittal, F. D. Smit, and J. V. Pilcher, Z. Phys. A336, 189 (1990)

[8] A. A. Cowley, A. van Kent, J. J. Lawrie, S. V. Förtsch, D. M. Whittal, J. V. Pilcher, F. D. Smit, W. A. Richter, R. Lindsay, I. J. van Heerden, R. Bonetti, and P. E. Hodgson, Phys. Rev. C43, 678 (1991).

[9] W. A. Richter, A. A. Cowley, R. Lindsay, J. J. Lawrie, J. V. Pilcher, S. V. Förtsch, R. Bonetti, and P. E. Hodgson, Phys. Rev. C46, 1030 (1992)

[10] Workshop on Multistep Direct Reactions, ed. R. H. Lemmer, World Scientific (1991), p. 159.

[11] W. A. Richter, A. A. Cowley, G. C. Hillhouse, J. A. Stander, J. W. Koen, S. W. Steyn, R. Lindsay, R. E. Julies, J. J. Lawrie, J. V. Pilcher, P. E. Hodgson. Phys. Rev. C49, 1001 (1994)

[12] W. A. Richter, A. A. Cowley, S. W. Steyn, J. A. Stander, J. W. Koen, J. J. Lawrie, J. V. Pilcher, G. C. Hillhouse, R. E. Julies, R. Lindsay, M. G. van der Merwe, and P. E. Hodgson, Phys. Rev. C54, 1756 (1996)

[13] S. W. Steyn, Ph. D. Thesis, University of Stellenbosch (1997)

[14] R. Bonetti, A. J. Koning, J. M. Akkermans, P. E. Hodgson, Phys. Rep. 247, 1 (1994).

[15] R. Bonetti, C. Chiesa, University of Milan, unpublished.

[16] P. Schwandt, H. O. Meyer, W.W. Jacobs, A. D. Bacher, S. E. Vigdor, M. D. Kaitchuck and T. R. Donoghue, Phys. Rev C26, 55 (1982).

[17] D. G. Madland, Los Aalamos National Laboratory Report LA-UR-87-3382, unpublished.

[18] W. A. Richter, R. Lindsay, private communication.

[19] W. A. Richter, Acta Phys. Slov. 45, 733 (1995)

[20] M. B. Chadwick, P. G. Young, D. C. George and Y. Watanabe, Phys. Rev. C50, 996 (1994)

[21] A. J. Koning, M. B. Chadwick, Phys. Rev. C56, 970 (1997)

[22] Ph.D. thesis: S. Karataglidis, Univ. of Melbourne, Australia ; P. J. Dortmans and K. Amos, Phys. Rev. C49, 1309 (1994); S. Karataglidis, P. J. Dortmans, K. Amos and R. de Swiniarski, Phys. Rev. C52, 861 (1995).

[23] J. Raynal, computer code DWBA91 (unpublished)

THE A~130 HIGHLY-DEFORMED REGION: NEW RESULTS AND GLOBAL DIFFERENTIAL LIFETIME MEASUREMENTS

M.A. Riley[1], F.G. Kondev[1*],
R.W. Laird[1], J. Pfohl[1 †], D.E. Archer[1,2], T.B. Brown[1‡], R.M. Clark[3],
M. Devlin[4], P. Fallon[3], D.J. Hartley[1 §], I.M. Hibbert[5 ¶], D.T. Joss[6],
D.R. LaFosse[4 ||], F. Lerma[4], P.J. Nolan[6], N.J. O'Brien[5], E.S. Paul[6],
D.G. Sarantites[4], R.K. Sheline[1], S.L. Shepherd[6], J. Simpson[7]
and R. Wadsworth[5]

1. Dept. of Physics, Florida State University, Tallahassee, Florida 32306, USA
2. Lawrence Livermore National Laboratory, Livermore, CA 94550 USA
3. Nuclear Science Division, LBNL, Berkeley, CA94720, USA
4. Dept. of Chemistry, Washington University, St. Louis, MO 63130, USA
5. Dept. of Physics, University of York, York Y01 5DD, UK
6. Oliver Lodge Laboratory, University of Liverpool, Liverpool L69 7ZE, UK
7. CLRC, Daresbury Laboratory, Daresbury, Warrington, WA4 4AD, UK

*Present address: Argonne National Laboratory, Argonne, IL 60439, USA.
†Present address: Sandia National Laboratories, Albuquerque, NM 87185, USA.
‡Present address: Chemistry Dept., Univ. of Kentucky, Lexington, KY 40506, USA.
§Present address: Dept. of Physics and Astronomy, Univ. of Tennessee, Knoxville, TN 37996, USA.
¶Present address: Oliver Lodge Lab., Univ. of Liverpool, Liverpool L69 7ZE, UK.
||Present address: Dept. of Physics and Astronomy, SUNY at Stony Brook, New York 11794, USA.

The Nucleus: New Physics for the New Millennium
Edited by Smit et al., Kluwer Academic / Plenum Publishers, New York, 2000.

Abstract: Two experiments have recently been performed in the A∼130 highly-deformed region using GAMMASPHERE and the Microball. A large number of highly-deformed structures have been observed (many of them for the first time). In addition, the quadrupole moments for a variety of configurations, including the $9/2^+[404]$ ($g_{9/2}$) proton, $1/2^+[660](i_{13/2})$ and $1/2^-[541]$ ($f_{7/2}$,$h_{9/2}$) neutron orbitals, were measured in a wide range of nuclei using the Doppler-shift attenuation method. While the involvement of the first two orbitals leads to quadrupole deformations that are comparable to those observed for the so-called superdeformed bands in this mass region, the β_2 values for structures that include the $1/2^-[541]$ neutron are found to lie intermediate between those observed for normally deformed and highly deformed bands. Deformation trends for the same configuration as a function of N are also established. New results on ^{135}Pm (N =75) are presented. Some unresolved questions and perspectives for the region are discussed.

1 INTRODUCTION

I (MAR) remember well the day that Andrew Kirwan and Paul Nolan discovered the first discrete superdeformed band in ^{132}Ce. It was 1984 and I was sitting next to them on the second floor of the Oliver Lodge Laboratory in Liverpool when the first gated spectra started being projected on their computer screen. The importance of this result was quickly realized [1] and a lifetime measurement confirming the large deformation $\beta_2 \approx 0.5$ [2] was later performed. These results came at a similar time to the discovery of the discrete superdeformed band in ^{152}Dy [3, 4] by the Liverpool-Daresbury-NBI collaboration in which John Sharpey-Schafer played such a leading role, along with Peter Twin and Barna Nyakó. These, and many other glorious discoveries at Daresbury Laboratory using the beloved TESSA series of Compton suppressed Ge spectrometers helped re-ignite the field of gamma-ray spectroscopy. This renaissance continues to this day and our distinguished host can be justly proud of his pivotal role in this revolution!

The recent studies of highly deformed (β_2=0.3−0.5), sometimes referred to as "superdeformed", structures in the region near mass A∼130 have revealed an important interplay between microscopic shell effects, such as the occurrence of large gaps in the nucleon single-particle energies, and the occupation of high-j low-Ω (intruder) orbitals in driving the nucleus towards higher deformation. Initially, it was thought that only the involvement of one or more $i_{13/2}$ neutrons could result in a strong polarization on the nuclear shape in this mass region [5]. However, it has been shown more recently that bands built upon the $9/2^+[404]$ ($g_{9/2}$) proton orbital in the odd-Z ^{131}Pr [6] and ^{133}Pm [7] isotopes, exhibit quadrupole deformations comparable to the values found for highly deformed structures which include $i_{13/2}$ neutrons. Furthermore, for nuclei below $N = 73$ where the occupancy of the $\nu i_{13/2}$ orbital is energetically unfavored, there are indications that bands involving the $1/2^-[541]$ ($f_{7/2}$,$h_{9/2}$) neutron may also be highly deformed [8, 9].

In order to elucidate the impact of the occupation of specific orbitals on the nuclear deformation, accurate lifetime measurements for a large number of bands in a variety of nuclei have been measured. In addition, we report on the observation of two high moment of inertia bands in the odd-Z $N = 75$ ^{135}Pm nucleus and discuss several unresolved issues in this mass region and some future perspectives.

2 DIFFERENTIAL QUADRUPOLE MOMENT MEASUREMENTS IN A~130 HIGHLY DEFORMED NUCLEI

While the quadrupole moment, Q_0, for some of the highly deformed bands in the A~130 region had been measured in the past using the Doppler-shift attenuation method (DSAM), conclusive comparisons between different nuclei were limited owing to systematic distinctions between experimental setups such as varying reactions and target retardation properties. Specifically, due to differences in the parameterization of the nuclear and electronic stopping powers, which act as an "internal clock" in the DSAM lifetime measurements, large variations in the measured Q_0 values have been reported for the same band. The absence of adequate experimental information on the time structure of the quasicontinuum sidefeeding contributions also results in an additional inaccuracy on the measured quadrupole moments. An excellent introduction to the DSAM technique and a detailed discussion of associated uncertainties with the method can be found in Ref. [10]. In the current work we have greatly reduced systematic problems by measuring the lifetime decay properties of a large selection of bands in different nuclei under nearly identical experimental conditions in terms of angular momentum input, excitation energy and recoil velocity profile. Furthermore, the high efficiency and resolving power of GAMMASPHERE made it possible in favorable cases, to greatly minimize the effect of sidefeeding on the measured quadrupole deformations, by gating on shifted transitions at the top of the band of interest, thus gaining some insight into the nature and time scale of the sidefeeding.

High-spin states in a wide range ($Z = 58 - 62$) of nuclei were populated after fusion of a ^{35}Cl beam with ^{105}Pd target nuclei. Thin and backed target experiments were performed at the 88-Inch Cyclotron at the Lawrence Berkeley National Laboratory with beam energies of 180 (thin target) and 173 MeV (backed target). The thin target consisted of an isotopically enriched ^{105}Pd foil with a thickness of 500 μg/cm^2. The backed target was a 1 mg/cm^2 thick ^{105}Pd foil mounted on a 17 mg/cm^2 Au backing. Emitted γ-rays were collected using the GAMMASPHERE spectrometer [11] consisting of 57 (thin target) and 97 (backed target) HPGe detectors. The evaporated charged particles were identified with the MICROBALL detector system [12], whose selection capabilities allowed a clean separation of the different charged particle channels.

The present work focuses on the properties of structures which involve the important $9/2^+[404]$ ($g_{9/2}$) proton, $1/2^+[660](i_{13/2})$ and $1/2^-[541]$ ($f_{7/2}$,$h_{9/2}$) neutron orbitals, in the odd-N (Z=60) ^{133}Nd (populated in the αp2n channel) and ^{135}Nd (3p2n) isotopes, and the odd-Z (Z=59) ^{130}Pr (2α2n), ^{131}Pr (2α1n) and ^{132}Pr (1α2p2n) nuclei. Typically more than about 50×10^6 (thin target) and 20×10^6 (backed target) events (of a fold ≥ 3) per particle gated channel were collected.

The backed target data were used to extract the quadrupole deformation using the centroid-shift technique in conjunction with the Doppler-shift attenuation method [10, 13]. This was done in two ways. In the first method, the data were sorted into two-dimensional matrices in which one axis consisted of "forward" (31.7° and 37.4°) or "backward" (142.6° and 148.3°) group of detectors and the other axis was any coincident detector. Spectra were generated by summing gates on the cleanest, fully stopped transitions at the bottom of the band of interest and projecting the events onto the "forward" and "backward" axes. These spectra were then used to extract the fraction of the full Doppler shift, $F(\tau)$, for transitions within the band of interest. In the second method, the data were sorted into a number of double-gated spectra which contained counts registered by particular group of detectors. Specifically, for the relatively strongly populated $\nu i_{13/2}$ bands in ^{133}Nd and ^{135}Nd, gates were also set on in-band "moving" transitions in any ring of detectors and data were incremented into separate spectra for events detected at "forward", "90°" and "backward" angles.

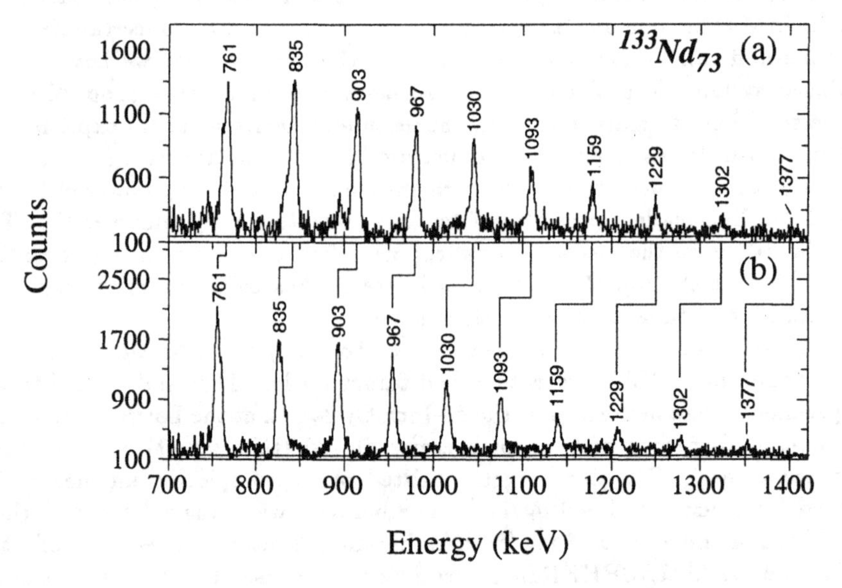

Figure 1 (a) Forward and (b) backward angle coincidence γ-ray spectra [32] for the $1/2^+[660]$ ($i_{13/2}$) band in ^{133}Nd formed from combinations of all double gates on in-band transitions from 345 keV up to 1228 keV. The peaks are labeled with the unshifted energies.

Sample spectra for the $\nu 1/2^+[660]$ ($i_{13/2}$) band in ^{133}Nd are shown in Fig. 1. It should be noted, that the implementation of the latter method made it possible to eliminate the effect of sidefeeding for states lower in the cascade.

In order to extract the intrinsic quadrupole moments from the experimental $F(\tau)$ values, calculations using the code FITFTAU [14] were performed. The

F(τ) curves were generated under the assumption that the band has a constant Q_0 value. In the modeling of the slowing process of the recoiling nuclei, the stopping powers were calculated using the 1995 version of the code TRIM [15]. The corrections for multiple scattering were introduced using the prescription given by Blaugrund [16]. Where appropriate, the sidefeeding into each state was taken into account according to the experimental in-band intensity profile using a rotational cascade of three transitions with the same Q_0 as the in-band states. It should be emphasized, that although the uncertainties in the stopping powers and the modeling of the sidefeeding may contribute an additional systematic error of 15–20% in the absolute Q_0 values, the relative deformations are considered to be accurate to a level of 5–10%. Such precision allows a clear differentiation in the Q_0 values to be made, which was used in turn as evidence for the involvement of specific orbitals within a band configuration.

1. Bands involving the $1/2^+[660]$ ($i_{13/2}$) neutron orbital: Collective structures built upon the $1/2^+[660]$ ($i_{13/2}$) intruder neuton orbital have been observed in the chain of odd-N (Z=60) Nd isotopes from ^{133}Nd up to ^{137}Nd [17, 18]. These bands have been connected to the normally deformed structures [18, 19, 20, 21, 22, 23], so that their spin, parity and excitation energy are unambiguously determined. In addition, the g-factor experiment performed in the case of ^{133}Nd [24] independently confirms the $\nu 1/2^+[660]$ configuration assignment. Quadrupole moment measurements were carried out previously using both the centroid-shift and lineshape DSAM techniques [25, 24, 26, 27, 28]. Lifetimes of low-spin members of the band in ^{135}Nd and ^{133}Nd were also measured via the Doppler-shift recoil-distance method [30, 31]. The F(τ) values and the corresponding quadrupole deformations deduced in the current work when gates were set on the stopped 409, 440, 513 and 603 keV transitions in ^{133}Nd, and 546 and 676 keV γ-rays in ^{135}Nd are shown in Figs.2(c) and 2(d). Our observations (which are of higher precision) are in agreement with the previously measured quadrupole deformations for the bands in ^{133}Nd [25], and ^{135}Nd [26]. The comparison of the intensity profiles for these two bands, shown in Figs.2(a) and 2(b), reveals that the sequence in ^{135}Nd is fed significantly from the side over a range of transitions for which the F(τ) values change very rapidly. Such a behavior led to speculations [26, 27], that the sidefeeding lifetimes could be as much as four times slower than the in-band levels which led to the deduction of a quadrupole deformation for ^{135}Nd which exceeded that of ^{133}Nd. Figs. 2(e) and 2(f) show our observations, when spectra gated on the Doppler-shifted in-band 1158, 1228, 1300 and 1377 keV γ-rays in ^{133}Nd, and 1146, and 1216 keV γ-rays in ^{135}Nd, were used.

We found a roughly 10% increase in the deformation of both these two bands, compared to values deduced when gates were set on stopped transitions. These results allow us to estimate that the sidefeeding lifetimes are only about 1.3–1.4 times slower than those for the in-band levels which is consistent with recent measurements by Clark *et al.* [33] for superdeformed structures in ^{131,132}Ce.

The present observations, together with the values for the band in ^{137}Nd (Q_0=4.0(5) [β_2=0.22(3)]) [25], indicate clearly for the first time that in the

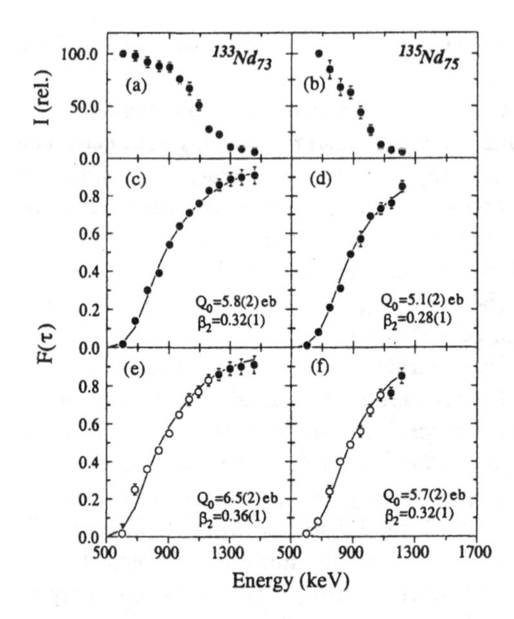

Figure 2 Intensity profiles (a) and (b), $F(\tau)$ values and corresponding quadrupole deformations deduced by gating on stopped transitions (c) and (d), and "moving" (solid) transitions (e) and (f) for the $1/2^+[660]$ ($i_{13/2}$) bands in ^{133}Nd and ^{135}Nd, respectively [32].

odd-N Nd nuclei there is a systematic decrease in the deformation of the $\nu i_{13/2}$ band as the neutron number increases [32]. Such an experimental trend is now in line with predictions by Total Routhian Surface and Ultimate Cranker calculations with pairing [25, 21], as well as by Cranked Nilsson-Strutinsky calculations [34] which do not include pairing.

2. Bands involving the $9/2^+[404]$ ($g_{9/2}$) proton orbital: Initially, a highly deformed band built upon the $9/2^+[404]$ ($g_{9/2}$) proton orbital was observed in ^{131}Pr by Galindo-Uribarri *et al.* [6]. The present work has established a value of Q_0=5.5(8) *eb* [β_2=0.32(5)] for this band which is much larger than Q_0=3.9(3) *eb* [β_2=0.23(2)] deduced for the normally deformed $\pi h_{11/2}$ structure in the same nucleus. Recently, Brown *et al.* [35] have observed a strongly coupled band in the neighboring odd-odd ^{130}Pr isotope which was suggested to include the $9/2^+[404]$ ($g_{9/2}$) proton orbital coupled to the $7/2^-[523]$ ($h_{11/2}$) neutron. The measured and calculated $F(\tau)$ values for this structure, as well as those for the normally deformed $\pi h_{11/2} \otimes \nu d_{5/2}$ band are shown in Fig.3(a). The results for the $\pi g_{9/2}$ and $\pi h_{11/2}$ configurations in ^{131}Pr, deduced from the current work, are presented in Fig.3(b). We report Q_0=6.1(5) *eb* [β_2=0.35(3)] for the $\pi g_{9/2} \otimes \nu h_{11/2}$ band [36], which is similar or perhaps slightly larger compared to the value for the $\pi g_{9/2}$ band in ^{131}Pr. Our observations for the quadrupole deformations of the $\pi g_{9/2}$ and $\pi h_{11/2}$ bands in ^{131}Pr are in agreement with the values reported by Galindo-Uribarri *et al.* [6]. It is notable that the deformation of structures that involve the $\pi 9/2^+[404]$ ($g_{9/2}$) configuration is comparable to those for the $1/2^{+}[660]$ ($i_{13/2}$) bands in the neighboring nuclei ^{133}Nd and ^{135}Nd isotopes thus confirming the important role played by the former orbital in building highly deformed structures in the region.

3. Bands involving the $1/2^{-}[541]$ ($f_{7/2}$,$h_{9/2}$) neutron orbital: Two decoupled bands, referred to as band 1 and band 2 in the current work, were identified in ^{130}Pr [37, 38] in agreement with the parallel work of Smith *et al.* [39].

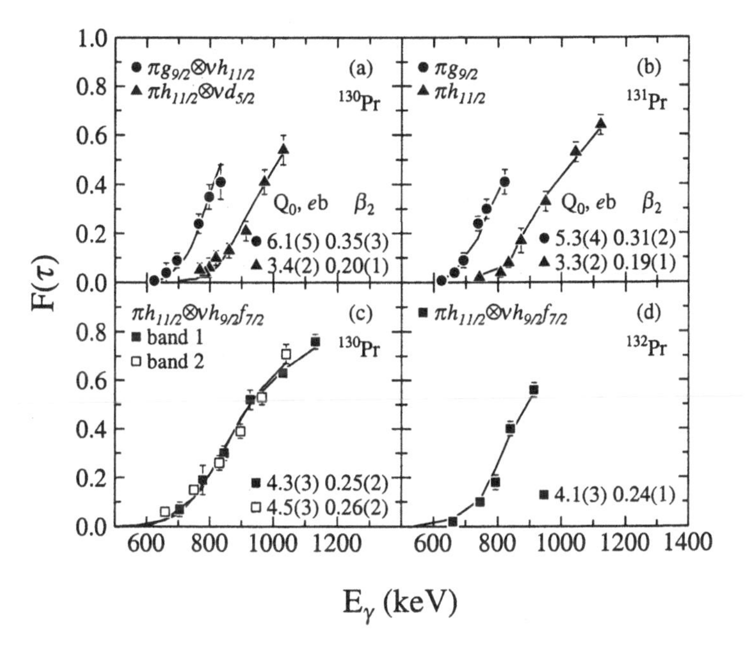

Figure 3 The experimental and calculated F(τ) values as a function of γ-ray energy for selected bands in ^{130}Pr, ^{131}Pr and ^{132}Pr Calculated curves as shown as solid lines and correspond to the best fit to the data [36, 38].

We confirm the previously reported decoupled band in ^{132}Pr [40, 41], but we propose different spin values compared to Ref. [41], and identify several additional in-band and inter-band transitions. These observations, together with the measured rotational alignments, band crossing properties and the orbitals expected near both the proton and neutron Fermi surfaces, led us to conclude that the configuration of these decoupled structures in ^{130,132}Pr includes the $1/2^{-}[541]$ ($f_{7/2}$,$h_{9/2}$) neutron orbital coupled to the $3/2^{-}[541]$ ($h_{11/2}$) proton. Such an interpretation is also supported by the measured quadrupole deformations, shown in Figs.3(c) and (d). Thus, the occupancy of the $1/2^{-}[541]$ ($f_{7/2}$,$h_{9/2}$) neutron orbital results in the observation of enhanced deformed bands for nuclei below $N = 73$. The corresponding growth in quadrupole deformation values, however, is not as large as those observed when the $\nu i_{13/2}$ or $\pi g_{9/2}$ orbitals are involved.

A summary of the results discussed above may be found in Table 1.

Table 1 Quadrupole moments and deformations for selected bands in ^{133,135}Nd and ^{130,131,132}Pr.

Nucleus	Configuration[(1)]	Q_0, eb [β_2] Expt.	Ref.
^{133}Nd	$\nu i_{13/2}$	5.8(2) [0.32(1)]	[32]
		6.5(2) [0.36(1)][(2)]	[32]
^{135}Nd	$\nu i_{13/2}$	5.1(2) [0.28(1)]	[32]
		5.7(2) [0.32(1)][(2)]	[32]
^{130}Pr	$\pi g_{9/2} \otimes \nu h_{11/2}$	6.1(4) [0.35(2)]	[36]
^{131}Pr	$\pi g_{9/2}$	5.3(4) [0.31(2)]	[36]
^{130}Pr	$\pi h_{11/2} \otimes \nu (f_{7/2}, h_{9/2})$ (band 1)	4.3(3) [0.25(2)]	[38]
	$\pi h_{11/2} \otimes \nu (f_{7/2}, h_{9/2})$ (band 2)	4.5(3) [0.26(2)]	[38]
^{132}Pr	$\pi h_{11/2} \otimes \nu (f_{7/2}, h_{9/2})$	4.1(3) [0.24(1)]	[38]
^{130}Pr	$\pi h_{11/2} \otimes \nu d_{5/2}$	3.4(2) [0.20(1)]	[38]
^{131}Pr	$\pi h_{11/2}$	3.3(2) [0.19(1)]	[38]

(1) $\pi g_{9/2}$: 9/2$^+$[404], $\pi h_{11/2}$: 3/2$^-$[541], $\nu h_{11/2}$: 7/2$^-$[523], $\nu d_{5/2}$: 5/2$^+$[402], $\nu(f_{7/2}, h_{9/2})$: 1/2$^-$[541], $\nu i_{13/2}$: 1/2$^+$[660]
(2) Deduced by gating above the level of interest, so that the effect of side feeding was eliminated. We may reasonably expect a similar increase of ~10% for the other highly deformed bands listed.

3 NEW RESULTS IN ^{135}Pm

A comprehensive new level scheme has been deduced for ^{135}Pm from our thin target experiment [42]. Two bands of high moment of inertia have been observed for the first time in this nucleus, see Fig. 4. Unfortunately no firm linking transitions to the normal deformed level scheme have been established yet. Thus the exact excitation energies and spins of the bands are not yet known. Their behavior in terms of moment of inertia and very small signature splitting is extremely similar to the established 9/2$^+$[404] ($g_{9/2}$) proton band in ^{133}Pm [7]. It would be tempting to make this same assignment to the bands in ^{135}Pm except that the expected strong dipole transitions between the bands are very weak and cannot be seen. Since such small signature splitting is not expected for any other proton orbital near the Fermi surface, the 9/2$^+$[404] orbital may still be involved, however the large g-factor would have to be quenched by other aligned quasiparticles. This possibility is also supported by other experimental differences with ^{133}Pm such as (i) that the decay of the bands into the low spin parts of the level scheme has not been observed, (ii) the frequency range (and thus spin range) of the bands ($\hbar\omega$ = 0.25 - 0.65 MeV) is quite different to ^{133}Pm ($\hbar\omega$ = 0.17 - 0.45 MeV) and (iii) the fact that yrast transitions up to the 805 keV ($31^-/2 \rightarrow 27^-/2$) can be seen, see Fig. 4. While several candidates exist, a preferred three quasiparticle configuration has not yet emerged satisfying all the observed experimental features.

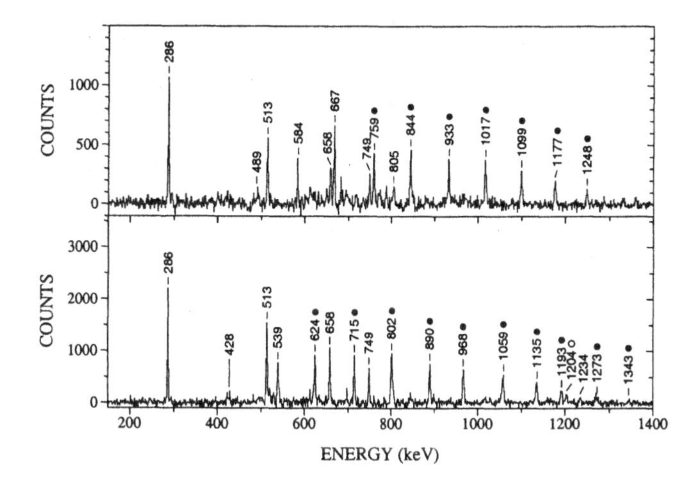

Figure 4 Coincidence γ-ray spectra [42] for new high moment of inertia bands in ^{135}Pm. The normal deformed yrast transitions have energies of 286, 513, 658, 749 and 805 keV. The solid dots indicate the transitions used in the sum of double gates in order to generate each spectrum.)

4 FUTURE PROSPECTS FOR THE REGION

While a large body of research exists pertaining to this highly deformed region and a lot of physics has been learnt, a huge amount of work remains to be done! In the following section a few interesting questions are posed. Obviously this is a short and incomplete list and one that will no doubt grow and self generate as new experiments are performed.

Firstly only a handful of highly deformed structures (mainly in Nd nuclei) have been linked successfully to the normal deformed low spin yrast structures. What about so many others such as, ^{130,132}Ce, ^{132,133}Pr, ^{135}Pm etc? Only in ^{134}Nd has an excited highly deformed band been linked. What about the large number of other excited highly deformed sequences? Even worse, in many cases the absolute excitation energies and spins of the normal deformed structures are unknown! All this information is crucial if we are to correctly identify the "real" configurations involved in building the highly deformed structures. Also remember that for many years we believed that in the even-even Nd isotopes the yrast highly deformed band involved a two-quasiparticle $i_{13/2}$ excitation but now we know this is simply not true [29]! In addition, the role of the $9/2^+[404]$ ($g_{9/2}$) proton orbital has emerged in recent years as being very important in creating highly deformed structures. However work remains in charting out the characteristics of bands involving this special orbital, both in terms of excitation energy, alignments, deformation, signature splitting, single and multiquasiparticle configurations etc over a broader range of N and Z.

While very complex level schemes can now be created for nuclei in this region which display beautiful shape co-existence features, there remains a major problem with regard to the existence of discrete band structures at high spin and high excitation energy. Even though the input angular momentum in many reactions can be as high as $\approx 75\hbar$, discrete structures, with the exception of a few low ($\leq 1\%$) intensity highly deformed sequences, are rarely observed above spin $\sim 35\hbar$. What happens to the vast majority of the intensity flow as it cascades down from the high spin regime? Are the different co-existing minima fed differently as a function of spin? Are different shapes fed over different timescales? How high in spin do discrete structures exist before they lose their own identity? Does this identity loss occur at similar excitation energies for different shapes?

The role of pairing, its stability or demise at high spin is something that still requires clarification. Many observed high spin alignments are not clearly explained. Are they paired or unpaired band crossings and/or shape changes? Another important question involves what happens to the highly deformed bands at high spin. Is it possible to follow them all the way up to spin 60-$70\hbar$ where it has been predicted that they will terminate [34]? How quickly is the collectivity lost? There are some indications perhaps of termination effects from the moment of inertia behavior in ^{133}Nd but a new generation of instrumentation and more refined lifetime measurements, where the detailed changes in quadrupole moment as a function of spin can be accurately determined, are required to substantiate this prediction.

One of the great hopes is to bridge the gap between the different islands of superdeformation. It seems possible in the near future that we be able to see the full flowering of the superdeformed landscape from ^{132}Ce to ^{143}Eu to ^{152}Dy. Further studies of Pm, Sm as well as the heavier Nd's are necessary to help form this important evolutionary picture. Also what happens to the deformation trends as one moves to lighter N and Z systems than described in this work?

The question of $\Delta I = 2$ bifurcation is again one of current interest following recent results near ^{149}Gd [43]. Tentative signs of this behavior in Ce nuclei was reported [44] but so far no further evidence of this curious and unexplained phenomenon has been reported in this mass region.

The A$\sim$130 nuclei are expected to be extremely stable against fission having some of the highest spin values possible for nuclear systems. Discrete states up to near $60\hbar$ have been observed in ^{132}Ce, but just how high can this limit go and how do nuclei behave at such a critical point? Do hyper-deformed nuclei with a 3:1 major to minor axis ratio exist at the highest spins? Calculations indicate that this region of nuclei may well be one of the best to search for such exotic structures [45].

5 SUMMARY

The new generation of gamma-ray spectrometers, such as GAMMASPHERE, are now allowing us to perform exquisitely sensitive nuclear structure measurements. In addition, the quadrupole moments of a number of bands, in several Nd and Pr nuclei, were measured using the Doppler-shift attenuation method as a part of our systematic study dedicated to understanding the properties of highly deformed structures in mass A~130 region. Differences in the observed deformations clearly demonstrate the important role played by the occupation of the $9/2^+[404]$ ($g_{9/2}$) proton, $1/2^+[660](i_{13/2})$ and $1/2^-[541]$ ($f_{7/2}$,$h_{9/2}$) neutron orbitals on the properties of the highly deformed bands. New lifetime measurements in Pm nuclei will be forthcoming soon. In ^{135}Pm, a new pair of high moment of inertia bands were introduced. Finally a brief selection of unresolved (until the the next century that is!) questions relating to the highly deformed A~130 region were discussed.

6 ACKNOWLEDGMENTS

Graduate school in Liverpool with John Sharpey-Schafer, the new TESSA arrays, the trips to NBI and the initial halcyon days of Daresbury Laboratory are very special memories to me (MAR). So too was the period in the late eighties during my second stay in the Liverpool group. I will always be indebted to John for many many things, his guidance, enthusiasm, encouragement, the fruitful collaborations and his close friendship throughout the last two decades. I am joined by all your former students when I make the following declaration, *"John Sharpey-Schafer: The best there is. The best there was. The best there ever will be!"* Thank you John!

In addition, the authors wish to thank the staff of the LBNL GAMMASPHERE facility and the crew of the 88″ Cyclotron for their assistance during these experiments. The software support of D.C. Radford and H.Q. Jin and the target making wizardry of Bob Darlington are greatly appreciated. Support for this work was provided by the U.S. Department of Energy under Contract No. DE–AC03–765F00098 and Grant No. DE–FG02–88ER40406, the National Science Foundation, the State of Florida and the U.K. Engineering and Physical Sciences Research Council. MAR and JS acknowledge the receipt of a NATO Collaborative Research Grant.

References

[1] P.J. Nolan *et al.*, *J. Phys.* G11, (1985) L17.

[2] A.J. Kirwan *et al.*, *Phys. Rev. Lett.* 58, (1987) 467.

[3] P.J. Twin *et al.*, *Phys. Rev. Lett.* 57, (1986) 811.

[4] M.A. Bentley *et al.*, *Phys. Rev. Lett.* 59, (1987) 2141.

[5] R. Wyss *et al.*, *Phys. Lett.* B215, (1988) 211.

[6] A. Galindo-Uribarri *et al.*, *Phys. Rev.* C50, (1994) R2655.

[7] A. Galindo-Uribarri *et al.*, *Phys. Rev.* C54, (1996) 1057.

[8] A. Galindo-Uribarri *et al.*, *Phys. Rev.* C54, (1996) R454.

[9] R. Wadsworth *et al.*, *Nucl. Phys.* A526, (1991) 188.

[10] P.J. Nolan and J.F. Sharpey-Schafer, *Rep. Prog. Phys.* 42, (1979) 1.

[11] I.Y. Lee, Nucl. Phys. A520, (1990) 641c and, R.V.F. Janssens and F. Stephens, Nuclear Physics News, 6 (1996) 9.

[12] D.G. Sarantites *et al.*, *Nucl. Instrum. Methods Phys. Res.* A381, (1996) 418.

[13] T.K. Alexander and J.S. Forster, *Advances in Nuclear Physics*, New York: Plenum Press, 1978, vol.10, pp.197.

[14] E.F. Moore *et al.*, *Phys. Rev.* C55, (1997) R2150.

[15] J.F. Ziegler, J.P. Biersack, and U. Littmark, *The Stopping and Range of Ions in Solids*, New York: Pergamon Press, 1985; J.F. Ziegler, (priv. comm.).

[16] A.E. Blaugrund, *Nucl. Phys.* 88, (1966) 501.

[17] R. Wadsworth *et al.*, *J. Phys. G: Nucl. Phys.* 13, (1987) L207.

[18] E.M. Beck *et al.*, *Phys. Rev. Lett.* 58, (1987) 2182.

[19] D. Bazzacco *et al.*, *Phys. Lett.* B309, (1993) 235.

[20] D. Bazzacco *et al.*, *Phys. Rev.* C49, (1994) R2281.

[21] M.A. Deleplanque *et al.*, *Phys. Rev.* C52, (1995) R2302.

[22] S. Lunardi *et al.*, *Phys. Rev.* C52, (1995) R6.

[23] C.M. Petrache *et al.*, *Nucl. Phys.* A617, (1997) 228.

[24] N.H. Medina *et al.*, *Nucl. Phys.* A589, (1995) 106.

[25] S.M. Mullins *et al.*, *Phys. Rev* C45, (1992) 2683.

[26] R.M. Diamond *et al.*, *Phys. Rev.* C41, (1990) R1327.

[27] C.M. Petrache *et al.*, *Phys. Rev.* C57, (1998) R10.

[28] C.M. Petrache *et al.*, *Phys. Lett.* B219, (1996) 145.

[29] C.M. Petrache *et al.*, *Phys. Rev. Lett.* 77, (1996) 239.

[30] P. Wilssau, *et al.*, *Phys. Rev.* C48, (1993) R494.

[31] S.A. Forbes, *et al.*, *Z. Phys.* A352, (1995) 15.

[32] F.G. Kondev *et al.*, *Phys. Rev.* C, (submitted) .

[33] R.M. Clark *et al.*, *Phys. Rev. Lett.* 76, (1996) 3510.

[34] A.V. Afanasjev and I. Ragnarsson, *Nucl. Phys.* A608, (1996) 176.

[35] T.B. Brown *et al.*, *Phys. Rev.* C56, (1997) R1210.

[36] F.G. Kondev *et al.*, *Eur. Phys. J.* A2, (1998) 249.

[37] F.G. Kondev *et al.*, *J. Phys. G.* 25, (1999) 893.

[38] F.G. Kondev *et al.*, *Phys. Rev.* C59, (1999) (in press) .

[39] B.H. Smith *et al.* (to be published); L.L. Riedinger *et al.* (priv. comm.).

[40] K. Hauschild *et al.*, *Phys. Rev.* C50, (1994) 707.

[41] S. Shi *et al.*, *Phys. Rev.* C37, (1988) 1478.

[42] J. Pfohl *et al.*, to be published.

[43] D.S. Haslip *et al.*, *Phys. Rev. Lett.* 78, (1997) 3447.

[44] A.T. Semple *et al.*, *Phys. Rev. Lett.* 76, (1996) 3671.

[45] T.R. Werner and J. Dudek, *Atomic Data and Nuclear Data Tables* 50, (1992) 179.

QUASI-DYNAMICAL SYMMETRY – A NEW USE OF SYMMETRY IN NUCLEAR PHYSICS

David J. Rowe

Department of Physics, University of Toronto
Toronto, ON 25S 1A7, Canada*

rowe@physics.utoronto.ca

Abstract: Dynamical symmetry has become an indispensable tool in nuclear structure physics. However, while its uses are far from exhausted, it appears that many collective phenomena are naturally described by an extension of the concept to *quasi-dynamical symmetry*. Finding ways to exploit this concept will present an important challenge for the future.

1 INTRODUCTION

Symmetry has long been used to simplify descriptions of physical systems. However, in recent years it has been applied extensively to dynamical as well as static systems and has become indispensable.

Consider, for example, the symmetries associated with rotations. Rotational invariance of the Hamiltonian for an isolated system implies that angular momentum is a good quantum number. It also means that states of non-zero angular momentum form degenerate multiplets. However, rotational invariance alone says nothing about the dynamics of a rotor. One can think of the rotational invariance of a Hamiltonian as a *static symmetry* whereas bands of rotational states are a manifestation of a *dynamical symmetry*.

Almost all nuclear models have dynamical symmetry [1]. This is because solvable models are generally concerned with the dynamics of a finite number of degrees of freedom that are associated with dynamical groups of transformations. One can then ask: "Does dynamical symmetry have the potential to

*Supported by the Natural Sciences and Engineering Research Council of Canada.

explain all the phenomena in nuclear physics we would like to understand?" The answer is almost certainly "no". However, I will try to convince you that, with an extension to *quasi-dynamical symmetry*, it has the potential to take us a long way. But first, I want to ask "what are the objectives of theoretical nuclear physics". According to Thomas Kuhn [2], one can only ask the question within the terms of reference of the paradigm in which we currently work.

2 PARADIGMS AND MODELS OF NUCLEAR PHYSICS

The fact that we divide physics into subfields, ranging from elementary particle physics to geophysics, is based on an assumption that the physical world has a hierarchical structure in which the objects of one subfield are the building blocks of the next. Thus, one can regard the substructure of the nucleon as being in the domain of elementary particle physics and model the nucleus as a system of interacting nucleons. Of course, the interactions between nucleons are of interest to both elementary particle and nuclear physics. However, nuclear structure theorists generally take the interactions between nucleons as measurable quantities and concern themselves with the many-body properties of the quantum system that results. The first problem is that measured nucleon-nucleon interactions are much too singular for standard many-body theory to apply. Thus, one needs an intermediate stage in which the quantum system of nucleons with strong repulsive-core interactions is reduced to a shell model with much weaker effective interactions. At the top of the hierarchy of nuclear models, cf. Fig. 1, are models with few degrees of freedom designed to explain experimental phenomena. Such models include macroscopic models, e.g., col-

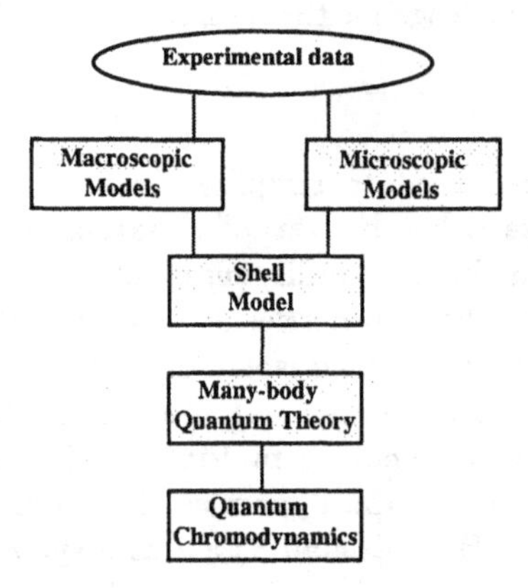

Figure 1 The hierarchy of nuclear models.

lective models, and microscopic models, e.g., pair-coupling models of nuclear superconductivity. My current interest is to learn how successful macroscopic collective models have their foundations in the shell model.

Fig. 2 shows some basic models that have been employed in nuclear structure physics. To a first approximation, the evolution in time is from the bottom

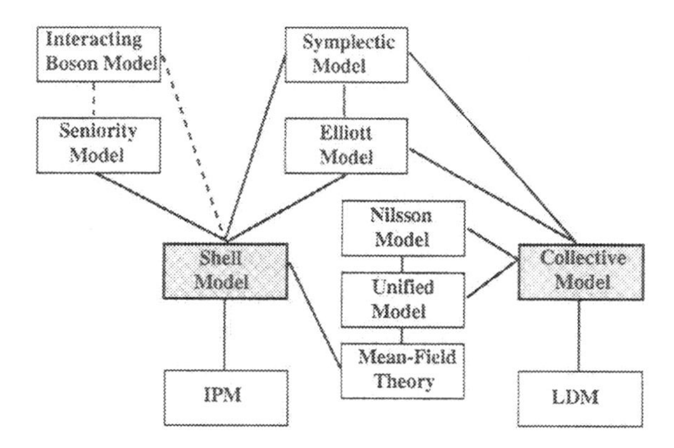

Figure 2 Basic models of nuclear structure.

upwards; related models are joined by lines. The most fundamental model is undoubtedly the Shell Model. However, since its full (untruncated) form is intractable, the shell model is really a formal framework upon which simpler models can hopefully be founded. The most physically intuitive and successful of models is surely the Collective Model. It demonstrates rather unequivocally that nuclei exhibit collective rotational and vibrational dynamics. Moreover, if it could be expressed in terms of interacting nucleons, i.e., realized as submodels of the shell model, it would acquire the potential to tell us about the character of nuclear matter as a quantum fluid. Elliott's SU(3) Model [3] and the Symplectic Model [4] go a long way towards achieving this objective but, as discussed below, not far enough.

It is interesting to recall the major evolutionary stages of the current nuclear structure paradigm. In the early fifties one had two competing models, the Independent-Particle (shell) Model and the Collective Model, which had evolved from the earlier Liquid Drop Model. The first involved all the nucleon degrees of freedom and was solvable because it assumed they could be decoupled so that the nucleons moved independently. The second considered only a few collective degrees of freedom. The two models were soon combined into a Unified Model which considered particles moving independently in a common field with collective degrees of freedom. However, beginning with Elliott's SU(3) model [3] and the seniority scheme [5] for coupling nucleons in pairs, a new miniparadigm, based on dynamical symmetry, was introduced. It is now known that dynamical symmetry is the unifying concept in terms of which all the models in the figure, as well as others, can be expressed.

3 DYNAMICAL GROUPS – SPECTRUM GENERATING ALGEBRAS

A dynamical symmetry for a classical system is a group of transformations of the phase space for the system having the property that all possible dynamical motions of the system are given by sequences of transformations of an initial starting point. Furthermore, starting from any initial state, it must be possible to reach all other states (i.e., all points of the phase space) by transformations contained in the dynamical group.

Given a dynamical group and a suitable Hamiltonian, the classical dynamics of a system is expressible in terms of the Lie algebra of infinitesimal generators of the group. Moreover, the model is quantized by constructing the unitary representation of the dynamical group. Thus, the Hilbert space for the quantized system is the carrier space of such a representation.

The infinitesimal generators of a dynamical group comprise a so-called *spectrum generating algebra*. In quantum mechanics, the elements of the SGA are identified with the physical observables of the model. Thus, a useful SGA for a model is one for which all the observables of physical interest are either elements of the SGA or simple functions of elements.

Dynamical symmetry has two important uses. In the first place it provides the mathematical apparatus for applications of a model; one can appeal to the representation theory of its SGA to work out the predictions of a model. A second use [1] is to give a model a foundation on a more fundamental model. This is a primary objective of the hierarchical paradigm.

Consider, for example, the problem of founding a collective model in the shell model. A SGA for the shell model is the infinite-dimensional algebra of one-body operators. Thus, if the collective model observables are expressed as one-body operators, the corresponding SGA becomes a subalgebra of the shell-model algebra. One can then seek irreps (irreducible representations) of the collective model within the space of the shell model. If we succeed in this, we can say that we have embedded the collective model in the shell model. More precisely, we have embedded a representation of the model in the shell model.

To embed a representation of a collective model in the shell model is to give the model a microscopic interpretation. Moreover, it means that the model can be derived from the shell model by restricting (i.e., truncating) the shell model space to the subspace that carries the corresponding irrep. Thus embedding and restriction are complementary (inverse) processes.

In general, a model has many inequivalent irreps. For example, a rotor model has different irreps with different moments of inertia. Thus, it will usually happen that the shell-model space is a direct sum (or direct integral) of many subspaces each of which carries an irrep of a collective model SGA. Moreover, a given irrep may occur (often infinitely) many times. Thus, the embedding of a model in the shell model is rarely unique. The model then has many microscopic interpretations, all compatible with the shell model, and the problem is to determine the most relevant. One way is to find the embedding that has lowest energy relative to some assumed shell-model Hamiltonian. By such methods, one can also select from among all irreps, not just among equivalent irreps, which realizations of a model are most appropriate for the low-energy spectrum of a given nucleus.

This perspective suggests an approach to nuclear structure theory in which one selects a phenomenological model, that describes the data reasonably well, and restricts a shell model calculation to the space of a number of lowest-lying irreps of the corresponding SGA. Even if one restricts to a single irrep, one gains much by proceeding in this way over simply fitting the data with the model. In the first place, one can (in principle) derive the adjustable parameters of the model from a microscopic Hamiltonian and, in the second place, one obtains microscopic wave functions. As a result, one can use the full artillery of shell model observables to learn more about the dynamical content of collective wave functions than can be inferred by model considerations alone. For example, one can compute the dynamical current flows associated with rotational states, about which the rotor model can say nothing because it has no current operator.

4 QUASI-DYNAMICAL SYMMETRY

To some extent, one can infer an appropriate embedding of a collective model in the shell model experimentally. For example, an irrep of the rigid-rotor model [1] is characterized by fixed values of intrinsic quadrupole moments and the expectation values of these moments in physical states can be measured to within experimental error. However, one can only determine the irrep in this way to within equivalence. Recall that any number of equivalent irreps of a group or Lie algebra can be mixed to form a new equivalent irrep. One can imagine that equivalent irreps of a collective model within a shell model space have common collective wave functions but different intrinsic structures. Thus, for example, one might choose a combination that minimizes the energy of a shell model Hamiltonian and thereby define a corresponding intrinsic state. A problem is that the intrinsic states of a physical rotor have vibrational fluctuations in the values of their intrinsic quadrupole moments. This means that an intrinsic state of a physical rotor is a linear combination of intrinsic states from different rigid-rotor irreps. We then describe the rotor as a soft rotor and, for such a rotor, the rigid-rotor dynamical group becomes a quasi-dynamical group.

We now give examples of what quasi-dynamical symmetry means in practice. The remarkable fact is that one can have huge mixing of similar irreps without significantly changing the predictions of a model. Thus, we discover that a model can often work well even when its dynamical symmetry is badly broken.

4.1 Spin-orbit mixing

It is known that the spin-orbit interaction mixes different SU(3) irreps. However, the spin-orbit interaction is a component of a (1,1) SU(3) tensor. Thus it can only directly mix a (λ, μ) irrep with irreps in the tensor product

$$(\lambda,\mu)\otimes(1,1) = 2(\lambda,\mu) + (\lambda+1,\mu+1) + (\lambda+2,\mu-1) + (\lambda-1,\mu+2)\,; \quad (1)$$

i.e., irreps with similar values of λ and μ. Moreover, for large values of $\lambda + \mu$, it turns out that the mixing is of a highly coherent nature [6, 7] and preserves SU(3) quasi-dynamical symmetry to a high degree of accuracy.

As an illustration, fig. 3 shows a spectrum that results from mixing SU(3)

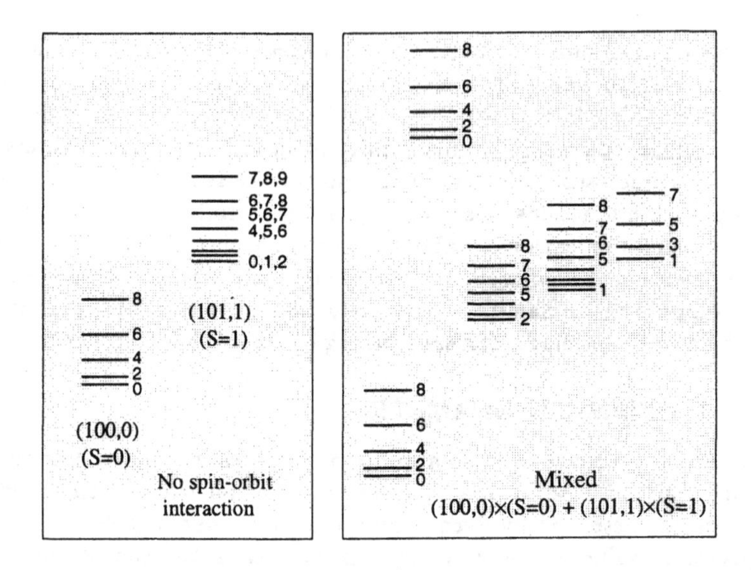

Figure 3 A spectrum that results from mixing SU(3) states with a spin-orbit interaction.

(100,0) spin-zero states and (101,1) spin-one states with a spin-orbit interaction. These are the kinds of SU(3) irreps needed for the description of rotational bands in rare-earth nuclei. The two $K=0$ bands that result are almost equal mixtures of states from the two irreps. However, they would be experimentally indistinguishable from pure irreps except for the appearance of small transitions between each other and the $K=2$ band that emerges. These effects are just of the type needed to explain the so-called beta- and gamma-vibrational bands.

In fact, SU(3) appears to be a better quasi-dynamical symmetry in heavy nuclei than in light. This is because the SU(3) quantum numbers are much bigger in heavy nuclei, the rotational motions are more adiabatic, and spin-orbit partners (i.e., single-particle states of spin $l \pm s$) are more widely separated.

4.2 Short-range pairing interactions

It is known that attractive short-range interactions favour nuclei with spherical shapes and superfluid properties whereas long-range forces favour deformed shapes and rotational spectra. It is also known that rotational states are given by an SU(3) model with quadrupole-quadrupole interactions [3] whereas spherical superfluid states are given by seniority coupling with an SU(2) quasi-spin SGA and pairing forces [5].

Many nuclei are observed to be both deformed and superfluid. But, since pairing forces mix SU(3) irreps and, conversely, quadrupole-quadrupole forces mix SU(2) irreps, neither SU(3) nor SU(2) can be a dynamical symmetry for a superconducting rotational nucleus. It turns out that SU(3) is an excellent quasi-dynamical symmetry [8].

To illustrate, consider a model with Hamiltonian

$$H(\alpha) = H_0 + \alpha V_{\mathrm{SU}(3)} + (1 - \alpha) V_{\mathrm{pairing}}\,. \tag{2}$$

This Hamiltonian can be diagonalized analytically when $\alpha = 0$ and 1. For $\alpha = 0$, it has an SU(2), quasi-spin, dynamical symmetry and a vibrational spectrum characteristic of a superfluid. For $\alpha = 1$, it has an SU(3) dynamical symmetry and a rotational spectrum. In general, it is diagonalizable on a computer by virtue of an Sp(3) dynamical symmetry which contains both SU(2) and SU(3) as subgroups. The spectrum of the lowest J = 0, 2, 4, 6 and 8 states is shown as a function of α in fig. 4. Details of the model are given in ref. [8].

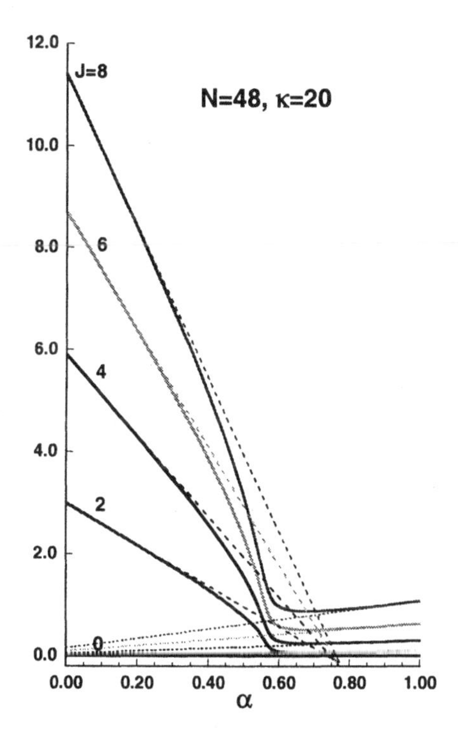

Figure 4 Energies of the lowest $J = 0, \ldots, 8$ states as a function of α for $H(\alpha)$.

The remarkable fact is that the spectrum remains superfluid for a sizable range of α and then flips to that of a rotor at a critical value of $\alpha \approx 0.58$. The sudden flip becomes increasingly sharp with increasing particle number indicating an approach to a phase transition. Even more remarkable is the behaviour of the wave functions. They are shown in fig. 5 for the lowest states of angular momentum J = 0, 2, 4, and, 8 as coefficients in an SU(3) $\supset$ SO(3) basis for four values of α. Below the critical value of α, the wave functions are complicated in the SU(3) basis. However, above the critical point, they become remarkably coherent. At $\alpha = 1$, the states belong to a single (32,8) irrep. For

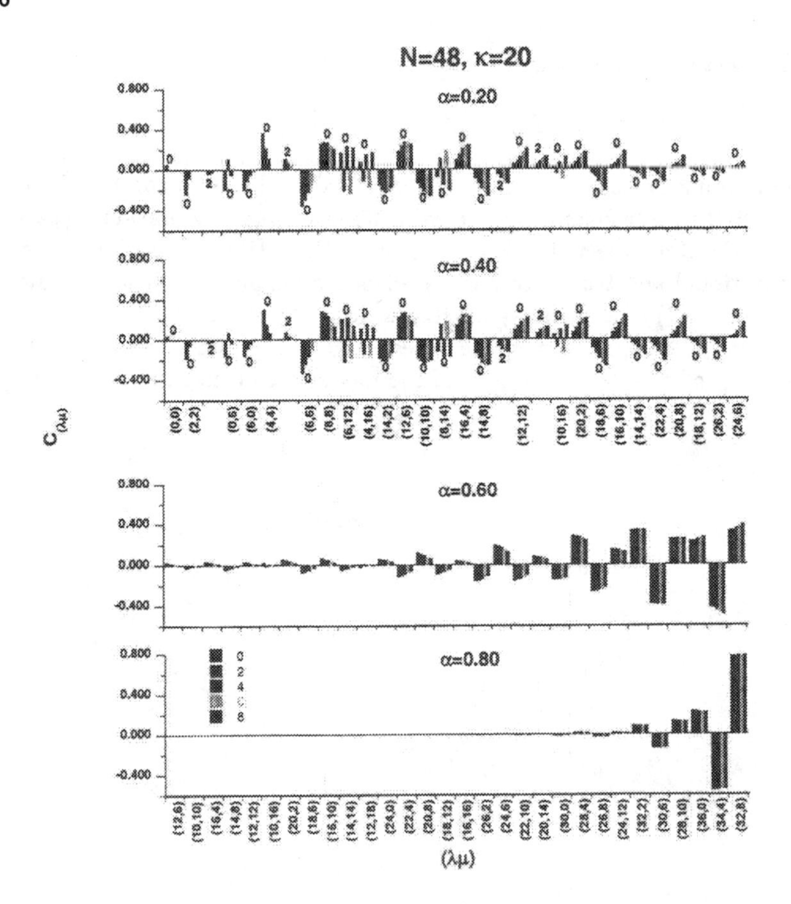

Figure 5 Wave functions in an SU(3) basis for the lowest-energy angular-momentum states of the Hamiltonian $H(\alpha)$ for four values of α.

smaller α, the pairing interaction mixes SU(3) irreps. But, for almost all α above the critical value, the mixing is such that the coefficients are of the form

$$|\Psi_{LM}\rangle \approx \sum_{\lambda\mu} C_{\lambda\mu}\, |(\lambda,\mu)K=0, JM\rangle\, ; \tag{3}$$

they are essentially independent of angular momentum. Thus, SU(3) is a very poor dynamical symmetry but an excellent quasi-dynamical symmetry for a large range of α above the critical value. The spectrum also remains in close agreement with the $J(J+1)$-dependence and ratios of B(E2)-transition rates characteristic of a rotor, even for $\alpha = 0.6$.

5 A SYMPLECTIC MODEL WITH DAVIDSON POTENTIALS

It is known [4] that the non-compact symplectic group Sp(3,R), sometimes called Sp(6,R), is a dynamical group for a microscopic model of rotations and quadrupole vibrations. We now use this model to infer the mixing of SU(3) irreps in the shell model due to collective couplings to higher major shells.

5.1 *Rotations of a diatomic molecule*

We consider first a model [9] of a diatomic molecule with a Hamiltonian for its relative motion given by

$$H = \frac{p^2}{2m} + \frac{1}{2}m\omega^2\left(r^2 + \frac{\varepsilon}{r^2}\right) . \tag{4}$$

Expressing lengths in harmonic-oscillator units, in which $\hbar/m\omega = 1$, gives

$$H = \frac{1}{2}\hbar\omega\left(-\nabla^2 + r^2 + \frac{\varepsilon}{r^2}\right), . \tag{5}$$

The potential $V(r) = \frac{1}{2}m\omega^2\left(r^2 + \frac{\varepsilon}{r^2}\right)$, known as the Davidson potential, is shown in fig. 6. It has a minimum at the point $\varepsilon = r_0^2$ and a width dependent

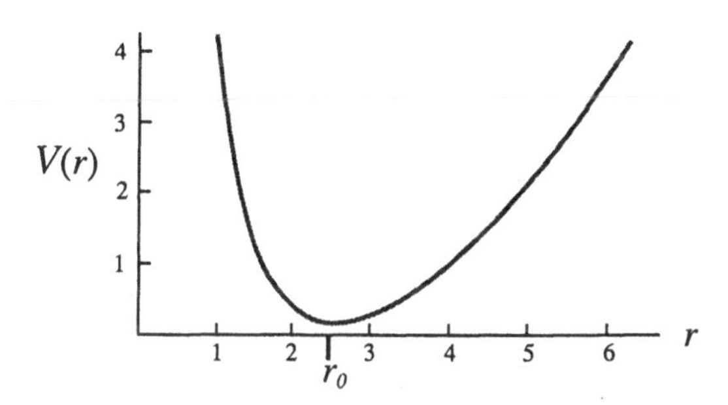

Figure 6 The Davidson potential $V(r)$.

on ω. Thus, ε and ω can be adjusted to give the equilibrium separation of the two atoms and the first vibrational excitation energy of the molecule.

The spectrum and wave functions of the model can be derived analytically [9] using the invariance of H under SO(3) rotations and an SU(1,1) dynamical symmetry. The group SU(1,1) has infinitesimal generators

$$Z_1 = -\nabla^2 + \frac{\varepsilon}{r^2}, \quad Z_2 = r^2, \quad Z_3 = -\tfrac{1}{2}\mathrm{i}(\mathbf{r}\cdot\nabla + \nabla\cdot\mathbf{r}), \tag{6}$$

which close under commutation to form an su(1,1) SGA. The energy levels are easy to derive because the Hamiltonian, $H = \frac{1}{2}\hbar\omega(Z_1 + Z_2)$, is an element of the su(1,1) algebra. They are given by

$$E_{nl} = [2n + 1 + \sqrt{(l + 1/2)^2 + \varepsilon}\,]\,\hbar\omega \tag{7}$$

or, expanded in inverse powers of ε,

$$E_{nl} = E_0 + 2n\hbar\omega + Al(l+1) - Bl^2(l+1)^2 + \dots , \tag{8}$$

with

$$A = \frac{\hbar\omega}{2\sqrt{\varepsilon}}, \quad B = \frac{1}{3\varepsilon} A. \tag{9}$$

Thus, for large values of ε, one obtains a relatively rigid rotor spectrum that is perturbed by centrifugal stretching for smaller values of ε. The ratios of dipole transition rates are likewise in accord with those of the corresponding rotor for large ε. More details can be found in ref. [9].

5.2 *Rotations of a many-particle nucleus*

For the rotations and quadrupole vibrations of an axially symmetric nucleus, consider a model with Hamiltonian [9]

$$H = H_0 + V(\beta, \gamma), \tag{10}$$

$$H_0 = \frac{1}{2m}\pi^2 + \frac{1}{2}m\omega^2\alpha^2, \quad V(\beta,\gamma) = \chi_1\left(\beta^2 + \frac{\varepsilon}{\beta^2}\right) + \chi_2\beta^3\cos 3\gamma, \tag{11}$$

where α is a monopole moment, β and γ are quadrupole deformation parameters and π is a suitable momentum. This model is embedded in the shell model by setting

$$\begin{gathered}\pi^2 \equiv \textstyle\sum_n p_n^2, \quad \alpha^2 \equiv \sum_n r_n^2, \\ \beta^2 \equiv \textstyle\sum_n Q_n \cdot Q_n, \quad \beta^3\cos 3\gamma \propto \sum_n [Q_n \times Q_n \times Q_n]^0,\end{gathered} \tag{12}$$

where Q_n is the quadrupole tensor for the n'th nucleon with components

$$Q_{n\nu} = r_n^2 Y_{2\nu}(\theta_n, \varphi_n). \tag{13}$$

The harmonic oscillator Hamiltonian H_0 has an SU(1,1) dynamical group with infinitesimal generators

$$Z_1 = \pi^2, \quad Z_2 = \alpha^2, \quad Z_3 = \frac{1}{4i\hbar}[\alpha^2, \pi^2] \equiv \tfrac{1}{2}\sum_n(\mathbf{r}_n \cdot \mathbf{p}_n + \mathbf{p}_n \cdot \mathbf{r}_n), \tag{14}$$

which satisfy the su(1,1) commutation relations

$$[Z_1, Z_2] = 4i\hbar, \quad [Z_3, Z_1] = 2i\hbar Z_2, \quad [Z_3, Z_2] = -2i\hbar Z_2. \tag{15}$$

The quadrupole moments, β and γ, are not expressible in terms of the su(1,1) operators. However, they can be expressed in terms of a larger algebra of observables, sp(3, R), viz. the algebra spanned by the 21 operators

$$K_{ij} = \textstyle\sum_n p_{ni}p_{nj}, \quad Q_{ij} = \sum_n r_{ni}r_{nj}, \quad S_{ij} = \frac{1}{2}\sum_n(r_{ni}p_{nj} + p_{nj}r_{ni}), \tag{16}$$

where r_{ni} and p_{ni}, with $i = 1, 2, 3$, are Cartesian components of $\mathbf{r}_n$ and $\mathbf{p}_n$, respectively.

It follows that Sp(3, R) is a dynamical group for the full Hamiltonian H. This means that we can use the representation theory of sp(3, R) to determine matrix elements of H and compute its spectrum. In doing this we have first to decide which nucleus we wish we consider and the appropriate irrep for its low-lying states. We must also choose suitable values for the parameters of H.

For the purposes of illustration, we choose ^{166}Er and, from experimental [10] and Nilsson model considerations [11], determine the appropriate Sp(3, R) irrep to be the one with lowest weight (328,250,250). The value of $\hbar\omega$ was set at $41A^{-1/3}$ MeV and ε was such that the Davidson potential has a minimum at a deformation $\beta = 0.35$. We set $\chi_2 = 0$ and fixed χ_1 by a self-consistency condition. The resulting spectrum is shown, in comparison with experiment and corresponding results for a rigid rotor, in fig. 7. The agreement is remarkable

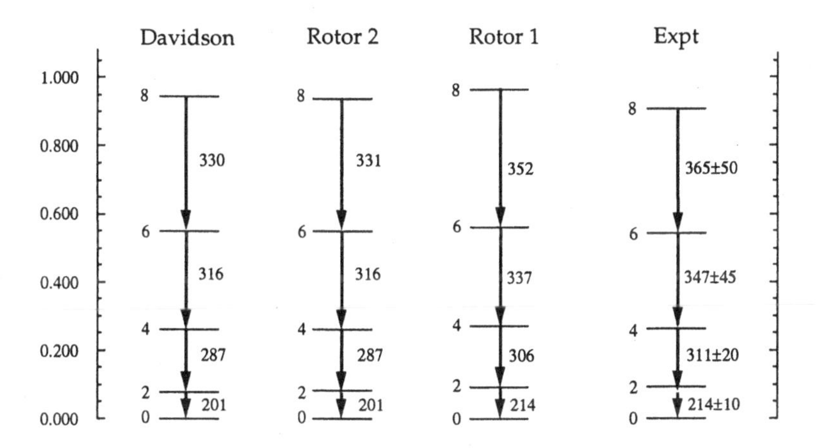

Figure 7 Low-lying energy levels and B(E2) strengths for ^{166}Er in the symplectic model with a Davidson interaction.

considering that the model uses the full many-nucleon kinetic energy without adjustable moments of inertia.

Although the Davidson potential preserves Sp(3, R) as a dynamical symmetry, it produces a huge mixing of SU(3) irreps from different major harmonic oscillator shells. This is seen in fig. 8 which shows the coefficients of the wave functions for the lowest J states in a U(3) basis. The coefficients are seen to be essentially independent of J for all states considered. Thus SU(3) emerges as a remarkably good quasi-dynamical symmetry.

From the results for the mixing of SU(3) irreps by the spin-orbit and pairing interactions, we anticipate that SU(3) will remain a good quasi-dynamical symmetry generally so long as the interactions remain within physical limits. We have cause to believe this simply because experiment shows that the observed yrast states in ^{166}Er follow a rotational sequence.

5.3 Application to a light nucleus

Application of the symplectic model to ^{20}Ne, with microscopic Brink-Boeker interactions [12, 13], gives the results shown in fig. 9. Note, however, that the

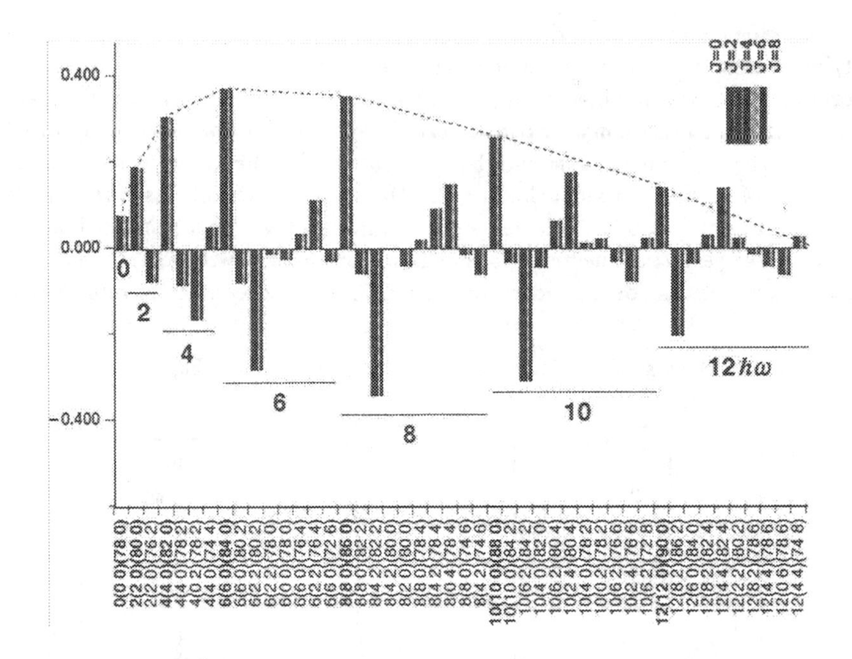

Figure 8 Wave functions in an SU(3) basis for the lowest-energy angular-momentum states of ^{166}Er in the symplectic model with a Davidson potential.

model predicts a continuation of the band to higher angular momentum states, albeit with a major gap between the 8^+ state and the lowest 10^+ state.

One can now enquire if SU(3) is a good quasi-dynamical symmetry for the ground state of ^{20}Ne. Fig. 10 shows the coefficients of some symplectic model wave functions in an SU(3) basis. It now appears that only for the $J=0$, 2 and 4 states is SU(3) a reasonably good quasi-dynamical symmetry. This result has to do with the contraction of SU(3) to a rotor algebra which occurs for angular-momentum values much less than $\lambda+\mu$. As a rule of thumb, quasi-dynamical SU(3) symmetry ceases to be good when J becomes comparable to $\frac{1}{2}(\lambda+\mu)$, where λ and μ are the lowest SU(3) labels for the band. For ^{20}Ne, the lowest SU(3) irrep (i.e., the one coming from the valence shell) is an (8,0) irrep.

6 SUMMARY

To appreciate the significance of quasi-dynamical symmetry, it is useful to review the evolution of a microscopic theory of nuclear collective motion.

6.1 Change of variable versus dynamical group methods

Early approaches were based on the presumption that success of a collective model implies a decoupling of collective degrees of freedom. For centre-of-mass motion and giant-resonance vibrations, it is possible to make a complete change

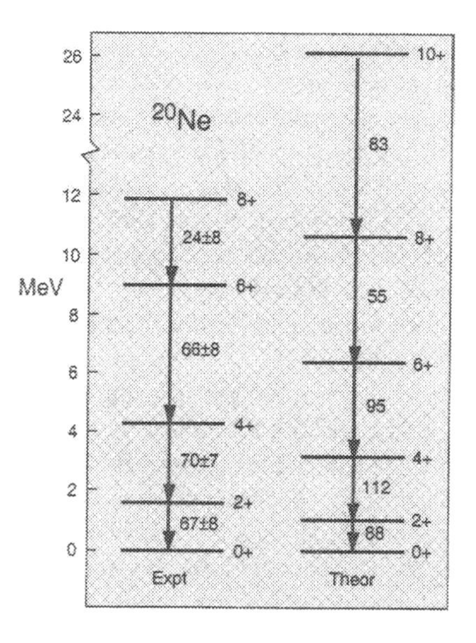

Figure 9 Low-lying energy levels and B($E2$) transition rates for ^{20}Ne calculated in the symplectic model.

from single-nucleon to collective and complementary intrinsic variables. However, while rotational and quadrupole vibrational variables were found [14], the complementary intrinsic variables were either not orthogonal or non-integrable. Both alternatives presented a dilemma. Non-orthogonal intrinsic coordinates meant that the collective and intrinsic motions could not be decoupled and we

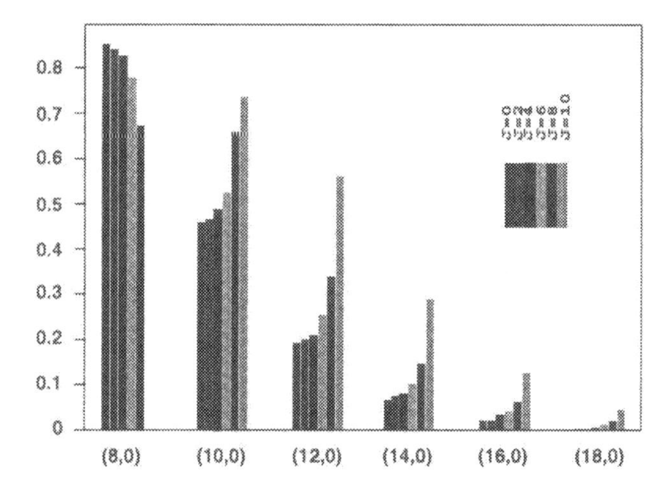

Figure 10 Wave functions in an SU(3) basis for the lowest-energy angular-momentum states of ^{20}Ne in a simple symplectic model calculation.

did not know, at the time, that one could handle the quantum mechanics of non-integrable coordinates with non-abelian gauge theory.

It was then learned that one could construct pure collective states with shell model wave functions within an irrep of a dynamical group of collective transformations (refs. given in [15]). This was a breakthrough. Since collective models have dynamical symmetries, it meant that the mathematical problem of embedding them in the shell model was largely solved. We could then focus on the physical problem of identifying which irreps or, as it turned out, combinations of irreps were most appropriate for the description of "physical" states.

A bonus of the algebraic (dynamical group) approach was the exposure of a common mistake in the quantization of a collective model. For example, it was customary to quantize a collective Hamiltonian, like that of of eqn. (11), by setting $\pi = -i\hbar\partial/\partial\alpha$. This endows the Hamiltonian H_0 with a Heisenberg-Weyl SGA. But it is wrong because it gives a lowest weight state (the ground state of H_0) with an energy $\frac{1}{2}\hbar\omega$ whereas the lowest weight state of the appropriate irrep of the su(1,1) SGA should have a very large ground-state energy (equal to the zero-point energy of the many-particle ground state). Thus, it would imply a very different su(1,1) representation and different model predictions. Similarly, the algebraic structure of the standard collective model needed adjustment to make it compatible with the many-nucleon structure of a nucleus; in this case the SGA that emerged was that of the symplectic model, sp$(3, R)$.

6.2 Sp$(3, R)$ as a dynamical group

Although changing to collective and intrinsic variables did not prove the most efficient way of deriving microscopic collective wave functions, the attempt was eminently worthwhile. From the vantage point of geometry and group theory, one discovered a remarkable system of spatial coordinates for a mass-A nucleus comprising three centre-of-mass coordinates, nine collective coordinates and $3A-12$ complementary coordinates [16, 17, 18]. The centre-of-mass coordinates are, orthogonal to the others. The nine collective coordinates are a coordinate system for a submanifold of the nuclear configuration space. The complementary coordinates alone did not form an integrable submanifold. However, they did when combined with three of the collective coordinates, We shall refer to this submanifold as the intrinsic submanifold. Thus, one obtained the insightful situation of two submanifolds of the full configuration space, one collective and one intrinsic, with a three-dimensional subspace in common.

Starting from a point in the common subspace, one can generate the common space by acting on the point with the SO(3) rotation group. If one acts on the starting point with a general linear group of transformations, GL$(3, R)$, one generates the collective submanifold and if one acts on it with an orthogonal group, O$(A-1)$, one generates the intrinsic space [17, 19]. The rotation group SO(3) that generates the common space, is a group of vortex rotations. Thus, it appears that one can think of vortex rotations as being either collective or intrinsic. Either way, they provide an essential coupling between the collective and intrinsic degrees of freedom.

This group theoretic/geometric perspective highlights the essential difficulty of constructing a decoupled collective subdynamics for the nucleus. It also suggests a solution. The important observation is that, excluding the centre-of-mass space, the nuclear configuration space is an orbit of the direct product group $\mathrm{GL}(3, R) \times \mathrm{O}(A)$.

It was also learned from an extension of this perspective that the many-nucleon phase space is a product of a centre-of-mass phase space and an orbit of the direct product group $\mathrm{Sp}(3, R) \times \mathrm{O}(A)$. Thus, one finds that, while the classical situation is complicated by the fact that the collective and intrinsic spaces overlap, the Hilbert space of spatial wave functions of the corresponding quantum system decomposes into a direct sum of $\mathrm{Sp}(3, R) \times \mathrm{O}(A)$ subspaces. What is more, it follows that $\mathrm{Sp}(3, R)$ is a dynamical group for pure collective motion. The group $\mathrm{Sp}(3, R)$ has a rich structure including, for example, a dynamical group for rigid rotations and, most important for interfacing the collective and independent-particle models, its Lie algebra $\mathrm{sp}(3, R)$ contains the many-nucleon kinetic energy and the harmonic oscillator shell model Hamiltonian. $\mathrm{Sp}(3, R)$ also contains Elliott's SU(3) group as a subgroup.

6.3 *The origins of quasi-dynamical symmetry*

The symplectic group, $\mathrm{Sp}(3, R)$, is successful in enabling one to derive rotational states with shell-model wave functions. But it fails to give low-energy beta- or gamma-vibrational excitations. The discovery of quasi-dynamical symmetry emerged during a search for a more general dynamical group that would give such states. We soon found that the more general group we were seeking did not exist and that the physics of low-energy vibrational excitations must be described by mixing $\mathrm{Sp}(3, R)$ irreps. However, we also found that, if the mixing occurred in the coherent manner of a quasi-dynamical symmetry, then the essential properties of the original dynamical symmetry would be preserved and give us the best of two worlds. The prototype of a quasi-dynamical symmetry is given by the softening of the rigid-rotor model to a soft rotor with vibrational excitations [20].

A rigid rotor has a well-defined dynamical group [21]. It is a subgroup of $\mathrm{Sp}(3, R)$ with irreps characterized by intrinsic states that are eigenstates of the quadrupole moment operators (i.e., with precisely-defined quadrupole moments). However, a physical rotor is *soft* and its intrinsic states have distributions of quadrupole moments; they are combinations of rigid-rotor intrinsic states. Nevertheless, if the effects of the Coriolis and centrifugal forces are negligible, each state of a soft-rotor band has a common intrinsic state and the intraband properties of the states become identical to those of a rigid-rotor band. The observable difference between rigid and soft rotors is that the latter have interband E2 transitions. We say that the dynamical group of the rigid rotor is a quasi-dynamical group for the soft rotor. Note that, although quasi-dynamical symmetry is only realized for a physical rotor when one neglects the Coriolis and centrifugal forces, it is nevertheless a precise concept.

The physics underlying quasi-dynamical symmetry has long been known in other language as the Born-Oppenheimer *adiabatic* approximation [22]. Recall that a non-relativistic system has translational and Galilean invariance. It also has rotational invariance but no rotational analog of Galilean invariance (because of Coriolis and centrifugal forces). Nevertheless, Coriolis and centrifugal effects are small for slow (adiabatic) rotations of a relatively rigid system. It follows that, in the adiabatic limit, rotational and intrinsic motion become decoupled. Thus, quasi-dynamical symmetry is simply an expression of the idea of adiabatic decoupling of collective degrees of freedom in the language of group theory where it becomes a potentially powerful computational tool.

6.4 SU(3) as a quasi-dynamical symmetry

It is known that SU(3) contracts [23] to the rigid-rotor algebra and, as a result, SU(3) states of large $\lambda + \mu$ and angular momentum $J \ll \lambda + \mu$ have all the properties of rotor model states. This, together with some properties of SU(3) Clebsch-Gordan coefficients, means that the concept of quasi-dynamical symmetry is well-defined for mixtures of large-dimensional SU(3) representations. Moreover, it follows that combinations of large-dimensional SU(3) irreps from any number of shells behave as soft-rotor bands provided the combinations are such that SU(3) is a quasi-dynamical symmetry. This is an important and significant result because of the observation that SU(3) is preserved as a quasi-dynamical symmetry by the dominant symmetry breaking interactions that destroy SU(3) as a full dynamical symmetry.

Acknowledgments

The author wishes to thank C.Bahri for many of the results shown.

References

[1] D.J.Rowe, *Prog. Part. Nucl. Phys.* 37 (1996) 265

[2] T.S. Kuhn, The structure of scientific revolutions, *Int. Encyclopedia of Unified Science.* Vol 2, no. 2 Univ. of Chicago Press,1970.

[3] J.P. Elliot, *Proc. Roy. Soc.*, A245 (1985) 1419.

[4] G. Rosensteel and D.J. Rowe, *Phys. Rev. Lett.*, 38 (1977) 10; D.J. Rowe, *Rep.Prog. Phys.* 48 (1985) 1419

[5] G. Racah, *Phys. Rev.*, 62 (1942) 438; G Racah and I. Talmi. *Physica.* 18 (1952) 1097; A.K. Kerman, *Phys. Rev..* 120, (1961) 300; A.K. Kerman, R.D. Lawson, and M.W. Macfarlane, *Phys. Rev..* 124 (1961) 162.

[6] J.P. Elliot and C. Wilsdon, *Proc. Roy. Soc.* A302 (1968) 509.

[7] P. Rochford and D.J. Rowe, *Phys. Lett..* B210 (1988) 5.

[8] C. Bahri, D.J. Rowe and W. Wijesundera, *Phys. Rev..* C58 (1998) 1539.

[9] D.J. Rowe and C. Bari, *J. of Phys.*, 31 (1998) 4947; C. Bahri and D.J. Rowe *in preparation.*

[10] M Jarrio, J.L. Wood and D.J. Rowe, *Nucl Phys.* A528 (1991) 409.

[11] J. Carvalho and D.J. Rowe, *Nucl. Phys..* A584 (1991) 1.

[12] M.G. Vassanji and D.J. Rowe, *Phys. Lett.* 127B (1983) 1.

[13] D.M. Brink and E. Boeker, *Nucl. Phys.* A91 (1967) 1.

[14] F.M.H. Villars, *Nuc. Phys.* 3 (1957) 240; S. Goshen and H.J. Lipkin *Ann. Phys., N.Y.* 6 (1959) 301; W. Scheid and W. Greiner, *Ann. Phys., N.Y.* 48 (1968) 493; D.J. Rowe,*Nucl. Phys.* A152 (1970) 273.

[15] D.J. Rowe, *Rep. Prog. In Phys.* 48 (1985) 1419.

[16] A.Y. Dzyublik, V.I. Ovcharenko, A.I. Steshenko and G.F. Filippov. *Sov. J. Phys.* 15 (1972) 487; W Zickendraht, *J. Math. Phys.* 12 (1971) 1663.

[17] P. Gulshani and D.J. Rowe, *Can. J. Phys.* 54 (1976) 970.

[18] D.J. Rowe and G. Rosensteel, *J. Math. Phys.* 20 (19 79) 465.

[19] D.J. Rowe and G. Rosensteel, *Ann. Phys., N.Y.* 126 (1980) 198,

[20] D.J. Rowe, P. Rochford and J. Repka, *J. Math. Phys.* 29 (1988) 572.

[21] H. Ui, *Prog. Theor. Phys.* 44 (1970) 153

[22] M. Born and J.R. Oppenheimer. *Ann. Phys.* 84 (1927) 457.

[23] E. Ínönü and E.P. Wigner, *Proc. Natl. Acad. Sci.* 39 (1953) 510.

PROMPT PARTICLE DECAYS OF DEFORMED SECOND MINIMA

D. Rudolph

Division of Cosmic and Subatomic Physics
Lund University, Box 118, S-22100 Lund
Sweden

dirk.rudolph@kosufy.lu.se

Abstract: High-spin states in neutron-deficient nuclei near doubly-magic ^{56}Ni were studied with heavy-ion fusion-evaporation reactions. Well deformed rotational structures in the second well of these nuclei were observed. Most interestingly, some of these bands were found to decay by prompt particle emission in competition to the expected γ decay-out. Recent results and perspectives of combined γ-ray and particle spectroscopy are presented.

1 INTRODUCTION

The advent of the latest generation of 4π Germanium detector arrays such as GAMMASPHERE [1] and EUROBALL [2] allow for more and more refined views into the world of nuclear structure. The high sensitivity of these devices, in particular if coupled to powerful ancillary detector systems, have given rise to numerous exciting, unprecedented, and surprising results [3]. The present contribution aims at one specific aspect of such array physics [4], namely the first observation of prompt discrete particle decays from deformed second minima. So far they have only been found in nuclei near the $N = Z = 28$ doubly-magic isotope ^{56}Ni.

Naturally, nuclei located in the vicinity of the proton drip line have a reduced Coulomb barrier. For them being in the ground state or in β-decaying isomeric states, this might lead to a competition between proton radioactivity and their (normal) β^+ decay. In fact, proton radioactivity was first observed in the decay of the $I^\pi = 19/2^-$ spin-gap isomer in ^{53}Co [5] illustrated on the right hand side of Fig. 1. While this type of proton decay is based on discrete-line spectroscopy, so-called β-delayed proton emission yields quasi-continuous

The Nucleus: New Physics for the New Millennium
Edited by Smit et al., Kluwer Academic / Plenum Publishers, New York, 2000.

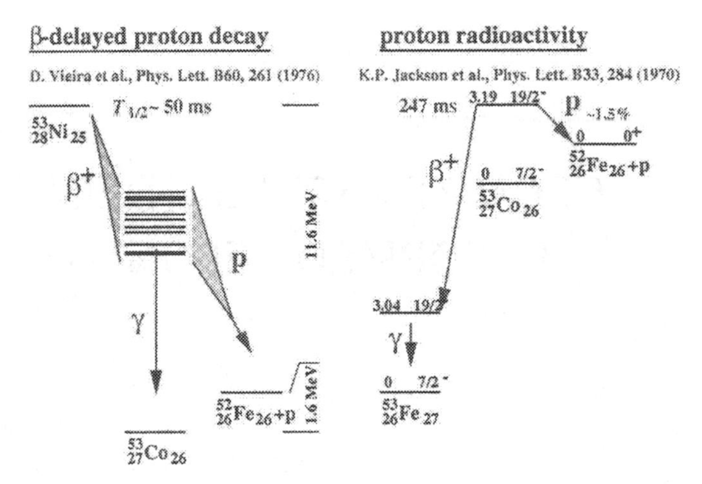

Figure 1 Sketch of previously known types of proton-emission processes encountered in low-energy nuclear structure studies: β-delayed proton decay from highly excited low-spin states (left) and the first observation of proton radioactivity (competing with β rays) in the case of the $19/2^-$ spin-gap isomer in ^{53}Co.

proton energy spectra. An example is shown on the left hand side of Fig. 1 [6]: Following the β^+-decay of ^{53}Ni low-spin states are populated in the regime of high excitation energy and level density in the daughter nucleus ^{53}Co which are unbound to prompt (typically s-wave) proton emission into the ground state of ^{52}Fe. In parallel, γ-rays might reach the ^{53}Co ground state. Note that the Q-window of the β^+ decay is large (13.2 MeV), and that the binding energy of ^{53}Co is small (1.6 MeV). More information on these early studies can be found, e.g., in a review article by J. Cerny and J.C. Hardy [7].

In the following years, more and more refined techniques were developed to map the proton drip-line and, hence, proton emitters in a somewhat heavier mass regime $A \sim 110 - 180$. These studies were mainly driven by the use of online recoil mass spectrometers at, e.g., GSI at Darmstadt, Daresbury Laboratory, or Argonne and Oak Ridge National Laboratories. Details of the latest (experimental) developments can be found in review articles by S. Hofmann [8], or P.J. Woods and C.N. Davids [9]. The experimental progress was accompanied by several continuously improved theoretical approaches – semiclassical WKB, Distorted Wave Born Approximation, two-potential approach – which were successfully applied to describe the properties of (spherical) proton emitting nuclei (see, e.g., Refs. [10, 11]). Recently, proton emission from the strongly deformed nuclei ^{131}Eu and ^{141}Ho was reported [12]. Here, the measured half-life of the activity can be used to tag the Nilsson orbits at the Fermi-surface and/or the size of the deformation. A new comprehensive theoretical description deals with emitted nucleons moving in deformed single-particle Nilsson levels [13, 14]. More information on theoretical aspects might be found in the contribution of N. Carjan [15].

Why is it of specific interest to study proton (or particle) radioactivity? E. Maglione and co-workers, referring to the deformed ground-state proton emitters, state [13] that

> ... proton decay can be used to probe small components of the deformed wave function in the mother nucleus, which would otherwise be very difficult, if at all possible, to measure.

S. Åberg and collaborators summarize as follows [11]:

> As studies of exotic nuclei progress, it will be very important to determine the empirical ordering of different single-particle states near and beyond the drip-line, to classify excited states, and, possibly, to investigate the competition between the proton radioactivity and other decay modes (such as gamma decay).

Here the competition between (discrete) particle and γ decays is explicitly mentioned, something which was studied more than 20 years ago by T. Døssing, S. Frauendorf, and H. Schulz [16], for the case of high-spin states:

> Future experiments on high angular momenta might reveal yrast traps or ... states ... from which the γ decay is strongly hindered. If such states exist above the critical angular momentum for particle emission then the nucleon will be emitted with just one energy, or a few possible energies, and this nucleon plus the following γ-decay cascade would tell a lot about the structure of the high spin state.

However, experimental evidence for these prompt and discrete exotic decays has been lacking until recently, but experiments at the large γ-ray detector facilities have by now revealed a plethora of states from which the γ decay in fact *is* strongly hindered. For example, the states at the bottom of superdeformed bands or second minima in the nuclear potential. These states and associated γ decays were recently established also in the light $A \sim 60$ regime (cf. Ref. [17]), and very recently we succeeded to observe for the first time a prompt ($\tau < 3$ ns) decay of a *well-deformed* excited rotational band in the $N{=}Z$ nucleus ^{58}Cu via emission of monoenergetic protons into a *spherical* excited state in ^{57}Ni [18]. In the following, a second case was established in the decay-out of a rotational band in the doubly magic nucleus ^{56}Ni [19], and there are indications for a third example in the neighbouring isotope ^{59}Cu [20]. Moreover, a 4% decay-out branch from the second minimum in ^{58}Ni constitutes the first observation of a prompt monoenergetic α radiation [21] into the spherical 2949 keV 6^+ yrast state in ^{54}Fe [22, 23, 24].

Figure 2 summarizes the proton decay of the band in ^{58}Cu. The irregularly spaced states in the first, spherical minimum, and the rotational band in the second minimum are shown in the upper left part. The spin values of the states in the first minimum were measured while those in the second minimum were inferred from the assigned configuration of the band, i.e., $\pi(g_{9/2}) \times \nu(g_{9/2})$, as well as the best estimates originating from partly measured spins of initial and final states of both γ and proton decay out of the band. Within the band angular distribution and correlation measurements clearly indicate quadrupole character for the γ-ray transitions [24].

Prompt Proton Decay
of a
Well-Deformed Band in ^{58}Cu

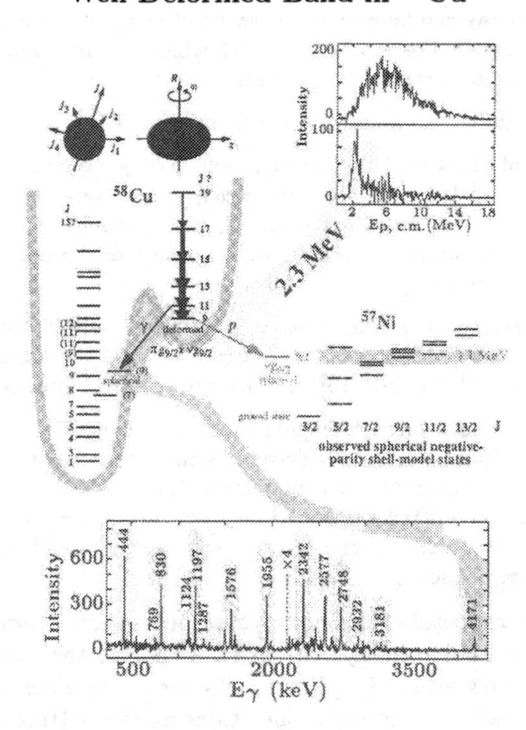

Figure 2 Overview of the prompt proton decay from the deformed band in the second potential well of ^{58}Cu into a spherical daughter state in ^{57}Ni. The γ-ray spectrum from the reaction ^{36}Ar+^{28}Si was gated by one detected alpha, proton, and neutron, and measured in coincidence with the second and third transition in the rotational band of ^{58}Cu (1197 and 1576 keV). It shows the other transitions in this band (830, 1955, 2342, 2748, and 3181 keV), the discrete linking transition at 4.2 MeV, and γ rays between states in the spherical minima of both ^{58}Cu (e.g., 444 keV) and ^{57}Ni (e.g., 1124 and 2577 keV). In the upper right part of the figure proton center-of-mass energy spectra are shown gated by one α particle, one or two protons, and one neutron, in prompt coincidence with the 444 keV ground-state transition in ^{58}Cu (top) and the 830 keV transition which feeds the proton decaying ^{58}Cu band head.

The γ-ray spectrum at the bottom of Fig. 2 implies that only the $I = 9/2$ state at 3701 keV excitation energy in the daughter nucleus ^{57}Ni is populated by the main branch observed at 2.3(1) MeV proton center-of-mass energy (cf. spectrum in the upper right part of Fig. 2). Many other levels with spin values close to 9/2 were observed in ^{57}Ni at similar excitation energies, and shell-model calculations indicate a high degree of mixing within these negative-parity states [24]. This fact plus the apparent selectivity of the proton decay clearly hint towards positive parity of the 3701 keV level, i.e., it is likely to reflect the $1g_{9/2}$ single-particle level with respect to the doubly-magic core ^{56}Ni. However, its parity yet remains to be determined. It should be noted that the measured FWHM of the proton peak can be attributed to the kinematic broadening rather than the intrinsic resolution of the CsI elements of MICROBALL [25]. More information related to the ^{58}Cu proton decay can be found in Refs. [18, 26].

Apparently, there are three major differences between the 'conventional' proton emission processes and these prompt discrete particle decays:

- The prompt particle decays compete with γ-radiation, while (ground-state) proton radioactivity competes with β^+ decay.
- The time scale is different by several orders of magnitude. Proton emitters possess typical half-lives in the micro- to millisecond range. β-delayed protons are *observable* on a similar time scale but, of course, the *intrinsic* decay times are much faster, i.e., in the attosecond regime. The prompt particle decays, however, seem to lie in the picosecond range.
- There is a drastic change of nuclear shape associated with the prompt particle decays – the initial states, situated at the bottom of rotational bands, have a deformation of $\beta_2 \sim 0.4$ while the final states are spherical shell-model states. For other particle decays, the shape of initial and final nuclear state is essentially the same.

2 EXPERIMENTS

Previously, mainly light-ion induced reactions have been used to populate excited states in ^{56}Ni [27, 28] and neighbouring nuclei. The experiment exploring the high-spin decays in this mass regime was performed at the 88-Inch Cyclotron at the Lawrence Berkeley National Laboratory. The fusion-evaporation reaction ^{28}Si(^{36}Ar,$xpynz\alpha$) at effective 136 MeV beam energy was used. The GAMMASPHERE array [1] comprised 82 Germanium detectors and the γ rays

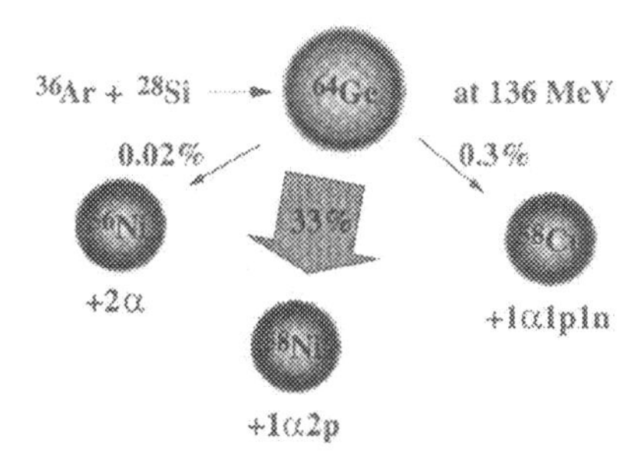

Figure 3 Experimental relative cross sections for the nuclei of interest.

were measured in prompt coincidence with the evaporated light particles to provide reaction channel selection. The charged particles were detected in the 95-element 4π CsI ball MICROBALL [25] (Fig. 4 left) while neutrons were measured in fifteen liquid scintillator neutron detectors. For details see, e.g., Refs. [18, 24].

Figure 3 provides the experimental relative cross-sections for the three residual nuclei of interest. As can be seen, the reaction dominantly populates excited states in the $1\alpha 2p$ channel ^{58}Ni. It absorbs no less than 1/3 of the reactions. On the one hand, this allows for restrictive and multiple gating procedures. On the other hand, one has to cope with an immense amount of information: While previously only five excited high-spin states have been reported for ^{58}Ni [29], a recent study at an intermediate-sized array revealed some 25 γ rays connecting states up to 8 MeV and spin $I \sim 10\ \hbar$ [30]. The (still incomplete) level scheme from our experiment contains more than 200 γ-ray transitions, including several rotational bands, possibly magnetic dipole bands, and a plethora of shell-model states, in total reaching up to 30 MeV excitation energy and spin $I \sim 30\ \hbar$. However, by exchanging one evaporated proton by a neutron (^{58}Cu+$1\alpha 1p1n$), the cross section drops by two orders of magnitude. This fact alone manifests that selective devices (here: neutron detectors) are vital to perform reliable and comprehensive spectroscopy at or even beyond the $N = Z$ line [4]. Interestingly, the cross section for the 2α channel is about a factor of

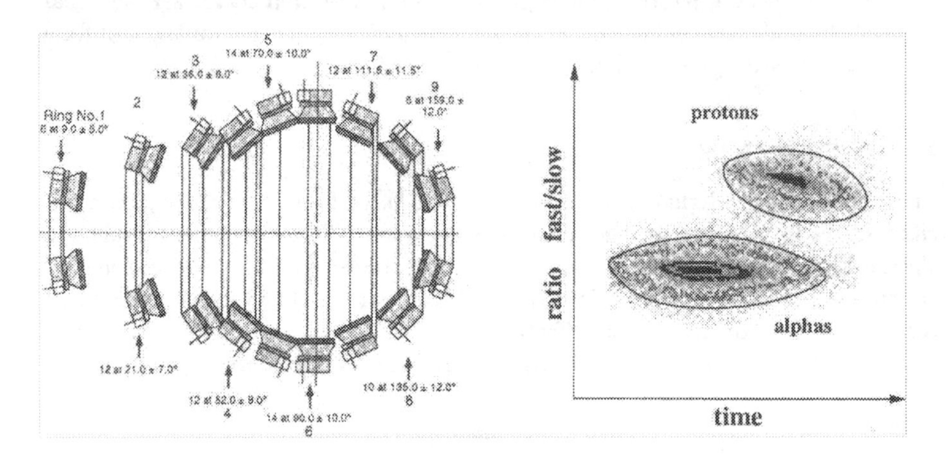

Figure 4 A schematic drawing of the 4π CsI array MICROBALL [25] is shown on the left hand side while a two-dimensional spectrum from the MICROBALL detector element 35 (ring 4) is shown on the right hand side. The leading edge discriminator time is plotted versus the charge ratio of the fast and the slow part ('particle identification') of the signal, both in arbitrary units. The spectrum is gated by two two-dimensional gates in similar matrices 'energy vs. time' and 'energy vs. ratio' for protons as well as α-particles. The third two-dimensional gate for each species, indicated by the solid lines in the plot, finally determines the unambigous particle identification.

ten too small as compared to *experimental* estimates gathered from this and other reactions in the mass region.

Protons and α particles were discriminated in the MICROBALL by pulse-shape techniques [25]. For the present analysis we employed a triple two-dimensional gating procedure for each particle species. They are set in the three two-dimensional spectra which can be created from the three signals from each MICROBALL element: time, energy, and the charge ratio between the early and late part of the pulse ('particle identification'). The right hand side of Figure 4 illustrates the final gates (solid lines) in a two-dimensional spectrum time vs. ratio. of MICROBALL detector element 35, situated in the fourth ring around 52° with respect to the beam axis. (52° in the laboratory frame corresponds to nearly 90° in the center-of-mass frame.) The spectrum is gated by two two-dimensional gates set in the matrices 'energy vs. time' and 'energy vs. ratio' for protons as well as α-particles. Clearly, the third gate unambiguously identifies either a proton or an α-particle. In addition, the energies and angles of the charged particles detected in the MICROBALL were used to evaluate the momenta of the individual recoiling nuclei. This kinematic correction results in a more precise Doppler-shift determination of the γ-ray energies, which is of particular importance in view of these light systems.

3 PROMPT ALPHA DECAY IN ^{58}Ni

A schematic overview of experimental evidence for the prompt discrete α decay from the deformed band in the second potential well of ^{58}Ni into the spherical 6^+ yrast daughter state in ^{54}Fe is illustrated in Fig. 5. On the right hand side a sketch of the relevant part of the level scheme of ^{58}Ni [21, 30] is shown, including a firm 4 MeV discrete γ-link from the strongly deformed states into the spherical minimum, as well as the α-decay branch into ^{54}Fe. It has a Q-value of 7.4 MeV inferred from the known masses.

The two γ-ray spectra on the left hand side of Fig. 5 are based on a $2\alpha2p$-gated $\gamma\gamma\gamma$ cube. The spectrum at the top shows the sum of three doubly-gated spectra within the well-known 411-1130-1408 keV $6^+ \rightarrow 4^+ \rightarrow 2^+ \rightarrow 0^+$ yrast sequence in the $2\alpha2p$ reaction channel ^{54}Fe. It reveals the sequence itself and a number of transitions connecting levels at higher excitation energies in ^{54}Fe [24]. However, it shows also one clear and relatively strong peak at 1663 keV. This is the same energy as that of the lowest transition in the deformed band in ^{58}Ni.

In fact, a spectrum gated by the 1663 keV line in a $2\alpha2p$-gated matrix clearly shows the three transitions in ^{54}Fe, the other transitions of the band in ^{58}Ni, but, even though weakly, also transitions connecting spherical low-spin states in ^{58}Ni. The latter occur due to pile-up in the MICROBALL detectors and/or imperfect particle gating. Therefore, we restricted the data analysis for α particles to the first four rings of MICROBALL for which the particle identification is unequivocal (cf. Fig. 4). This caused a drop of the α-detection efficiency from some 65 % to $\epsilon_\alpha = 51(1)$ %, which is affordable due to the high cross-section of the ^{58}Ni+$1\alpha2p$ channel. To further reduce possible contaminations

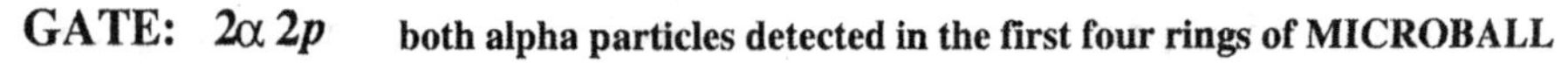

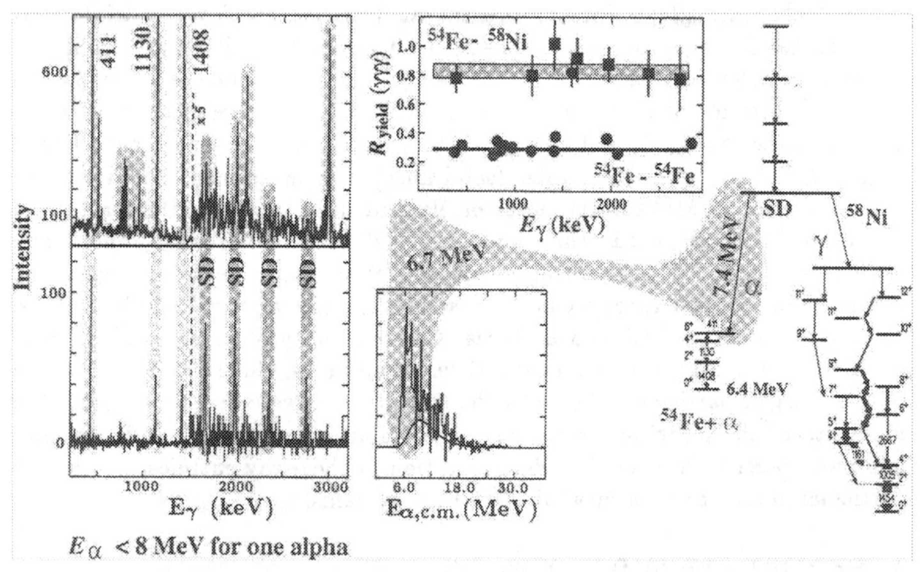

Figure 5 Overview of the prompt discrete α-decay from the deformed band in the second potential well of ^{58}Ni into the spherical 6^+ yrast daughter state in ^{54}Fe. On the right hand side the sketch of the decay schemes of the two nuclei indicate a firm 4 MeV γ link between the SD band and the spherical states in ^{58}Ni and the α-decay branch into ^{54}Fe. The two γ-ray spectra on the left hand side are based on a $2\alpha2p$-gated $\gamma\gamma\gamma$ cube. It shows the sum of three doubly-gated spectra within the $6^+ \rightarrow 4^+ \rightarrow 2^+ \rightarrow 0^+$ yrast sequence in ^{54}Fe (top) and the sum of nine doubly-gated spectra between this sequence and the lowest three transitions in the SD band in ^{58}Ni (bottom). For the latter at least one α-particle must have had an energy less than 8 MeV. The same γ-gating procedure was applied for the α center-of-mass energy spectrum yielding a peak structure at 6.7 MeV. The ratio of γ-triples yields is explained in the text.

an additional upper energy cut of 8 MeV for at least one of the two detected α particles was inferred. The striking result is the γ-ray spectrum at the bottom left of Fig. 5. It corresponds to the sum of nine doubly-gated spectra between the yrast cascade in ^{54}Fe and the lowest three members of the deformed band in ^{58}Ni. No other transition from either ^{58}Ni or ^{54}Fe is visible! By using identical γ-ray *double*-gating conditions and an overall $2\alpha2p$ particle gate but projecting on the α-particle center-of-mass energy the sum spectrum in the middle of Fig. 5 can be created. It clearly exhibits a peak structure at 6.7(2) MeV. The full curve corresponds to a normalized total projection of the α-particle energy axis. The missing 0.7(2) MeV with regard to the Q-value can be attributed to the kinetic energy which is transferred to the ^{54}Fe recoil in the course of the decay.

The ratio $R_{\rm yield}(\gamma\gamma\gamma) = Y(2\alpha 2p, E_\alpha < 8$ MeV$)/Y(2\alpha 2p)$ is plotted as function of γ-ray energy in the middle of Fig. 5. Again, both α particles had to be detected in the first four rings of MICROBALL. According to the peak in the α-energy spectrum this ratio should be close to unity for transitions associated with the α decay. Contrary, yields of transitions of ^{54}Fe in spectra doubly-gated with two transitions of ^{54}Fe should be reduced to $R_{\rm yield} = 1 - 0.83^2 = 0.31$, because 83 % of the total α-energy projection have energies in excess of 8 MeV. This estimate is in perfect agreement with the observation. It should be noted that these ratios provide one of the most stringent tests of the particle decays, especially if no concident γ-rays were present in the daughter nucleus [26].

By determining the yields of the $2^+ \to 0^+$ transitions (or other significant transitions between low-spin states) in spectra from $1\alpha 2p$-gated matrices or cubes which are (doubly-)gated on transitions in the deformed minimum of ^{58}Ni a (preliminary) α-branching ratio of $b_\alpha = 4(2)$ % can be estimated. Other experimental information includes the angular distribution of the emitted α ($\Delta l \sim 8 - 10\ \hbar$), and the investigation of additional fractional Doppler shifts yields an average quadrupole moment of $Q_t \sim 2.0$ *e*b for the band.

A naive α-decay scenario is based on the fact that all the deformed bands in the $A \sim 60$ region are built on an underlying four particle-four hole (4p-4h) structure across the $N = Z = 28$ gap. Hence, these four particles (two protons and two neutrons) are moving in identical orbits and, of course, might have a good chance of preforming an α-cluster. On the other hand, the orbits involved are low-spin orbits, and the *deformed* ^{56}Ni core (4p-4h with respect to the *spherical* one) is again doubly magic. A more sophisticated theoretical approach arises from an idea presented in the early 80's (at that time purely academic). Catara and co-workers calculated the probability distributions $P(r, R)$ for two identical particles which were moving in single-particle orbits outside a closed core and coupled to spin $I = 0$. (r is the distance between these two particles, R their center-of-mass distance to the core.) It turns out that for high-j orbits with $\Delta l = 1$ and opposite parity a distinct maximum of the probability distribution occurs at small r but large R, i.e., a clustering at the surface of the nucleus [31]. The calculations were performed for the $h_{11/2}$ and $i_{13/2}$ orbits, but there is a similar pair present in our case, namely $f_{7/2}$ and $g_{9/2}$. Detailed calculations are in progress.

4 ROTATIONAL BANDS AND THE SHELL MODEL

The rotational bands in the vicinity of ^{56}Ni are based on 4p-4h excitations across the magic $N = Z = 28$ shell gap between the $1f_{7/2}$ orbit and the so-called upper *pf* shell ($2p_{3/2}$, $1f_{5/2}$, and $2p_{1/2}$ orbits). In the Nilsson scheme, the crossing between the up-sloping [303]7/2 and the down-sloping [321]1/2 orbits opens a second distinct energy gap at quadrupole deformations $\beta_2 \sim 0.3$–0.5. Taking ^{56}Ni as an example, this configuration is reflected by the sequence of Routhians (calculated at $\beta_2 = 0.4$) labelled $4^0 4^0$ in Fig. 6(a). The levels below the *deformed* shell gap at $N = Z = 28$ are filled, and none of the (quasi)particles has been excited into the $\mathcal{N} = 4$ $1g_{9/2}$ intruder orbit [440]1/2.

The Routhians in Fig. 6(e) were calculated with a Skyrme-Hartree-Fock (HF) approach and the cranking approximation [32, 33]. The downsloping high-j low-Ω [440]1/2 neutron (and proton) orbit creates an energy gap of several MeV at frequencies $\hbar\omega \approx 1.0$ MeV. For ^{56}Ni the excitation of one neutron and one proton [4^14^1 in Fig. 6(a)] costs about 5 MeV of excitation energy at low spins, but it is predicted to become the yrast configuration beyond spin $I = 20\ \hbar$. Due to the expected shape driving effect of the $1g_{9/2}$ particles the calculated quadrupole deformation increases from $Q_t = 1.65$ eb for the 4^04^0 band to 1.89 eb for the 4^14^1 band. The odd-odd $N = Z = 29$ nucleus ^{58}Cu has one proton and one neutron in addition to ^{56}Ni. Consequently, it forms a deformed "doubly-magic" core at $\beta_2 \sim 0.4$ – the yrast 4^14^1 band is very well separated from other collective structures by a gap of some 4 MeV [18]. Moving

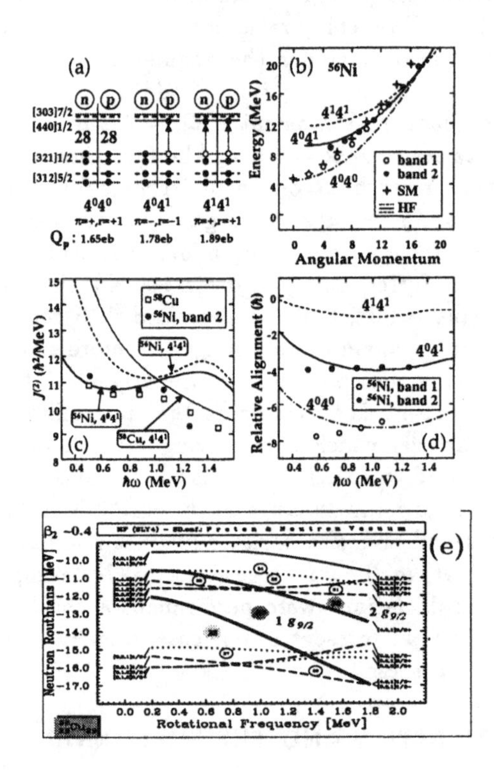

Figure 6 Properties of deformed bands near ^{56}Ni according to the cranked HF-SLy4 model. (a) Schematic diagram of single particle configurations corresponding to the calculated deformed minima The (π, r) quantum numbers of single-particle Routhians are: $(+,-i)$ - solid line; $(-,i)$ - dotted line; $(-,-i)$ - dashed line. (b) Energies of the 4^04^0, 4^04^1, and 4^14^1 bands compared to the full fp 6p-6h shell-model calculations (SM) and experimental bands 1. (c) Calculated dynamic moments of inertia of bands 4^04^1 and 4^14^1 in ^{56}Ni and the 4^14^1 band in ^{58}Cu. (d) Relative alignments, $i_{\rm rel} \equiv I(^{56}\text{Ni}) - I(^{58}\text{Cu})$, for deformed bands in ^{56}Ni. (e) Routhian diagram for ^{58}Cu.

to even larger deformations, the second [440]1/2 Routhian also becomes quickly favoured, leading to a superdeformed "doubly magic" core at $\beta_2 \sim 0.5$ for the $N = Z = 30$ system, i.e., ^{60}Zn [34].

In case of the 4^04^0 band in ^{56}Ni it was possible to compare the results from mean-field methods to those from state-of-the-art large-scale shell-model calculations. The result is shown in the E_x-I plane of Fig. 6(b). The shell-model calculations were done using the KB3 interaction introduced by the Madrid-Strasbourg group [35], and up to six particles were allowed to be excited from the $1f_{7/2}$ orbit into the upper pf shell. This leads to basis dimensions of some 25 million which is close to the present computational limits [36]. The overall agreement of the shell-model and the experimental results are striking though a closer look into Fig. 6(b) reveals some irregularities in the level spacing due to the truncation.

Table 1 Experimental properties of some of the rotational bands in the second well in the $A \sim 60$ region (status February 1999).

Isotope	$I_{\min}$ ($\hbar$)	$I_{\max}$ ($\hbar$)	$E_{x,max}$ (MeV)	*Max. rel. pop.* (%)	*Decay out* (%) γ_{total}	γ_{discrete}	p,α	$E_{p,\alpha}$ (MeV)	*Ref.*
^{56}Ni	(5)	(17)	20	18	82(5)	82(5)	18(5)	2.5(1)	[19]
^{58}Ni	16^+	26^+	29	1.5	96(2)	~ 40	4(2)	6.7(2)	[21]
^{58}Cu	(9)	(23)	23	30	33(5)	33(5)	67(5)	2.3(1)	[18]
^{59}Cu	$25/2^+$	$57/2^+$	32	30	> 95	~ 60	< 5	< 3	[20]
^{60}Zn	8^+	30^+	34	34;60	> 95	37(3)	< 5		[34]

In the mean-field calculations, this band can be ascribed to the 4^04^0 configuration with respect to a closed *deformed* core of ^{56}Ni. However, its alignment is predicted too large because pairing correlations were not considered [cf. Fig. 6(b)]. By promoting one particle across the deformed gap from the [321]1/2 to the [440]1/2 Routhian, one obtains the 4^14^0 and 4^04^1 bands which have negative parity and odd spins. Their energies agree well with an excited rotational band in ^{56}Ni, and the dynamic moment of inertia is consistent up to $\hbar\omega \approx 1.1$ MeV. However, a predicted band crossing between the [321]1/2 and [312]5/2 Routhians around $\hbar\omega \approx 1.5$ MeV has not been observed yet [21] [see Fig. 6(c)]. Next to its 50 % proton decay branch into the ground state of ^{55}Co [21] this band has also nearly identical transition energies to the band in ^{58}Cu. Assuming a band-head spin $J = 9$ for the band in ^{58}Cu the 4^04^1 configuration of ^{56}Ni has a relative alignment close to $-4\ \hbar$. This is consistent with the angular momentum proposed for the band in ^{56}Ni as illustrated in Fig. 6(d). The calculated 4^14^1 band in ^{56}Ni carries $i_{\text{rel}} \approx -0.5$, which is difficult to accommodate experimentally. Therefore, we associate a probably mixed 4^14^0 and 4^04^1 structure with the excited band. Unfortunately, the predicted dynamic moments of inertia of bands 4^04^1 in ^{56}Ni and 4^14^1 in ^{58}Cu differ both at low spins and at high spins which makes it difficult to understand their near identicality. While an accidental degeneracy cannot be ruled out, the possibility exists that the disagreement between experiment and theory has its roots in the microscopic input of the calculations, e.g., the treatment of pairing correlations or the inadequacy of today's effective interactions.

Another aspect of the bands in the second minimum are population and depopulation mechanisms. For some selected bands in the mass $A \sim 60$ region the numbers are compiled in Table 6. (Note that some of the nuclei are still under analysis.) Two features are striking: Firstly, all bands but the one in ^{58}Ni are much more strongly fed than common superdeformed (SD) bands in other mass regions. For the preferred reaction for ^{60}Zn, i.e., ^{40}Ca(^{28}Si,2α), the yield of the band is as high as 60 % of the 1004 keV $2^+ \rightarrow 0^+$ ground-state transition [34]. Secondly, the percentage of observed, discrete γ decay out amounts to typically 50 % of the total γ decay out, and it is comprised in some 5 (mainly $\sim$ 3-5 MeV stretched $E2$) transitions for ^{58}Ni, ^{59}Cu, and ^{60}Zn. In the other two cases, ^{56}Ni and ^{58}Cu, the combined discrete γ and particle decays can account for the complete yield of the respective band. For comparison, in the case of ^{194}Pb a total of twelve discrete links (mainly of stretched $E1$ type) account for only 21 % of the intensity of the band [37, 38, 39], for ^{194}Hg five transitions make up 5 % [40, 41]. Whether the differences in yields are simply related to the differences in level density, or whether the dominance of $E2$ decay-out transitions in the $A \sim 60$ region are related to the closeness of the $N = Z$ line remains to be investigated in more detail.

5 SUMMARY AND PERSPECTIVES

Clearly, the mass $A \sim 60$ region comprises a large variety of nuclear struture effects. First of all, the vicinity to the (soft) doubly-magic core ^{56}Ni provides an ideal testing ground for the spherical shell-model. Experimentally, a plethora of states in the first minimum are currently evolving, and the theoretical challenge seems to be not only the deduction of proper single-particle energies and effective two-body matrix-elements for the full fp model space but to include the $1g_{9/2}$ shell in one way or another. The spherical $1g_{9/2}$ states have been identified in the $A = 59$ nuclei and likewise in ^{57}Ni. Of course, this implies an additional dramatic increase in the basis dimensions for the conventional shell-model studies, i.e., a truncation in the number of particles either crossing the N,Z = 28 gap and/or being lifted into the $1g_{9/2}$ orbit will be unavoidable (at present). The Shell-Model Monte-Carlo (SMMC) [42] and Quantum Monte-Carlo Diagonalisation (QMCD) [43, 44] methods, however, are capable to bypass these huge dimensions as they trace only the most important components of the wave functions. Hence, they appear to be the favourable tools for near-future full $fpg_{9/2}$ shell-model studies.

At high spins ($I \sim 10 - 15$) collective structures dominate the experimental level schemes. Next to the strongly or superdeformed bands in the second minimum there are also a series of bands ranging from normally deformed $E2$ cascades over strongly coupled bands with about equally strong $M1$ and $E2$ transitions to plain $M1$ sequences. Experimentally, the knowledge is somewhat

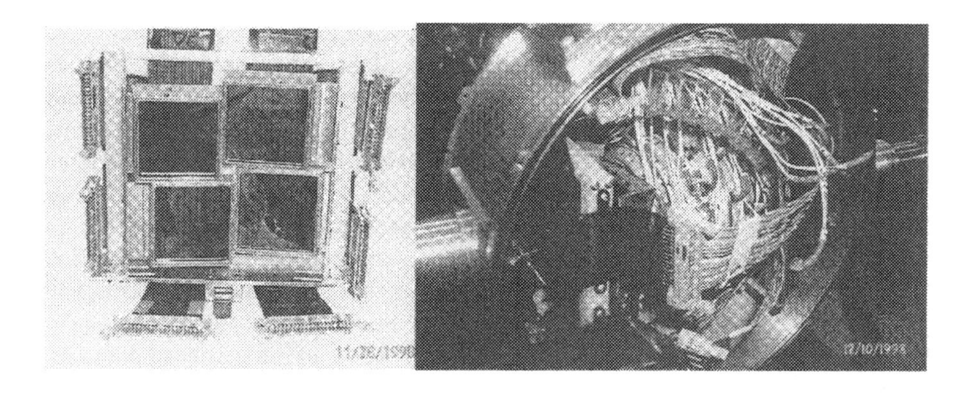

Figure 7 Photograph of the four ΔE-E Silicon strip detector telescopes used in a recent experiment at GAMMASPHERE (left). Each detector has 16 strips providing a total of 256 pixels per telescope. The coupling of this system to the MICROBALL (the first three rings of which were replaced) is shown at the right. Due to shadowing effects in the corners some 800 pixels were active.

scarce as compared to the spherical shell-model states, namely the spins and parities of some of the bands (and in few cases even the excitation energies) are yet to be determined. Mainly aiming at the electromagnetic decay properties of the γ-rays we recently performed a backed-target EUROBALL experiment entering the same compound nucleus as the first run at GAMMASPHERE. On the theoretical side the HF(B) results on the strongly deformed bands indicate a significant effect on the choice of the effective interaction used [18]. More detailed investigations [cf. Fig. 6(c),(d)] are underway. In addition, the normally deformed and/or strongly coupled bands offer the unique opportunity to compare and relate conventional or Monte Carlo shell-model calculations to approaches based on mean field theory.

Most interstingly, however, prompt monoenergetic proton and alpha decay lines were observed for the first time in the decay from high-spin states which are associated with a deformed secondary minimum in the potential. At present, the drastic change of shape in the course of the decay (all the daughter states are spherical) seems to prevent a satisfactory theoretical description and, hence, prediction. The impact of these discrete decay-out transitions on models describing the tunneling process from the second to the first minimum remains to be determined. Experimentally, of course, there is the quest for more cases, in particular in other mass regions, and to collect a more comprehensive picture of the process. A steep yrast line of the first minimum seems to be important, as well as a closeby drip-line (weak binding). The known decays also take place near or at the presumed band-head of the respective band. Hence, other favourable regimes are the $A \sim 80$ and $A \sim 90$ nuclei while, despite their relatively large excitation energy, the SD bands in the $A \sim 150$ region seem to be less good candidates. In the $A \sim 60$ region it is important not only to verify the decays in dedicated, independent experiments but to investigate the details of the decays such as precise branching ratios, decay times and, hence, strengths as well as potential fine structures. The combination of the above mentioned EUROBALL experiment together with a new experiment at GAMMASPHERE should provide that information. The latter recently took place at Argonne National Laboratory using again the reaction $^{36}Ar+^{28}Si$ but with the first three rings of MICROBALL (28 detector elements) replaced by four ΔE-E Silicon strip detector telescopes with a total of some 800 active pixels (cf. Fig. 7). As mentioned earlier the Doppler broadening is the decisive quantity for *in-beam particle spectroscopy*. Even for the new, much more dedicated set-up this limits the proton energy resolution to some 150-200 keV.

There is surely more than one puzzle to be solved in the next millenium!

Acknowledgments

First of all, I would like to thank the chairman as well as the Swedish Natural Science Research Councils for having provided me with the opportunity to join the stimulating conference at such a pleasant location. All the best to NAC and the country for the future! I am indebted to E. Caurier, D.J. Dean, J. Dobaczewski, P.-H. Heenen, W. Nazarewicz, F. Nowacki, A. Poves, and W. Satula for the effort they put into the theoretical description and understanding of the spherical and deformed states in the mass region, and letting me use their results. C. Baktash deserves credit for many invaluable discussions. Without the perfect and persistent work of the collaborators from Washington University, namely M. Devlin, D.R. LaFosse, and last but not least D.G. Sarantites, this work would not have been possible at all. Thanks to C.E. Svensson for helping me out with some of his transparencies and results. Finally, I want to thank M.J. Brinkman, R.M. Clark, P. Fallon, H.-Q. Jin, R. Krücken, I.-Y. Lee, R. MacLeod, A.O. Macchiavelli, L.L. Riedinger, and C.-H. Yu and the operating crew of the 88-Inch Cyclotron for their assistance during the experiments, and D. Balamuth, S. Freeman, M. Leddy, and C.J. Lister for organizing, providing and setting up the neutron detector array at GAMMASPHERE. Many thanks to C. Fahlander and H. Ryde for carefully reading the manuscript. This research was supported in part by the Swedish Natural Science Research Councils.

References

[1] I.-Y. Lee, Nucl.Phys. A520, 641c (1990).

[2] EUROBALL III, A European γ-ray facility, Eds. J. Gerl and R.M. Lieder, GSI 1992.

[3] GAMMASPHERE, The Beginning ... 1993-1997, Science Highlights booklet, Ed. M.A. Riley, (1998); http://www-gam.lbl.gov.

[4] C.J. Lister, this proceeding.

[5] K.P. Jackson, C.U. Cardinal, H.C. Evans, N.A. Jelley, and J. Cerny, Phys. Lett. B33, 281 (1970).

[6] D.J. Vieira, D.F. Sherman, M.S. Zisman, R.A. Gough, and J. Cerny, Phys. Lett. B60, 261 (1976).

[7] J. Cerny and J.C. Hardy, Annu. Rev. Nucl. Sci 27, 333 (1977).

[8] S. Hofmann, in *Nuclear Decay Modes*, Ed. D.N. Poenaru, IOP Publishing, Bristol, 1996, p. 143ff.

[9] P.J. Woods and C.N. Davids, Ann. Rev. Nucl. Part. Sci. 47, 541 (1997).

[10] V.P. Bugrov and S.G. Kadmensky, Sov. J. Nucl. Phys. 49, 967 (1989).

[11] S. Åberg, P.B. Semmes, and W. Nazarewicz, Phys. Rev. C56, 1762 (1997).

[12] C.N. Davids *et al.*, Phys. Rev. Lett. 80, 1849 (1998).

[13] E. Maglione, L.S. Ferreira, and R.J. Liotta, Phys. Rev. Lett. 81, 538 (1998).

[14] E. Maglione, L.S. Ferreira, and R.J. Liotta, Phys. Rev. C59, R589 (1999).

[15] N. Carjan, this proceeding.

[16] T. Døssing, S. Frauendorf, and H. Schulz, Nucl. Phys. A287, 137 (1977).

[17] C.E. Svensson *et al.*, Phys. Rev. Lett. 79, 1223 (1997).

[18] D. Rudolph *et al.*, Phys. Rev. Lett. 80, 3018 (1998).

[19] D. Rudolph *et al.*, submitted to Phys. Rev. Lett.

[20] C. Andreoiu *et al.*, to be published.

[21] D. Rudolph *et al.*, submitted to Phys. Rev. Lett.

[22] J. Styczen *et al.*, Nucl. Phys. A327, (1979) 295.

[23] J. Huo, H. Sun, W. Zhao, and Q. Zhou, Nucl. Data Sheets 68, 887 (1993).

[24] D. Rudolph, C. Baktash, M.J. Brinkman, M. Devlin, H.-Q. Jin, D.R. LaFosse, L.L. Riedinger, D.G. Sarantites, and C.-H. Yu, Eur. Phys. J. A4, 115 (1999).

[25] D.G. Sarantites *et al.*, Nucl. Instrum. Meth. A381, 418 (1996).

[26] D. Rudolph, in *Proceedings Nuclear Structure '98*, Gatlingburg, TN, Ed. C. Baktash, APS, 1999.

[27] J. Huo, Nucl. Data Sheets 67, 523 (1992).

[28] J. Blomqvist *et al.*, Z. Phys. A 322, 169 (1985).

[29] M.R. Bhat, Nucl. Data Sheets 80, 789 (1997).

[30] S.M. Vincent *et al.*, J. Phys. G, in press.

[31] F. Catara, A. Insolia, E. Maglione, and A. Vitturi, Phys. Rev. C29, 1091 (1984).

[32] J. Dobaczewski and J. Dudek, Comp. Phys. Commun. 102, 166 (1997); 102, 183 (1997); to be published.

[33] E. Chabanat *et al.*, Nucl. Phys. A627, 710 (1997).

[34] C.E. Svensson *et al.*, submitted to Phys. Rev. Lett.

[35] A. Poves and A. Zuker, Phys. Rep. 70, 235 (1981).

[36] E. Caurier, shell-model code, Strasbourg (1990).

[37] M.J. Brinkman *et al.*, Phys. Rev. C53, R1461 (1996).

[38] A. Lopez-Martens *et al.*, Phys. Lett. 380B, 18 (1996).

[39] K. Hauschild *et al.*, Phys. Rev. C55, 2819 (1997).

[40] T.L. Khoo *et al.*, Phys. Rev. Lett. 76, 1583 (1996).

[41] G. Hackman *et al.*, Phys. Rev. Lett. 79, 4100 (1997).

[42] S.E. Koonin, D.J. Dean, and K. Langanke, Phys. Rep. 278, 1 (1996), and references therein.

[43] M. Honma, T. Mizusaki, and T. Otsuka, Phys. Rev. Lett. 75, 1284 (1995).

[44] M. Honma, T. Mizusaki, and T. Otsuka, Phys. Rev. Lett. 77, 3315 (1996).

UNIFIED DERIVATION OF SELF-CONSISTENT MULTIPOLE INTERACTIONS IN NUCLEI

Hideo Sakamoto

Faculty of Engineering, Gifu University, Gifu 501-1193, Japan

sakamoto@cc.gifu-u.ac.jp

Abstract: The present status of our research on self-consistent effective interactions in nuclei is discussed. By means of the field coupling method, we present a general prescription to derive effective interactions which are consistent with single-particle potentials. Using this method, doubly-stretched multipole interactions in rotating nuclei are constructed. The iso-vector components of effective interactions are discussed in terms of the nuclear self-consistency.

1 INTRODUCTION

It is well recognized that due to the self-sustained, strongly-bound and close-packing system, interactions between nucleons in the nucleus may be quite different from the bare nucleon-nucleon interactions in free space, and are generally dependent on the state of the system. To understand the dynamical aspects of nuclear many-body systems, by the use of the concept of *nuclear self-consistency* as an important guiding principle, we have developed a theory of self-consistent effective interactions in nuclei which is simple and yet effective for investigating anharmonicities in fundamental modes of motion.

We applied the framework to improve the conventional multipole interaction model so as to satisfy the nuclear self-consistency rigorously even if the system is deformed. As a result, the doubly-stretched multipole interaction model was proposed and has been successfully applied to low-lying collective states and giant resonances in deformed nuclei [1, 2]. The theory of self-consistent effective interaction has been extended for a system with a velocity dependent mean potential. It has been shown that the self-consistent velocity dependent effective interaction plays a role to recover the local Galilean invariance and eliminate various unphysical effects arising from the spurious velocity dependence of the

The Nucleus: New Physics for the New Millennium
Edited by Smit et al., Kluwer Academic / Plenum Publishers, New York, 2000.

mean potential. By use of this method, the origin of the multipole pairing interaction has been clarified [3].

Recently, by means of the field coupling method, we have formulated a general prescription to derive effective interactions which are consistent with single-particle potentials [4]. In this paper we intend to use the prescription to derive self-consistent effective interactions in rotating nuclei and to examine the isovector component of effective interactions.

2 EFFECTIVE INTERACTIONS CONSISTENT WITH SINGLE-PARTICLE POTENTIALS

Here we will briefly summarize the unified derivation of the self-consistent effective interactions [4]. Let us consider a 2^λ-pole collective oscillation mode which is characterized by the variational displacement of the nucleonic field variable as $\mathbf{r} \to \mathbf{r} + \delta\mathbf{r}$ with $\delta\mathbf{r} = -\sum_\mu \alpha^*_{\lambda\mu} \nabla Q_{\lambda\mu}$ $(\lambda \geq 1)$. The collective velocity field associated with this mode is irrotational and incompressible. We will require that nucleonic velocities entering into the velocity-dependent single-particle potential are to be measured relative to the collective velocity $\mathbf{v}(\mathbf{r})$ so that the potential is to be invariant under the local Galilean transformation $\mathbf{p} \to \mathbf{p}+\delta\mathbf{p}$ with $\delta\mathbf{p} = M\mathbf{v}(\mathbf{r})$.

Then the variation of the average one-body potential δV and that of the density distribution $\delta\rho$ produced by the oscillation are determined from the conditions $V(\mathbf{r} + \delta\mathbf{r}, \mathbf{p} + \delta\mathbf{p}) = V_0(\mathbf{r}, \mathbf{p})$ and $\rho(\mathbf{r} + \delta\mathbf{r}, \mathbf{p} + \delta\mathbf{p}) = \rho_0(\mathbf{r}, \mathbf{p})$. These conditions provide relations between δV and $\delta\rho$ through the displacement vectors $\delta\mathbf{r}$ and $\delta\mathbf{p}$ as

$$\delta V(\mathbf{r}, \mathbf{p}) = \delta V_r + \delta V_p = -\delta\mathbf{r} \cdot \nabla V_0 - \delta\mathbf{p} \cdot \nabla_p V_0,$$
$$\delta\rho(\mathbf{r}, \mathbf{p}) = \delta\rho_r + \delta\rho_p = -\delta\mathbf{r} \cdot \nabla\rho_0 - \delta\mathbf{p} \cdot \nabla_p \rho_0.$$

The field couplings δV_r and δV_p can be expressed as

$$\delta V_r = \kappa_\lambda \sum_\mu \alpha^*_{\lambda\mu} F_{\lambda\mu}, \quad F_{\lambda\mu} = (\nabla Q_{\lambda\mu} \cdot \nabla V_0)/\kappa_\lambda,$$
$$\delta V_p = \tilde{\kappa}_\lambda \sum_\mu \dot{\alpha}^*_{\lambda\mu} \tilde{F}_{\lambda\mu}, \quad \tilde{F}_{\lambda\mu} = M(\nabla Q_{\lambda\mu} \cdot \nabla_p V_0)/\tilde{\kappa}_\lambda.$$

Then the effective two-body interactions whose Hartree field coincide with these field couplings are derived as

$$H^{(\lambda)}_{int} = \frac{\kappa_\lambda}{2} \sum_\mu F^\dagger_{\lambda\mu} F_{\lambda\mu} + \frac{\tilde{\kappa}_\lambda}{2} \sum_\mu \tilde{F}^\dagger_{\lambda\mu} \tilde{F}_{\lambda\mu}.$$

The coupling constants are determined from self-consistency conditions

$$\alpha^*_{\lambda\mu} = A < F^\dagger_{\lambda\mu} >, \quad \dot{\alpha}^*_{\lambda\mu} = A < \tilde{F}^\dagger_{\lambda\mu} > .$$

Notice here that the above formulation applies not only to spherical nuclei regarding $\mathbf{r}$ as the ordinary coordinate but also to deformed nuclei regarding it as the doubly-stretched coordinate.

3 EFFECTIVE INTERACTIONS IN ROTATING NUCLEI

We start from the cranked harmonic oscillator (CHO) Hamiltonian

$$\tilde{h} = \frac{1}{2m}p^2 + \frac{m}{2}(\omega_x^2 x^2 + \omega_y^2 y^2 + \omega_z^2 z^2) - \Omega l_x' = \sum_{i=1}^{3} \hbar\omega_i \left(c_i^\dagger c_i + \tfrac{1}{2}\right),$$

where the stretched angular momentum l_x' is used instead of l_x for simplicity. The CHO Hamiltonian is diagonalized by means of the canonical transformation $c_1 = c_x$, $\begin{pmatrix} c_2 \\ c_3 \end{pmatrix} = W \begin{pmatrix} c_y \\ c_z \end{pmatrix}$, $W = \begin{bmatrix} u & -v \\ v & u \end{bmatrix}$. As a saturation condition for the rotating system of nucleons, we postulate $\omega_1\omega_2\omega_3 = \omega_0(\Omega)^3$, which guarantees constancy of the volume energy or constancy of the Fermi energy in the presence of a change in the shape of the rotating system.

We now introduce pseudo-coordinates q_i and conjugate momenta p_i as $[q_i, p_j] = i\hbar\delta_{ij}$, $q_i = \sqrt{\hbar/m\omega_i}q_i'$, $p_i = \sqrt{\hbar m\omega_i}p_i'$ with $\begin{pmatrix} q_2' \\ p_3' \end{pmatrix} = W \begin{pmatrix} y' \\ p_z' \end{pmatrix}$, $\begin{pmatrix} q_3' \\ p_2' \end{pmatrix} = W \begin{pmatrix} z' \\ p_y' \end{pmatrix}$. Then we can bring the total CHO Hamiltonian in the form

$$\tilde{H} = \sum_{n=1}^{A} \tilde{h}_n, \quad \tilde{H} = \sum_{i=1,2,3} \left(\frac{1}{2m}p_i^2 + \frac{m}{2}\omega_i^2 q_i^2\right) \equiv T + V_0,$$

where the potential V_0 is spherical in terms of the doubly-stretched pseudo-coordinates $q_i'' = (\omega_i/\omega_0)q_i$.

Now we will apply the field coupling method to this potential. As the variational displacement vector for the multipole collective oscillation modes, we assume $\mathrm{q}'' \to \mathrm{q}'' + \delta\mathrm{q}''$, $\delta\mathrm{q}'' = \sum_{\lambda\mu} \kappa_{\lambda\mu}\alpha_{\lambda\mu}''^{*}\nabla'' Q_{\lambda\mu}''$, which is a natural extension of the irrotational and incompressible model to the deformed rotating system. Then the self-consistent effective interactions for this system are derived as

$$H_{int} = -\frac{1}{2}\sum_{\lambda\mu} \chi_\lambda^{self} Q_{\lambda\mu}''^{\dagger} Q_{\lambda\mu}'', \quad \chi_\lambda^{self} = \frac{4\pi}{2\lambda+1}\frac{m\omega_0^2}{A\left\langle (q'')^{2\lambda-2}\right\rangle_0}.$$

Notice that $Q_{\lambda\mu}''$ is defined in terms of the pseudo-coordinates.

4 ISO-VECTOR COMPONENT OF EFFECTIVE INTERACTIONS

For a two-component system with $\mathring{\omega}_n \neq \mathring{\omega}_p$, the nuclear self-consistency condition is violated because of the isospin dependence of the one-body field

$$V_0 = \frac{M}{2}(\mathring{\omega}_\tau r'')^2, \quad \mathring{\omega}_\tau = \mathring{\omega}_0(1+\tau\eta) = \begin{cases} \mathring{\omega}_n \\ \mathring{\omega}_p \end{cases}, \quad \tau = \begin{cases} +1 & \text{for n} \\ -1 & \text{for p} \end{cases}.$$

Here $\eta = \theta(N-Z)/A$ and θ is a model dependent numerical factor. Then even if we consider a purely iso-scalar variational displacement

$$\delta\mathrm{r}'' = \frac{\chi_\lambda}{\lambda M\mathring{\omega}_0^2}\nabla''(\alpha_\lambda \cdot Q_\lambda'') = \frac{\chi_\lambda}{\lambda M\mathring{\omega}_\tau^2}\nabla''(\alpha_\lambda \cdot \tilde{Q}_\lambda''),$$

there appears an iso-vector component in the induced field coupling as

$$\delta V = -\chi_\lambda(\alpha_\lambda \cdot \tilde{Q}''_\lambda) = -\chi_\lambda(1+\tau\eta)^2(\alpha_\lambda \cdot Q''_\lambda).$$

As a result, the self-consistent multipole interactions in this case become

$$H_\lambda = -\frac{\chi_\lambda}{2}(\tilde{Q}''_\lambda \cdot \tilde{Q}''_\lambda), \quad \widetilde{Q_{\lambda\mu}}'' = (\frac{\mathring{\omega}_\tau}{\mathring{\omega}_0})^2 Q''_{\lambda\mu}.$$

Though there is an isospin dependent scaling factor in the operator, these interactions are essentially iso-scalar ($T=0$) character.

For the interactions of essentially iso-vector ($T=1$) character, it is useful to construct them so that the $T=0$ and $T=1$ modes are maximally decoupled with each other. For this purpose, we introduce the so-called *reduced isospin* $\tau^{(\lambda)}$ and express the $T=1$ interactions as

$$H_\lambda^{(T=1)} = \frac{\chi_\lambda^{(T=1)}}{2}(\tau^{(\lambda)}\tilde{Q}''_\lambda \cdot \tau^{(\lambda)}\tilde{Q}''_\lambda).$$

From the decoupling condition, the reduced isospin is determined as

$$\tau^{(\lambda)} = \begin{cases} +2Z < r^{2\lambda-2} >_p /A < r^{2\lambda-2} > & \text{for n} \\ -2N < r^{2\lambda-2} >_n /A < r^{2\lambda-2} > & \text{for p} \end{cases} .$$

For the dipole case, our reduced iso-vector operator guarantees the translational invariance of the $N \neq Z$ system.

Acknowledgments

The author is grateful to the late Dr. T. Kishimoto for many useful discussions at the early stage of the present work. Thanks are also due to Drs. T. Kammuri, T. Marumori and K. Matsuyanagi for their continuous encouragement. This research was partially supported by the Grant-in-Aid of the Ministry of Education, Science, Sports and Culture No. 09740195.

References

[1] T. Kishimoto, J. M. Moss, D. H. Youngblood, J. D. Bronson, C. M. Rozsa, D. R. Brown and A. D. Bacher, Phys. Rev. Lett. 35, 552 (1975).

[2] H. Sakamoto and T. Kishimoto, Nucl. Phys. A501, 205 (1989); A501, 242 (1989).

[3] H. Sakamoto and T. Kishimoto, Phys. Lett. 245B, 321 (1990).

[4] T. Kubo, H. Sakamoto, T. Kammuri and T. Kishimoto, Phys. Rev. C 54, 2331 (1996).

THE RECOIL SHADOW ANISOTROPY METHOD

E. Gueorguieva[1,2], C. Schück[1], A. Minkova[2], M. Kaci[3], Ch. Vieu[1], B. Kharraja[1,4], J.J. Correia[1] and J.S. Dionisio[1]

[1]C.S.N.S.M. CNRS-IN2P3, F-91405 Orsay, France
[2]Department of Atomic Physics, University of Sofia, Sofia, 1164, Bulgaria
[3]Institut Galilée, University Paris XIII, F-93430, Villetaneuse, France
[4]Department of Physics, University of Notre Dame, Notre Dame, Indiana 46556

gueorgui@csnsm.in2p3.fr

Abstract: The present work reports on a new method for identifying isomers in the nanosecond range and measuring their half-lives. The Recoil Shadow Anisotropy Method (RSAM) can be applied to experiments using thin targets and gamma multidetector arrays including collimated composite detectors. Lifetime measurements can thus be performed at the same time as standard spectroscopic measurements, without any additional device. This method was first developed for data obtained with Eurogam-II and a number of isomers with half-lives between 0.5 and 18 ns have been measured so far.

1 INTRODUCTION

In heavy-ion reactions using thin targets, the γ rays issued from delayed transitions are emitted in flight at some distance from the target. Thus, they are seen through different solid angles by the individual crystals of the composite Ge detectors (see figure 1). The shadow imposed by the collimators results therefore in a sizeable decrease of the delayed γ-ray intensities in some of the individual detectors (e.g. A and C with respect to B and D in figure 1). Obviously no similar effect is observed for prompt transitions, emitted while the nucleus is still at the target position. Therefore the comparison of the counting rates in the individual Ge crystals allows one to distinguish prompt from delayed transitions and thus to locate isomeric levels in a known level scheme.

Furthermore, lifetime measurements can be performed by using an appropriate calibration of this shadow effect. Since this effect depends only on the

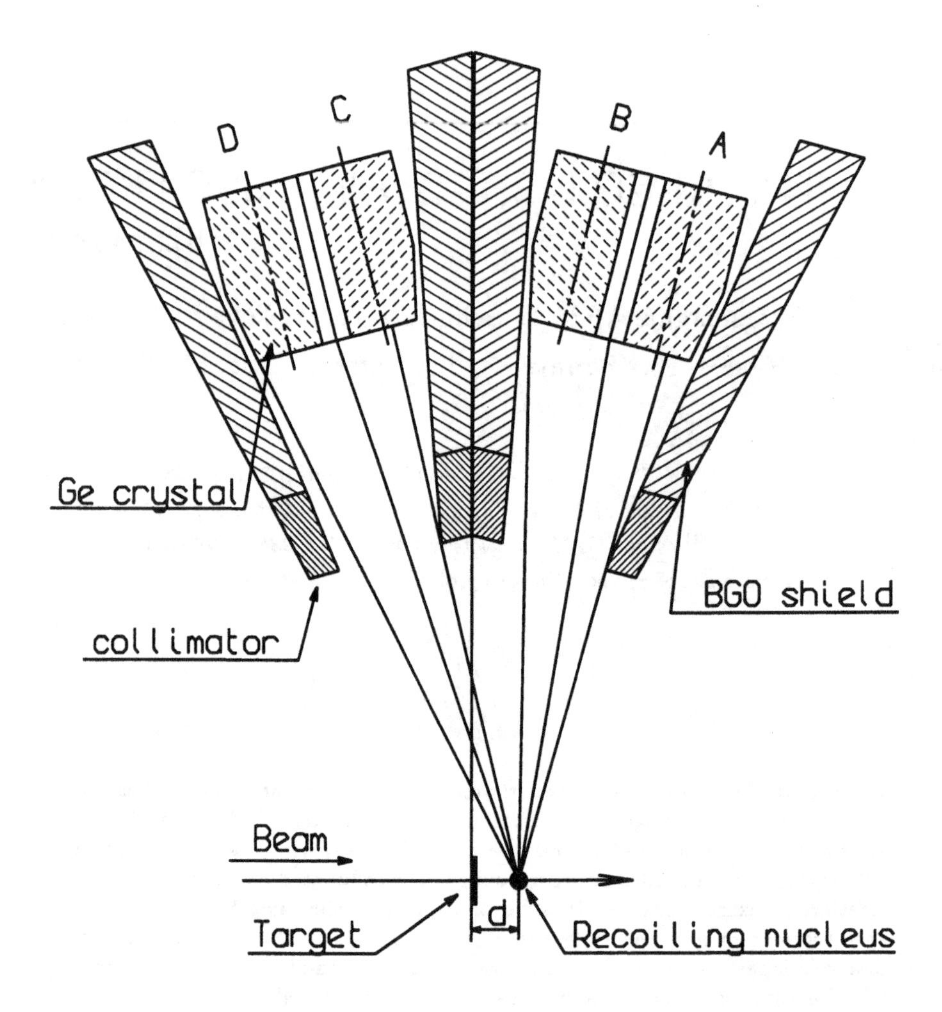

Figure 1 Sketch of two clover detectors showing the collimator shadow effect on the different clover rings of the Eurogam-II array.

geometry of the multidetector array, it is clear that it is a characteristic feature of the array and that the same calibration can be used in any other experiment performed with this array.

2 THE METHOD

This method was first applied to data obtained with the Eurogam-II array, operating at Strasbourg. It included 30 large Ge detectors (74%) arranged in four rings at 22°, 46°, 134° and 158° and 24 composite "clover" detectors located in two rings, close to 90°. Each of the clover detectors contained four individual coaxial Ge crystals (20%) [1], which resulted in four rings of individual Ge crystals at 71°, 80°, 100° and 109° with respect to the beam direction. Tungsten collimators prevented the detection of direct gamma-rays by the BGO suppressors, defining a sharp solid angle focused on the target position. It is clear that the shadow effect depends on the mean distance from the target to where the delayed γ-rays are emitted. In order to perform a quantitative analysis of this dependence, we have introduced the anisotropy parameter, A_γ, as a measure of the collimator shadow effect, defined for every γ-ray as the ratio:

$$A_\gamma = \frac{I_\gamma(80^\circ) - I_\gamma(71^\circ) + I_\gamma(109^\circ) - I_\gamma(100^\circ)}{I_\gamma(80^\circ) + I_\gamma(71^\circ) + I_\gamma(109^\circ) + I_\gamma(100^\circ)} \tag{1}$$

where $I_\gamma(\theta)$ is the gamma-ray intensity in the detectors located at angle θ ($A_\gamma = 0$ for prompt transitions). An isomeric state can be easily located by comparing the the experimental "difference" and "sum" spectra, defined as:

Difference spectrum: [Sp(80°)+Sp(109°)] - [Sp(71°)+Sp(100°)]

Sum spectrum: [Sp(80°)+Sp(109°)] + [Sp(71°)+Sp(100°)]

Indeed, only delayed transitions remain in the difference spectrum, while all transitions (prompt and delayed) are present in the sum spectrum as illustrated in figure 2, showing the difference and sum spectra corresponding to the decay path of the I^π=31/2$^-$ isomer ($t_{1/2}$=6.1ns) in ^{191}Au [2].

Moreover according to equation (1), the experimental anisotropies A_γ can be extracted for any γ ray as the ratio between the corresponding peak areas in the difference and sum spectra.

2.1 Anisotropy calibration curve

For a specific array, the anisotropy parameter can be calculated as a function of the recoil distance $v.t_{1/2}$ parameter: (v is the nuclear recoil velocity and $t_{1/2}$ the isomeric half-life)

$$A(v.t_{1/2}) = \int_0^{6.t_{1/2}} A_{inst}(x).\epsilon_{clo}(x).\lambda e^{-\lambda t} dt \tag{2}$$

x (x = v.t) being the distance of the nucleus from the target position at the time of γ-ray emission (the upper limit of the integral $6.t_{1/2}$ accounts for more than 98% of the isomeric decay). The calculated anisotropy curve $A(v.t_{1/2})$ for Eurogam-II is presented in figure 3.

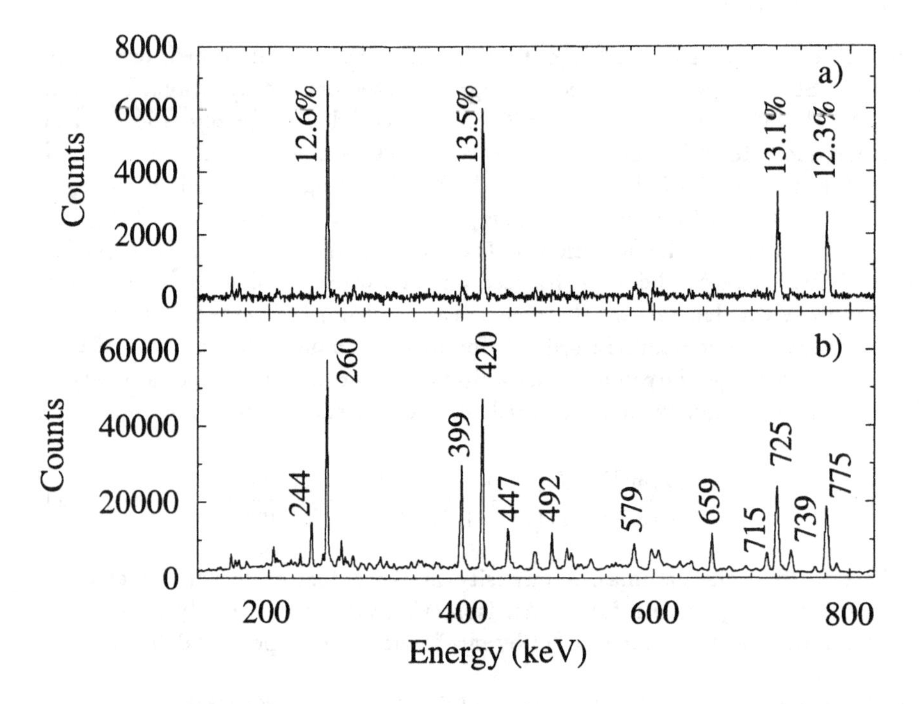

Figure 2 "Difference" (a) and "sum" (b) spectra gated on the 301.keV transition in ^{191}Au. The experimental anisotropies for the delayed transitions are included in (a).

2.2 Test of the RSAM method

A number of previously known isomeric lifetimes have been independently measured in the ^{191}Au, ^{194}Hg, ^{148,149}Gd, ^{193}Pb and ^{194}Pb nuclei, produced in five different experiments at Eurogam-II. The experimental anisotropies A_γ were measured according to formula 1 (the nucleus recoil velocity being deduced from the Doppler shifts of selected γ-lines at forward and backward angles). Excellent agreement has been obtained between the measured anisotropies and the calculated curve (see figure 3). Therefore, the $A(v.t_{1/2})$ curve can be used as a calibration curve for measuring the half-lives of new isomers.

2.3 Search for new isomers

We have applied this new technique in order to search for new isomeric states. So far, five new isomers have been identified in ^{190}Au, ^{192}Au and ^{190}Hg. Their half-lives, between 0.5 and 7 ns, have been determined [3]. Recently the 20^+ isomers in ^{190}Au and ^{192}Au nuclei have been measured in two independent experiments using the delayed coincidence method applied to I.C. measurements [3]. The half-life obtained, $t_{1/2}$=6.8±0.3 ns and $t_{1/2}$=5.4±0.3 ns, are in excellent agreement with the RSAM values.

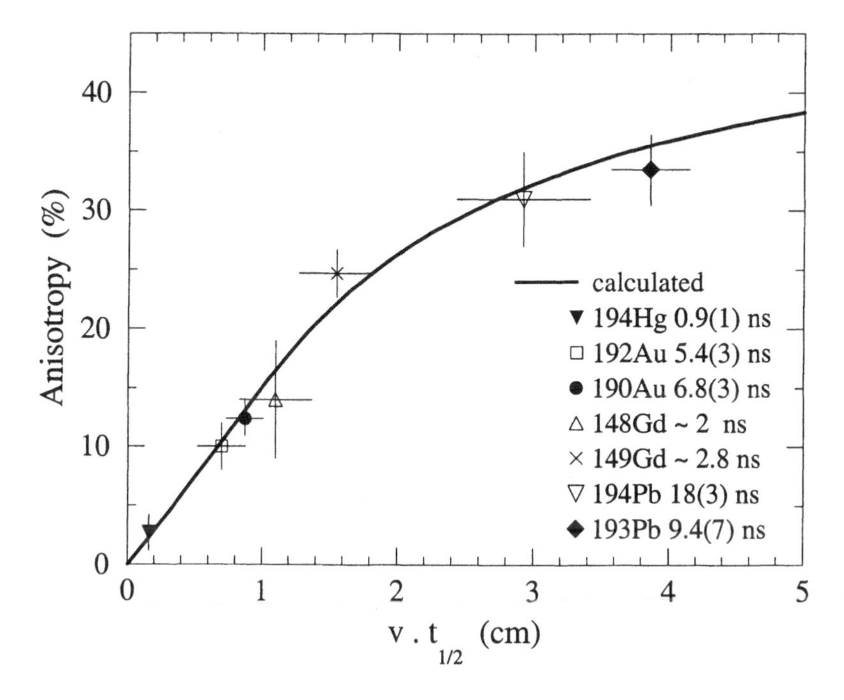

Figure 3 Calculated calibration curve $A_\gamma(v.t_{1/2})$ for the Eurogam-II array compared with the extracted anisotropies and $v.t_{1/2}$ parameters for some known and new isomers.

3 PROPERTIES OF THE RSAM

For a given isomeric state, the γ-ray anisotropies, A_γ, do not depend on the energies or multipolarities of the delayed transitions, neither on the gating transitions (so long as all the intensity passes through the isomeric level). Depending on the recoil velocities, half-lives between 0.5 ns and $\sim$ 20 ns can be measured.

4 CONCLUSION

In summary, we have developed a new method, the Recoil Shadow Anisotropy Method (RSAM) for performing nanosecond lifetime measurements. The half-lives of several previously known isomers have been re-measured and five new isomeric levels have been discovered. This technique can be applied to other multidetector arrays, such as Euroball, Gammasphere, Afrodite, Exogam, etc.

Acknowledgments

We are particularly grateful to all our colleagues of the Eurogam-II collaboration for making their data available to us.

References

[1] G. Duchêne et al. Workshop on Large γ Detector Arrays, Chalk River, AECL 10613 (1992)

[2] V. Kölschbach et al., Nucl. Phys. A439, 189 (1985).

[3] E. Gueorguieva et al., to be published.

ALPHA PARTICLE EMISSION IN THE INTERACTION OF 400 MeV ^{12}C WITH ^{59}Co AND ^{93}Nb

E. Gadioli, M. Cavinato, E. Fabrici, E. Gadioli Erba,
C. Birattari, I. Mica, S. Solia
Dipartimento di Fisica, Universitá di Milano, Istituto Nazionale di Fisica Nucleare,
Sezione di Milano, via Celoria 16, 20133 Milano, Italia
T.G. Stevens, S.H. Connell, J.P.F. Sellschop
Schonland Research Centre for Nuclear Sciences,
University of the Witwatersrand, Johannesburg, South Africa
G.F. Steyn, S.V. Förtsch, J.J. Lawrie, F.M. Nortier
National Accelerator Centre, Faure, South Africa
A.A. Cowley
Department of Physics, University of Stellenbosch, South Africa

Abstract: Inclusive double differential cross sections of α particles emitted in the interaction of 400 MeV ^{12}C ions with ^{59}Co and ^{93}Nb were measured at several emission angles between 10° and 140°. These spectra are satisfactorily reproduced in a priori calculations by considering the contributions of the *spectator* α particles produced in ^{12}C break-up, the *participant* α particles which have been re-emitted with most of their energy after incomplete fusion reactions, the pre-equilibrium α particles which are mainly produced in complete fusion reactions, and the α particles which evaporated from the equilibrated nuclei produced at the end of the fast stage of the de-excitation process.

1 INTRODUCTION

Studies of ^{12}C-induced reactions [1, 2] have shown that incomplete fusion of α particles and ^{8}Be fragments produced in the break-up of the projectile contributes significantly to the reaction cross section. These processes lead to the emission of a large number of α particles. At very forward angles the dominant contribution is ascribable to *spectator* α particles, which have a mean energy corresponding approximately to the beam velocity. At larger emission angles, a large fraction of the yield is contributed by *participant* α particles which undergo incomplete fusion (both as single units and as correlated pairs constituting ^{8}Be fragments) followed by re-emission with only a slight reduction of

The Nucleus: New Physics for the New Millennium
Edited by Smit et al., Kluwer Academic / Plenum Publishers, New York, 2000.

their initial energy [3, 4]. One also observes the emission of pre-equilibrium α particles produced by the coalescence of excited nucleons, as well as low-energy α particles which evaporated from the equilibrated nuclei produced at the end of the de-excitation process.

Thus far, few systematic measurements of the inclusive double differential cross sections of α particles emitted in ^{12}C-induced reactions were performed. The resulting lack of data, both as a function of the incident ^{12}C energy and the target nucleus mass, makes it difficult to confirm or disprove previous assumptions regarding the dominant reaction mechanisms. Exclusive measurements may in certain cases provide a more stringent test of some aspects of the theory. However, the measurement of α particles in coincidence with heavy residues detected at very forward scattering angles might not select the reaction mechanism unambiguously. At high incident energies, for example, the angular distributions of residues produced in complete and incomplete fusion processes largely overlap. In addition, it may be difficult to detect heavy residues with low energies, which will be produced abundantly [4]. More accurate results could perhaps be obtained in $\alpha-\gamma$ coincidence measurements such as those discussed in [2]. However, such measurements of all the contributions ascribable to a particular reaction mechanism might be rather complicated since many different residues can be produced, of which some may have a very low yield.

Because of these considerations, we decided to measure the inclusive cross sections of α particles produced in the interaction of ^{12}C with two nuclei having notably different mass and charge, namely ^{59}Co and ^{93}Nb. This is also in line with our previous investigations [3, 4], where our main interest was to obtain *comprehensive* information on all the contributing reaction mechanisms. Preliminary results obtained at an incident ^{12}C energy of 400 MeV are presented.

2 EXPERIMENTAL SET-UP

The ^{12}C beam of 400 MeV nominal energy and a 200 MeV α-particle beam (which was used to calibrate the detectors) were supplied by the cyclotron facility of the National Accelerator Centre, Faure, South Africa [5]. The targets and detectors were mounted inside a 1.5 m diameter scattering chamber. Two complementary detector assemblies were employed. These were conventional charged-particle telescopes, each consisting of a stack of two Si surface-barrier detectors and a NaI stopping detector. This arrangement gives two ΔE-E combinations for particle identification in each telescope, corresponding to lower and higher energy regions. In one of the telescopes the Si detectors had thicknesses of 30 μm and 500 μm respectively, while different thicknesses of 100 μm and 2000 μm were used in the second telescope. This ensured that the gaps in the measured spectra associated with dead layers in each telescope did not overlap. By combining the data of the two telescopes, complete energy spectra could be obtained for each scattering angle from a threshold of about 7 MeV up to the maximum ejectile energies. The detector assemblies were mounted on opposite sides of the beam axis in the same reaction plane, utilizing the two rotatable arms of the scattering chamber.

Data were acquired at angles ranging from 10° to 140°. The overall systematic uncertainty of the absolute cross section values is estimated to be less than 10%. A selection of the measured spectra is shown in Fig. 1

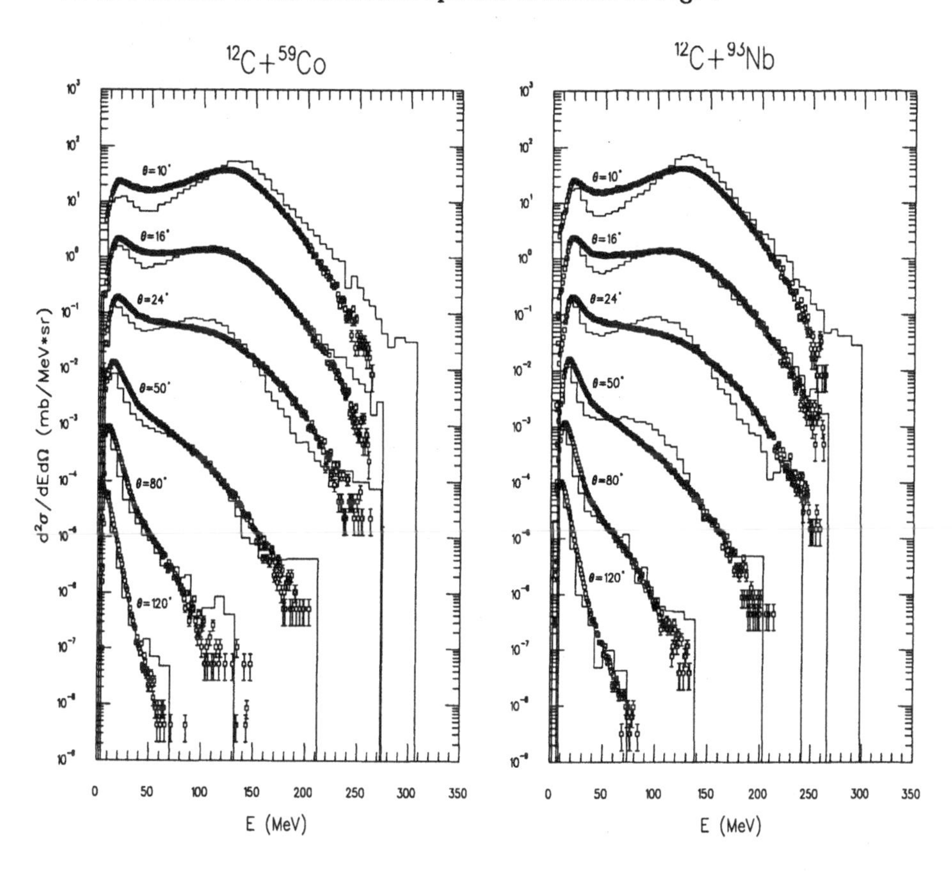

Figure 1 A selection of experimental (squares) and calculated (histograms) double differential α-particle spectra in the interaction of ^{12}C with ^{59}Co and ^{93}Nb, at an incident energy of 400 MeV. The individual spectra at the lab. emission angles as indicated have been scaled down in decades with increasing angle for purposes of display.

3 THEORY AND CALCULATIONS

Up to incident energies of about 400 MeV, only a few mechanisms need to be considered in the interaction of ^{12}C with nuclei [2, 3, 4, 6]. These are the complete fusion of ^{12}C with the target nucleus; the incomplete fusion of α particles from projectile break-up (both as single entities and as pairs when they appear as ^{8}Be); and the transfer of single nucleons from the projectile to the target nucleus. In addition to the *spectator* α particles, one should observe a significant contribution of *participant* α particles with slightly reduced energy in incomplete fusion reactions. In fact, since only high angular momentum waves contribute to incomplete fusion, these interactions occur in a low density region of the nucleus. Thus, the probability for α particle re-emission with most of their initial energy after only a few interactions with target nucleons may become large. In addition to contributions from projectile break-up, pre-equilibrium α particles may be produced by coalescence of excited nucleons during the thermalization of the composite nuclei produced in the primary projectile-target interactions. Eventually, the equilibrated nuclei produced at the end of the fast stage of the de-excitation cascade may evaporate a significant number of low-energy α particles.

In our calculations, the incomplete fusion cross sections are evaluated with the generalized critical angular momentum model [2, 4]. The resulting values are given in Table 1, together with the cross sections for complete fusion and nucleon transfer reactions.

Table 1 Cross sections (mb) of the mechanisms contributing to the total reaction cross section for the interaction of ^{12}C with ^{59}Co and ^{93}Nb (CF = complete fusion; BeICF = ^{8}Be incomplete fusion; αICF = α particle incomplete fusion; pT = proton transfer; nT = neutron transfer.)

Mechanism	^{59}Co	^{93}Nb
CF	315.3	383.2
BeICF	343.1	365.3
αICF	666.7	794.9
pT	1027.9	1060.3
nT	514.0	530.1

3.1 Break-up α particles

The double differential cross sections of the α particles and ^{8}Be fragments which are produced in ^{12}C break-up are evaluated in the Serber approximation. In this approximation it is assumed that the break-up transition matrix is given by the Fourier transform of $\psi(r)$, the wave function for the relative motion of the α particle and the ^{8}Be fragment in ^{12}C [7, 8]. The following expression is used for $\psi(r)$, which is obtained by matching a square well wave function with a repulsive core to a Yukawa wave function:

$$\psi(r) = A\,\frac{\sin K(r-b)}{r} \quad \text{for } b \leq r \leq R+b,$$
$$\psi(r) = C\frac{e^{-r/R_o}}{r} \quad \text{for } r > R+b, \tag{1}$$

where b and R are, respectively, the core and the matching radius, $K = \sqrt{2\mu(V_o - B)}/\hbar$, V_o is the square well depth and $R_o = \hbar/\sqrt{2\mu B}$. The corresponding Fourier transform is given by

$$\psi(\mathrm{p}) = \frac{4\pi A(\hbar c)^2}{(2\pi\hbar)^{3/2}} \frac{1}{pc} \{ \frac{1}{(\hbar K c)^2 - (pc)^2} [pc \ \sin KR \ \cos \frac{pc(b+R)}{\hbar c}$$
$$- \hbar K c \ \cos KR \ \sin \frac{pc(b+R)}{\hbar c} + \hbar K c \ \sin \frac{pcb}{\hbar c}]$$
$$+ \frac{\sin KR}{(\frac{\hbar c}{R_o})^2 + (pc)^2} [pc \ \cos \frac{pc(R+b)}{\hbar c} + \frac{\hbar c}{R_o} \ \sin \frac{pc(R+b)}{\hbar c}]\}, \tag{2}$$

where $\mathrm{p} = \mathrm{p}_S - (M_S/M_P)\mathrm{p}_P$, and p_S, p_P, M_S and M_P are the momenta and the masses of, respectively, the spectator fragment and the projectile. Multiplying for the three-body phase space factor [8], assuming the target mass $\mathrm{M}_T \gg \mathrm{M}_P$, one finally obtains for the double differential cross section of the spectator fragment

$$\frac{d^2\sigma}{dEd\Omega} \propto p_S \ p_{part} \ |\psi(\mathrm{p})|^2, \tag{3}$$

where p_{part} is the momentum of the participant fragment.

In evaluating the projectile momentum, one must take into account that during break-up the projectile has been slowed down by the Coulomb barrier that acts between it and the target nucleus. After the break-up, the spectator fragments are accelerated by the Coulomb repulsion of the residue.

3.2 *α particles which are re-emitted after incomplete fusion*

The evaluation of this contribution is according to a procedure used in previous studies of α-particle induced reactions [9, 10] and is still rather tentative. The results of the calculation are expected to depend quite sensitively on the geometry of the incomplete fusion process, which occurs at large impact parameters in ^{12}C-induced reactions and involves, initially, low density regions of the target nucleus. In the case of α-particle induced reactions, where the incident α particles interact with the whole nuclear volume, the time evolution of the cascade of α-nucleon interactions and the probability of α-particle re-emission during the course of this cascade may be evaluated on the basis of the statistical competition of the processes which may occur. These processes are α-nucleon collisions which lead to a more complex state (whose decay rates are obtained from the α-nucleus imaginary optical potential) and the emission of α particles into the continuum (whose decay rates are evaluated from the detailed balance principle). However, in ^{12}C-induced reactions the α particle may be re-emitted after a smaller number of collisions with target nucleons and with a larger probability simply because it may scatter toward even more peripheral nuclear regions. Since the nucleon density in the peripheral region is small, further α-nucleon interactions become progressively more unlikely and α-particle emission therefore more likely. Our present aim is to generalize our previous model to describe this process. However, for the present study we have not attempted to evaluate *a priori* both the average number of collisions and the emission probability. We have deduced the most suitable values for these quantities from a fit to the experimental cross sections, by considering both the previous activation data [3, 4] and the double differential cross sections of the

α particles emitted at angles between 20° and 80°, where this process seems to be dominant.

To simulate the α-nucleon interaction in the peripheral nuclear region, we evaluated the double differential cross sections of the re-emitted α particles with the procedure discussed in [9, 10] by using, in a local density approximation, reduced values of the nucleon and the α particle Fermi energies. The predicted angular distribution of the α particles after a single nucleon collision can be expressed by

$$\frac{d^2\sigma}{dEd\Omega} \propto e^{-(\theta/\Delta\theta)^2}, \tag{4}$$

where $\Delta\Theta = 2\pi/k\Delta R$, k is the α particle wave number and $\Delta R = c\ A^{1/3}$ is the thickness of the nuclear surface region where the α-nucleon collisions are expected to occur. The values of c, which depend on the α particle energies before and after the interaction as well as the nucleon and α particle Fermi energies, vary from 0.75 to 1.1 fm. The angular distribution of the α particles after a given number of collisions is evaluated by means of a Monte Carlo procedure by assuming that, after each collision, equation (4) yields the α particle angular distribution with respect to its direction before the collision.

3.3 Pre-equilibrium α particles

The double differential cross sections of the pre-equilibrium α particles are calculated with the Boltzmann master equation theory [11, 12, 13, 14, 15]. However, the theory discussed in these references deals with pre-equilibrium emission in complete fusion reactions, while in the present case we also have to expect a significant contribution from the incomplete fusion processes in which the fusing fragment is not re-emitted. The evaluation of this contribution is also rather tentative and is made by assuming that the angular and energy dependences of the pre-equilibrium α particles emitted in incomplete fusion reactions are the same as in complete fusion reactions. Their yield is scaled by the ratio of the participant to the projectile mass. This procedure seems to be appropriate for pre-equilibrium nucleons emitted in complete and incomplete fusion of light heavy ions [16]. This approximation predicts that the major contribution to the pre-equilibrium α particle spectra is expected to come from complete fusion reactions ($\sim 2/3$ of the total). The comparison of spectra of pre-equilibrium α particles emitted in complete fusion reactions induced by projectiles of significantly different mass, namely ^{16}O and ^{32}S, at comparable incident energies per nucleon [17], suggests that the previous assumption might lead to an overestimation of the pre-equilibrium yield by a factor of $\sim$1.5. This should lead to an overestimation of at most 12 % for the total pre-equilibrium α particle yield, which is considered to be acceptable. It must also be stressed that we have neglected (as is usually done) the possibility that before emission the α particles may interact with the nucleons of the composite nucleus, thereby losing part of its energy.

3.4 *Evaporated α particles*

At all the emission angles one observes a relatively large yield of low energy α particles which have evaporated from the intermediate equilibrated nuclei. These nuclei are produced at the end of the fast thermalization phase, after the emission of the break-up fragments and the pre-equilibrium particles. The evaluation of this contribution is quite standard and has been done using conventional parameters [18].

All the above contributions to the double differential spectra have been calculated with a Monte Carlo procedure. It is assumed that these contributing events can be described as a joint probability sequence of independent events, which are evaluated once the energy and the angular dependence of the emitted α particles are known. The energy and the emission angle of each emitted particle are evaluated in the centre of mass system of the emitting nucleus and converted to the laboratory system.

4 RESULTS AND DISCUSSION

The comparison of the predicted and experimental α particle spectra at the most forward angles, where the contribution of the spectator α particles (in the ^{8}Be incomplete fusion) and of the α particles produced in the dissociation of the spectator ^{8}Be nuclei (in the α particle incomplete fusion) dominates, suggests a very small value for the core radius b. In the calculations shown in Fig. 1, a value of b=0 and a well depth V_o=21 MeV were adopted, with a corresponding square well radius of 1.67 fm. The corresponding root mean separation of the α particle and the ^{8}Be is approximately $< r^2 >^{1/2}$= 0.74 fm. This roughly corresponds to a radius of about 2.17 fm if ^{12}C is considered to be composed of three tightly packed α particles. The break-up contribution also depends on the state in which ^{8}Be is produced when ^{12}C breaks up. The best agreement with the data is obtained by assuming that ^{8}Be is mainly produced in the ground state.

At angles between about 20° and 80° and emission energies exceeding about 50 MeV, the major contribution to the measured spectra is that of the α particles which are re-emitted after an incomplete fusion. In the case of the reaction ^{12}C + ^{59}Co, this contribution has been calculated using values of 5 and 20 MeV, respectively, for the nucleon and α particle Fermi energies in the composite nucleus. For the average number of α-nucleon collisions before re-emission, we have used values of 3 and 2, respectively, in the cases of ^{8}Be and α particle incomplete fusion. For the corresponding α particle re-emission probabilities we used values of 0.5 and 0.8, respectively [4]. The same parameters were used in the case of ^{12}C + ^{93}Nb, except for the nucleon and α particle Fermi energies in the composite nucleus, for which we adopted values of 10 and 40 MeV, respectively. This is not unreasonable since incomplete fusions may occur in regions of lower nuclear density in the case of ^{59}Co (as compared to ^{93}Nb) due to its smaller size.

At still larger angles the contributions of pre-equilibrium and evaporated α particles dominate, and these spectra are reasonably well reproduced both in shape and absolute value. The pre-equilibrium contribution to the double differential cross sections have been evaluated as discussed in [11, 12] using the *high energy approximation* discussed in [13].

The comparison of the theoretical predictions with the measured data (see Fig. 1) shows that the main features of the experimental spectra are quite satisfactorily reproduced, even though the measured spectra are less pronounced in the region of the break-up peak. This may be a consequence of the approximations made in order to simplify the calculations. Among these are the neglect of final state interactions of the spectator α particles and the semiclassical approximations adopted to model the energy and angular distributions of the participant α particles after each α-nucleon interaction. But the overall agreement is such that we may confidently conclude that the dominant reaction mechanisms have been identified correctly. The success of the present analysis nevertheless points to further refinements that may be introduced in the theoretical model.

References

[1] H. C. Britt and A. R. Quinton, Phys. Rev. **124** (1961) 877.
[2] K. Siwek-Wilczynska et al., Nucl. Phys. **A330** (1979) 150.
[3] E. Gadioli et al., Phys. Lett. **B394** (1997) 29.
[4] E. Gadioli et al., Nucl. Phys. **A641** (1998) 271.
[5] J. V. Pilcher et al., Phys. Rev. C **40** (1989) 1937.
[6] C. Birattari et al., Phys. Rev. **C54** (1996) 3051.
[7] R. Serber, Phys. Rev. **72** (1947) 1008.
[8] N. Matsuoka et al., Nucl. Phys. **A311** (1978) 173.
[9] E. Gadioli and E. Gadioli Erba, Z. Phys. **A299** (1981) 1.
[10] E. Gadioli, E. Gadioli Erba and M. Luinetti, Z. Phys. **A321** (1985) 107.
[11] I. Cervesato et al., Phys. Rev. **C45** (1992) 2369.
[12] M. Cavinato et al., Z. Phys. **A347** (1994) 237.
[13] C. Brusati et al., Z. Phys. **A353** (1995) 57.
[14] M. Cavinato et al., Phys. Lett. **B405** (1997) 219.
[15] M. Cavinato et al., Nucl. Phys. **A643** (1998) 15.
[16] E. Fabrici et al., Z. Phys. **A338** (1991) 17.
[17] R. Wada et al., Phys. Rev. **C39** (1989) 497.
[18] P. Vergani et al., Phys. Rev **C48** (1993) 1815.

DECAY OUT OF THE YRAST SUPERDEFORMED BAND IN ^{191}Hg

S.Siem,[1,2] P.Reiter,[1] T.L.Khoo,[1] T.Lauritsen,[1] M.P.Carpenter,[1] I.Ahmad,[1] H.Amro,[1,4] I.Calderin,[1] T.Døssing,[3] S.M.Fischer,[1] U.Garg,[5] D.Gassmann,[1] G.Hackman,[1] F.Hannachi,[6] R.V.F.Janssens,[1] B.Kharraja,[5] A.Korichi,[6] A.Lopez-Martens,[6] E.F.Moore,[4] D.Nisius[1] and C.Schuck[6]

1 Argonne National Laboratory 2 Univ. of Oslo, 3 The Niels Bohr Institute, 4 North Carolina State Univ., 5 Notre Dame Univ., 6 C.S.N.S.M, Orsay *

siem@sun0.phy.anl.gov

Abstract: The excitation energies and spins of the yrast superdeformed band in ^{191}Hg have been determined by analyzing the quasicontinuum spectrum connecting the superdeformed and normal-deformed states. The results from this analysis, combined with that given by one-step decay lines, give confident assignments of the spins and energies of the yrast superdeformed band in ^{191}Hg.

1 INTRODUCTION

The yrast superdeformed (SD) band in ^{191}Hg was the first SD band to be discovered in the A 190 region [1]. There are now more than 175 known SD bands in the A=150 and 190 regions, but only 3 bands in ^{194}Hg[2] and ^{194}Pb[3] have known spins and excitation energies through one-step linking transitions. It is important to obtain these quantities for an odd-A SD band since that gives information on the relative pair correlation energies in normal-deformed (ND) and superdeformed (SD) states, thereby providing a stringent test for theory. So far, the main information has come from J^2 moments of inertia of the SD bands.

*This research is supported in part by the U.S. Dept. of Energy under contract No. W-31-109-ENG-38. S. Siem acknowledges a NATO grant through the Research Council of Norway.

The Nucleus: New Physics for the New Millennium
Edited by Smit et al., Kluwer Academic / Plenum Publishers, New York, 2000.

An alternative way of setting limits on the spins and excitation energies of SD bands is to analyze the quasicontinuum decay spectra. It can complement results in those cases where one or two decay pathways are known. In most cases, where one-step transitions are not observed, it is the only alternative at present. The method of analyzing the quasicontiniuum spectrum to extract the spin and excitation energy of a SD band has been successfully tested[4] in the case of SD band 1 in ^{194}Hg, where the spins and excitation energies are known from several one-step γ ray transitions connecting the SD band with known ND states.

The Quasi-Continuum Method

Superdeformed states in ^{191}Hg were populated using the ^{174}Yb(^{22}Ne,5n)^{191}Hg-reaction. The experiment was performed using the GAMMASPHERE Ge-detector array, which had 96 detectors at the time of the experiment. The 120 MeV beam was provided by the 88" Cyclotron at Lawrence Berkeley National Laboratory. The 3.1mg/cm^2 ^{174}Yb target had a 6.8mg/cm^2 ^{197}Au backing to stop the recoil nuclei before the decay-out γ-rays are emitted. A total of 2x10^9 triple γ-coincidence events were collected. A method has been developed at Argonne[5, 4] to isolate the quasicontinuum γ-spectrum connecting the SD and ND states. First, the data are sorted with double gates on SD transitions to ensure clean spectra. The background subtraction is done using the FUL method[6]. Corrections are done for summing effects and for neutron interactions in the detectors, which is mostly in the forward detectors. The spectra are then unfolded to eliminate contributions from Compton scattered γ rays and corrected for the detector full-energy efficiency. The area of the spectrum is now normalized to multiplicity by requiring the peak corresponding to the most intense transition in the SD band to have unit area. The discrete peaks below 1 MeV are removed so that the remaining spectrum contains the sought-after decay-out component, plus feeding components of statistical, quadrapole and M1/E2 nature. To extract the decay-out spectrum we subtract away the feeding components. The feeding statistical spectrum is obtained from a Monte Carlo simulation for ^{192}Hg and renormalized to the high-energy portion of the ^{191}Hg spectrum to account for the different entry conditions in 4 and 5-n channels. Then the quadrapole and dipole feeding components can be extracted by angular distribution analysis in the center-of-mass system. The low-energy region of the decay-out spectrum cannot be extracted due to the presence of the larger E2 and M1/E2 feeding components. The decay-out spectrum below 800keV in Fig. 1 represents a guess.

Monte Carlo simulations suggest that each decay-out γ ray removes $0.5(2)\hbar$ of spin, on the average. By multiplying the average energy and spin removed by each γ ray with the multiplicity, the energy and spin removed by the decay-out component is found to be: ΔE = 3.4MeV and ΔI = $2.0\hbar$. From the peak areas of ND transitions, the mean spin and energy at which the cascade enters the known ND level sheme is found. The energy and spin of the level fed by the 351keV SD transition is determined to be E_{exit}= 5.75 $\pm$ 1MeV and

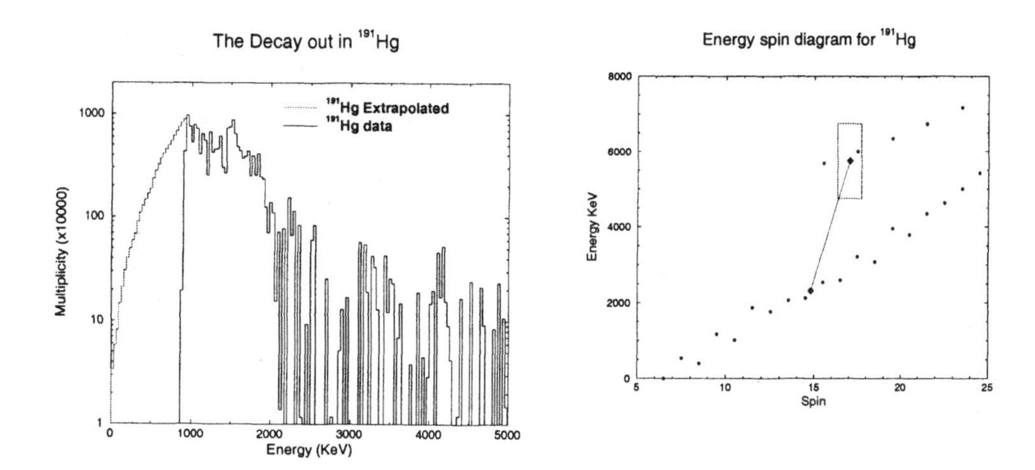

Figure 1 a)The continuum spectrum connecting the superdeformed and normal-deformed states. b)The spin and excitation energy of the level feed by the 351keV SD transition from the quasi continuum analysis (diamond); the box represents the uncertainty. The circles represent results from the one-step lines.

I_{exit}= 17.0 ± 0.7 $\hbar$ see Fig. 1. Contributions to the uncertainty come from the calculated feeding statistical spectrum, the normalization to multiplicity and the extrapolation of the decay-out spectrum to lower γ-energies.

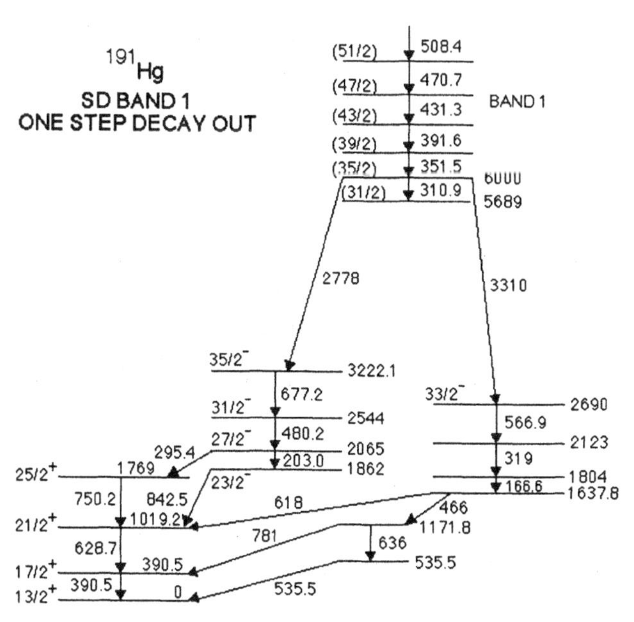

Figure 2. Partial level sheme, showing the one-step decay pathways connecting SD and ND levels.

One-step transitions

Two transitions, at 2778keV and 3310keV, that connect SD and ND states in one step have been assigned[7] on the basis of coincidence data; both place the decay-out level at 6000 keV (see Fig. 2). The angular distribution coefficient of the stronger one-step line (2778 keV), A2= 0.57± 0.48, is consistent with a $\Delta I = 0$ assignment, suggesting a 35/2 $\hbar$ spin assignment for the decay-out level. We rule out the possibility of it being an E2 transition, because that would require M3 multipolarity for the other one-step transition. The spin is consistent with an assignment[8] of a $j_{15/2}$ particle configuration. However, the experimental data do not allow for a parity assignment. A $j_{15/2}$ particle configuration assignment, would require the one-step lines to be M1 transitions. From neutron capture experiments it is known that 8 MeV E1 transitions are about 5 times more likely than M1 transitions. However, in ^{191}Hg the one-step lines have significantly lower energy ≈3 MeV. In fact, M1 transitions around this energy are observed both in neutron capture and also from the decay of the SD band in ^{194}Pb.

Conclusion

The spins and excitation energies of the yrast SD band in ^{191}Hg have been determined by analyzing the quasicontiniuum spectrum connecting the SD and ND states. The results are in good agreement, within the error bars, with that given by the one-step linking transitions. The level fed by the 351 KeV SD transition has $E_x = 6.000$ MeV and $I = 35/2\ \hbar$, i.e. at the main point of decay, the SD state is 2.778 MeV above the yrast line. Extrapolated to 13/2, the spin of the ground state, the excitation energy of the SD band is 4.74 MeV. It is important to obtain these quantities in an odd-even nucleus since that gives information on the relative pair correlation energies in ND and SD states, thereby providing a stringent test for theory. Information on pair quenching in excited states will be obtained by comparing the experimental and theoretical [9] decay-out spectra from SD bands in even-even and odd-even Hg nuclei.

References

[1] F. Moore *et al.* Phys. Rev. Lett (1990)

[2] T. L. Khoo *et al.* Phys. Rev. Lett 76 (1996)1583

[3] A. Lopez-Martens *et al.* Phys. Lett. 380B (1996)18

[4] T. Lauritsen *et al.* Heavy ion Physics, 6(1997)229, Proc. Symp. on Nucl. Structure at the Limits, ANL(1996).

[5] R. G. Henry *et al.* Phys. Rev. Lett 73 (1994)777

[6] B. Crowell *et al.* Nucl. Instr. & Methods A235 (1995)p.575

[7] P. Reiter *et al.* Phy. Div. Ann. Report for 1997, ANL-98/24 page 38

[8] M. P. Carpenter *et al.* Phys. Rev. C 51(1995)

[9] T. Døssing *et al.* Phys. Rev. Lett. 75 (1995) 1276.

SPECTROSCOPY AT THE HIGHEST SPINS IN NORMAL DEFORMED NUCLEI

J. Simpson

CLRC, Daresbury Laboratory, Daresbury, Warrington, WA4 4AD, UK

Abstract: Very high spin states (50-60$\hbar$) have been observed in the transitional nuclei ^{156}Dy and $^{158-162}$Er using the γ-ray spectrometers Eurogam, Euroball and Gammasphere. The lowest energy positive parity bands in ^{156}Dy are found to evolve smoothly towards oblate shape in contrast to the sudden change observed in the neighbouring nucleus ^{158}Er. In ^{159}Er, the decay path of the positive parity yrast bands fragments above $I^{\pi}=\frac{85}{2}^{+}$ and favoured single particle states are identified at $\frac{89}{2}^{+}$ and $\frac{101}{2}^{+}$. In ^{160}Er the yrast structure is found to remain collective up to and beyond 50$\hbar$ with no sign of band termination. In ^{161}Er three bands are observed well above spin 50$\hbar$. In the yrast even parity, positive signature $(+,+\frac{1}{2})$ band a crossing is observed near spin 50$\hbar$ which is interpreted an an unpaired band crossing involving the rearrangement of the occupation of single neutron orbitals. In the negative parity bands there is evidence for a crossing involving single proton orbitals and possible band termination. These data on band termination, unpaired band crossings and the competition between these phenomena at these very high spins will be discussed.

1 INTRODUCTION

Gamma-ray spectroscopy has provided an enormous quantity of information on the behaviour and structure of atomic nuclei. This information is most intriguing and is challenging our understanding of nuclear phenomena when the nucleus is stressed to its limits, for example, by spinning the nucleus to very

The Nucleus: New Physics for the New Millennium
Edited by Smit et al., Kluwer Academic / Plenum Publishers, New York, 2000.

large values of angular momentum. In order to probe this limit a very efficient and extremely sensitive instrument is required. This instrument is the large multi-detector gamma-ray spectrometer. Presently nuclear physicists are at a most interesting and exciting period with the operation and exploitation of the latest generation of very efficient γ-ray spectrometers, such as Euroball [1] and Gammasphere [2]. These spectrometers, with total photopeak efficiencies up to 10% at 1.3 MeV allow an unprecedented study of the properties of the atomic nucleus with sensitivities up to or better than 10^{-5} of the production cross-section. The improvement in sensitivity with time is schematically illustrated in figure 1 where the fraction of the reaction channel that can be studied in discrete line spectroscopy is plotted as a function of nuclear spin. The performance and description of these arrays and the techniques involved are well documented in some recent review articles [3, 4].

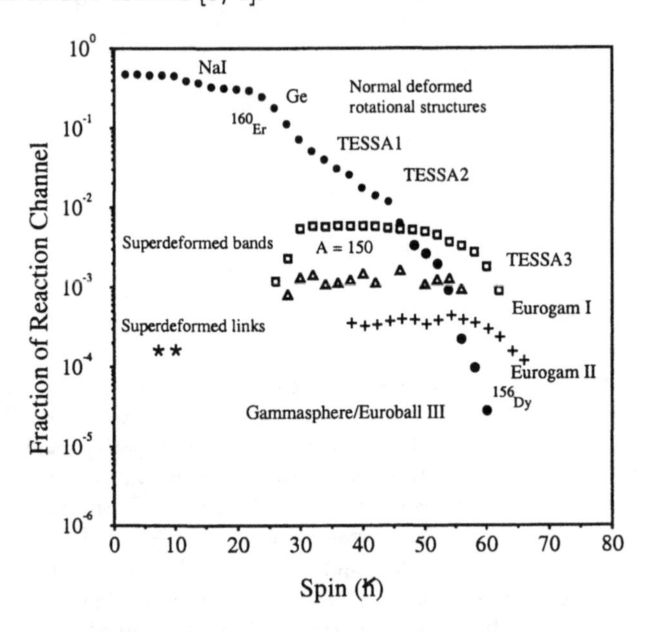

Figure 1 Plot of the observed fraction of reaction channel as a function of spin. Various arrays are indicated at their approximate level of sensitivity. The intensity distribution for selected high spin phenomena are plotted. Normal deformed states to spin ≈ 50$\hbar$ in ^{160}Er [5] and spin 60$\hbar$ in ^{156}Dy [6] (•), the yrast superdeformed band in ^{152}Dy (□) [7] and excited superdeformed bands in ^{151}Tb (△) [8] and ^{152}Dy (+) [9]. The links between the superdeformed bands and the normal deformed structures in the mass = 190 region are also shown (∗)

This very high sensitivity is obtained by collecting, in high-fold coincidence, as much as possible of the radiation emitted from a nucleus as it decays in a large array of escape-suppressed Ge spectrometers (ESS). Each ESS comprises a high resolution Ge detector surrounded by an escape suppression shield, usually of BGO. Composite Ge detectors, that have more than one Ge crystal in the same cyrostat, and segmented Ge detectors, that have their outer contact electrically divided, are now commonplace in these arrays.

In this contribution I will discuss some of the recent results from the spectroscopy of normal deformed (not superdeformed) nuclei to the very highest spins.

2 SPECTROSCOPY AT THE HIGHEST SPINS IN ^{156}Dy AND ^{161,162}Er.

The region of transitional nuclei around N=90 have been particularly rich in nuclear structure phenomena at very high spin. Most notable are the observation of band terminations and the observation of unpaired band crossings at very high spin indicating the decline of static nuclear pair correlations. The light A$\sim$160 Dy and Er nuclei form isotopic chains in which the highest spin states in normal deformed nuclei have been observed. Two recent experiments have been performed on ^{156}Dy and ^{161,162}Er that have revealed new information on the properties of nuclei at the highest spins.

The nucleus ^{156}Dy was populated at very high spin and excitation energy in the reaction ^{124}Sn(^{36}S,4n) at a beam energy of 165 MeV. The Gammasphere array [2], which consisted of 93 escape suppressed spectrometers, was used to detect the de-exciting γ rays. A total of 1.3 x 10^9 events were collected, which were unfolded off line into $\approx$ 2.5 x 10^{10} γ^3 coincidences.

The nuclei ^{161,162}Er were populated at very high spin using the reaction ^{130}Te(^{36}S,xn) at a beam energy of 170 MeV. The target consisted of two stacked foils of ^{130}Te, each of thickness 0.5 mg cm^{-2}. The Euroball array [1] was used to detect the de-exciting γ rays. For this experiment the array consisted of 14, 7 element Cluster detectors [10], 26, 4-element Clover detectors [11] and 30 single crystal Ge detectors [12]. A total of 2.0 x 10^9 events were collected, which were unfolded into $\approx$ 2 x 10^{10} γ^3 coincidence events. In both experiments the γ^3 events were analysed by David Radford's software analysis package levit8r and the γ^4 events using the Radford hypercube [13].

The analysis of the ^{156}Dy data has enabled the most comprehensive level scheme of any nucleus above spin 40 to be established [6]. Thirteen sequences are observed above spin 45 with several approaching and one possibly going beyond spin 60! These are the highest spin discrete states observed to date in a normal deformed nucleus.

The excitation energy of the (parity,signature) (+,0) bands in ^{156}Dy observed to high spin are plotted with respect to a rigid rotor reference energy in figure 2. They are compared with the cranked Nilsson-Strutinski calculations, similar to those described in [14]. The behaviour of the highest spin states in the $(+,0)_1$ and $(+,0)_2$ is particularly interesting. The highest spin states gain significant energy compared with the rigid rotor. This is characteristic of the phenomena of smooth or unfavoured band termination, typical in the mass $\approx$ 110 region [15]. In soft band termination a particular configuration slowly evolves from a prolate ($\gamma = 0°$) shape to a fully aligned oblate ($\gamma = 60°$) shape. In general the final terminating state is unfavoured with respect to a rigid rotor, see figure 2(a). The comparison between the calculations and experiment for ^{156}Dy is

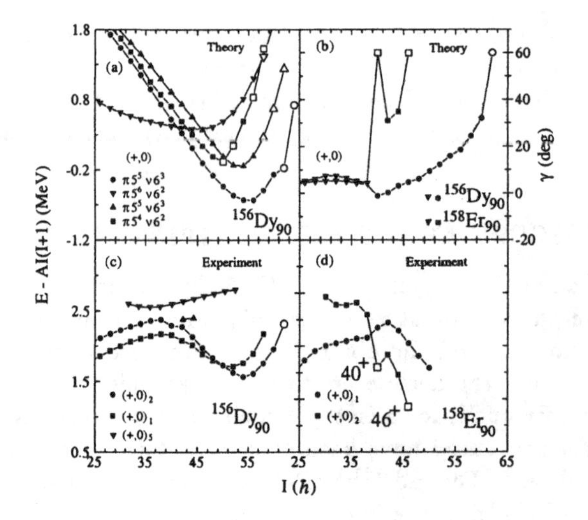

Figure 2 (a) Calculated excitation energy minus a rigid rotor reference as a function of spin for the (+,0) bands in ^{156}Dy [6]. (b) Calculated evolution with spin of the γ deformation for ^{156}Dy and ^{158}Er. (c) and (d) Experimental excitation energy of the (+,0) bands in ^{156}Dy and ^{158}Er, [16].

very impressive. For example the calculations predict that the $(+,0)_1$ band is the $\pi 5^5\ \nu 6^3$ configuration. This configuration evolves smoothly with spin from $\epsilon_2 \approx 0.19$, $\gamma \approx 0^0$ at 40^+ (prolate), to $\epsilon_2 \approx 0.16$, $\gamma \approx 16^0$ at 54^+ (triaxial) and eventually to $\epsilon_2 \approx 0.10$, $\gamma = 60^0$ at 62^+ (oblate) as shown in figure 2(b). The final state at 62^+ is a fully aligned band termination state. In addition, band termination states are also established in the negative parity bands at 52^- and 53^- [6].

The observation of an unfavoured band termination in ^{156}Dy is in stark contrast to the classic example of favoured band termination in the neighbouring nucleus ^{158}Er [16], see figure 2(b) and (d) and figure 3. In this case the oblate terminating states are very favoured and a rapid transition from prolate to 'oblate' shape is observed along the yrast line. These particularly favoured states have been observed in a range of Er nuclei from 156 to 159 [18]. One of the motivations for the Euroball experiment to observe very high spin states in ^{161,162}Er, was to investigate if band termination could be identified in the heavier Er nuclei. In ^{158}Er the fully aligned band terminating state at 46^+ has the configuration $[\pi(h_{\frac{11}{2}})^4]_{16^+} \otimes\ [\nu(i_{\frac{13}{2}})^2(h_{\frac{9}{2}})^3(f_{\frac{7}{2}})^3]_{30^+}$ [19, 20, 16]. If three neutrons are added to this configuration into the most energetically favourable neutron orbitals then fully aligned, terminating states can be predicted in ^{161}Er. In the negative parity bands for example such states are predicted at $\frac{109}{2}^-$ and $\frac{111}{2}^-$ with the configurations $[\pi(h_{\frac{11}{2}})^4]_{16^+} \otimes\ [\nu(i_{\frac{13}{2}})^4(h_{\frac{9}{2}})^3(f_{\frac{7}{2}})^4]_{\frac{77}{2}^-}$ and $[\pi(h_{\frac{11}{2}})^4]_{16^+} \otimes\ [\nu(i_{\frac{13}{2}})^4(h_{\frac{9}{2}})^4(f_{\frac{7}{2}})^3]_{\frac{79}{2}^-}$, respectively. However, recent calculations [21] predict that such states in ^{161}Er with three more valence neutrons

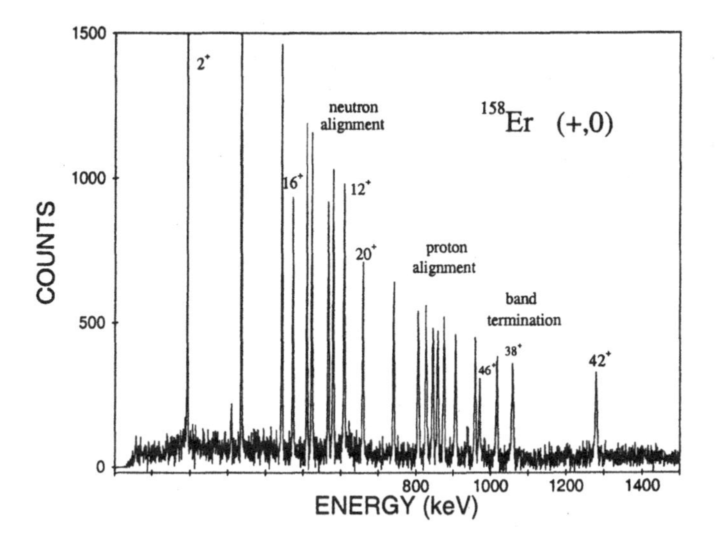

Figure 3 The classic band terminating spectrum of the (+,0) band in ^{158}Er obtained with the Eurogam spectrometer [17]. The spectrum is in coincidence with the $44^+ \rightarrow 42^+$ transition.

are at least 0.5 MeV above the yrast states up to spin 60 and instead the bands observed move into the unpaired regime.

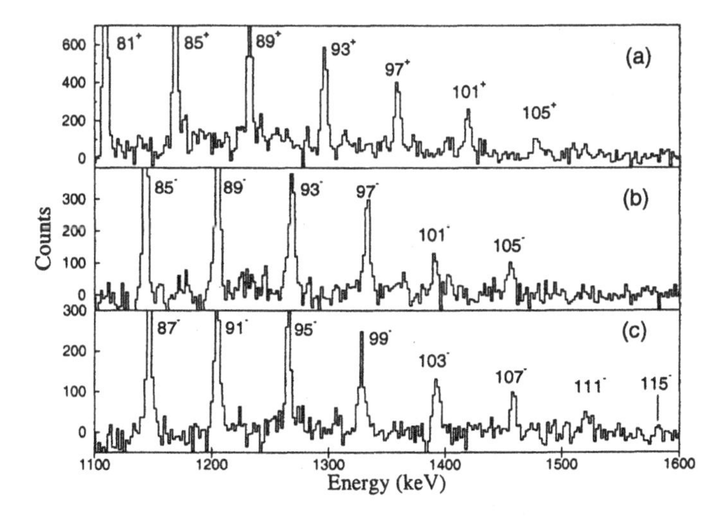

Figure 4 Four-fold coincidence spectra of the (a) $(+,+\frac{1}{2})$, (b) $(-,+\frac{1}{2})$ and (c) $(-,-\frac{1}{2})$ bands in ^{161}Er. These spectra were obtained by summing many spectra each with the requirement that there were three other coincident transitions in the band. The spectra are labelled by 2I, where I is the spin of the decaying state.

There has been a persistent search for definitive evidence of the transition from the paired to unpaired phase in nuclear matter in high spin nuclear studies. One of the most convincing pieces of experimental evidence is the observation of unpaired band crossings. Such crossings are between bands based on different single particle configurations. The rotational frequencies at which such crossings occur are specific and highly dependent on the single particle spectrum of states and are not usually expected to occur in several rotational bands at the same frequency as is the case in the paired regime. Prior to this recent experiment on Euroball only one unpaired band crossing has been observed in the $(-,1/2)$ sequence in ^{159}Er at $\hbar\omega = 0.56$ MeV [22].

Preliminary analysis of the Euroball data on ^{161}Er has established three bands to $\frac{105}{2}^-$, $\frac{109}{2}^+$ and $\frac{123}{2}^-$ and the $(0,+)$ band in ^{162}Er to 56^+, see figure 4 and see figure 7. The data on ^{161}Er are particularly exciting with evidence for

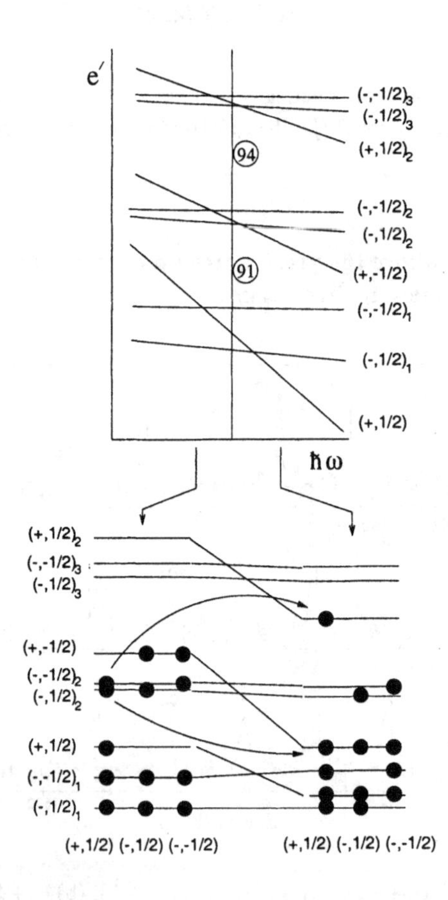

Figure 5 A schematic illustration of the occupation of single-neutron orbitals for the lowest energy configurations in ^{161}Er at low and high rotational frequency. The exchange of particles in the $(+,+\frac{1}{2})$ configuration is indicated.

the highest spins and unpaired crossings involving bands with different single neutron and single proton orbitals.

The top of the $(+,+\frac{1}{2})$ band shows evidence for a band crossing at $\approx \frac{101}{2}^{+}$. This crossing can be explained by extending the schematic model that was used to explain the unpaired band crossing in ^{159}Er [22] as a crossing between bands based of different single-neutron configurations. Figure 5 shows the neutron orbitals expected to be occupied at both low and high rotational frequency for lowest energy configurations in ^{161}Er. Using this model the first unpaired band crossing is predicted to occur in the $(+,+\frac{1}{2})$ band when it is energetically favourable to occupy the $(+,-\frac{1}{2})$ [651] and $(+,\frac{1}{2})_2$ [642] orbitals rather than two odd parity N = 5 orbitals $(-,-\frac{1}{2})$ and $(-,+\frac{1}{2})_2$. This particular crossing is blocked in both the negative parity bands since the odd parity orbitals are already occupied. Recent cranking calculations [21] for the $(+,+\frac{1}{2})$ states, however, predict a crossing at about this spin that involves the exchange of configuration involving both neutron and protons, see figure 6.

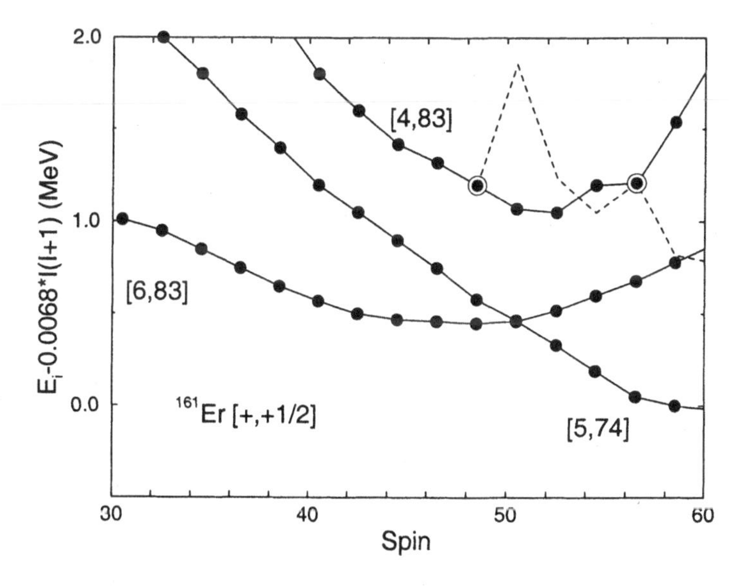

Figure 6 Cranked Nilsson-Strutinski calculations [21] of the energy of the lowest $(+,+\frac{1}{2})$ states in ^{161}Er as a function of spin. The bands are labelled by particles outside the ^{146}Gd core as [A,CD]. A is the number of $h_{\frac{11}{2}}$ protons, respectively and C and D are the number of $(h_{\frac{9}{2}}/f_{\frac{7}{2}})$ and $i_{\frac{13}{2}}$ neutrons, respectively.

In the $(-,-\frac{1}{2})$ band there is tentative evidence for a crossing at $\frac{119}{2}^{-}$. The calculations predict an unpaired band crossing at this spin involving a change in just the proton configuration. If this crossing is confirmed then this data provides direct evidence for the quenching of static pairing (neutron and proton) correlations at high spin.

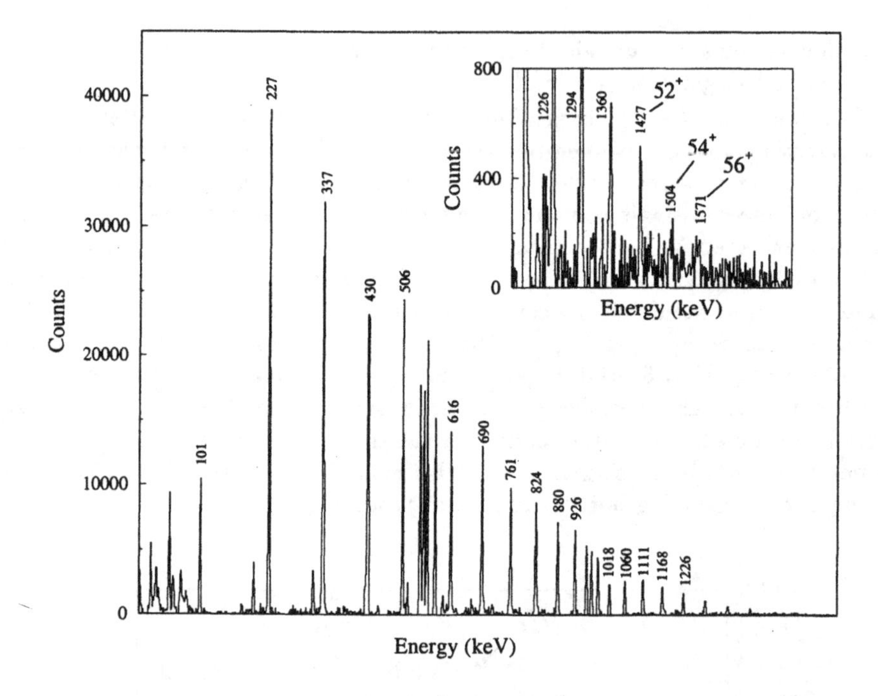

Figure 7 Spectrum of the (+,0) band in ^{162}Er. These spectra were obtained by summing many spectra each with the requirement that there were two other coincident transitions in the band. The insert shows the highest spin region.

In ^{162}Er the (+.0) bands is established up to 56^+ with no sign of a band crossing. This is consistent with the interpretation involving the exchange of just neutrons since the negative signature [651] $(+,-\frac{1}{2})$ orbital is already occupied.

3 SUMMARY

The new generation of very high efficiency arrays has given us the possibility to study nuclear structure phenomena at the very highest spins approaching the fission limit. The data on ^{156}Dy and ^{161,162}Er are the first to show discrete states in normal deformed nuclei up to spin $60\hbar$. A rich variety of competing structures, terminating bands and unpaired bands are observed. This is clearly an area that will blossom in the next few years.

4 ACKNOWLEDGEMENTS

I would like to thank all those who I have worked with over the years. Thanks to Mark Riley, Filip Kondev, Andrew Bagshaw and John Lisle, the main players in the ^{156}Dy and Er experiments. Special thanks to JFSS, who I have the pleasure to work with for over 20 years, for his continuing enthusiasm and encouragement to continue this mission to seek out the highest spins.

References

[1] J. Simpson, Z. Phys. A358 (1997) 139.

[2] I.Y. Lee, Nucl. Phys. A 520 (1990) 641c.

[3] C.W. Beausang and J. Simpson, J. Phys. G: Nucl. Part. Phys. 22 (1996) 527.

[4] P.J. Nolan, F.A. Beck and D.B. Fossan Ann. Rev. Nucl. Part. Sci. 45 (1994) 561.

[5] J. Simpson *et al.*, J.Phys. G 13 (1987) L235.

[6] F.G. Kondev *et al.*, Phys. Lett. B 437 (1998) 35.

[7] P.J. Twin *et al.*, Phys. Rev. Lett. 57 (1986) 811.

[8] Th. Byrski *et al.*, Phys. Rev. Lett. 64 (1990) 1650.

[9] P.J. Dagnall *et al.*, Phys. Lett. B335 (1994) 313.

[10] J. Eberth, Prog. Part. Nucl. Phys. 28 (1992) 495, J. Eberth *et al.*, Nucl. Instrum. Methods A 369 (1996) 135

[11] G. Duchene *et al.*, to be published, F.A. Beck *et al.*,91994) Proc Conf. on Physics form Large γ-ray detector Arrays (Berkeley) LBL 35687, CONF 940888, UC 413, p 154.

[12] C.W.Beausang *et al.*, Nucl. Instrum. Methods A 313 (1992) 37

[13] D.C. Radford, Nucl. Instrum. Methods Phys. Res. A 361 (1995) 297.

[14] I. Ragnarsson *et al.*, Phys. Src. 34 (1986) 651.

[15] I. Ragnarsson *et al.*, Phys. Rev. Lett. 74 (1995) 3935.

[16] J. Simpson *et al.*, Phys. Lett. B327 (1994) 187.

[17] J.F.Sharpey-Schafer, collected works of J.F.Sharpey-Schafer

[18] F.G. Kondev *et al.*, J. Phys. G. in press.

[19] J. Simpson *et al.*, Phys. Rev. Lett. 53 (1984) 648.

[20] P.O. Tjom *et al.*, Phys. Rev. Lett. 55 (1985) 2405.

[21] A. Afansjev and I. Ragnarsson private communication.

[22] M.A. Riley *et al.*, Phys. Rev. Lett. 55 (1987) 553.

GENERALIZED IMPULSE APPROXIMATION APPLIED TO QUASIELASTIC POLARIZED PROTON SCATTERING

B.I.S. van der Ventel, G.C. Hillhouse and P.R. de Kock

Department of Physics,
University of Stellenbosch,
Stellenbosch, 7600, South Africa

Abstract: Quasielastic $(\vec{p},\vec{n})$ and $(\vec{p},\vec{p}\,')$ scattering is employed to study the effect of the surrounding nuclear medium on the nucleon-nucleon interaction. This is done by calculating the associated polarization transfer observables $D_{i'j}$, within the framework of the Relativistic Plane Wave Impulse Approximation (RPWIA) and using a general Lorentz invariant representation of the nucleon-nucleon scattering matrix.

Quasielastic proton-nucleus scattering is assumed to be scattering from mainly one nucleon of the target nucleus. It therefore provides a way to study the effect of the surrounding nuclear medium on the NN interaction. This is done through the calculation of complete sets of spin observables, denoted by $D_{i'j}$ where j (i') refers to the initial (final) polarization. Deviations of the $D_{i'j}$'s from the free values could be viewed as resulting from a medium modification of the NN interaction by the nuclear medium. Calculations are done within the framework of the Relativistic Plane Wave Impulse Approximation (RPWIA) in which the medium effect is contained entirely in the effective nucleon masses

The Nucleus: New Physics for the New Millennium
Edited by Smit et al., Kluwer Academic / Plenum Publishers, New York, 2000.

M_1 and M_2 for the projectile and target nucleons respectively [1]. The average values for M_1 and M_2 can both be calculated microscopically from relativistic optical potentials [2]. However, we consider them presently as free parameters which can be varied over a range of physically acceptable values,

$$(0.5; 0.5) \leq (\tfrac{M_1}{M}; \tfrac{M_2}{M}) \leq (1.0; 1.0)$$

with step size of 0.1, and where M denotes the free nucleon mass. *The aim is to extract an optimal set which best describes all experimental polarization transfer observables for a given target and laboratory energy at the centroid of the quasielastic peak for* $(\vec{p}, \vec{p}\,')$ and $(\vec{p}, \vec{n})$ scattering. The calculations are done at the quasielastic peak (using free two-body kinematics, and ignoring recoil effects) since there it is assumed that the scattering is dominated by a single-step process and is therefore the closest to true free NN scattering. For the nucleon-nucleon scattering matrix, $\hat{F}$, we have employed both the standard (but incorrect) five-term parameterization in terms of the Fermi covariants (the SPVAT form or the IA1 representation) [1] as well as a (more correct) general Lorentz invariant expansion (the IA2 representation) of $\hat{F}$ [3].

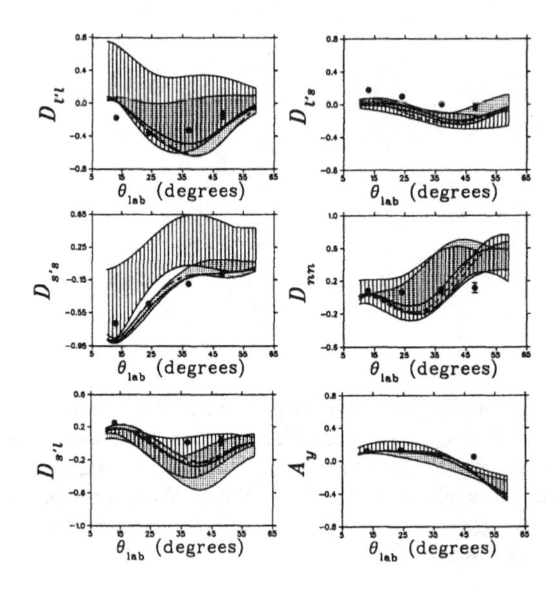

Figure 1 Values of $D_{i'j}$ versus θ_{lab} for $^{40}Ca(\vec{p}, \vec{n})$ at $T_{lab} = 200$ MeV. Solid and dashed lines represent the calculations with optimal effective mass values in respectively the IA2 and IA1 representations. For each, the band denote the range of values which result from varying $\frac{M_1}{M}$ and $\frac{M_2}{M}$ over the full range. The straight line hatch pattern denotes the IA1 model; the dotted hatch pattern the IA2 model. The long-dash–short-dash lines represent the free mass values. Data (at $\theta_{lab} = 13°, 24°, 37°, 48°$) are from Ref. [4].

For the results in Fig. 1 ($T_{lab} = 200$ MeV) the $(\vec{p}, \vec{n})$ observables show a strong variation with respect to angle. For D_{nn} the variation of the theoretical calculation is too strong to reproduce the data which is quite flat and it is clear that no combination of effective masses can remedy this. The variation of the theoretical calculation with respect to angle in A_y is not so strong and the optimal curves describe the data very well, except at $\theta_{lab} = 48^\circ$. A very interesting observable is $D_{s's}$, since it shows a *range of angles for which the IA1 and IA2 effective mass bands do not overlap.* This observable is sensitive to whether one uses the IA1 or IA2 representation of $\hat{F}$. The IA2 optimal curve describe the data for $D_{s's}$ very well. For all observables, at least one of the data points, (measured at $\theta_{lab} = 13^\circ$, 24°, 37° and 48° [4]) misses the IA2 band. The data point at $\theta_{lab} = 13^\circ$ lies within the IA1 effective mass band and would have led to the incorrect conclusion that this point can be predicted by the use of an effective mass. The correct IA2 representation, however, clearly shows that this is not the case (since it does not lie within the IA2 effective mass band) and indicates that other effects must be included at that particular angle.

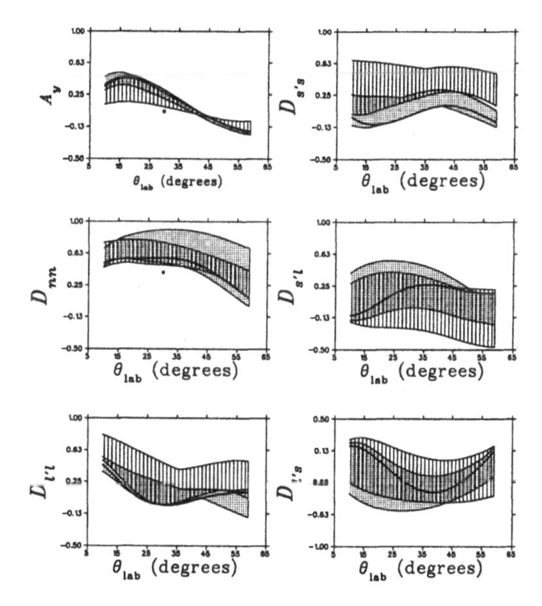

Figure 2 Values for $D_{i'j}$ versus θ_{lab} for $^{40}Ca(\vec{p}, \vec{p}\,')$ at $T_{lab} = 200$ MeV, with the solid line representing the free mass calculation. Data (at $\theta_{lab} = 30^\circ$) are from Ref. [5].

In Fig. 2 we consider the reaction $^{40}Ca(\vec{p}, \vec{p}\,')$ at $T_{lab} = 200$ MeV. The data point at $\theta_{lab} = 30^\circ$ [5] lies outside both IA1 and IA2 bands for D_{nn} and A_y. No effective mass combination can describe these two observables. No data exists, however, for the other observables and therefore it would be extremely helpful to measure a *complete set and at a few more angles* for this target and energy. Even though the IA2 repesentation is the most advanced treatment of medium effects to date, it still fails to describe all observables, and therefore other effects like multiple scattering and distortions have to be included.

References

[1] C.J. Horowitz and D.P. Murdock, Phys. Rev. C37, 2032 (1988).

[2] G.C. Hillhouse and P.R. de Kock, Phys. Rev. C49, 391 (1994)

[3] J.A. Tjon and S.J. Wallace, Phys. Rev. C36, 1085 (1987).

[4] C.L. Hautala, *Measurement of Polarization Observables in the Quasielastic Region on ^{nat}Ca and ^{nat}Pb using the $(\vec{p},\vec{n})$ Reaction at 200 MeV*, Ph.D Thesis, Ohio University (1998), unpublished.

[5] D.S. Carman, *AIP Conference Proceedings of the Eight International Symposium on Polarization Phenomena in Nuclear Physics*, edited by Edward J. Stephenson and Steven Vigdor, (American Institute of Physics, New York, 1995), p. 445.

EXOTIC NUCLEI PRODUCED AND PROBED USING LASERS AND ION TRAPS

P. Van Duppen

Instituut voor Kern- en Stralingsfysica University of Leuven B-3001 Leuven, Belgium

Abstract: Online isotope separators (ISOL) have produced beams of exotic nuclei for several decades. Continuous efforts in the development and refinement of the targets and ion sources and the detection systems has allowed detailed studies of far-unstable nuclei. Recent implementation of resonant photo ionisation and of ion manipulation using electromagnetic ion traps enlarges the research possibilities around ISOL systems substantially. In this contribution we will highlight these developments and point out their capabilities and limitations.

1 INTRODUCTION

One of the main goals of nuclear physics studies is to understand the manifestation of the strong, weak and electromagnetic interaction in the atomic nucleus. Experimental and theoretical investigations over several decades have resulted in models that describe the nuclear-structure properties of nuclei, that are situated around the line of stability, in a satisfactory way. However, once one goes away from the line of stability towards the proton- or neutron drip line, the situation changes dramatically. The balance between protons and neutrons of these so-called exotic nuclei is so far from equilibrium - i.e. from the Z/N-ratio in the valley of stability - that it is questionable if nuclear models, valid close to stability, survive at these boundaries. Simple extrapolations of our understanding towards these regions often fail to reproduce the observed phenomena

The Nucleus: New Physics for the New Millennium
Edited by Smit et al., Kluwer Academic / Plenum Publishers, New York, 2000.

and surprises show up. Experiments with these close-to-drip line nuclei will reveal critical information to guide the developments of theoretical models and to test their applicability. Furthermore, many of these still unknown nuclei are lying on nucleosynthesis paths; the reaction-, decay- and ground-state properties of some of these nuclei can play a key role in the development of reliable nucleosynthesis scenarios.

Exotic nuclei are extremely difficult to produce and to study because of their low production cross section, the short half life involved and the overwhelming production of unwanted species in the same target. In essence there are two complementary ways to produce and prepare exotic nuclei for study: Isotope Separation On-Line (ISOL) eventually followed by post-acceleration, and In-Flight Separation (IFS). Online isotope separation have in the past played an important role in research related to exotic nuclei. Recently two main technical developments, resonant laser ionisation and ion traps, have been implemented in several ISOL systems. It is the aim of this contribution to discuss these developments that have boosted the research programs at ISOL systems and to give the reader a flavour of the physics program one might expect for the new millennium.

2 RESONANT LASER IONISATION

Resonant photo ionisation, whereby the atoms are stepwise excited by laser photons until they reach the continuum, is a very efficient and selective process. With commercially available laser systems high efficiency (close to 100%), ultimate element selectivity (eventually isotope selectivity) and high universality (one can ionise atoms of almost all elements) can be reached. For an introduction in the field of resonant ionisation the reader is referred to the following references [1, 2].

The principle of laser ionisation has been successfully implemented in an on-line ion source for the production of beams of radioactive ions in two different ways [3]. At the ISOLDE facility, the laser light is shone into a hot cavity that is connected to the target container [4, 5]. The atoms that diffuse from the target material and enter the hot cavity are ionised through interactions with the laser light and are subsequently subtracted from the source. Several elements (Li, Be, Mg, Mn, Ni, Cu, Zn, Ag, Cu and Sn) have been laser ionised at ISOLDE with efficiencies between 3 and 20 %. A schematic drawing of the laser source is shown in Fig. 1.

Different experiments have been performed over the last few years making use of laser ionisation: decay study of neutron-rich Ag (up to ^{129}Ag) [6] and Mn nuclei (up to ^{69}Mn) [7], decay study of neutron-deficient ^{58}Zn, production of a ^{7}Be target, and laser spectroscopy of ^{11}Be. At the LISOL facility, a different approach is followed. The reaction products are thermalised in a gas cell and photo ionised before they reach the exit hole of the gas cell [8]. The latter approach uses the "thin target" technique as the reaction products have to recoil out of the target which makes the total production rates limited. But it has the advantage that, for certain "refractory type" elements the long diffusion

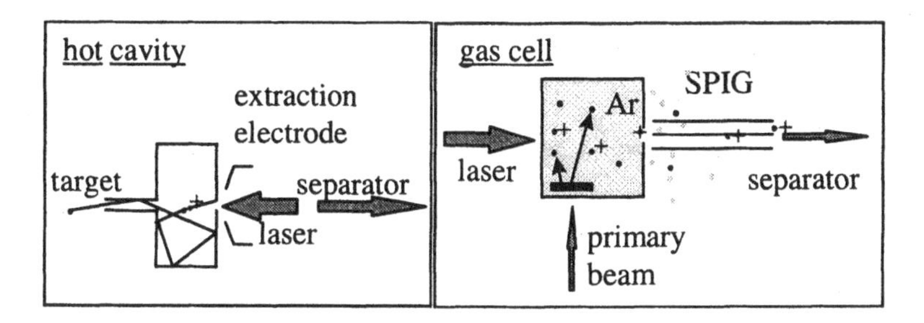

Figure 1 Schematic drawing of the two approaches of a laser ion source at ISOL systems. On the left hand side the hot-cavity approach [4] is shown while on the right hand side the gas-cell approach [8] is sketched.

times in the "thick target" approach are avoided. In this way the two techniques are complementary. Results have been obtained using light-ion induced fusion and fission reactions. For example the β-decay of $^{68-74}$Ni [9] and $^{67-70}$Co [10] has been studied revealing nuclear structure information around the semi-magic N=40 shell gap.

3 ION TRAPS

Using a combination of electric and/or magnetic fields, ions can be trapped for an extended period of time. An overview on the principles of ion traps and the research program performed with them can be found in ref. [11]. The capability to confine a cloud of ions in space for an extended period of time opens up a whole series of possibilities for manipulating radioactive ions and for performing high-precision measurements. The Penning traps coupled to the ISOLDE separator (so-called ISOLTRAP) has been operational for several years and mass measurements of exotic nuclei have been performed [12].

The ISOLTRAP is a tandem of traps. Starting with an RF-trap (Paul trap) with buffergas cooling to make short beam pulses from the quasi continuous ISOLDE radioactive ion beam and to cool the ions prior to injection in the cooling and purifying second trap. This second trap (Penning trap) aims at further cooling the ion cloud and this in a mass selective way [13]. The technique makes use of the fact that the different motions (axial oscillations, cyclotron motion and magnetron motion) in a Penning trap are involved in the cooling process. The mass resolving power (R=M/ΔM) obtained in the second trap can reach up to 10^5. In this way a bunch of purified and cooled ions are delivered to the third trap (Penning trap) which is constructed to obtain a high mass resolving power. In this trap the ions are subject to RF radiation. If the RF frequency equals the cyclotron frequency the ions heat up. The ion cloud is subsequently ejected from the trap and the time of flight is measured. The time of flight versus the RF frequency then displays a resonance (minimum time of flight at the resonance frequency). The mass resolving power is in first order linearly dependent on the time the ions are subjected to the RF power. For RF-irradiation times of 1 second a typical resolving power of about 10^6 is obtained. This translates in an accuracy for the mass determination of the order of 10^-7. For A=100 this corresponds to an accuracy of about 10 keV. Note that this accuracy doesn't change if one goes away from the line of stability as can be nicely seen in Figure 2 of Ref. [12].

These mass measurements, that are complementary to mass measurements performed for example in the ESR storage ring at GSI [14], will certainly bring interesting new results in the next millennium Besides the possibility to perform precise measurements, ion traps offer the possibility to manipulate and cool the ions in an efficient and effective way. Behind the laser ion source at LISOL [8] a sextupole ion guide (SPIG) has been installed [15] (see fig. 1). A picture of the SPIG is shown in fig. 2.

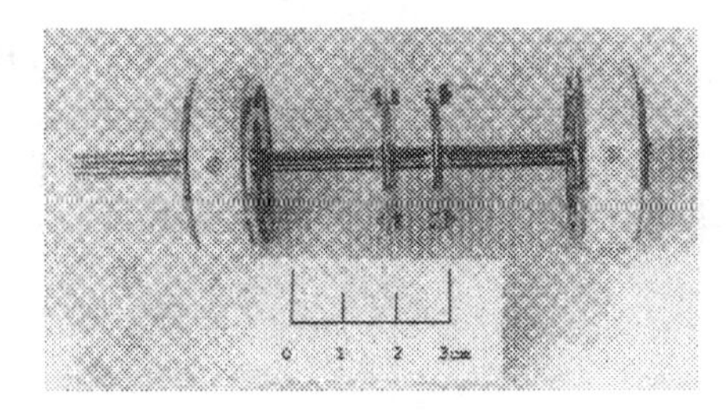

Figure 2 Photo of the sextupole ion guide (SPIG). The six rods are connected three by three and the RF power is applied via the two centre metal rings. The isolators (white teflon cylinders) are carefully shielded to avoid charging.

This RF ion trap guides the ions from the high pressure zone to the low pressure zone prior to acceleration. The ions are cooled at the same time by buffer gas collisions. For example, after installation of the SPIG the mass resolving power of the separator increased by a factor of five [15]. Using a trap for cooling of secondary beams after mass separation has also been proposed.

A 60 keV radioactive ion beam will be injected into the ion trap. For this project a Penning trap (as for the REX-ISOLDE project) as well as an RF-trap is proposed [12]. This technique of radioactive ion trapping and cooling is very promising but requires much technical development. Therefore a group of research institutes and universities (JYFL-Jyväskylä, GSI-Darmstadt, CERN-Geneva, LMU-Munich, GANIL/LPC-Caen, KULeuven) have joined efforts in the so-called EXOTRAPS network aimed at dealing with the specific problems related to trapping and cooling of exotic nuclei [16]. Besides mass measurements and ion manipulation, traps are also planned to be used for fundamental interaction studies like e.g. measurements of the β-ν correlation after β-decay of a radioactive atom that is not imbedded in a solid matrix.

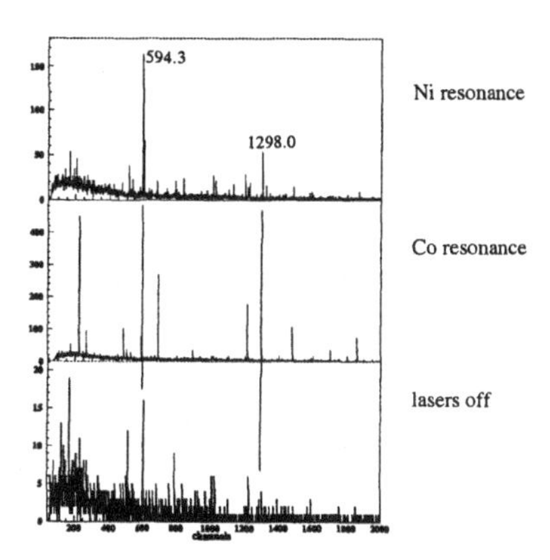

Figure 3 β-gated γ-ray spectra obtained at mass 69 when the lasers were tuned on Co resonance (top), Ni resonance (middle) and off (bottom). Energies of the γ-lines are in keV.

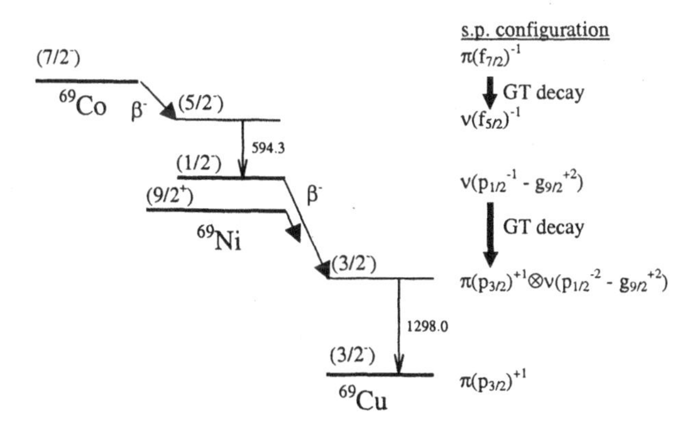

Figure 4 Schematic drawing showing the main β-decay branch for the decay of ^{69}Co - ^{69}Ni - ^{69}Cu.

4 SPECTROSCOPY ALONG THE Z=28 LINE

At LISOL decay studies of $^{68-74}$Ni [9] and $^{67-70}$Co [10] produced with the laser ion source combined with the SPIG have been performed. To show the strength of the method we show in Fig. 3 the γ-spectra obtained at mass 69 using a proton induced fission reaction on ^{238}U. From these spectra one can immediately identify the γ-rays that belong to ^{69}Co or ^{69}Ni. The dominant 594.3 keV transition present only in the Co on-resonance spectrum belongs to the β-decay of ^{69}Co and can be identified, using the selectivity of the ($\pi f_{7/2} \rightarrow \nu f_{5/2}$) Gamow-Teller β decay, as the $\nu(f_{5/2}^{-1}\text{-}g_{9/2}^{+2}) \rightarrow \nu(p_{1/2}^{-1}\text{-}g_{9/2}^{+2})$ transition in ^{69}Ni. The 1298.0 keV line belongs to the β-decay of the low-spin isomer in ^{69}Ni [17] whose half life could be determined to be 3.5 (5) s. We notice that this isomer is populated via the strong 594.3 keV transition when the lasers were tuned to Co on-resonance. When the lasers where tuned to Ni on-resonance the transition is still present in the spectrum indicating direct population of the low-spin isomer in the fission reaction. This 1298.0 keV transition itself could be identified as belonging to the decay of a $3/2^-$ excited state to the $3/2^-$ ground state in ^{69}Cu. From the selective Gamow-Teller decay ($\nu p_{1/2} \rightarrow \pi p_{3/2}$) one can conclude that the excited state must have a $\pi p_{3/2} \otimes \nu p_{1/2}^{-2}\text{-}\nu g_{9/2}^{2}$ configuration as main component of its wave function. Note that the first excited 0^+ state in ^{68}Ni is situated at 1770 keV [18] with a possible $\nu p_{1/2}^{-2}\text{-}\nu g_{9/2}^{2}$ configuration. Fig. 4 shows a very schematic decay scheme showing only the main β- and γ-decay branches. More details can be obtained in ref. [19]. These measurements do not only give information on the proton and neutron single-particle states and half lives of neutron-rich nuclei along the Z=28 closed shell, but shed light on the importance of the N=40 sub-shell closure and on extrapolations towards the doubly magic nucleus ^{78}Ni.

5 CONCLUSION

In this contribution we have described new technical developments that are implemented at on-line isotope separators. Laser ionisation allows the production of relative pure ion beams of exotic nuclei while ion traps offer the possibility to perform high precision measurements as well as ion manipulation to improve the beam quality of the low-energy radioactive ion beam. These issues, once fully developed, will initiate the exploration of new areas of the chart of nuclei, close to the drip lines, that were not accessible before. Furthermore laser ionisation and ion manipulation are of prime importance for the production of post-accelerated radioactive ion beams. If the planned experiments with these low-intensity (sometimes very low intensity) radioactive ion beams are to become a success, one needs pure beams of very good ion optical qualities. Laser ionisation combined with cooling (eventually mass selective cooling) makes it possible to fulfil these conditions. Resonant photo ionisation will produce beams in an element selective way and the cooling guaranties an efficient acceleration process (only little loss of intensity). Both developments will shape the nuclear-structure research of exotic nuclei in the next millennium.

References

[1] V.S. Lethokov, Laser Photoionization Spectroscopy (Academic Press, Or-

lando, 1987)

[2] G.S. Hurst and M.G. Payne, Principles and Applications of Resonance Ionization Spectroscopy (Hilger, London, 1988)

[3] P. Van Duppen, *Nucl. Instr. and Meth.* B126, 66 (1997)

[4] V.N. Fedoseyev, *et al.*, *Nucl. Instr. and Meth.* B126, 88 (1997)

[5] J. Lettry, *et al.*, *Rev. of Scien. Instr.* 69, 761 (1998)

[6] K.L. Kratz *et al.*, Proc. of the Int. Conf. "Fission and Properties of Neutron-Rich Nuclei", ed. J.H. Hamilton and A.V. Ramaya, World Scientific, 586 (1998)

[7] M. Hannawald *et al.*, *Phys. Rev. Lett.* 82 (1999)

[8] Y. Kudryavtsev *et al.*, *Nucl. Instr. and Meth.* B114 (1996) 350

[9] S. Franchoo *et al.*, *Phys. Rev. Lett.* 81 (1998) 3100

[10] L. Weissman *et al.*, *Phys. Rev. C* (1999) in print

[11] Proc. Nobel Symposium 91 on "Trapped Charged Particles and Related Fundamental Physics", Lysekill, Sweden, 1994, Phys. Scripta T59 (1995)

[12] G. Bollen, *Nucl. Phys.* A616 (1997) 457c

[13] M. König *et al.*, *Int. J. Mass Spectrom. Ion. Proc.* 142 (1995) 95

[14] H. Geissel, *Nucl. Phys.* A616 (1997) 316c

[15] P. Van den Bergh *et al.*, *Nucl. Instr. and Meth.* B126 (1997) 194

[16] http://www.jyu.fi/ armani/exotraps/frames.htm

[17] R. Grzywacz *et al.*, *Phys. Rev. Lett.* 81 (1998) 766

[18] M. Bernas *et al.*, *J. Phys. Lett.* 45 (1984) L851

[19] W. Mueller *et al.*, to be published

MEASURING QUADRUPOLE MOMENTS IN A RECOIL-SHADOW GEOMETRY USING THE LEMS METHOD

K. Vyvey[1], G. Neyens[1], D.L. Balabanski[2], S. Ternier[1], N. Coulier[1], S. Teughels[1], G. Georgiev[1] and R. Coussement[1]

[1] Instituut voor Kern- en Stralingsfysica, University of Leuven, B-3001 Leuven, Belgium
[2] Faculty of Physics, St. Kliment Ohridsky University of Sofia, BG-1164 Sofia, Bulgaria

Abstract: The applicability of the LEMS technique using a recoil-shadow configuration, is demonstrated. As a test case the ^{69}Ge($I^{\pi} = \frac{9}{2}^{+}, \tau = 4\mu s, \mu = -1.0011 n.m.$) released out of a ^{nat}Fe foil with a recoil velocity of 1.6(5)%c has been studied. The magnetic field induced by the atomic electrons at the place of the ^{69}Ge nucleus is found to be 2200(600)T.

1 INTRODUCTION

The LEMS method (Level Mixing Spectroscopy) has proven to be a powerful tool to study quadrupole moments of high spin isomers [1, 2, 3, 4, 5]. In these experiments the change of the nuclear spin orientation due to a combined electric quadrupole and magnetic dipole interaction is measured as a function of the magnetic field strength. The electric field gradient (EFG) is provided by in-beam implantation of the reaction recoils in a suitable host and the magnetic field by a superconducting magnet. Detection limitations are the main restriction for the applicability of the method : The number of holes in the magnet and thus the number of detector positions is limited so that a lot of the usual methods of cleaning the spectra cannot be applied due to the low counting rates. The only way to reduce the background radiation at this point, is the use of timing, which is not efficient for short life times (lower than 300 ns). Therefore, the improvement of the peak to background ratio by using the recoil-shadow configuration, i.e. separating target and host and shielding the detectors from the target, has been investigated. Improvements of the peak to background ratio up to a factor of 5 can be reached this way[6]. A complicating factor for the application of this technique is that often the recoiling nuclei are not fully stripped, with as a consequence a loss of the nuclear spin orientation due to the interaction between randomly oriented electron spin $\overline{J}$ and the nuclear spin $\overline{I}$ during the flight through vacuum. In this paper a theoretical study and first experimental results on the change in the anisotropy due to the pure $\overline{I}.\overline{J}$ interaction and the combined $\overline{I}.\overline{J}$ + LEMS interaction will be presented.

2 LEVEL MIXING AND $\overline{I}.\overline{J}$ INTERACTIONS

The formalism of nuclear level mixing has been discussed extensively in Ref. [1]

2.1 The LEMS technique

The LEMS method has been extensively treated in Ref. [1]. Here only the relevant features to the experiments described in this paper are mentioned. After production and orientation in a fusion-evaporation reaction the isomers of interest are caught in a suitable host, often a polycrystal, where they are submitted to a combined electric quadrupole and magnetic dipole interaction. The magnetic field is oriented parallel to the beam axis, i.e. the symmetry axis of the initial orientation. A time-integrated measurement of the anisotropy of the γ-radiation as a function of the magnetic field strength B is performed. In a LEMS curve three regimes can be distinguished. At zero magnetic field only the quadrupole interaction is present and the initial orientation is decreased to the hard-core value for the anisotropy. At high magnetic fields (several Tesla), the electric quadrupole interaction is negligible compared to the Larmor precession of the isomeric spins around $\overline{B}$. As the precession axis coincides with the initial orientation axis, the initial anisotropy is measured. At intermediate fields there is a competition between the quadrupole and the magnetic interaction and a smooth change from the hard-core anisotropy to the initial full anisotropy takes place. This part of the LEMS curve is sensitive to the ratio of the quadrupole interaction frequency $\nu_Q = \frac{eQV_{zz}}{h}$ to the magnetic moment μ of the isomer. So, if the magnetic moment is known, the quadrupole interaction frequency can be deduced. All experiments have been performed at the CYCLONE cyclotron in Louvain-la-Neuve, Belgium.

3 THE $\overline{I}.\overline{J}$ INTERACTION

If there is a distance between target and host, the $\overline{I}.\overline{J}$ interaction influences the anisotropy as a function of the magnetic field strength, if the nucleus is not fully stripped. At zero magnetic field the nuclear spin $\overline{I}$ and the electron spin $\overline{J}$ combine to a total angular momentum $\overline{F}$ around which both precess. As the electron spin is randomly oriented the net result of this precession is a lowering of the orientation of the nuclear ensemble. The orientation can be restored by applying a high enough magnetic field parallel to the orientation axis, so that both, electron and nuclear spin precess around this magnetic field. The Hamiltonian of the interaction is given by $H = a\overline{I}.\overline{J} + \overline{\mu}_J.\overline{B} - \overline{\mu}_I.\overline{B}$ with $a = \frac{\mu_I \langle B_{hf}(0) \rangle}{IJ\hbar^2}$, $\overline{\mu}_I$ the nuclear magnetic moment, $\overline{\mu}_J$ the electronic magnetic moment and B_{hf} the hyperfine field induced by the atomic electrons. So, if μ_I is known, B_{hf} can be deduced out of a and vice versa [7].

An experiment using the ^{nat}Fe(^{16}O,2pn)^{69}Ge reaction with a beam energy of 65 MeV has been performed. The target thickness of 1.57mg/cm^2 is thin enough to release all recoils, which have been stopped in a 1.78 mg/cm^2 thick Ni host at a distance of 5 cm of the ^{nat}Fe target. A 50 μm thick Ta foil served as beam stopper. The high-Z beam stopper has been chosen in order to decrease the background radiation. The Ni foil has been heated up to 450°C, a temperature at which it is in paramagnetic phase. Since Ni is cubic, no EFG and thus no quadrupole interaction is present unless defects are trapped by the probe nuclei. An experiment applying a direct production in a ^{nat}Ni(^{12}C,2pn)^{69}Ge reaction proves that indeed no defects are created, as the analysis of the 398 M2 keV transition results in a flat curve for the anisotropy as a function of B (fig. 1a). So, if a change in the anisotropy is observed, it is entirely due to the $\overline{I}.\overline{J}$ interaction during recoil time of the ^{69}Ge ions (order of 5 ns).

The $\overline{I}.\overline{J}$ decoupling curve for the ^{69}Ge(I=9/2) isomers released out of the Fe foil with a velocity of 1.6(5)%c is shown in fig. 1b. Notice that the recoiling ^{69}Ge atoms can occupy several electron states, so that the fit results for a and J are a kind of average over all possible states. The values for $\nu_{IJ} = \frac{\hbar}{2\pi}a = 1.5(4)$ GHz and J = 2.5 result in a hyperfine field of 2200(600)T at the place of the nucleus. An external field of 0.8 T is necessary to decouple the $\overline{I}.\overline{J}$ interaction.

4 THE COMBINED LEMS + $\overline{I}.\overline{J}$ INTERACTION

If there is a recoil distance between target and host and an EFG is present in the host, the $\overline{I}.\overline{J}$ interaction and the LEMS interaction take consecutively place. Remember $\overline{I}.\overline{J}$ interaction is only active as long as the atom is free. As our goal is to measure quadrupole moments, we are interested in the influence of the $\overline{I}.\overline{J}$ interaction on the LEMS curve. A computer program has been developed to describe the influence of these interactions on the angular distribution, by calculating the perturbation factors. The details of the calculations will be published elsewhere.

As a test case an experiment has been performed on the ^{69}Ge($I^\pi = 9/2^+$,τ=4μs, μ=-1.0011 n.m, Q=100(20)efm^2) isomer recoiling into a Pt host. Again a

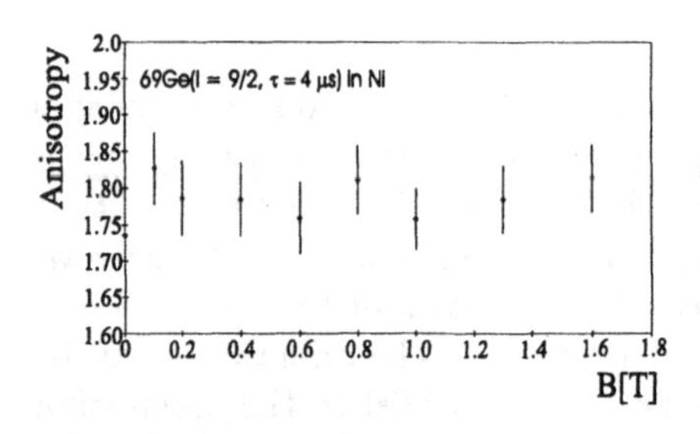

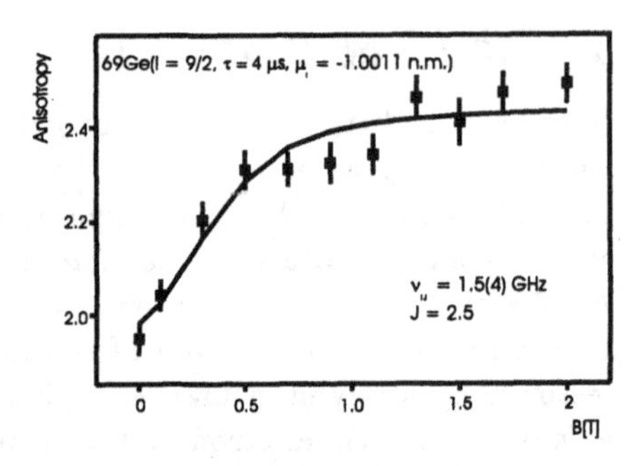

Figure 1 (a) LEMS curve for the ^{69}Ge(I=9/2) isomer in Ni at 450°C, using a direct production in the Ni. The flat curve proves that no defects in the Ni are present. (b) $\overline{I}.\overline{J}$ decoupling curve for the ^{69}Ge(I=9/2) isomers recoiling out a ^{nat}Fe foil with a velocity of 8(3)%c. The frequency ν_{IJ} is equal to $\frac{\hbar}{2\pi}a$.

^{nat}Fe(^{16}O,2pn)^{69}Ge reaction with a beam energy of 65 MeV has been used. In a first experiment the ^{nat}Fe was evaporated on the Pt, in a second one the Fe foil was placed at a recoil-distance of 6 cm. The target thickness was twice 1.57mg/cm^2. Also Pt is cubic, but other experiments [8] have shown that defects are easily created in this host, resulting in a defect associated EFG. At zero recoil distance a pure LEMS interaction takes place. Out of the amplitude of the LEMS curve can be derived that 55% of the Ge nuclei end up in a defect associated site. 24(5)% of the Ge isomers is interacting with a smaller EFG (ν_{Q_1} = 6.4(1.5) MHz), 31% with a bigger one (ν_{Q_2} = 18(2) MHz). Fig. 2 shows that the curve for a 6 cm recoil distance has a larger amplitude, caused by an extra lowering in the anisotropy due to the $\overline{I}.\overline{J}$ interaction. In a first

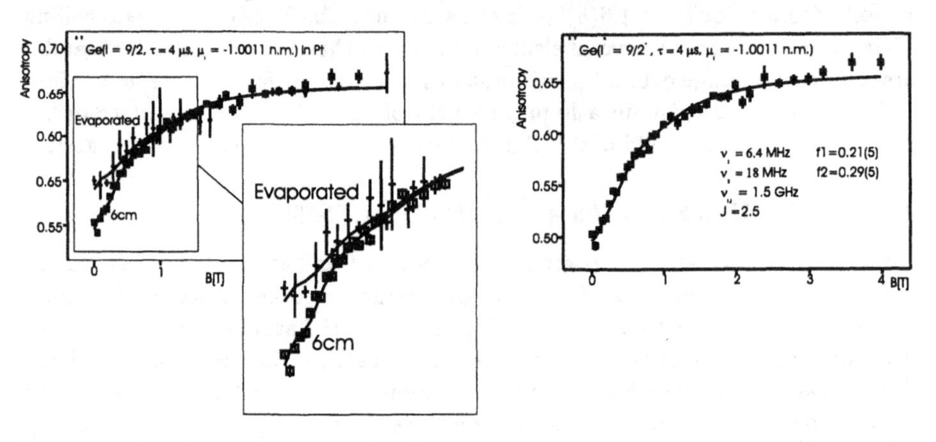

Figure 2 (a) LEMS curves for the ^{69}Ge(I=9/2) isomer in Pt at zero and 6 cm recoil-distance. No $\overline{I}.\overline{J}$ interaction has been taken into account. The fit results can be found in table 1. (b) Combined LEMS+$\overline{I}.\overline{J}$ fit for the 6 cm recoil distance.

Table 1 Fit results for the ^{69}Ge(I=9/2) isomer in Pt with a zero and a 6 cm recoil-distance. For a zero recoil distance only the pure LEMS interaction takes place. For a 6 cm recoil distance also the $\overline{I}.\overline{J}$ interaction is present, which has been taken into account in ^{b}, but not in ^{a}.

d	ν_{Q_1}(MHz)	ν_{Q_2}(MHz)	f_1	f_2	ν_{IJ} (GHz)	J
0	6.4(1.5)	18(2)	0.24(5)	0.31(5)	-	-
6 cm ^{a}	5.0(0.8)	17(4)	0.50(8)	0.40(5)	not in fit	not in fit
6 cm ^{b}	6.4	18	0.21(5)	0.29(5)	1.5	2.5

fit, the $\overline{I}.\overline{J}$ interaction has not been taken into account in order to verify the influence of the $\overline{I}.\overline{J}$ interaction on the quadrupole interaction frequencies. The same quadrupole frequencies within the error bar have been found (Table 1). Only the fractions differ as they are directly connected to the amplitude of the LEMS curve. In a second fit, both the LEMS interaction and the $\overline{I}.\overline{J}$ interaction have been taken into account. The $\overline{I}.\overline{J}$ fit parameters have been fixed by the experiment with the Ni host, the quadrupole frequencies by the experiment with zero recoil distance. Now also the fractions of nuclei submitted to an EFG agree and a slightly better χ^2 (3%) has been found.

5 CONCLUSION

The experiments on Ge(Pt) reveal that the influence on the quadrupole frequencies due to the $\overline{I}.\overline{J}$ coupling is negligible. Only in exceptional cases, when the $\overline{I}.\overline{J}$ interaction strength is very strong compared to the quadrupole interaction strength, the $\overline{I}.\overline{J}$ interaction should be taken into account. In this case a combined LEMS + $\overline{I}.\overline{J}$ fit results in the correct quadrupole frequencies and the $\overline{I}.\overline{J}$ fit parameters can be fixed in an experiment where the nuclei of interest recoil into a cubic host.

References

[1] F. Hardeman, G. Scheveneels, G. Neyens, R. Nouwen, G. S'heeren, M. Van den Bergh and R. Coussement *Phys. Rev. C* 43,130(1991).

[2] G. Scheveneels, F. Hardeman, G. Neyens and R. Coussement *Phys. Rev. C*43,2566(1991).

[3] F. Hardeman, G. Scheveneels, G. Neyens, R. Nouwen, G. S'heeren, M. Van den Bergh and R. Coussement *Phys. Rev. C* 43,514(1991).

[4] G. Neyens, R. Nouwen, G. S'heeren, M. Van Den Bergh and R. Coussement *Nucl. Phys.* A555,629(1993).

[5] K. Vyvey, G. Neyens, D.L. Balabanski, S. Ternier, N. Coulier, S. Teughels, G. Georgiev, R. Wyckmans, R. Coussement, M. Mineva, P.M. Walker,

A.P. Byrne, G.D. Dracoulis and P. Blaha *accepted in Journ. of Phys. G, Proceedings on the Int. Conf. on Nuclear Structure at the extremes* Lewes, (1998).

[6] K. Vyvey, G. Neyens, S. Ternier, N. Coulier, S. Michiels, R. Coussement, D.L. Balabanski and A. Lépine-Szily *Acta Polonica* B28,329(1997) ,*Proceedings on the XXXI Zakopane Schools of Physics* Zakopane.

[7] A. Little, H.C. Jain, S.M. Lazarus, T.K. Saylor, B.B. Triplett and S.S. Hanna *Hyp. Int.* 8,318(1980).

[8] S. Ternier, G. Neyens, K. Vyvey, J. Odeurs, N. Coulier, S. Michiels, R. Coussement, D.L. Balabanski and R. Kulkarni *Nucl. Instr. Meth.* B140,235(1998).

HALF-LIFE FOR THE rp-PROCESS WAITING POINT NUCLIDE ^{80}Zr USING DELAYED GAMMA TAGGING

J. Ressler,[1] A. Piechaczek,[2] W. Walters,[1] A. Aprahamian,[3] J. Batchelder,[4] C. Bingham,[5] D. Brenner,[6] T. Ginter,[7] C. Gross,[4] R. Grzywacz,[5] D. Kulp,[8] B. MacDonald,[8] W. Reviol,[5] K. Rykaczewski,[9] J. Stone,[1] M. Wiescher,[3] J. Winger,[10] and E. Zganjar.[2]

[1] Department of Chemistry, University of Maryland, College Park, MD 20742
[2] Physics Department, Louisiana State University, Baton Rouge, LA 70803
[3] Department of Physics, University of Notre Dame, Notre Dame, IN 46556
[4] Oak Ridge Institute for Science and Education, Oak Ridge, TN 37831
[5] Department of Physics, University of Tennessee, Knoxville, TN 37996
[6] Department of Chemistry, Clark University, Worcester, MS 01610
[7] Department of Physics, Vanderbilt University, Nashville, TN 37235
[8] School of Physics, Georgia Institute of Technology, Atlanta, GA 30332
[9] Physics Division, Oak Ridge National Laboratory, Oak Ridge, TN 37831
[10] Physics Dept., Mississippi State University, Mississippi State, MS 39762

Abstract: A 4.1(5) s half-life has been measured for ^{80}Zr at the Holifield Heavy Ion Research Facility of Oak Ridge National Laboratory. Using the inverse fusion reaction $^{24}Mg(^{58}Ni,2n)^{80}Zr$, nuclides of interest were produced and separated by the Recoil Mass Spectrometer prior to implantation onto the tape of a Moving Tape Collector. The ^{80}Zr half-life was then determined by observing delayed gamma rays depopulating the 4 μs isomer at 312 keV in the daughter ^{80}Y, a technique referred to as Delayed Gamma Tagging, or DGT.

The Nucleus: New Physics for the New Millennium
Edited by Smit et al., Kluwer Academic / Plenum Publishers, New York, 2000.

The light isotopes of elements with $Z < 50$ are thought to be synthesized in the astrophysical rp-process. This rapid proton capture mechanism has been discussed in detail by Schatz *et al.* [1]. Due to the proton instability of ^{81}Nb, ^{80}Zr lies at a particulary critical point in the rp-process. Further nucleosynthesis beyond this point must "wait" for ^{80}Zr to decay or undergo di-proton capture. Calculations for the ^{80}Zr half-life have resulted in a wide range of values due to rapid nuclear structure changes in the $N = Z = 40$ region. Differences in calculated half-lives are a consequence of uncertainties in both the Q_{EC} and the estimated log ft values for the ^{80}Zr decay.

^{80}Zr nuclei were produced using a 195 MeV ^{58}Ni beam from the Holifield tandem accelerator to bombard a 500 μg/cm^2 thick enriched ^{24}Mg foil at the target postion. Reaction fragments were subsequently separated by A/Q ratios through the Recoil Mass Spectrometer, and fragments with a single A/Q peak associated with $A = 80$ were implanted onto the tape of a Moving Tape Collector.

The experiment was designed assuming the 0^+ ground state of ^{80}Zr would populate one or more 1^+ levels in the daughter ^{80}Y. These levels would then depopulate, at least in part, to the recently identified 4 μs 2^+ isomer at 312 keV [2], emitting a single 84 keV gamma ray. Coincidence events were sought in which a time-to-amplitude converter (TAC) was started by a β^+-particle, annihilation gamma, or prompt gamma decay from a 1^+ level, and stopped by an 84 keV gamma from the 2^+ isomer. In this manner, events unique to the decay of ^{80}Zr could be identified despite numerous prompt coincidence events from the daughter and other nuclides with the same A/Q ratio. This method is referred to as Delayed Gamma Tagging, or DGT.

Three plastic scintillators were used to identify positrons from the ^{80}Zr decay, and three clover HPGe detectors were used to detect annihilation and prompt gamma radiations. An event in any one of these six detectors provided the TAC start pulse. A low energy Ge detector with dimensions 70 mm x 30 mm detected the 84 keV gamma rays for the TAC stop pulse. From an analysis of isomeric decay events, a half life of 4.1(5) s for ^{80}Zr was determined.

The spectrum of gamma rays observed in the clover detectors coincident with delayed isomeric transitions revealed a clear peak at 311 keV. We assign this line at 311 keV as one of the primary 1^+ levels populated in the beta decay of ^{80}Zr which depopulates to the 2^+ isomer at 312 keV. Therefore, a strongly populated 1^+ level is is paced at 623 keV in ^{80}Y.

The half-life of ^{80}Zr is slightly shorter than predicted in some astrophysical models. Hence, this nuclide is less of a waiting point, permiting synthesis of heavier nuclides in stellar events with shorter time duration and less extreme proton flux and energies [3].

The credibility of the 4.1(5) s half-life for ^{80}Zr would have been difficult to establish without the DGT method because the daughter ^{80}Y has an isomer with a 5 s half-life and a production cross section at least 100 times larger.

This work was supported by the U. S. Department of Energy and the U. S. National Science Foundation.

References

[1] H. Schatz *et al.*, Phys. Reports C 294, 167 (1998).
[2] J. Döring *et al.*, Phys. Rev. C 57, 1159 (1998).
[3] M. Wiescher *et al.* in: " ENAM 98", *AIP Conf. Proc.* 455, 819 (1998).

NUCLEAR STRUCTURE AND THE rp-PROCESS IN X-RAY BURSTS

Michael Wiescher

Department of Physics
University of Notre Dame
Notre Dame, IN 46556, USA *
wiescher.1@nd.edu

Abstract: Nuclear Astrophysics is concerned with the study of nuclear processes at stellar temperature and density conditions and its influence on nucleosynthesis and energy generation in stars and stellar explosions. Detailed understanding of nuclear processes and improved observational data allow a highly sophisticated analysis of the hydrodynamic conditions in static and dynamic stellar processes. In my talk I will concentrate on one particular nucleosynthesis scenario, the explosive hydrogen burning which is associated with accretion processes on the surface of a neutron star. These events have been observed over the last ten years as X-ray bursts. The nuclear processes driving the explosion are characterized by a sequence of fast proton capture reactions and subsequent β-decays near the proton drip linc up to $Z\approx50$. I will show how the nuclear structure characteristics along the process path determines the ignition and the timescale of the explosion and how it influences the luminosity curve of the burst and the subsequent abundance distribution in the crust of the neutron star. I summarize the present need for reaction and structure data for neutron deficient isotopes necessary for a better interpretation of the characteristics of X-ray bursts.

1 INTRODUCTION

The abundances of the elements and isotopes have been originated in a multitude of nucleosynthesis processes that occurred over the history of our uni-

*Funding provided by NSF grant PHY94-07194 and PHY94-02761.

verse. Primordial nucleosynthesis in the third minute of the Big Bang caused the formation of light isotopes below mass A=8, ^{2}H, ^{3}He, ^{4}He, ^{7}Li. After the formation of stars, further nucleosynthesis processes take place in static stellar burning processes as well as in explosive burning processes in the final stages of stellar life. The products of these processes are mixed with the interstellar material by solar wind induced mass losses from the surface of the star, or by ejection due to explosive events like novae or supernovae. The presently observed abundance distribution is based on a multitude of observational data, mainly from solar and stellar photospheric and meteoritic abundances. This curve reflects the nucleosynthesis history and the chemical evolution of our universe.

Historically the field of nuclear astrophysics has been concerned with the interpretation of the observed abundance distributions and with the formulation and description of the contributing nucleosynthesis processes. Each of these nucleosynthesis processes can be characterized by a specific signature in the resulting abundance distribution. The charged particle reactions in the hydrogen, helium and carbon burning stages of stellar evolution are mainly responsible for the production of elements below mass 56. High abundances typically correspond to nuclei with high binding energy, in particular to nuclei with closed shell structure. On the other side neutron induced reactions which take place in the s- and the r-process are responsible for the formation of heavier elements. Again the high abundance peaks observed in the isotopic abundance distribution are explained by the high stability of closed shell nuclei.

Improved observational instrumentation and spectroscopic detection methods allow detailed observation and analysis of the surface abundances of stars. This is of particular importance for the analysis of the abundances at the surface of deep-convective stars which are characterized by the nucleosynthesis products of the shell burning processes. An analysis of these abundances in connection with a detailed understanding of the correlated nucleosynthesis can reveal deeper knowledge of the associated convective processes in the star itself.

Also important is the analysis of the abundances in the material ejected in stellar explosions. The observed abundances reflect directly the nucleosynthesis conditions during the explosive event. A direct comparison between the predicted and observed abundances will yield information about the temperature, density and hydrodynamical conditions in the explosive event. Another critical signature is the light curve which is directly associated with the energy release and the radiative energy loss during the explosion. This structure of the light curve reveals important information about the energy production rate in the explosion as well as the time structure of the energy release.

In the following sections I will discuss at a specific example the nuclear reaction and nuclear structure implications for nucleosynthesis and energy generation. I will concentrate on the explosive hydrogen burning as it is expected to take place in X-ray bursts which have been frequently observed over the last twenty years and are charceterized by their particular time structure of the luminosity curve. I will concentrate specifically on the nuclear physics aspects for the ignition of the thermonuclear runaway, which trigger the explosion, the freeze out of the runaway at the peak of the explosion, and the nucleosynthesis during the cooling phase of the X-ray burst.

2 THE IGNITION OF X-RAY BURSTS

The current interpretation of type-I X-ray bursts [1] is based on accretion processes in a close binary system. Accretion takes place from the filled Roche-Lobe of an extended companion star onto the surface of a neutron star. Typical predictions for the accretion rate vary from 10^{-10} to 10^{-9} $M_\odot$ y^{-1}. The accreted matter is continuously compressed by the freshly accreted material until it reaches sufficiently high pressure and temperature conditions to trigger the ignition of thermonuclear reactions [2, 3, 4]. Since the accreted material is hydrogen rich, the initial thermonuclear burning is dominated by the pp-chains and the CNO cycles which slowly heat the layer of the accreted material. However, neither the pp-chains nor the hot CNO cycles can cause a thermonuclear runaway [3] because of their temperature independent energy generation rate. The thermonuclear runaway requires an instability, where the temperature sensitivity of the nuclear energy generation rate $\dot{\epsilon}$ exceeds the temperature sensitivity of the cooling rate $\dot{\epsilon}_{\rm cool}$:

$$\frac{{\rm d}\dot{\epsilon}}{{\rm d}T} > \frac{{\rm d}\dot{\epsilon}_{\rm cool}}{{\rm d}T}. \tag{1}$$

Therefore, the explosive nuclear burning occurs as soon as temperature sensitive reaction sequences like the triple α process or the break-out of the CNO cycles start to contribute significantly to the nuclear energy release. The conditions at which that happens affect the amount of material that can be accreted between bursts and therefore influences burst intervals and burst luminosities.

2.1 *The role of break-out reactions from the hot CNO cycles*

At the typical high densities of $\rho \approx 10^5 - 10^6$ g/cm^3 and temperatures of $T \geq 10^8$ K in the accreted layer, the ^{14}O(α,p)^{17}F, ^{15}O(α,γ)^{19}Ne, and ^{18}Ne(α,p)^{21}Na reactions may cause a break-out from the β-limited hot CNO cycles. The present estimates of the reaction rates [5] indicate that at rising temperatures the ^{15}O(α,γ)^{21}Na reaction will be the first break-out reaction to occur. The associated energy release is shown in figure 1 as a function of temperature for a typical density of 10^6 g/cm^3 assuming that the first waiting point is ^{24}Si [8]. For comparison figure 1 also shows the energy generation rate from the CNO cycles and from the triple α process. Figure 1 indicates that at a temperature of 0.3 GK the energy production rate becomes temperature dependent. This is the ignition temperature for the thermonuclear runaway at a density of 10^6 g/cm^3. At these conditions, the energy generation from the break-out process is comparable to the triple α reaction and both processes trigger the thermonuclear runaway. At temperatures above the ignition temperature the break-out process becomes the dominant energy source and determines the heating time scale. Again, this depends critically on the associated reaction rates. If the rate for the $^{15}O(\alpha,\gamma)^{19}Ne$ reaction would be significantly smaller than suggested [5, 6], the onset of its large energy contribution would be delayed compared to the ignition by the triple α reaction resulting in a slower raise of the burst temperature.

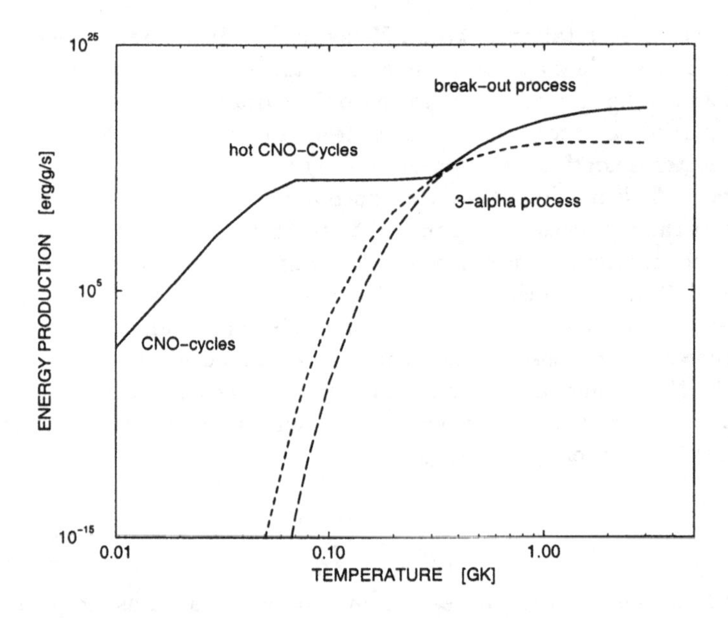

Figure 1 Energy generation in hot hydrogen/helium burning as a function of temperature calculated for solar hydrogen and helium abundances at a density of ρ=10^4 g/cm^3. The energy generation at lower temperatures is determined by the cold and hot CNO cycles. At higher temperatures the break-out flux from the CNO cycles dominates the energy generation. The energy release from the triple α reaction contributes significantly in the temperature range around T≈0.3 GK.

2.2 The ignition of the thermonuclear runaway

To demonstrate the influence of the break-out reactions, the on-set of the thermonuclear runaway at electron degenerate conditions has been calculated self-consistently in a one mass zone X-ray burst model [7, 8]. For this model hydrostatic equilibrium is maintained by keeping the total pressure P in the burning zone constant. The total pressure is determined by the ion pressure, the pressure of the degenerate electron gas, and the radiation pressure. Figure 2 shows the temperature and density profile as a function of time in the burning-zone during the thermonuclear runaway and the cooling phase of the burst. The pre-burst phase is dominated by the hot CNO cycles. The energy production and the nucleosynthesis are shown in figure 3. The two peaks in the energy production are caused by the conversion of the initial abundance of ^{12}C into ^{14}O (1b) – which depends sensitively on the $^{13}N(p,\gamma)^{14}O$ rate – and on the conversion of ^{16}O into ^{15}O by two subsequent proton capture reactions (1a) – which depends on the $^{17}F(p,\gamma)^{18}Ne$ rate. Because of the slow decay of the ^{14}O and ^{15}O isotopes the CNO process is halted and the energy production drops.

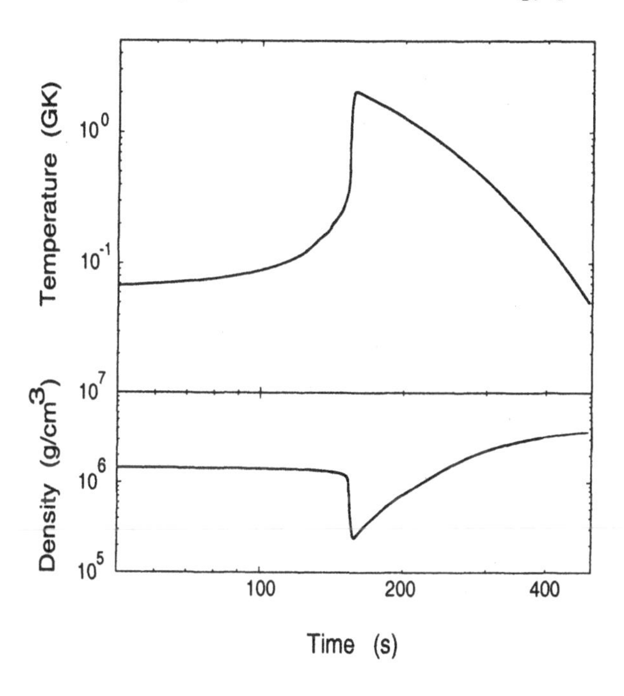

Figure 2 Temperature and and density as function of time for the thermonuclear runaway in a simplified X-ray-burst model. For details see text.

During this phase the main reaction flow is confined to the CNO cycles. The thermonuclear runaway itself is initiated by the triple α reaction and the $^{15}O(\alpha,\gamma)^{19}Ne$ reaction. Figure 3 indicates the onset of the triple α reaction by the additional increase of the ^{14}O abundance after depletion of ^{13}N and the onset of the $^{15}O(\alpha,\gamma)^{19}Ne$ reaction by the rapid depletion of ^{15}O. Shortly after ignition the temperatures are high enough for the $^{14}O(\alpha,p)^{17}F$ and the $^{18}Ne(\alpha,p)^{21}Na$ reactions to establish a continuous reaction flow from helium into the rp-process.

The onset of the thermonuclear runaway depends strongly on the rates of the break-out reactions $^{15}O(\alpha,\gamma)^{19}Ne$ and $^{18}Ne(\alpha,p)^{21}Na$. Both rates are based on structure information for the low energy alpha unbound states in the compound nucleus ^{19}Ne [6, 9] and ^{22}Mg [10]. Extensive studies have been performed recently at the Louvain la Neuve radioactive beam facility to measure the rate for $^{18}Ne(\alpha,p)^{21}Na$ [11, 12]. The data indicate good agreement with the predictions at high temperatures, while at low temperature conditions the experimental rate is substantially smaller. This might be explained by the lack of low energy data and additional experiments are clearly neccessary. To demonstrate the importance of the break-out reactions, the above described calculations have

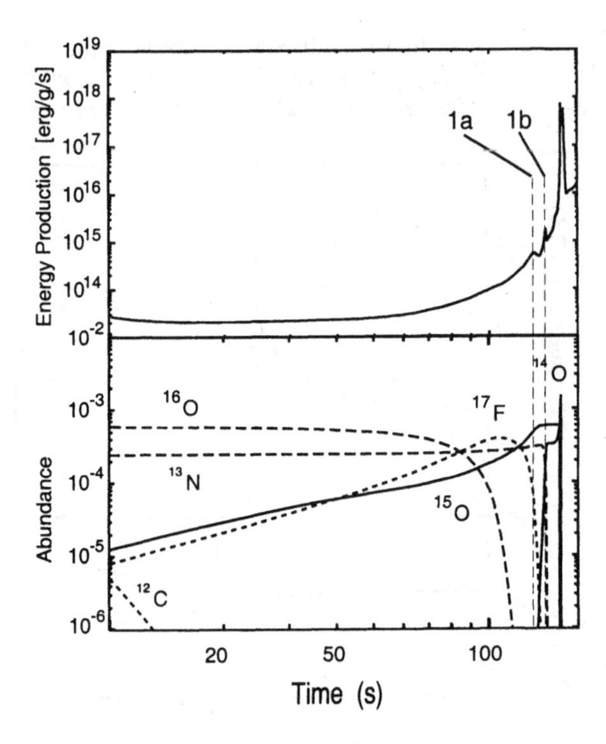

Figure 3 Energy production and nucleosynthesis in the early phase of an X-ray burst. Peak 1a and 1b in the upper part of the figure correspond to the burning of ^{17}F and ^{13}N which is shown in the lower part of the figure.

been repeated with substantially reduced reaction rates for both break-out processes. Figure 4 shows the consequences for the nucleosynthesis and the energy production. No break-out occurs and the bulk of the material remains in the CNO mass range, the remaining increase in luminosity is explained by the feeding of the CNO region via the triple alpha process and the associated increase in energy generation from the hot CNO cycles.

Sufficiently large reaction rates for $^{15}\mathrm{O}(\alpha,\gamma)^{19}\mathrm{Ne}$ and $^{18}\mathrm{Ne}(\alpha,\mathrm{p})^{21}\mathrm{Na}$ are required to trigger the thermonuclear runaway via the rp-process.

3 THE THERMONUCLEAR RUNAWAY IN THE rp-PROCESS

The rp-process has been first suggested by Wallace and Woosley [13] as the dominant nucleosynthesis process in explosive hydrogen burning at higher temperature and density conditions. The process is characterized by a sequence of fast proton capture reactions and a subsequent β-decay close to the drip line. In the following I want to discuss the waiting point concept for the rp-process which explaines the time structure for the energy production and defines the reaction path at high temperature conditions.

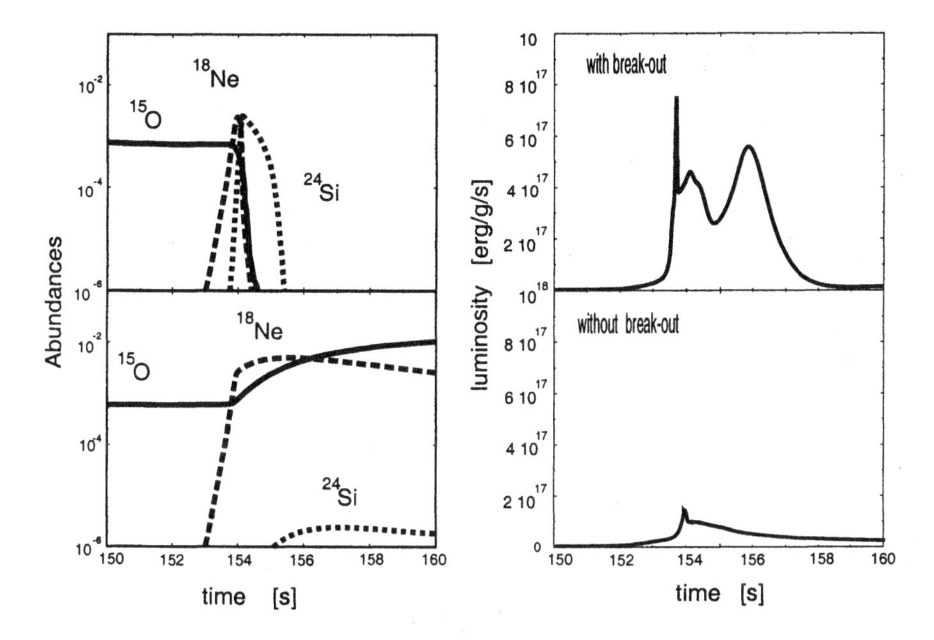

Figure 4 The left hand part of the figure compares the nucleosynthesis based on the estimated break-out reaction rates with the nucleosynthesis based on largely reduced rates. The right hand side shows the corresponding energy production.

3.1 Waiting points in the rp-process

At high temperature conditions $T \geq 8 \cdot 10^8$K the reaction path and the reaction flow for the rp-process can be characterized by the waiting point concept. Waiting point nuclei are typically even-even nuclei which are characterized by low or negative Q-values for subsequent proton capture due to pairing energy contributions. If $Q \leq 24kT$, the proton capture is strongly surpressed because of the inverse photodisintegration of the compound nucleus, if $Q \leq 0$ the compound nucleus is proton unbound and decays immediately back to the target nucleus. In these cases the reaction path is determined by a (p,γ)-(γ,p)- and/or by the (p,γ)-(p)-equilibrium, respectively.

The main impedance for the reaction flow are the lifetimes of the waiting point nuclei along the reaction path. In the range of $sdf_{7/2}$-shell nuclei with $10 \leq Z \leq 28$ the proton drip line is characterized the proton unbound T_z=-5/2 nuclei. The T_z=-3/2 nuclei are typically particle stable. In this mass range proton capture on T=2 nuclei is therefore largely inhibited and the time scale for the process is therefore mainly determined by the lifetimes of the T=2 isotopes near the drip line.

For higher Z nuclei, $28 \leq Z$ in the range of the $p_{3/2}$-shell T_z=-3/2 nuclei are proton unbound and the drip line is determined by the T=1 even-even nuclei. The life time of these isotopes is relatively short and the process proceeds fast towards higher Z-nuclei. The main waiting points are ^{55}Ni and ^{56}Ni since the reaction flow is delayed by the ^{56}Ni(p,γ)–^{57}Cu(γ,p) and the ^{55}Ni(p,γ)–^{56}Cu(γ,p) equilibrium which causes considerable enrichment in ^{55}Ni and in particular in ^{56}Ni. In the case of ^{56}Ni the effective lifetime depends strongly on the proton capture rates on ^{56}Ni and on ^{57}Cu while the electron capture remains rather slow [14]. Figure 5 shows the effective life time for ^{56}Ni as a function of temperature for a typical density of ρ=10^6 g/cm^3. While initially the lifetime drops with increasing temperature due to the proton capture on ^{56}Ni, for temperatures above T$\approx$1.5·10^9 K the effective life time of ^{56}Ni increases again due to the photodisintegration of ^{57}Cu. This increase is reduced by the effects of the second proton capture process on ^{57}Cu.

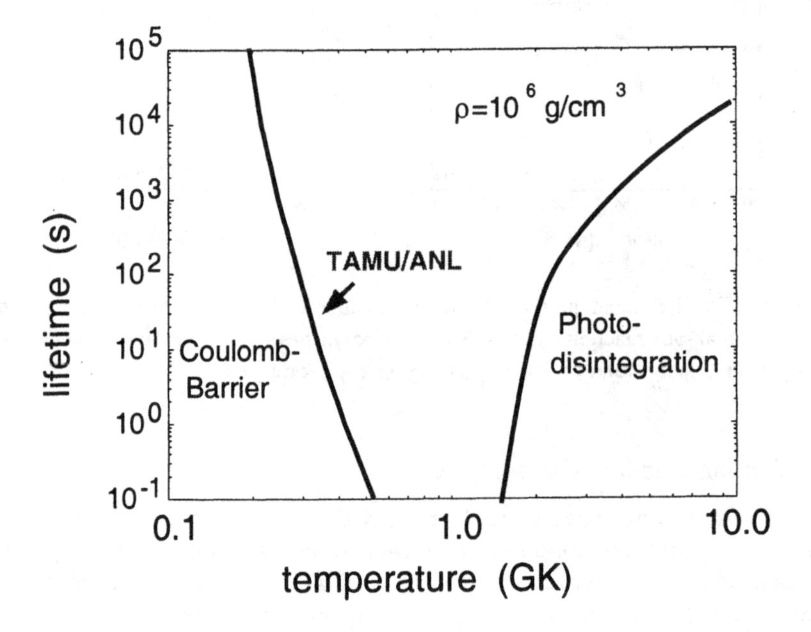

Figure 5 The effective lifetime of ^{56}Ni as a function of temperature calculated for a density of ρ=10^6 g/cm^3.

Therefore, only within a relatively small temperature window between $\approx$ 0.3·10^9 and 2.·10^9K and for relatively high densities $\rho \geq 10^5$ g/cm^3 fast ($\leq$100 s) rp-process nucleosynthesis beyond ^{56}Ni is possible. Therefore, the effective lifetime depends sensitively on the reaction rates involved. The rate for ^{56}Ni(p,γ)^{57}Cu is based on the measurements of excitation energies of proton unbound states in ^{57}Cu [15] and of the single particle structure of these states [16], the rate of ^{57}Cu(p,γ)^{58}Zn is entirely based on Hauser Feshbach assumptions. In both cases more data are clearly desirable.

3.2 The time structure of the burst

The following section will focus on the rp-process characteristics calculated selfconstistently for temperature conditions during the thermonuclear runaway in an one mass zone X-ray burst model. Figure 6 shows the energy production and the associated nucleosynthesis over the duration of the X-ray burst.

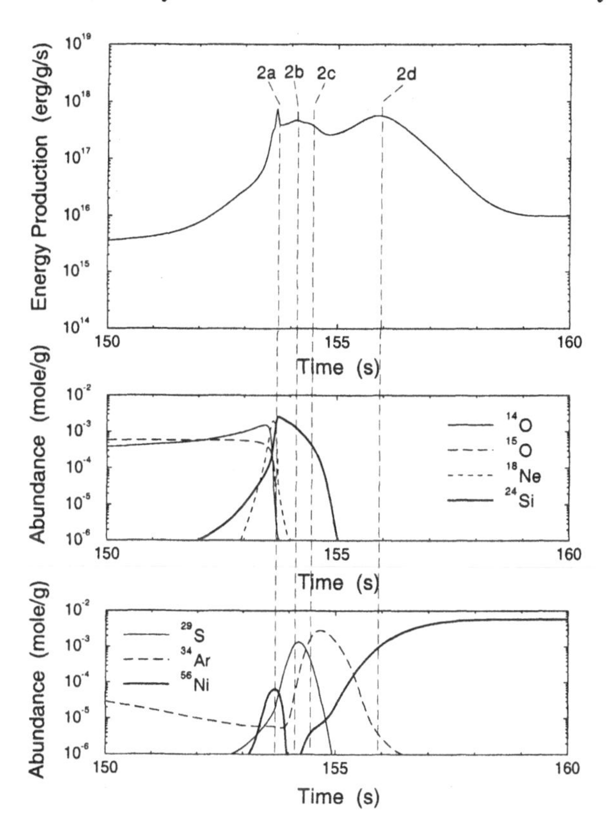

Figure 6 The energy production and the nucleosynthesis during the thermonuclear runaway of the X-ray burst.

The energy production curve exhibits a pronounced structure which is directly correlated with the above discussed waiting point concept of the rp-process [8, 14]. As discussed in the previous section the burst is initiated by the conversion of the hot CNO waiting point isotopes ^{14}O, ^{15}O, and ^{18}Ne by the αp-process to ^{24}Si. The T=2 nucleus ^{24}Si is a waiting point since ^{25}P is proton unbound. After the decay of ^{24}Si the process is reignited at higher temperatures by proton and α capture reactions leading to the production of the next waiting point isotopes ^{29}S and ^{34}Ar. The rapid increase in temperature allows subsequent α-capture to bridge these waiting points and leads to a rapid conversion of the waiting point isotopes to ^{56}Ni. At this time a continuous reaction flow converts the accreted He-abundance via the αp-process and the rp-process to ^{56}Ni. The reaction flow, integrated over the duration of the burst is shown in figure 7.

Since most of the reaction path is characterized by (α,p) reactions considerably more helium is burned than hydrogen. At this point, peak temperatures

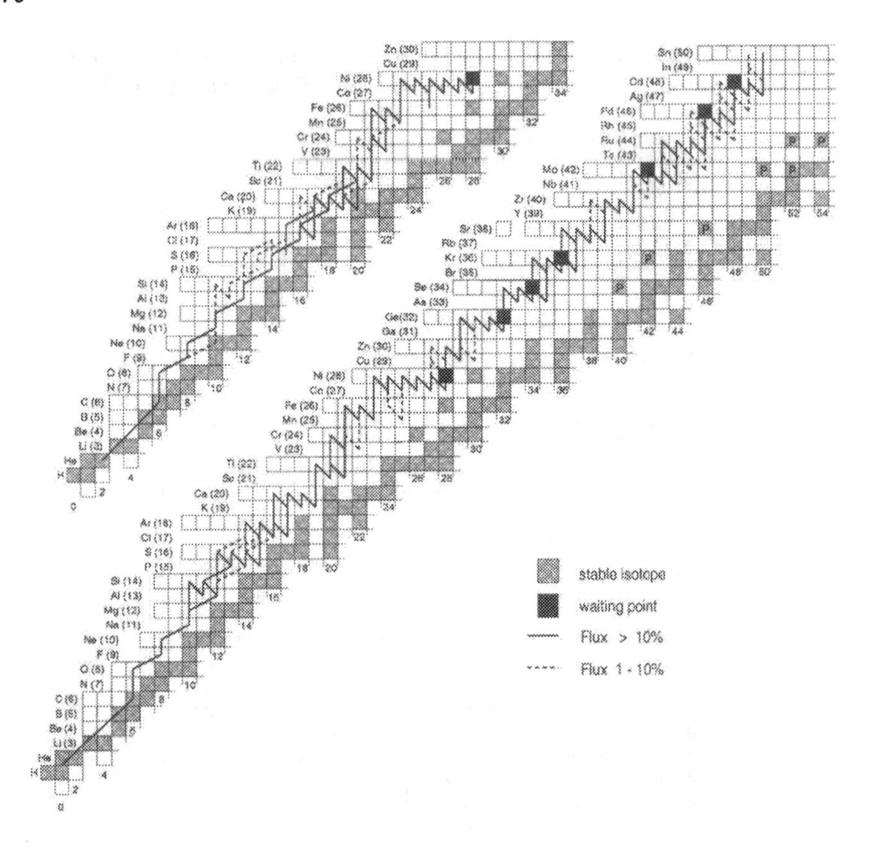

Figure 7 The αp- and rp-process reaction path during the ignition and the thermonuclear runaway phase of the X-ray burst. The reaction path up to ^{38}Ca is characterized by a sequence of α capture reactions. Also shown is the reaction flux in the cooling phase of the X-ray burst. Here the overall flux is dominated by the rp-process, only for very light Z nuclei (Z$\leq$14) is the Coulomb barrier low enough for α capture reactions to compete with the proton capture and the β-decay.

of $T \approx 2.0 \cdot 10^9$ K have been reached and further processing is halted by the ^{56}Ni(p,γ)-(γ,p) equilibrium [14]. The energy production drops rapidly while most of the initial heavy isotope abundances as well as a large fraction of the initial helium remains stored in the waiting point nucleus ^{56}Ni (see figure 6). The drop in energy production causes a slow down in the temperature increase just before the peak temperature is reached.

At these peak temperature conditions ^{56}Ni has a lifetime of approximately 100 s because the proton capture is largely balanced by the inverse photodisintegration of ^{57}Cu (see figure 5). However, with the decrease in temperature and the parallel increase in density the ^{56}Ni(p,γ)-(γ,p) falls out of equilibrium and the effective lifetime of ^{56}Ni decreases down to a fraction of a second versus proton capture at temperatures $T \approx 1.5 \cdot 10^9$ - $1 \cdot 10^9$ K as shown in figure 5. For temperatures below $0.6 \cdot 10^9$ K the lifetime increases again rapidly because of the temperature dependence in the reaction rate for $^{56}Ni(p,\gamma)^{57}Cu$. The small lifetime of ^{56}Ni within this temperature window allows the ignition of the last phase of the energy burst by proton capture on ^{56}Ni. This phase is characterized by nucleosynthesis via the rp-process beyond ^{56}Ni which operates during the cooling period until the Coulomb barriers prevent any further proton capture processes.

4 THE ENDPOINT OF THE rp-PROCESS

Figure 7 also shows the reaction flow integrated over the duration of the cooling phase. Notice that due to the lower temperature the αp-process is only dominant below sulfur, at higher masses the reaction path is characterized by the rp-process pattern leading up to ^{100}Sn. Figure 8 shows the details of the burst structure during this cooling phase of the burst. Several peaks in the energy production are due to the depletion of ^{56}Ni and the further processing towards the N=Z waiting point nuclei ^{64}Ge and the subsequent nucleosynthesis towards ^{68}Se and ^{72}Kr. These waiting point nuclei originally have been suggested as possible termination points for the rp-process because of their rather long β-decay lifetime [17, 18, 19]. Yet, it has been shown [8, 14] that the effective lifetime of these nuclei is actually shortened by two sequential proton capture reaction which bridge the barely bound isotope ^{65}As [20] as well as the unbound isotopes ^{69}Br [18, 19] and ^{73}Rb [19, 21]. High density and temperature conditions in the accreted layer during this phase allow two-proton capture processes to become prevalent. It should be pointed out that the two-proton capture rates used in this study depend heavily on the reliability of predicted masses for the relevant isotopes [14, 22]. In the final phase ^{68}Se and ^{72}Kr are eventually converted to heavier isotopes with masses $A \geq 72$. The most abundant emerging isotopes are the N=Z nuclei ^{80}Zr, ^{92}Cd, and ^{96}Ru as indicated in figure 8.

4.1 *Beta decay of ^{80}Zr and the nucleosynthesis in the mass 80 range*

Very limited experimental information is available about nuclear reaction and nuclear decay processes for nuclei along the rp-process path in the mass 76 to 100 range. The calculations are entirely based on model predictions for masses, lifetimes, and cross sections [14]. The nucleosynthesis predictions may therefore carry considerable uncertainties. This is particularly the case for the mass A=80 region where the nuclear structure and the correlated capture and decay processes are characterized by large deformations. The specific example I want to concentrate on are the uncertainties in the β-decay of ^{80}Zr. The observed high production of ^{80}Zr (and subsequently of ^{84}Mo) in the rp-process model calculations is mainly due to the long lifetime of $T_{1/2}$=6.85 s which has been predicted by QRPA model calculations for a prolate ground state deformation of ϵ_2=0.383 [14]. Several attempts to verify the lifetime predictions

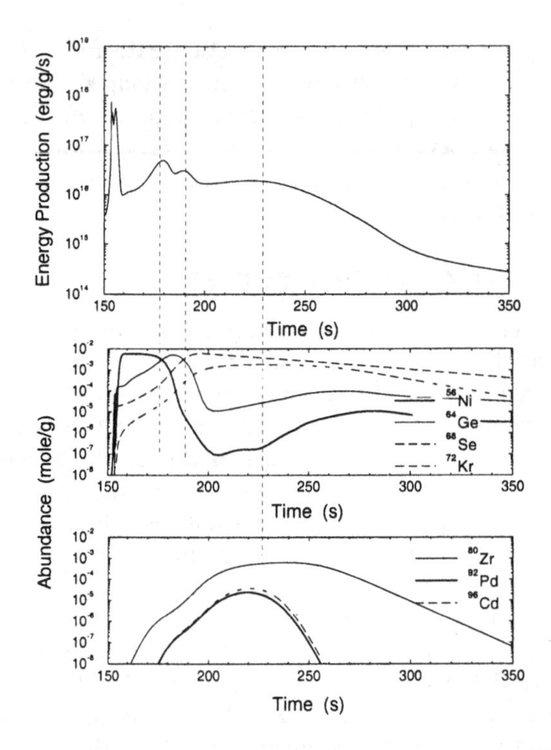

Figure 8 The figures shows the energy production and nucleosynthesis associated with the destruction of the waiting point nuclei during the cooling phase of the X-ray burst.

experimentally failed [1], but led to the identification of an isomeric $J^{\pi}=1^{-}$ state in ^{80}Y at 228 keV excitation energy. This isomeric state β/EC decays with a 19 % branching to low excited states in ^{80}Sr with a half life of 4.7 s [23]. The half life of the ^{80}Y ground state was determined to be $T_{1/2}$=30.1 s. In the aforementioned calculations the β-decay lifetimes of ^{80}Zr and ^{80}Y were based on the ground state decay only. However, β-decay processes in high temperature scenarios also involve β-decay of thermally excited states [14]. Their lifetime can be significantly different from the lifetime of the ground state due to the higher Q_{β}-values. For the highly deformed nucleus ^{80}Zr, the excitation energy of the first 2^{+} state is rather low, $E_x \approx$290 keV [24], therefore the decay through the thermally excited state is initiated at temperatures above $\approx$ 1 GK. The allowed β-decay will populate 1^{+} and/or 2^{+} states which are predicted at an excitation range $E_x \approx$300-600 keV [23], possibly to the 2^{+} (6 μs) isomeric state at 312 keV which decays by 84 keV γ-emission to the 1^{-} isomeric state at 228 keV [23, 25]. The decay pattern of ^{80}Zr is demonstrated in figure 9.

[1] A recent measurement at the Oak Ridge RMS separator has been successful, the data indicate a halflife of $\approx$4 s which reduces the predicted ^{80}Zr abundances by 30 % (J.J. Ressler, private communication and to be published).

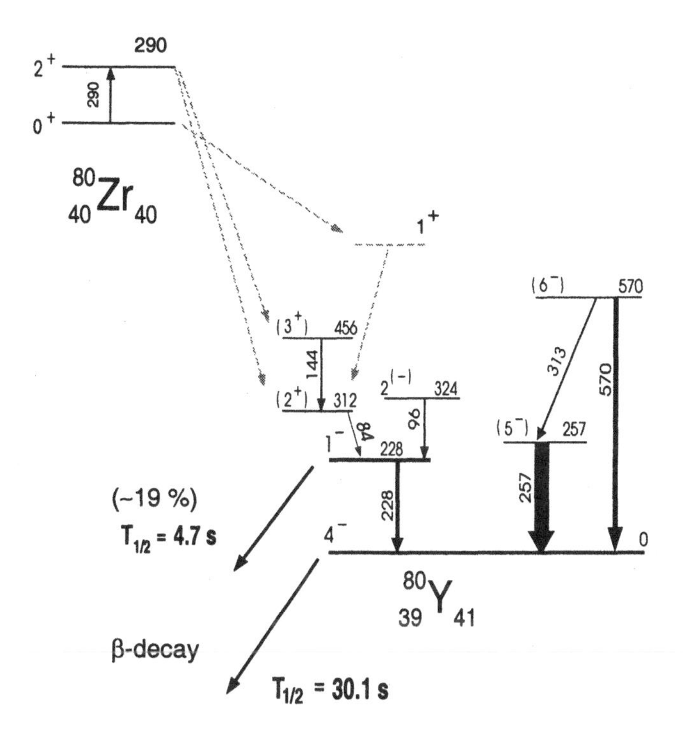

Figure 9 The level scheme of ^{80}Y and the proposed decay routes of the ground state and the first excited state in ^{80}Zr at high temperature conditions.

A QRPA calculation for such a decay suggests an effective half life of ^{80}Zr of $\approx$3 s for the temperature conditions in the cooling phase of the X-ray burst.

4.2 *The A=80 abundance in X-ray bursts*

In the following, I will discuss the implications of the above described decay patterns for the nucleosynthesis of ^{80}Zr and for the final abundance of ^{80}Kr after the freeze ot of the burst.

Figure 10 shows the development of the abundances of the N=Z isotopes ^{56}Ni, ^{64}Ge, ^{68}Se, and ^{80}Zr calculated on the assumption of a pure ground state decay of ^{80}Zr. In comparison the figure also shows the abundances calculated with the addition of a fast decay sequence through thermal excitation of ^{80}Zr with the subsequent β^+ decay to the isomeric states in ^{80}Y.

The figure clearly shows a factor four reduction of the ^{80}Zr abundances while the abundances of the other isotopes remain unaffected. For the formation of ^{80}Kr in the freeze out phase the decay of all A=80 neutron deficient isotopes have to be considered. Assuming only ground state decay, significant enrichment can be observed for ^{80}Y along the rp-process path due to its long half-life of 30.1 s. The feeding of the isomeric states in ^{80}Y by the decay of ^{80}Zr reduces the effective half life of ^{80}Y significantly and subsequently reduces its abundance

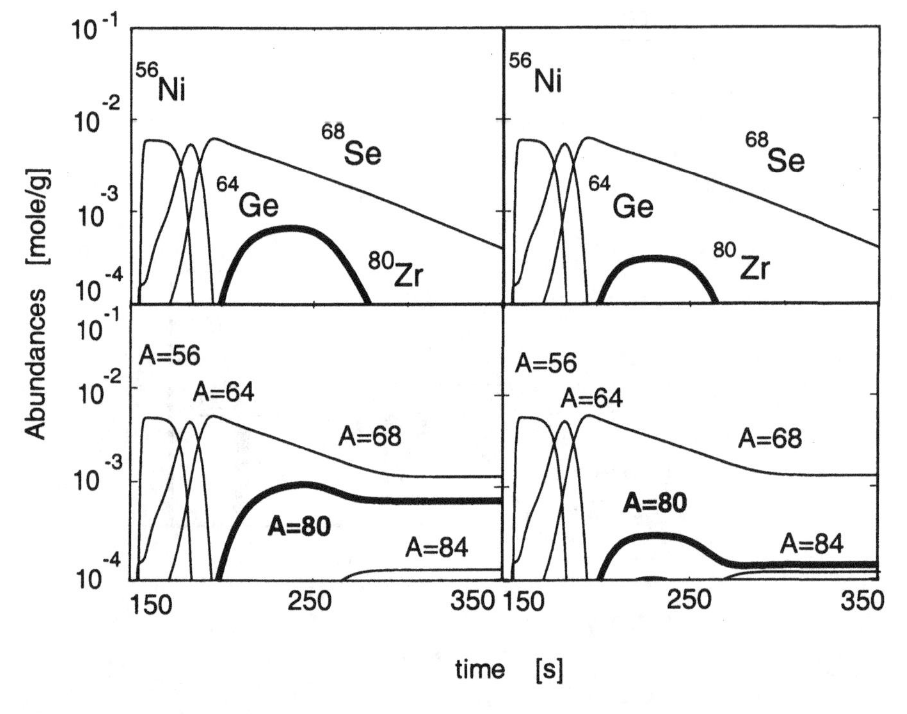

Figure 10 The two top figures show the development of the abundances for ^{56}Ni, ^{64}Ge, ^{68}Se, and ^{80}Zr as a function of time. The figure on the top left shows abundances based on pure ground state decay of ^{80}Zr while the one on the right side includes the decay via the excited states. The lower part of the figure shows the total abundances for mass A=56, 64, 68, 80, and 84 nuclei calculated at the same conditions.

during the cooling phase. This is demonstrated in figure 10 comparing the development of the total A=80 abundance for the two discussed cases. Including thermal excitation of ^{80}Zr with subsequent decay through the isomeric states of ^{80}Y reduces the A=80 abundance by about one order of magnitude compared to ground state decay only. Additional experimental information about masses of the involved isotopes, their β-decay patterns, and the β-feeding and decay of isomeric states in the mass A=80 range are necessary to perform a more reliable and consistent study of the rp-nucleosynthesis in this mass range.

5 CONCLUSION

It has been shown within the framework of a simplified hydrostatic X-ray burst model that the rp-process is sufficiently fast to process light mass material from the CNO range up to the mass A=100 region. The time structure of the nucleosynthesis process and the associated energy production depends on the effective lifetimes of the waiting point nuclei along the rp-process path. These are in turn determined by the temperature and density conditions in the burning zone. The main delay for the ignition of the thermonuclear runaway are the break-out reactions of the hot CNO cycles. The main impedance for the rp-process nucleosynthesis is the ^{56}Ni(p,γ)-(γ,p) equilibrium which can only be bridged in the cooling phase of the burst. For the subsequent rp-process nucleosynthesis up to the mass A=100 range the dominant impedances are the long effective lifetimes of ^{64}Ge, ^{68}Se, and ^{72}Kr. These lifetimes are determined by the temperature and density dependent two-proton capture reaction rates and by the β-decay of their ground states and possibly their thermally excited states. The last waiting point is the largely deformed N=Z nucleus ^{80}Zr which cannot be depleted by two-proton capture since ^{81}Nb is highly proton unbound [14]. It was shown in this study that the effective lifetime of ^{80}Zr and therefore its abundance can be sufficiently reduced by thermal population of its first excited state and its subsequent β-decay feeding the short-lived isomeric 1^- level in ^{80}Y [23]. To further investigate possible X-ray burst nucleosynthesis contributions to the galactic abundances, more detailed theoretical studies are clearly necessary in order to evaluate the various mass loss effects out of the gravitational potential of the neutron star.

Acknowledgments

I want to thank A. Aprahamian, J. Döring, and J. Görres from the University of Notre Dame, L. Bildsten from the University of California at Berkeley, W. Bradfield-Smith and A. Shotter from the University of Edinburgh, W. Galster and P. Leleux from the Universite de Louvain la Neuve, H. Schatz from GSI Darmstadt, F.K Thielemann from the Universität Basel, and W. Walters from the University of Maryland for their collaboration in the many various projects which led to the here presented summary of X-ray burst nucleosynthesis.

References

[1] W. Lewin, J. van Paradijs, and R. Taam. *Space Sci.Rev.*, 62:233, 1993

[2] A. Ayasli, and P. Joos. *Astrophys.J.*, 256:637, 1982

[3] R. Taam. *Ann.Rev.Nucl.Sci.*, 35:1, 1985

[4] R. Taam, S.E. Woosley, T. Weaver, and D. Lamb. *Astrophys.J.*, 413:324, 1993

[5] K.I. Hahn, *et al. Phys.Rev.* C, 54:1999, 1996

[6] Z.Q. Mao, H.T. Fortune, and A.G. Lacaze. *Phys.Rev.* C, 53:1197, 1996

[7] L. Bildsten *The many Faces of Neutron Stars* ed A Alpar, L Buccheri and J Van Paradijs (Dordrecht, Kluwer) in press

[8] M. Wiescher, H. Schatz, and A. Champagne. *Phil.Trans.Roy.Soc.*, 356:1, 1998

[9] K. Langanke, M. Wiescher, W.A. Fowler, and J. Görres. *Astrophys.J*, 301:629, 1986

[10] J. Görres, M. Wiescher, and F.K. Thielemann. *Phys.Rev.* C, 51:392, 1995

[11] W. Bradfield-Smith, *et al. Proc. of the Int. Workshop XXVI on Gross Properties of Nuclei and Nuclear Excitations, Hirschegg, Austria*, ed M Buballa *et al* (Gesellschaft für Schwerionenforschung, Darmstadt) 364, 1998

[12] A.N. Ostrowski, *et al. J.Phys.G: Nucl.Part.Phys.*, 24:1553, 1998

[13] R. Wallace, and S.E. Woosley. *Astrophys.J.Suppl.*, 45:389, 1981

[14] H. Schatz, *et al. Phys.Rep.*, 294:168, 1997

[15] X.G. Zhou, *et al. Phys.Rev.* C, 53:982, 1996

[16] E. Rehm, *et al. Phys.Rev.Lett*, 80:676, 1998

[17] R.K. Wallace, and S.E. Woosley. *Proceedings of Accelerated Radioactive Beam Workshop, Parksville, Canada*, eds. L. Buchmann, J. D'Auria, (TRI-UMF, Vancouver), 1985

[18] B. Blank, *et al. Phys.Rev.Lett*, 74:4611, 1995

[19] R. Pfaff, *et al. Phys.Rev.*C, 53:1753, 1996

[20] J. Winger, *et al. Phys.Rev.*, 48:3097, 1993

[21] M.F. Mohar, *et al. Phys.Rev.Lett*, 66:1571, 1991

[22] P. Möller, J.R. Nix, D. Myers, and W.J. Swiatecki. *At.Data. Nucl.Data. Tab.*, 59:185, 1995

[23] J. Döring, *et al. Phys.Rev.* C, 57:1159, 1998

[24] C.J. Lister, *et al. Phys.Rev.Lett.*, 59:1270, 1987

[25] P.H. Regan, *et al. Acta Phys.Pol. B*, 28:431, 1997

MAGNETIC DIPOLE BANDS IN ^{190}Hg

A.N. Wilson

Department of Physics
University of York
Heslington
York YO10 5DD

Abstract: A series of experiments aimed at studying high-spin states in ^{190}Hg have been carried out over the past few years using the large γ-ray arrays Eurogam II, Gammasphere and Euroball III. The data obtained during these experiments has allowed a considerable expansion of the level scheme for the states in the first minimum. In particular, three new, high-spin structures have been observed at excitation energies around 5.5 MeV, with band-head spins of $\approx 17\hbar$. One of these is an irregular sequence of competing M1 and E2 transitions, similar to the magnetic dipole bands already known in heavier Hg isotopes. In the others, which have more regular energy-spacings, no ΔI= 2 transitions have been observed. Two of these structures (the irregular cascade and the stronger of the two dipole-only cascades) have been linked to the low-lying states and where possible polarisation and DCO measurements have established the spins and parities of the levels, confirming in the case of the regular structure that the in-band γ rays are indeed M1 transitions. Lower limits can be set on the B(M1)/B(E2) ratios and a comparison made with the predictions of various models. In order to explain the lack of visible E2 transitions in the high statistics data sets involved, it is necessary to postulate both a very small deformation (not predicted by Total Routhian Surface calculations) and a very large B(M1) strength, perhaps produced by a configuration similar to that responsible for the shears bands in the Pbs. Such structures are unexpected in the light Hg isotopes and present an excellent test for the predictive power of the currently favored models.

The Nucleus: New Physics for the New Millennium
Edited by Smit et al., Kluwer Academic / Plenum Publishers, New York, 2000.

1 INTRODUCTION

The study of high-spin structures consisting of very strong M1 transitions has been the subject of much experimental and theoretical effort over recent years. The existence of "shears" bands in the Pb isotopes (and more recently in nuclei around $A \approx 110$) has given rise to a picture of an alternative rotational mode in nuclei with a very small deformation. This model has been applied with great success to the structures in the Pb [?, ?] and Sn [?] isotopes. In essence, the shears picture describes the generation of angular momentum through the gradual alignment of a pair of initially perpendicular angular momentum vectors, one associated with a high–j, low–Ω particle excitation and the other with a high–j, high-Ω hole excitation. Rotational–like cascades of M1 transitions are created as the angle between the vectors closes step-by-step. Such a scenario requires a low deformation; if the nucleus were significantly deformed, the "classical" rotation of the core would induce alignment of the orbitals through the Coriolis effect and the shears mechanism would be washed out.

Cascades of competing M1 and E2 transitions have also been seen in several Hg isotopes; however these structures have not been thought to arise from the same phenomenon. In fact, their properties are adequately accounted for by "standard" CSM calculations and a more normal coupling of the particle angular momenta. The B(M1)/B(E2)s measured in these bands are $\approx 1-10$ $(\mu_N/eb)^2$; this is in contrast with the larger values associated with the dipole bands observed in the Pb isotopes, which are generally around 20–50 $(\mu_N/eb)^2$.

Four new dipole bands, three associated with ^{190}Hg and one with ^{188}Hg, have been observed in data from recent high-statistics experiments using the Eurogam II and Gammasphere spectrometers. Two of the bands in ^{190}Hg and the one in ^{188}Hg have some very unusual properties, including surprisingly regular energy spacings and extremely large B(M1)/B(E2) ratios.

2 THREE DIPOLE BANDS IN ^{190}Hg

Three new structures consisting of cascades of M1 transitions have been observed in data obtained using the Eurogam II, Gammasphere and Euroball III arrays. Two of the bands (DB1 and DB2) are shown in figures ??(a) and (b); despite the very high statistics in the spectra, there is no evidence for the presence of E2 cross-over transitions. This places a lower limit on the B(M1)/B(E2) values of around 50 $(\mu_N/eb)^2$. The third structure (DB3) is highly irregular and consists of both M1 and E2 transitions. It is similar to the bands already known in the heavier Hg isotopes such as ^{192}Hg [?, ?], although it extends to lower spin and excitation energy. Figure ??(c) shows the partial level scheme obtained for ^{190}Hg including DB1 and DB2; also shown are the spins and excitation energies of three levels in DB3. These three are the lowest level observed, a level which appears to mark a change in structure and the highest level observed. These have been included to illustrate the similarity in I, E_x between all three bands.

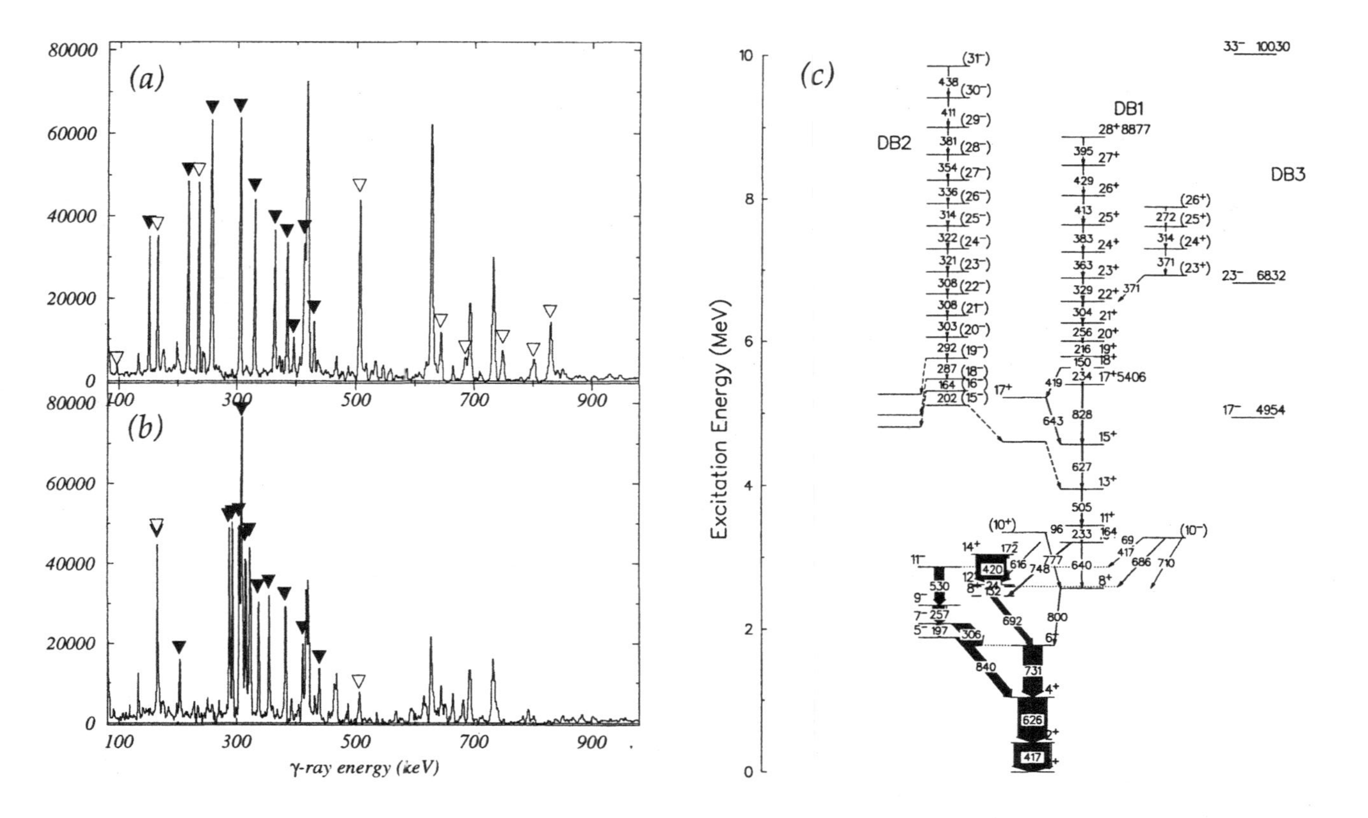

Figure 1 Spectra showing (a) DB1 and (b) DB2 in ^{190}Hg. The spectra are double-gated on the cleanest transitions in the bands; in-band transitions are marked with filled triangles. Transitions identified as the decay from the bands to the known states are marked with open triangles. Part (c) shows the partial level scheme deduced from the coincidence data.

DCO ratios have been extracted where possible and the dipole nature of the in-band transitions confirmed. In addition, the presence of Clover detectors in the Eurogam II array has allowed a measurement of the polarisation of the γ rays in DB1: the results show unambiguously that the transitions are magnetic, rather than electric, in character.

3 NULL AND NEW RESULTS IN OTHER Hg ISOTOPES

Following the observation of DB1 and DB2 in ^{190}Hg, attempts have been made to find similar structures in other Hg isotopes. A search of the low energy spectrum (100 — 600 keV) of ^{192}Hg revealed nothing similar, despite even greater statistics than those obtained while studying ^{190}Hg. However, when data on ^{188}Hg obtained by Fallon et al with the Gammasphere array was searched, a new structure which appears very much like DB1 was found. Although there remain some uncertainties as to the ordering of the γ rays in this structure, a large part of the decay to the previously known levels in ^{188}Hg has been established, allowing the relatively confident assertion that the bandhead spin and excitation energy are $I \approx 18\hbar$ and $E_x \approx 5.4$ MeV. This band consists of 9 transitions, with energies 120, 251, 249, 309, 317, 350, 365, 415 and 490 keV.

4 DISCUSSION

There are some problems in explaining the origin of the regular bands described above. If these structures were shears bands, based on the same configurations as those which support the bands in the Pb isotopes, then the promotion of two protons into the $i_{13/2}$ orbitals is required. An inspection of the Nilsson levels reveals that such excitations are possible, but only at significant degrees of (oblate) deformation. This creates a tension between the two requirements of the shears mechanism: just as the appropriate particle–hole excitations are needed in order to achieve the perpendicular coupling of the angular momentum vectors, so also a very small deformation ($\beta_2 \leq 0.1$) is required to ensure that collective rotation is not a more favorable means of generating angular momentum. In addition to this, if these new bands were simply analogous to the shears bands observed in the Pb isotopes, one would expect to see them across the Hg chain and not limited to the neutron deficient isotopes, as now appears to be the case.

If one looks to the standard CSM calculations, there are several available configurations which could give rise to M1 bands. These involve a pair of high-j $i_{13/2}$ quasineutrons coupled to various possible high-K proton excitations, involving either $h_{9/2}$, $h_{11/2}$, $i_{13/2}$ orbitals or some combination of the three. These configurations have been invoked in the interpretion of the dipole bands observed in the heavier Hg isotopes [?, ?]. Using the Donau and Frauendorf model, one expects such configurations to give rise to B(M1) values ranging from 1 μ_N^2 for the $\nu(i_{13/2})^2_{j=12}\pi(h_{11/2})^2_{j=10}\pi(h_{9/2})^2_{K=8}$ configuration to $\approx 8-9$ μ_N^2 for the $\nu(i_{13/2})^2_{j=12}\pi(h_{9/2})^2_{K=11}$ configuration. Although this last configuration does allow a large B(M1), if one makes the (reasonable) assumption of a deformation of approximately $\beta_2 = -0.15$, it would only result in a B(M1)/B(E2) of around 20 $(\mu_N/eb)^2$. In order to achieve the large values associated with DB1 and DB2 (the lower limit placed by the lack of observable E2 transitions is B(M1)/B(E2)$\geq 50 - 60$ $(\mu_N/eb)^2$), the deformation would have to decrease by a factor of at least two, in which case the excitation to the appropriate orbitals becomes more and more energetically expensive.

To conclude: three surprisingly regular structures, composed only of magnetic dipole transitions, have been observed in ^{188,190}Hg. It remains an open question as to what type of configuration these bands are built upon. Neither the magnetic rotation scenario nor the standard Donau and Frauendorf formalism seem to provide appropriate frameworks within which to explain their behaviour. The observation of such bands poses a challenge to the accepted models.

References

[1] R.M.Clarke et al. Nucl. Phys. A 562, 121 (1993)

[2] G.Baldsiefen et al. Nucl. phys. A 574, 521 (1994)

[3] D.G.Jenkins et al. Phys. Lett. B 428, 23 (1998)

[4] Y.Le Coz et al. Z Phys A 348, 87 (1994)

[5] P.Willsau et al. Z Phys A 355, 129 (1996)

III Closing Speaker

SUMMARISING "THE NUCLEUS"

G.D. Dracoulis

Department of Nuclear Physics
R.S.Phys.S.E, Australian National University
Canberra, A.C.T. 0200, Australia
George.Dracoulis@anu.edu.au

Abstract: Reflections on conference summaries and nuclear physics.

Finish me off

I knew this would be an interesting, satisfying and very pleasant week, I just wasn't sure what I would do at the end. I am still not sure, but since the organisers left it to me to choose the format, I half-decided some time ago that I would not summarise what was said, to avoid the perennial problem of awarding geurnseys to special friends, or enemies, or important laboratories etc. I was confident that given the quality of the participants and the broad range of physics covered in the conference programme, the interactions I was likely to have during the week, repulsive and attractive, would surely trigger some ideas.

As a precaution, I consulted my handbook on public speaking [1] which reminded me of one of Mark Twain's many personal aphorisms:

> *It usually takes me more than three weeks to prepare a good impromptu speech.*

So forewarned, I did some homework before arriving. And I also took the second precaution of attending all sessions. I can report that there were, appropriately, 60 presentations, excluding this one.

Memory plays tricks

The first piece of homework was in the form of a Gedanken experiment, where I tried to recall the Summary talks of others. What was it that stuck ? After a lot of deep thought, all that sprang to mind was a few snippets from two talks, out of the several hundred I had conscientiously (and consciously) attended.

One was part of a masterly presentation by Walter Greiner at the 1978 International Conference on Nuclear Interactions, held in Canberra (about 10,000 miles due east) . I was one of the locals elevated to give an invited paper and the Summarizer, who I am sure meant well by being courteous to local hosts, duly showed one of my overheads of a nuclear level scheme, but he showed it as in Figure 1, back-to-front and upside-down ! When audience murmuring caught his attention, he said [1], "oh, it doesn't matter, they all look much the same." In doing so, he lost the chance of making a friend and missed the point, a point I want to make again later. That is,

Nuclei are not all the same !

They are, in fact all different and each one has properties that are its alone.

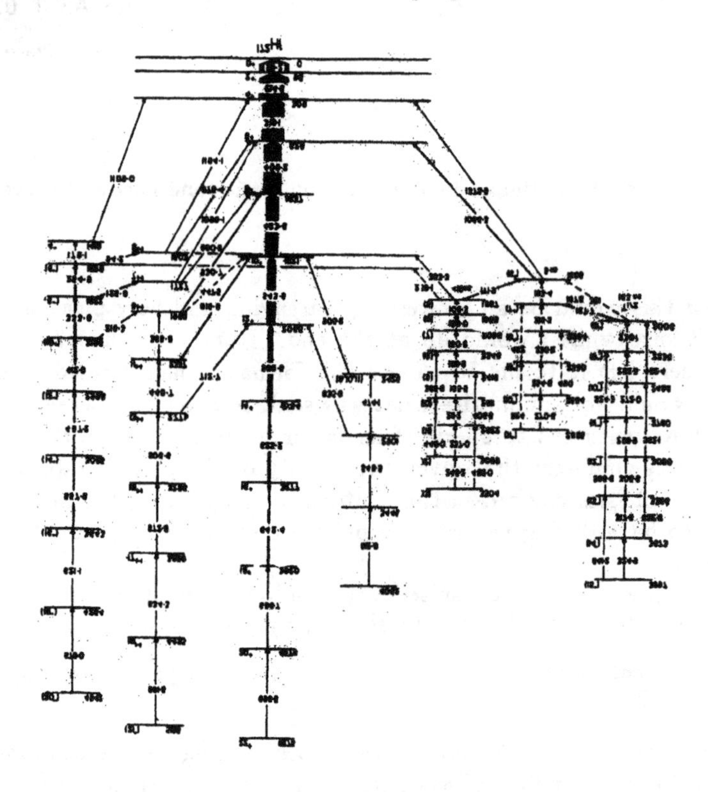

Figure 1 The level scheme of ^{172}Hf as seen down under.

The other fragment that sprang to mind was from Morinaga's *Summerly* talk, as he referred to it, at the International Conference on Nuclear Structure, held in Tokyo in 1977. He got my laugh with a catalogue on the subject of "

[1]at least this is roughly what I remember

the L-R Degeneracy " which included a list of common (risque) problem words which went ...

.

.

Election

.

.

etc

Well if that was all I could remember after 30 years of conference attendance, this approach obviously wasn't going to get me far, so I did some more homework and started reading published conference summaries, or at least those I could easily get my hands on, going back to Vickie Weisskopf at the Kingston Conference in 1960.

This was a fascinating and absorbing excercise and one I recommend to all of you here, old and young, as a way of catching a glimpse of the personalities and the physics – a lot of what we take for granted about nuclear structure but was not understood at the time; the great changes that have occurred and the persistent vitality of the field; even the occasional rediscovery of the wheel.

As a check on my memory, I looked up Morinaga's talk at Tokyo and found no mention of the "L-R Degeneracy" although he does indeed refer to his own summerly talk . So much for the written record !

What he did mention was

- the no-show of Superheavies
- whether we need to include quarks in describing nuclei
- the cost of travel
- the meaning of big conferences
- the young

Some of these are recurring themes which I will get back to later. He also made the point that there were

> *...no contributions on mere numbers as there used to be at the early stages of Nuclear Physics. This applies to two serious contributions from myself (Morinaga) that were not chosen for oral presentation !* [2]

The present organisers might like to note that I also had three serious submissions to the present conference, which weren't deemed suitable for oral presentation, and I am equally miffed. Well, worrying about Conference summaries turns out to be an old pastime and there were some comforts in the thoughts of the eminent speakers that went before.

[2] in the interests of brevity, and for the sake of and effect, I have occasionally paraphrased this and other quotes that follow. As always, the serious scholar should return to primary sources.

Others have been here before

Rudolf Peierls at the International Conference on Nuclear Physics in Paris, in 1964 (on the occasion of the 30th anniversary of the discovery of artificial radioactivity) said something like:

> *I am more sorry than you are that Victor Weisskopf couldn't make it because he was supposed to give the summary !*

Ben Mottelson at Gatlinburg in 1966 said:

> *In order to escape the universal problem of conference summarizers – of selecting some contributions and ignoring others – I would rather ...talk about my own .*

Gerry Brown at Tokyo in 1967 laid it on the line:

> *Let me frankly tell that* I do not *propose to summarize what was said, but what was* not *said. I hope this will discourage people from asking me to give summary talks in the future !*

... Amen.

Eugene Wigner at Montreal in 1969 gave a very serious and conscientious summary but he did let his guard slip near the end when he said :

> *... I must admit that there were a few presentations which were too difficult for me to follow; I am looking forward to reading them in print.*

Now there's a frightening prospect !

John Schiffer at Munich in 1973 put it this way:

> *(... on the three techniques used in giving a summary.) One is to enumerate as much as possible of what was said then go through the list and pick out items to make nasty comments about – known as the Gerry Brown technique !*

Good mates ?

Frank Stephens at the Oak Ridge High-Spin Conference in 1982 added a more human touch when he revealed that he had nightmares about the prospect:

> *...I'm standing before a very serious audience who expect me to say something intelligent and not only do I have absolutely no idea what to say, I don't even know what the subject is !*

Stuck In the Middle With You

I would to mention a few things that came up during the week, including the place of Nuclear Physics, which was commented on by the Minister in her opening address, and on the question of what is fundamental, and whether we need to worry about sub-nucleon structure.

Let's put the nucleus where it belongs, in the middle (Figure 2). In many ways, Nuclear Physics is self-contained but it has connections to numerous fields, fundamental and applied, some of which it spawned. It is central to many of them. There was peripheral discussion during the week on the ques-

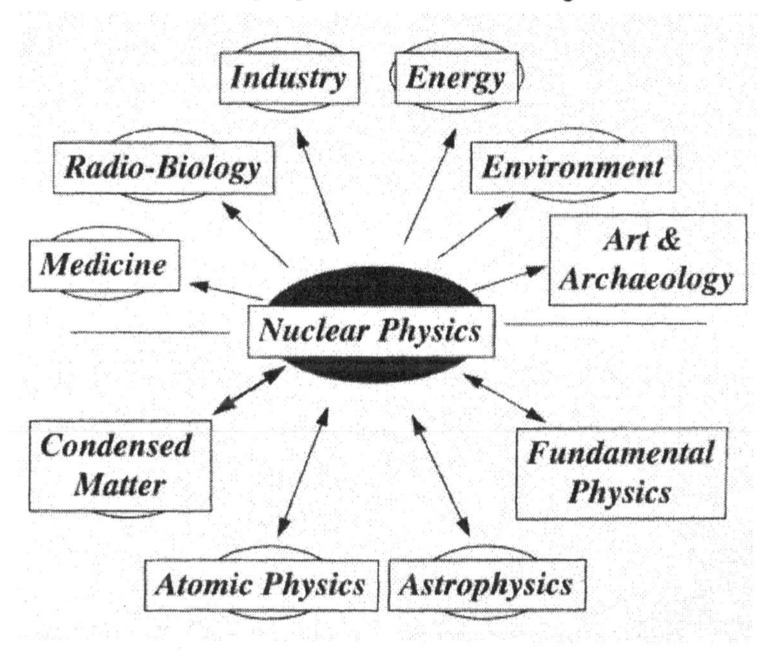

Figure 2 Nuclear Physics is in the middle. Adapted from [8]

tion of what is fundamental. Let me add a few comments (most of which I have borrowed from vocal advocates of "small science" usually directing (anti-reductionist) criticisms towards particle physicists) .

What is Fundamental

- The Fundament is the base. *That from which others can be derived.*
- But each layer of the physical hierarchy can be successfully represented (in having fundamental foundations) while remaining *largely decoupled* from the other layers .
- Each layer functions with its own "elementary" entities or concepts which in general are *not simply derived* from those of the "lower" level, but constructed in creative efforts .
- In the nucleus, the constituent nucleons might be all the same, but their combinations are not. It's the interactions between nucleons that are

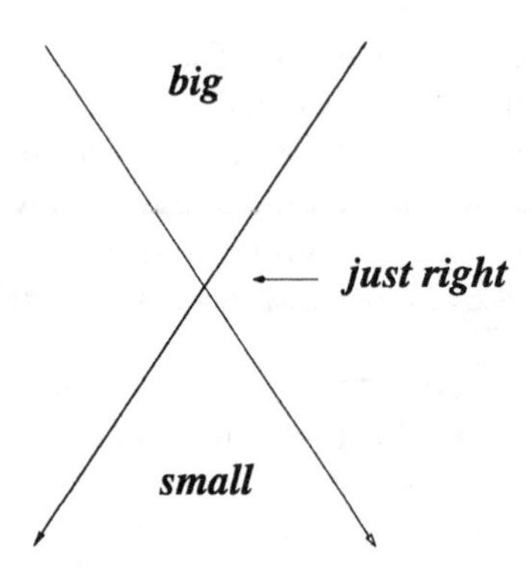

Figure 3 The nuclear scale.

important, not just that they make up the nucleus, or what the nucleons themselves are made up of ....

Of course, there is no doubt that there are universal concepts which can be used across the boundaries of the different levels, and productive analogies which can be made between systems, to illuminate some problem, in Nuclear Physics or wherever. But while such analogies might throw light on the subject, they rarely provide a solution. This is usually because the scales are different and the scale defines what will be important.

I like to see the nuclear scale as in Figure 3. Not too big, not too small, but just right. That might seem a little flippant, but the nuclear scale is peculiar and it makes the nucleus special.

What's special about the Nucleus ?

It

- has a few particles but a mean field
- shows collective (boundary and shape) effects
- has a structure sensitive to individual orbits
- has collective and intrinsic motions observable through individual and interfering quantum states
- exhibits internal contradictions – pairing– coherence length in the smallest superfluid object known is larger than the object itself
- has symmetries (identicality of constituent nucleons)
- is cold and isolated – in the ground state at room temperature,

and the scales of the competing motions are similar. A typical vibrational amplitude $<\beta^2>^{1/2}\sim 0.2$ for example is similar to the deformation, the inequality of major and minor axes, $\delta \sim 0.2$. The characteristic energies of vibration (1 MeV), single-particle excitation (2 MeV) and , for example, pairing (1 MeV) are often comparable. These are strange bodies where the dynamics is poised, almost in conflict.

Quarks, fame and humility

The question about whether we need to worry about sub-nucleon structure to describe nuclei (or their interactions) came up indirectly a number of times in this conference, and has done many times before. If Denys Wilkinson were here he could expound on the subject but let me quote from his presentation at the Brazil Conference in 1989 where he addressed just this question and came to the conclusion that:

> *The traditional view that the nucleus is composed of neutrons and protons must be supplemented by explicit exchange currents and (after an analysis covering 50 pages) it is concluded that Nuclear Physics below 1 GeV/c can be conducted without reference to quarks.*

This is disappointing in a way but it is neither good, nor bad. It's just the way it is.

This brings me in roundabout way to my favourite paper. Back in the late '70s there was an apparently successful attempt to calculate nuclear binding energies and other properties using a type of quark exchange model, which I for one, found attractive and interesting. The author of the model was proud of it and I recall he used to begin his talk (or perhaps end his talk) with a list of collaborators, a list which was a blank sheet. My favourite publication is the paper reproduced in Figure 4, by Phil Elliott. Not because it was critical, but because it has an eleven word abstract, is about one hundred words long, contains one reference and produced six citations for the original author ! A model of brevity which I often wish I could emulate.

I am showing this to make a point about the importance of both publication and of humility. John Sharpey-Schafer used to beat the drum about the writing of papers – a crucial part of the discipline and training of researchers in Nuclear Physics. When you send something to a journal for publication you are, one way or another, taking a chance and exposing yourself. Rarely are you 100% sure of your ground but you have to do it because the process itself forces you to be as sure as possible and still come to a conclusion in a finite time. You have to proceed with the best of intentions (and sometimes with the hope of becoming famous) and accept that there is a chance of being wrong, in print, for all to see.

That brings me to my second-favourite referee's report, reproduced in Figure 5 . As you read this you have to imagine me opening my eagerly awaited mail in front of the pigeon-holes in the corridor of our Department office block, with the expression on my face changing from one of hopeful anticipation to one of embarassment and even shame.

I thought, my god what stupid mistake have I made now ! In those few seconds I vowed to stop the arrogance, to be kind to all of Gods' creatures

L.B: 1C Nuclear Physics A324 (1979) 349 © North-Holland Publishing Co., Amsterdam

A COMMENT ON A QUARK EXCHANGE MODEL

J. P. ELLIOTT
School of Mathematical and Physical Sciences, University of Sussex, Brighton, England

Received 16 February 1979

Abstract: Attention is drawn to an error in a paper by [illegible].

A recent paper [1]) by [illegible] proposes a quark exchange model of nuclear structure according to which the binding energies of nuclei are calculated by counting the number of "quark exchange bonds". In subsect. 3.3 when discussing the structure of ^{4}He he states that the class (4) of the symmetric group S_4 has three elements whereas in fact it has six elements. Correction of this error would introduce a factor $\frac{1}{2}$ into eq. (3.33) of [illegible]'s paper which relates the binding energies of ^{4}He and ^{3}H and would appear to destroy much of the agreement with experiment claimed by [illegible]. For this reason, if for no other, one must therefore have serious doubts about [illegible]'s model while at the same time not wishing to question those relationships which are inherent to other established models, such as the α-particle model, and which are incorporated within [illegible]'s paper.

Reference

1) [illegible], Nucl. Phys. A308 (1978) 381

349

Figure 4 An exemplary paper. [2]

including my colleagues, and to abandon all ambition and just work quietly and humbly in the background. Of course, that only lasted for a few seconds as I realised that the font looked a little odd, and the wording seemed unlikely. As the truth dawned, I felt relief and I got a little mad. Later I got even.

Nuclear states, Serendipity and the Devil

I would like to return to the point that

every nucleus is different

Much of what is behind new directions in Nuclear Structure, as in many fields, is aimed at higher sensitivity, not just because the easy cases have been

THE PHYSICAL REVIEW

AND

PHYSICAL REVIEW LETTERS

EDITORIAL OFFICES - BOX [illegible] - RIDGE, NY [illegible]

Telephone [illegible]

FAX [illegible] Telex [illegible]

Cable Address [illegible]

[illegible]

14 February 1990

Dr. G.D. Dracoulis
Department of Nuclear Physics
Australian National University
P.O. Box 4
Canberra 2601, AUSTRALIA

Re: Shape difference implied by quenched Coriolis interaction in ^{175}Os

By: G.D. Dracoulis and B. Fabricius CNJ414

Dear Dr. Dracoulis:

The above manuscript has now been returned from the second referee who stated that it is the most stupid thing he has ever read since cold fusion.

Please refrain from sending us any future manuscripts for publication in 'The Physical Review' or 'Physical Review Letters'.

Sincerely yours,

Gerard J. Drake
Associate Editor
Physical Review C

(PUBLICATIONS OF THE AMERICAN PHYSICAL SOCIETY)

053-G0AD

Figure 5 My second-favourite referee's report.

done, but because each level of sensitivity may expose a new layer of understanding.

Sensitivity can give you

- New Nuclei
- New Effects
- New Shapes
- New Motions
- New Couplings

We are always at the mercy of nuclear reaction mechanisms which make some nuclei at any time innaccessible, and some states difficult to populate, but its important to remember that

Weakly populated states are not unimportant states

One argument we often make as a community to justify the push towards radioactive beams and new nuclei is that only about 40% of those nuclei predicted to be stable to particle decay are known. That argument by itself sounds rather weak, unless you put it within the context that every nucleus is different. The next nucleus you characterize might have similar general structures to others, but its quantum states, and possibly its dynamical motions, will be unique. It might, because of the particular disposition of those states, contain the clues to an understanding that is generally applicable, but is simply not demonstrated in any other nucleus.

This somewhat random characteristic is one reason why it is is possible for young researchers to make a substantial contribution, despite the maturity of the field.

I would like to try and illustrate that, and the importance of interpreting the detail, with an example taken from our own work on the level scheme of ^{179}W published a few years ago. The situation when we began the study (part of a general characterisation of metastable states in deformed nuclei) is illustrated in Figure 6. The rate of decay of a strongly populated 5-quasiparticle isomer with a projection K = 35/2, predominantly into the rotational band based on the $7/2^{-}[514]$ neutron orbital, had been difficult to understand [3, 4]. Although the connecting 610 keV E2 transition ($\lambda = 2$) was retarded, giving a lifetime of 710 ns for the 3349 keV state, it was not retarded nearly enough. As one group put it [4] it was

> ***impossible to understand that the half life can be as short as 710 ns for a transition that changes the K-quantum number by more than 14 units***

The measured strength of 10^{-4} single particle units was a factor of 10^{8} faster than the expected value of about $[10^{-1}]^{\Delta K - \lambda} = 10^{-12}$ single particle units. For many years, the prevailing view was this was an (unexplained) result of a chance degeneracy.

Our remeasurement of the level scheme using techniques such as time - correlations which gave us an order of magnitude of sensitivity, exposed another layer of detail which showed that the decay was more complicated, as shown partly in Figure 7 . There were close-lying states at spins $31/2^{-}$, $33/2^{-}$ and $35/2^{-}$ from three structures, the $7/2^{-}[514]$ band, the $35/2^{-}$ isomer, and an intervening band based on a $23/2^{-}$ state. Interference between these states was perturbing the energies and branches.

Understanding the branching of the $31/2^{-}$ state turned out to be crucial for a number of reasons, particularly because the branch out what appeared to be a member of the interceding band to the $27/2^{-}$ member of the $7/2^{-}$ band was so dominant. The out-of-band B(E2) dominated (at 180 units of intensity) to the point that the in-band transition (of 4 units) was just observable,

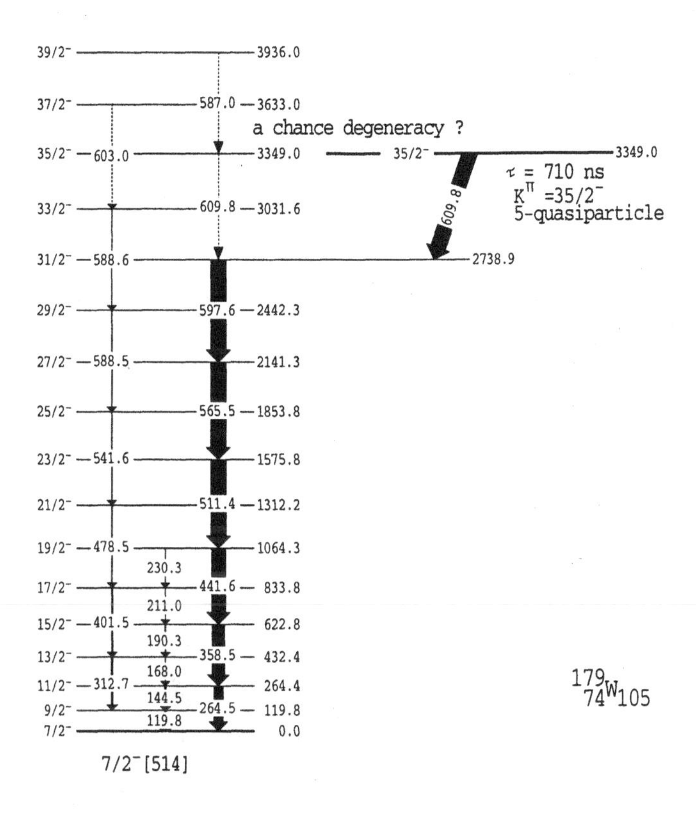

Figure 6 The decay of the $35/2^-$ isomer in ^{179}W as known in 1983.

$$\frac{B(E2)_{598}}{B(E2)_{477}} = \frac{180}{4}\left(\frac{477}{598}\right)^5 = 14.5,$$

apparently a result of destructive interference. In terms of the mixing amplitudes in the $31/2^-$ states and the $27/2^-$ states defined in Figure 8, the individual $B(E2)$ values were of the form

$$B(E2) = [\alpha\gamma < 1|\mathrm{E2}|1 > +\beta\delta < 2|\mathrm{E2}|2 >]^2$$

where the unperturbed $E2$ components have their usual K-dependence such that $B(E2)_{unp.} = \frac{5}{16\pi}e^2Q_0^2 < J_i 2K0|J_f K >^2$. Assuming the same deformation in both bands and associating configuration $|1>$ with $K = 7/2$ and configuration $|2>$ with the intervening band of unknown K, each $B(E2)$ will depend on the mixing and the ratio of Clebsch-Gordan (CG) coefficients as

$$B(E2) \propto \left[\alpha\gamma + \beta\delta \times \frac{< K >_{CG}}{< K = 7/2 >_{CG}}\right]^2$$

Complete mixing in the upper states (not a necessary assumption but a good approximation in this case) requires a mixing matrix element of about 17 keV which can be used to determine all amplitudes and therefore the branching ratio.

For $K = 23/2$, equal to the value of the spin of the bandhead, we get

$$\frac{B(E2)_{598}}{B(E2)_{477}} = ft\left[\frac{0.707(0.990) - 0.707(-0.142) \times \sqrt{0.19}}{0.707(0.142) - 0.707(0.992) \times \sqrt{0.19}}\right]^2 = 13.3$$

close to the observed value. In fact, the interference depends sensitively on the K-projection of the intervening band as illustrated in Figure 9, constraining K to a value $\sim 23/2$ and effectively localising the angle of the total spin with respect to the deformation axis, and and a tilt angle of about 46°, close to the value predicted by Frauendorf for what he had called Fermi-aligned rotation

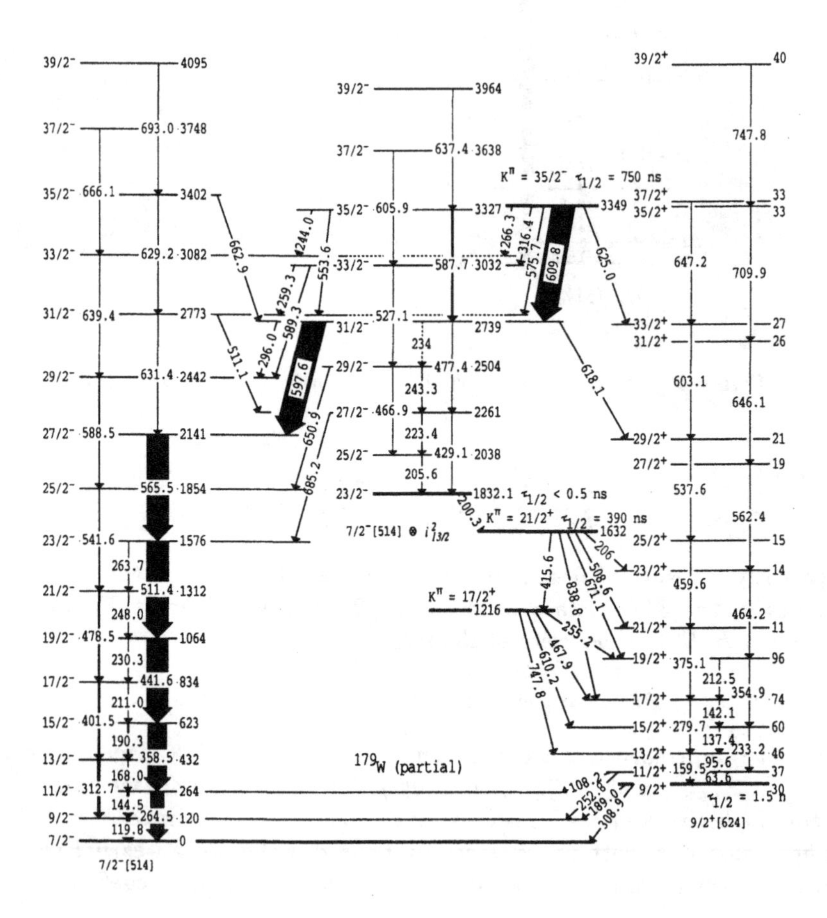

Figure 7 Partial scheme for the ^{179}W isomer from the work of Walker *et al* [11]

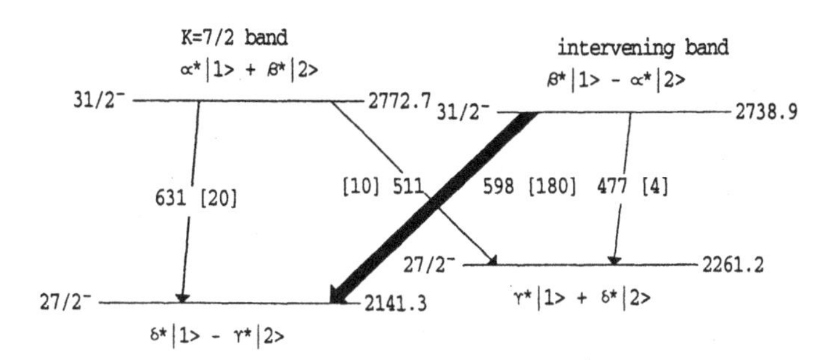

Figure 8 Mixing amplitudes and interference in ^{179}W .

[5]. This coupling mode could only arise in re gions where the deformation coupling (controlled by the pairing) is balanced by the rotation coupling. It

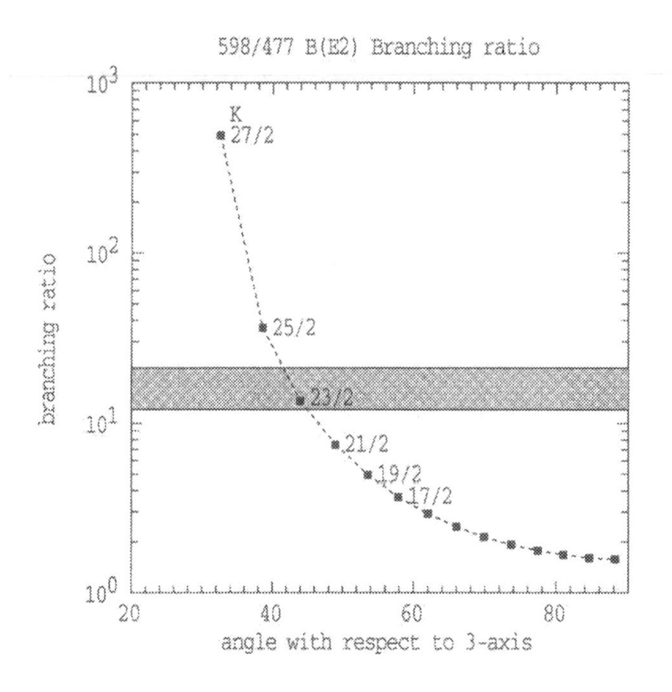

Figure 9 Branching ratios in ^{179}W as a function of the K-value of the intervening band.

was predicted to show the unusual characteristics of a simultaneous aligned angular momentum, but also a good - K value, and no signature splitting. Such a motion was the complement of the well known s-band alignment where a pair of $i_{13/2}$ neutrons align with rotation so that their individual projections cancel, viz $\Omega_1 - \Omega_2 \sim 0$. The new "$t$"-band (as it is now called [6]) would correspond to the situation where both spins localise on the same side of the nucleus with respect to the 1-axis, so that their projections add; $\Omega_1 + \Omega_2 \sim 8$.

So we have in the present case a band that corresponds to a combination of the $7/2^+[633]$ and $9/2^+[624]$ neutrons, and the spectator $7/2^-[514]$ neutron, to give

$$K = [7/2^+ + 9/2^+] + 7/2^- = 23/2^-.$$

That's a long story but the point is that the attention to detail resulted in both the resolution of a problem with an apparent anomalous hindrance (it is no longer ΔK=14 and the decay proceeds from 5- to 3- to 1-quasiparticle states) and the discovery of primary and direct evidence for a new form of rotational motion. All because of the interference between states which fortuitously fall close enough together to mix. A close degeneracy impossible to predict. And any nucleus could hold such clues.

Towards the future

As I said above, our difficulty is that we are at the mercy of reaction mechanisms, at least at present, which do not allow us to choose exactly where we want to begin in a particular nucleus. But I believe we have already seen one aspect of the future unfolding in the last few years in the coming together of many of the techniques and the practitioners of subfields which functioned previously in isolation.

The Coming Together of

- In-beamers
- β-decayers
- Laserers
- Trappers
- Separators
- Fissioners
- Fragmenters
- . etc.

What we didn't hear a lot about during the conference were the reaction mechanisms, particularly heavy ion reactions, which underly much of the effort in spectroscopy and which are vital for future progress. I had in mind the increasing realisation that heavy ion reactions, particularly at energies close to the Coulomb barrier are surprisingly sensitive to the structure of the target and projectile [7].

Weighty subjects

Could these in some way be tuned to produce nuclei that would otherwise not survive the competing processes of fission . One example is the suggestion of Iwamoto *et al* [8] that higher order shapes oriented in a way to allow themselves to hug each other closely (Figure 10) could provide a window to the fusion of very heavy nuclei, providing the initial nuclei exist. In this case they don't

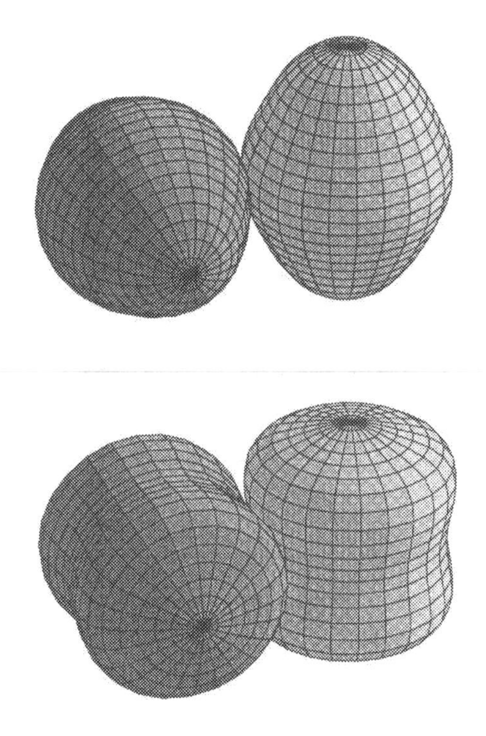

Figure 10 Heavy nuclei with exaggerated shapes in collision with particular relative orientations. The upper pair have ϵ_4 deformations of the same sign as stable ^{150}Nd nuclei, the lower pair with ϵ_4 and ϵ_6 chosen to allow a close "hug". (Adapted from [8].)

(as yet) and despite tremendous efforts by experimenters, there has been slow progress towards the holy grail of finding Superheavies. It is as if we are held back by the worries of Amos de Shalit who, at the Heidelberg Heavy Ion Conference in 1969 warned of the dangers of producing nuclei with $\bar{\nu} \sim 10$ with the rhetorical plea

> ***can we physicists do anything at this stage to prevent catastrophic results that may come out of mass production of such heavy nuclei. We...(should) give the matter the most serious thought before we proceed full speed..***

He was perhaps simultaneously too pessimistic, and too optimistic. We are progressing but mass production seems like a long way off !

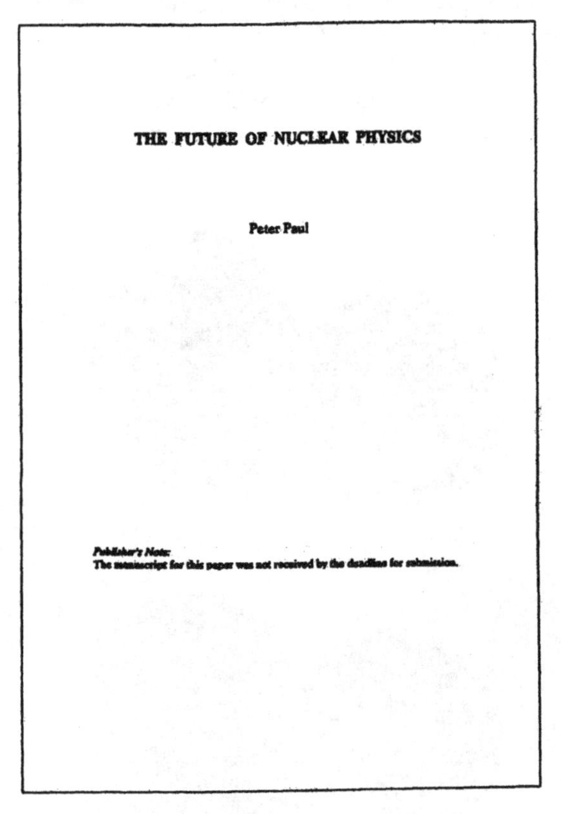

THE FUTURE OF NUCLEAR PHYSICS

Peter Paul

Publisher's Note:
The manuscript for this paper was not received by the deadline for submission.

Figure 11 Self explanatory. [9]

Contrasts

I won't attempt to cover the status of heavy ion reactions but more committed rapporteurs would do well to consult history again. If you have the time read D Allan Bromley's paper in the 1974 Nashville conference on Reactions Between Complex Nuclei which, in his usual comprehensive fashion included

1 Chinese proverb

65 Printed pages

62 Figures, and

160 References

For lighter reading on the future try the report of Peter Paul's summary at the 1989 INPC Conference in Brazil which I have reproduced in Figure

11. There was of course a lot more in that presentation than an undelivered manuscript (a development I can sympathise with) might suggest but the future does not, in any case, lie in the pronouncements of a few wise old men, or even foolish young men, like me. It lies within the Nucleus. If the Nucleus doesn't have any more secrets, or if we don't have any more ideas about where secrets are hidden, or cunning ideas about techniques, experimental and theoretical, to dig those secrets out, then it won't have a future.

Bouncing back

But history argues against that. The resilience of Nuclear structure is a confirmation of its underlying richness. Many predictions have been made in the past, often by people who *knew* that the nuclear system would become bland as one cranked up some variable. I picture this as in Figure 12 where the nuclear well would just become more shallow, and broader, and the characteristic states, indistinct. But that is not the truth of the matter: pairing did not disappear above 10 $\hbar$, structure is not washed out at high excitation energy, nuclear wave functions are not randomly mixed, the whole thing is not a blob. Co-existence phenomena, multiple minima in the nuclear potential well are not

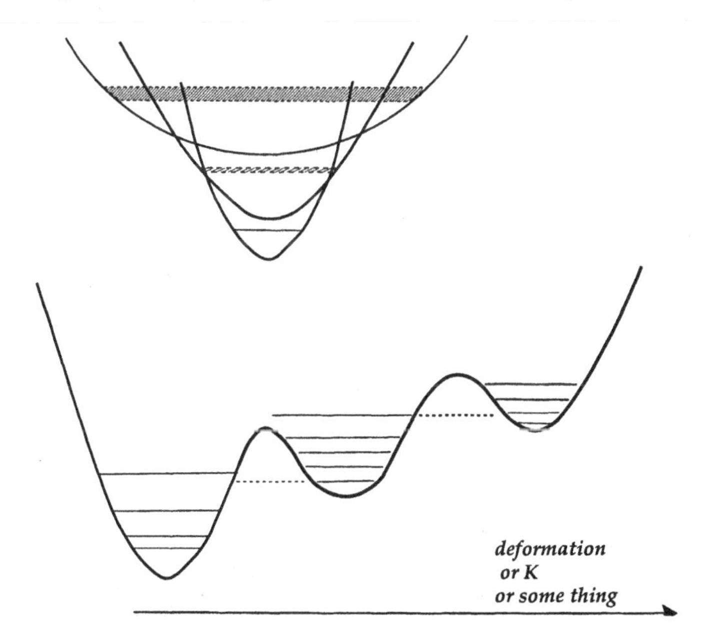

Figure 12 Two views of the Nuclear Landscape.

unusual, they are, as a function of all sorts of variables, ubiquitous.

The nucleus is apparently a lot more resilient than many of us. I am confident that if we exploit our opportunities, with new techniques, exotic beams, ordinary beams, whatever, the field will prosper. Providing we look after the new generations. That is where I want to say something about John Sharpey-Schafer. I hope he will forgive me if I don't mention his triumphs in physics, but as well as those, he has made a serious mark in the field by nurturing young talent – by having confidence in his young colleagues, by expressing that confidence and by making space for it to flourish. And by infecting them with enthusiasm. The nuclear physics community owes him a debt of gratitude.

NUCLEAR STABILITY
RULES
O.K !

BY
N. FEATHER, F.R.S.

CAMBRIDGE
AT THE UNIVERSITY PRESS
1952

Figure 13 Nuclear Stability Rules. O.K ! *I have taken some liberty with the original title.*

On behalf of the organisers let me thank all those who made the effort to share their results and ideas. On behalf of the participants, thank you to the organisers. It's been a stunning week ! David Inglis in his summary of the Rutherford Jubilee Conference held in Manchester in 1961 managed to include a quote from Darwin, a feat I won't try to top, but I will quote Inglis's quote of Weisskopf, who said in his closing remarks for the Kingston conference of the previous year,

> The nucleus is so small, and it has so few parts, and still it shows a tremendous variety of phenomena. What a marvellous invention ! It is worth devoting a lifetime to it.

Inglis thought this was an understatement.

My final indulgence is to recommend a historic textbook to you (Figure 13), one which John may well have has on his bookshelf, although its title might look a little different.

Nuclear Stability Rules. *O.K.*!

References

[1] Louise Bostock *Speaking in Public. Collins Pocket Reference Series* 1994. HarperCollins (Glasgow)

[2] Reprinted from Nuclear Physics A, **A324** (1979) 349 J.P. Elliott, "A comment on a Quark exchange model" (Page 349 only) Copyright (1979), with permission from Elsevier Science

[3] *Connection between backbending and high-spin isomer decay in* ^{179}W F.M. Bernthal *et al* Phys. Lett. 74B (1978) 211

[4] J. Pedersen *et al* Phys. Scr. T5 (1983) 162

[5] *Spin alignment in heavy nuclei* S. Frauendorf Phys. Scr. 24 (1981) 349

[6] S. Frauendorf Nucl. Phys. A557 (1993) 259c

[7] *Measuring Barriers to Fusion* M. Dasgupta, D.J. Hinde, N. Rowley and A.M. Stefanini, Annu. Rev. Nucl. Part. Sci. 48 (1998) 401

[8] *Collisions of Deformed Nuclei: A path to the far side of the superheavy island* A. Iwamoto, P. Moller, J.R. Nix and H. Sagawa Nucl. Phys. A596(1996)329

[9] Proc. of the 1989 Int. Nucl. Phys. Conf. Sao Paolo, Brasil August 20-26,Volume 2, Invited Papers page 807 (World Scientific)

[10] *Impact and Applications of Nuclear Science in Europe: Opportunities and Perspectives* NuPECC Report, December 1994

[11] *Resolution of the* ^{179}W *- Isomer Anomaly: Exposure of a Fermi-aligned s-band* P.M. Walker, G.D. Dracoulis, A.P. Byrne, B. Fabricius, T. Kibédi and A.E Stuchbery, Phys. Rev. Lett. 67 (1991) 433

AUTHOR INDEX

SUBJECT INDEX

CPSIA information can be obtained
at www.ICGtesting.com
Printed in the USA
LVHW061728210620
658633LV00030B/1522